NEW MEDIA and PUBLIC RELATIONS

PETER LANG
New York • Washington, D.C./Baltimore • Bern
Frankfurt am Main • Berlin • Brussels • Vienna • Oxford

NEW MEDIA and PUBLIC RELATIONS

Sandra C. Duhé, EDITOR

PETER LANG
New York • Washington, D.C./Baltimore • Bern
Frankfurt am Main • Berlin • Brussels • Vienna • Oxford

Library of Congress Cataloging-in-Publication Data

New media and public relations / edited by Sandra C. Duhé.
p. cm.
Includes bibliographical references.
1. Public relations. 2. Internet in public relations.
I. Duhé, Sandra C.
HM1221.N47 659.20285'4678–dc22 2006025283
ISBN 978-1-4331-0124-3 (hardcover)
ISBN 978-0-8204-8801-1 (paperback)

Bibliographic information published by **Die Deutsche Bibliothek**.
Die Deutsche Bibliothek lists this publication in the "Deutsche Nationalbibliografie"; detailed bibliographic data is available on the Internet at http://dnb.ddb.de/.

Cover design by Joshua Hanson

The paper in this book meets the guidelines for permanence and durability of the Committee on Production Guidelines for Book Longevity of the Council of Library Resources.

29 Broadway, 18th floor, New York, NY 10006
www.peterlang.com

Printed in the United States of America

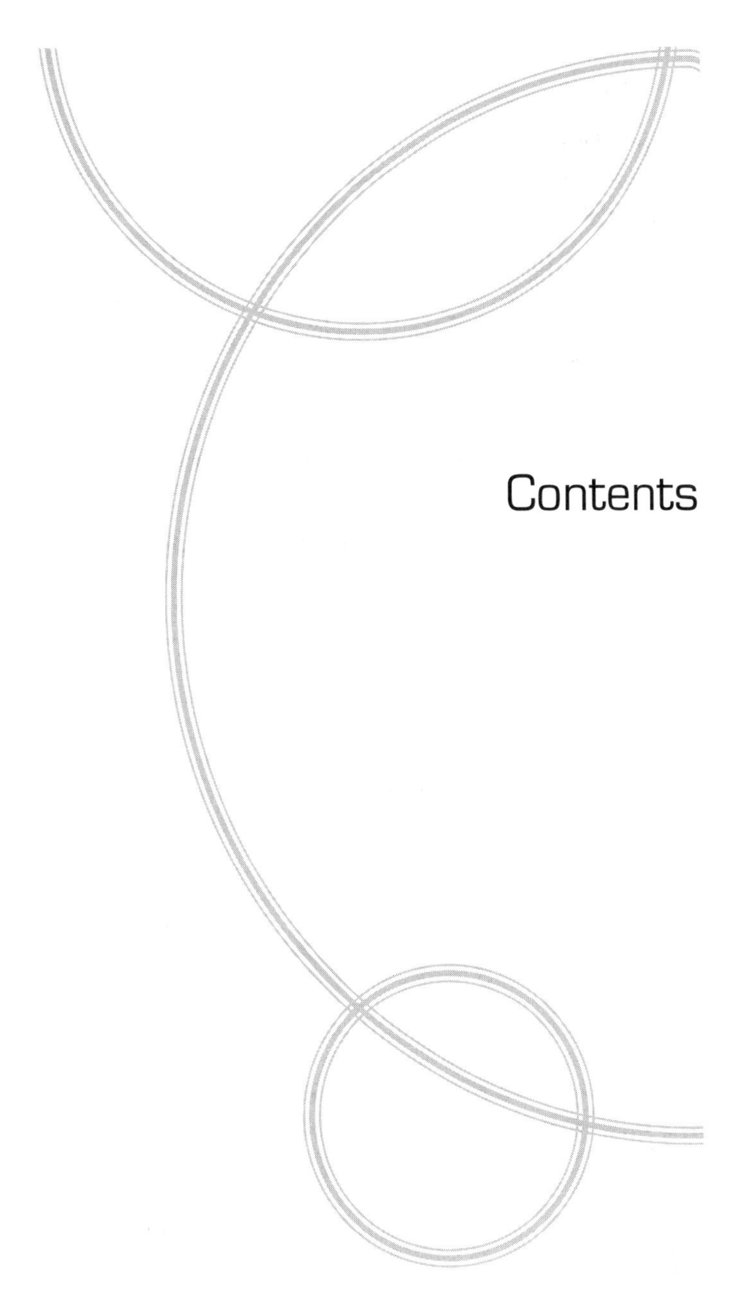

Contents

Editor's Note

Sandra C. Duhé

The genesis of this book traces back to a conversation in fall 2005 with my friend and colleague Ty Adams. Ty has authored or edited several books on Internet-mediated communication and suggested we put our heads together to compile an anthology on how new media are impacting the practice of public relations. I was just beginning my second year of teaching at the University of Louisiana at Lafayette after spending fourteen years working in the corporate sector. Needless to say, I was still learning about the lingo and pace of academic life, but I decided to take Ty up on his offer. A few months later, Ty was approached with another attractive book offer and graciously retreated from our project. But he encouraged me to press on, and I'm so glad that I did.

It has been a privilege for me to be a part of this endeavor. I am truly fortunate to have had the opportunity to work with so many bright, dedicated scholars and practitioners from around the globe and play some part in gaining a wider audience for their work. The countless hours spent reviewing and organizing these chapters have not only reaffirmed my decision to leave corporate life for the classroom, but also reignited my hopes for the advancement of our chosen field of study and practice.

It's an understatement to say the Internet has impacted the way we communicate as individuals, groups, and organizations. With increasing speed and expanding reach, we can collaborate faster, cast our research nets broader, and share information at a rate that boggles the mind. Large, hierarchical entities tend to lag behind smaller, more nimble organizations in the adoption and application of new media technologies. Consequently, dominant voices are now accompanied, challenged, and sometimes overshadowed by voices previously marginalized.

New media have facilitated a more open, transparent, and interactive society. Organizations and their publics now have more equitable footing on which to check each other's activities and motives. Public relationships are a critical component of this communication conundrum.

The following chapters are divided into two major parts. Part I focuses on how new media have impacted conventional thinking in public relations, primarily from a theoretical point of view. Here, authors revisit established models in light of new technologies and offer new lenses through which public relations can be studied. These exploratory ideas provide promising areas for further research and refinement and are offered in that spirit.

Part II, which comprises the bulk of the chapters, examines how new media are being used in public relations practice. These chapters are organized under unifying themes that will appeal to a broad range of reader interests: relationship and reputation management; power, resistance, and social change; a variety of niche applications; and crisis communication. Using an accessible writing style, the authors provide a rich array of scholarly and practitioner insights into the potential and challenges of building mediated relationships.

Included in the chapters are numerous references to Web sites, blogs, and other electronic publications. Sites are referred to in the past tense, not because they are no longer active (most should be), but rather because content is likely to have changed since the book went to press. Readers are encouraged to visit the URLs provided to experience firsthand what the authors discuss.

New Media and Public Relations is designed to show advanced undergraduate and graduate students as well as practitioners how new media are being used to interact with an organization's stakeholders and to spark new research and applications in this critical, ever-changing area. Each chapter includes thought-provoking discussion questions and suggested readings intended to expand on the authors' thoughts and lead to additional development of these ideas.

I owe a great deal of gratitude to the seasoned and up-and-coming authors who entrusted me with their contributions to this book; I am honored to be in their company. My appreciation extends to Mary Savigar, Phyllis Korper, Damon Zucca, Bernadette Shade, Meredith Ackroyd, and many others behind the scenes at Peter Lang; Katie Delahaye Paine; and the external reviewers: Each of you invested considerable time and provided needed support to get this book to market. Thank you, above all, for sharing my enthusiasm for this project. My heartfelt thanks go to my wonderfully supportive husband

and family who thought they would see more of me once I left corporate life but who have found, alas, that I'm busy as ever (and loving it). Last, but certainly not least, thanks, Ty, for giving me my wings.

Sandy Duhé
March 2007

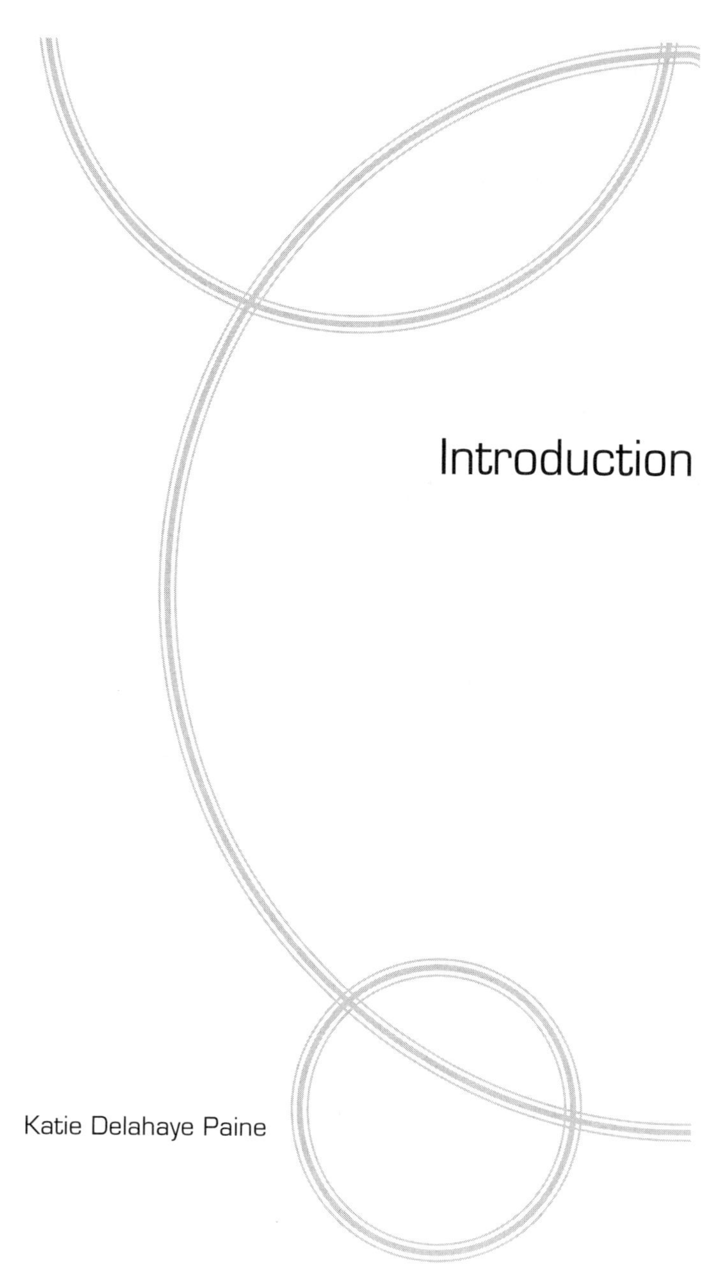

Introduction

Katie Delahaye Paine

Every once in a great while, times and technology converge to change the way we receive and distribute news. The fifteenth century had the printing press, the nineteenth century had broadsheets, and, of course, the twentieth century brought TV, radio, and the Internet. Just as those inventions defined their times, so the rise of new media, in all of its parameters and permutations, will define the twenty-first century. The explosion of consumer-generated media has turned every teenager and tourist with a camera phone into a journalist. From neighborhood to international nonprofit, organizers are using the speed and ease of the Internet to generate support for their causes. Organizations that used to claim that you can't measure public relations are now evaluating their public relations programs in hard dollar terms, such as sales and revenue. The notion of hiding bad news behind a public relations offensive now seems as antiquated as the manual typewriters that the equally antiquated press release used to be typed on.

The implications for twenty-first century public relations practitioners are all at once far reaching, terrifying, and enormously exciting.

First and foremost, public relations practitioners need to realize that managing the media is no longer an option. The media is now anyone with a cell phone, a laptop, a digital camera, or a tape recorder. With the advent of Wikipedia, YouTube, and, of course, blogs, the concept of the media as a gatekeeper or content provider is laughable. Whether it is sending back images from the front lines in the Middle East, to transmitting the scores of a Little League game seconds after the last inning, few people are waiting for the news to arrive on their doorstep. Instead, they are creating their own news stories, images, and sounds to accompany any given event.

If controlling the flow of information is no longer an option, managing your relationships is a necessity. With no one person or editor in control, your ability to influence anyone is entirely dependent on the credibility of your message and the reputation of your organization. That reputation and credibility is formed by the relationships you have established with your publics.

Some might see this era as one in which "the lunatics are in charge of the asylum." Others, however, will see the opportunity. So many assumptions are being challenged on a daily basis that many public relations professionals want nothing more than to throw up their hands and run for the safety of the last century. In reality, every generation that experiences enormous changes in media must be prepared to be more flexible, innovative, and creative than ever before. Today's public relations practitioners are no exception. The problem is that with all the opportunity comes a whole new array of tools, techniques, options, and problems that need to be examined.

New Media and Public Relations is the instruction manual, guidebook, map, and global positioning system (GPS) that twenty-first century communicators need to successfully navigate this new landscape. It is a comprehensive manual, from which readers will learn how to better understand the complex nature of relationships that the new media world brings together. It blows away many of the established theories of public relations and puts them back together in a way that is relevant and applicable in the new environment. Furthermore, it explains the trends and tendencies that are emerging within the new media framework so savvy public relations professionals can study up and take advantage of those trends. Whether your media is wikis or blogs, whether your mission is to raise money or sell product, and whether you're in the middle of a crisis or planning a new product launch, *New Media and Public Relations* offers insights and detailed case studies that explain how these new theories should be put into practice.

Katie Delahaye Paine, CEO
KDPaine & Partners, LLC
March 2007

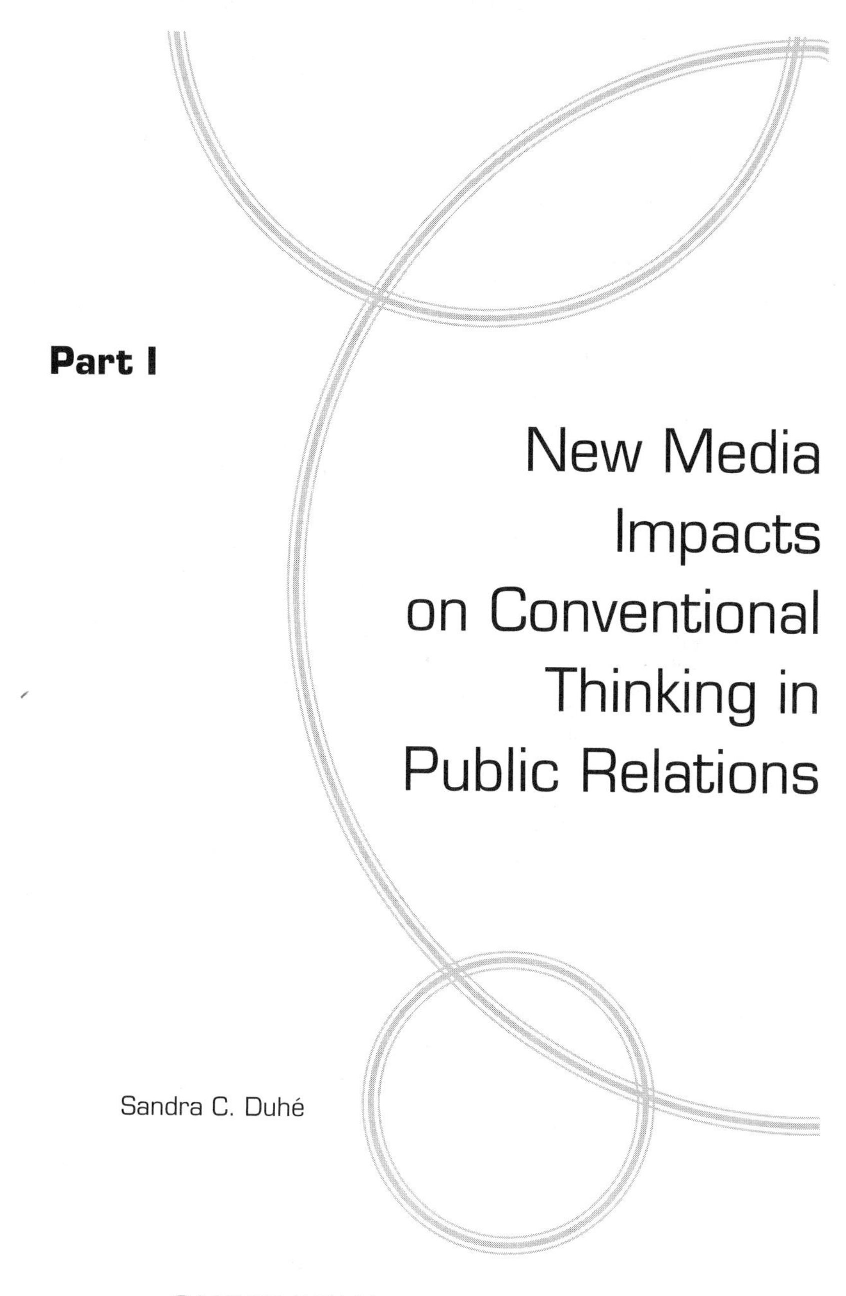

Part I

New Media Impacts on Conventional Thinking in Public Relations

Sandra C. Duhé

OVERVIEW

The important thing in science is not so much to obtain new facts as to discover new ways of thinking about them.

Sir William Bragg (1862–1942), recipient of the 1915 Nobel Prize in Physics

Change is inherent to advancing a field of study. The practice of public relations has been explored, defined, and tested using a variety of scientific lenses, including mass communication, organizational behavior, cognition, and culture. Beyond providing us with a framework for ongoing investigation, our present set of models and theories has, more importantly, helped to establish public relations as a recognized and worthwhile area of inquiry. For these accomplishments, we owe our gratitude to those scholars and practitioners who have preceded us.

We cannot afford to rest on our laurels, however. For, without change, there is no growth. Intellectual curiosity leads to change but not without risk to the pioneers behind it. As we expand, challenge, and attempt to modernize conventional thinking in public relations, we run the risk of being wrong. We risk having our ideas questioned, our credibility tested, and our work reversed. Such is the way of scholarship. Without the risk takers, we become stagnant.

The authors of these first five chapters embody the spirit of Sir William Bragg's observation that the essence of science is innovation. While honoring the contributions of traditional models, we seek new ways of thinking about them in light of emerging technologies.

Bey-Ling Sha offers a reconfiguration of the four models of public relations, first introduced by Grunig and Hunt over twenty years ago, into more modern (and measurable) strategic and tactical dimensions of practice. Zoraida Cozier and Diane Witmer challenge us to move away from outdated models about communicating with publics and adopt a structurationist perspective that recognizes the role of institutions, identities, and ideologies in the life cycle of public participation. Jennifer Bartlett updates current thinking to recast the impersonal influence of media in a symmetrical, as opposed to an asymmetrical, role in relationship management. My own chapter examines organizational transparency through the lens of complexity theory, which casts doubt upon the efficacy of command-and-control thinking in mediated stakeholder relations. Finally, Damion Waymer calls for a reconceptualization of the spiral of silence hypothesis, noting that the Internet has effectively increased the share of voice for minority-opinioned publics.

These chapters build on previous work to provide new ways of thinking about public relations scholarship and practice. They are offered not as a destination, but as a starting point for further research, validation, and debate. May this journey continue.

Dimensions of Public Relations

Moving beyond Traditional Public Relations Models

Bey-Ling Sha

With changes in technology and the explosion of media channels, classic public relations models can no longer adequately describe public relations practice, especially as the relationships between organizations and their stakeholders have grown increasingly complex and malleable. For over a generation, public relations students have learned about Grunig's models of public relations: press agentry, public information, two-way asymmetrical, and two-way symmetrical (J. E. Grunig & Hunt, 1984). But, in recent years, the venerated models of public relations have been reconceptualized as seven dimensions of public relations behavior that span both the strategic and tactical aspects of public relations practice. This chapter examines ways in which public relations program planners in the research and evaluation phases of the four-step process can measure stakeholders' perceptions of an organization's communication efforts across the new dimensions of public relations. Understanding these new measurements can help public relations practitioners to plan more effective programs to enhance organizational relationships with strategic publics.

A generation of public relations students and practitioners have learned about the four models of public relations: press agentry, public information, two-way asymmetrical, and two-way symmetrical (cf. J. E. Grunig & Hunt, 1984). Now that the Grunig and Hunt textbook that introduced the models to public relations is nearly a quarter-century old, the time has come to move on from these classic articulations to explore modern dimensions in public relations. Even J. E. Grunig (2001) himself has urged that we continue to push forward in the study of public relations behaviors of organizations and their publics.

To that end, this chapter examines ways in which public relations program planners in the research phase of the four-step process can measure stakeholders' perceptions of an organization's communication efforts across the new dimensions of public relations.[1] Understanding these new measurements can help public relations practitioners to plan more effective programs to enhance organizational relationships with strategic publics.

Background on Public Relations Models

The original four models of public relations were meant to elucidate descriptive, historical, and normative aspects of our field.

Descriptive aspects. The first model of public relations—press agentry—described a practice in which public relations practitioners sought publicity and press coverage for their clients, often at the expense of accuracy and truth. The second model—public information—portrayed public relations practitioners as *journalists-in-residence* who disseminated accurate information about their organizational employers to target publics (cf. J. E. Grunig & Hunt, 1984).

Third, the two-way asymmetrical model—sometimes called *scientific persuasion*—illustrated a practice in which organizations sought feedback from their publics, but only for the purpose of better persuading those stakeholders to see the organization's point of view. Finally, in the two-way symmetrical model of public relations, organizations engaged in two-way communication with their publics both to change the publics, as well as possibly to change the organization itself (cf. J. E. Grunig & Hunt, 1984).

The four models of public relations have been applied to public relations practice around the world, even though some scholars pointed to our need for "context specific, inductive and ground-up approaches" that generate theory from native cultures rather than imposing theory on those cultures (Bardhan & Patwardhan, 2004). Despite the drawbacks of doing so, researchers have examined the models of public relations in Brazil (Penteado, 1996), China (Chen, 1996; McElreath, Chen Azarova, & Shadrova, 2001), Costa Rica (Gonzalez & Akel, 1996), Greece (J. E. Grunig, L. A. Grunig, Sriramesh, Huang, & Lyra, 1995), southern India (Sriramesh, 1992), Russia (McElreath et al., 2001), South Africa (Holtzhausen, Petersen, & Tindall, 2003), South Korea (Kim, 2003; Kim & Hon, 1998), Romania (VanSlyke Turk, 1996), the United Kingdom (Gregory, 2003), Taiwan (Huang,

1990), Thailand (Ekachai & Komolsevin, 1996), and the United Arab Emirates (Badran, Turk, & Walters, 2003).

Historical aspects. The articulation of four models of public relations was useful not only in describing public relations practices in the United States and other countries, but also in illustrating historical aspects of our field (J. E. Grunig & Hunt, 1984), at least with respect to the United States. Even though several authors' conceptualizations of public relations history dates back to ancient times and Aristotelian efforts at persuasion (cf. Cutlip, Center, & Broom, 2006; J. E. Grunig & Hunt, 1984), the press agentry model encapsulates public relations in the early modern era, with efforts by press agents such as P.T. Barnum to gain notoriety for their clients (Cutlip, 1994; J. E. Grunig & Hunt, 1984). By the early 1900s, the nature of public relations efforts began to change, as organizations responded to muckraking journalism with renewed efforts to tell their version of the truth (Cutlip et al., 2006), thus ushering in the public information model of the practice, of which Ivy Lee was one embodiment (Cutlip, 1994; J. E. Grunig & Hunt, 1984).

With the advent of the two World Wars, the U.S. government launched a massive effort to determine whether the American public would support its country's entrance into what were initially European conflicts, as well as what persuasive messages would work best in shoring up that support (J. E. Grunig & Hunt, 1984). These efforts to persuade the public using social scientific methods, a practice for which Edward Bernays, in particular, was widely recognized, underscored the two-way asymmetrical model of public relations (Cutlip, 1994; J. E. Grunig & Hunt, 1984).

By the end of World War II, a new model of public relations had evolved, grounded in the belief that organizations had to be open to changing their own behaviors and not just interested in changing those of their publics. This attitude was articulated by Arthur Page's principle of "proving it with action" (cf. Cutlip, Center, & Broom, 2006), meaning that organizations had to act on their convictions, rather than merely talking about them. The belief in the importance of an organization's being open to change is reflective of the two-way symmetrical approach, in which both organizations and publics adjust to each other (cf. Cutlip & Center, 1978; J. E. Grunig & Hunt, 1984). One pioneer of this approach was Earl Newsom (Cutlip, 1994).

The traditional models of public relations, with their corresponding historical periods, key figures, and organizational attitudes toward their publics, are summarized in Figure 1. Missing from that visual summary, however, is an elucidation of the normative aspects of the public relations models.

Normative aspects. The two-way symmetrical model, primarily by itself, but also in conjunction with the two-way asymmetrical model, has been called the ideal model of ethical public relations (J. E. Grunig, 1992; J. E. Grunig & L. A. Grunig, 1996), as well as the model most indicative of excellence in public relations practice (J. E. Grunig, 1992; L. A. Grunig, J. E. Grunig, & Dozier, 2002). Also, Dozier, L. A. Grunig, and J. E. Grunig (1995) explicated that excellent organizations in fact practice a "contingency model" of public relations, meaning that knowing how to use *both* the asymmetrical and symmetrical models of public relations formed "the core of communication excellence" (p. 50). The contingency model reflected in several ways the "mixed-motive model" first articulated by Murphy (1991), as well as the continuum approach first recommended by Hellweg (1989).

The normative aspects of the two-way symmetrical model were not without controversy. For example, the two-way symmetrical model was supposed to be most effective in enabling organizations to deal with activist publics (Anderson, 1992; L. S. Grunig, 1986; Huang, 1990; Sha, 1995), as well as for activists themselves in certain situations (J. E. Grunig & L. Grunig, 1997). However, other scholars argued that this model was an altogether inappropriate strategy for activists (Rodino & DeLuca, 1999). Given the controversies surrounding the models of public relations, the time had clearly come for some serious consideration of reconceptualizing these descriptive, historical, and normative classifications of public relations practice.

Figure 1: Comparison of Descriptive and Historical Aspects

Public Relations Models	Historical Periods	Organizational Attitudes	Example Practitioners
Press Agentry	Pre-Seedbed Era (pre-1900s)	Public-Be-Damned	P.T. Barnum
Public Information	Seedbed Era (1900–1917)	Public-Be-Informed	Ivy Lee
Two-Way Asymmetric	World War I (1917–1919)		
	Booming Twenties (1919–1929)	Mutual Understanding	Edward Bernays
	Roosevelt and World War II (1930–1945)		
Two-Way Symmetric	Postwar Boom (1946–1964)	Mutual Adjustment	Arthur Page and Earl Newsom
	Period of Protest and Empowerment (1965–1985)		
	Digital Age and Globalization (1986–Present)		

Note. Public relations models as explained in J. E. Grunig and Hunt (1984). Historical periods as articulated in Cutlip, Center, and Broom (2006). Organizational attitudes as articulated by Edward Bernays and Arthur Page (cited in Cutlip et al., 2006). Example practitioners as identified in Cutlip (1994); Cutlip et al. (2006); and J. E. Grunig and Hunt (1984).

Articulating New Dimensions of Public Relations Practice

Toward the end of the last century, the study of public relations models had begun to evolve. Scholars no longer tried to fit public relations practice into one of the four static models, which had received criticism for being unrealistic and idealistic (cf. Cancel, Cameron, Sallot, & Mitrook, 1997; Leichty & Springston, 1993; Murphy, 1991; Yarbrough, Cameron, Sallot, & McWilliams, 1998). In some ways, the evolution away from the models of public relations toward the dimensions of public relations behavior was perhaps inevitable, as scholars reconsidered the fundamental aspects of the former.

In his own account of the evolution of the public relations models, J. E. Grunig (2001) articulated four dimensions: one-way/two-way, symmetrical/asymmetrical, interpersonal/mediated, and ethical/unethical. These same four dimensions are isolated by L. A. Grunig, J. E. Grunig, and Dozier (2002) in their latest articulation of a theoretical framework for excellent public relations, in which they reworked quantitative data from the Excellence study (cf. J. E. Grunig, 1992) in support of the first three dimensions. Also, one should note that the dimension of mediated/interpersonal communication is similar to what Hallahan (2001) proposed: an alternative continuum of mass communication—interpersonal communication.

Thus, depending on the researcher, from four to seven dimensions of public relations behaviors were reconceptualized and then measured (e.g., J. E. Grunig, 2000, 2001; J. E. Grunig & L. A. Grunig, 1996; L. A. Grunig, J. E. Grunig, & Dozier, 2002; Huang, 1997, 2004b; Rhee, 1999; Sha, 1999a, 1999b).[2] This chapter's purpose is to clarify these dimensions so that scholars can use them in developing public relations theory and so that public relations practitioners can use them to enhance strategic planning efforts as part of the four-step process. Specifically, we will look at both conceptual and methodological aspects of seven dimensions of public relations that span both strategic and tactical aspects of public relations planning.

Public relations strategy

Public relations strategy is about actions that an organization can take to accomplish its goals and objectives (Cutlip, Center, & Broom, 2006). On the strategic side, four public relations dimensions reflect organizational approaches to problem-solving or organizational worldviews about the management of relationships with stakeholders. These dimensions may be considered strategic in the sense that they lay the foundation for the manner in which an organization might try to achieve a public relations objective: two-way communication, symmetrical communication, ethical communication, and conserving communication.

Two-way communication. The direction of communication underlies one part of the initial conceptualization of the four models of public relations (cf. Dozier, L. A. Grunig, & J. E. Grunig, 1995; J. E. Grunig & Hunt, 1984). Defined here, two-way communication measures whether information flows from the organization to the public and vice versa, or whether the organization puts out information without seeking any input or feedback.

Symmetrical communication. The second characteristic underlying the original four public relations models had been the purpose of communication, specifically, whether the organization was interested in changing the public only or changing both the public and itself (cf. Dozier, L. A. Grunig, & J. E. Grunig, 1995; J. E. Grunig & Hunt, 1984). When an organization sought to change only its target publics, it was said to be asymmetrical, whereas symmetrical communication denoted an organization that was willing to make mutual adjustments in its relationship with stakeholders. As discussed above, the concept of symmetrical communication has generated much scholarship in public relations. Two additional public relations dimensions reflecting organizational strategy have grown out of the concept of symmetrical communication: ethical communication and conservation.

Ethical communication. Ethical communication in essence looks at the extent to which an organization behaves with honesty, accuracy, and other ethical considerations in its public relations efforts. In some recent research, the dimensions of symmetrical and ethical public relations behaviors were collapsed, since statistical analysis showed that only one factor was extracted from the items that constructed the symmetrical and ethical dimensions (cf. Huang, 2004a). Across three data sets reported in Huang (2004a), the merged dimension of symmetrical/ethical communication had construct measurement reliabilities of 0.75, 0.74, and 0.76, as well as alphas of 0.75, 0.72, and 0.71, respectively. These data offer empirical support for the conceptualization of symmetrical communication as being inherently the most ethical of the original models of public relations (cf. J. E. Grunig & L. A. Grunig, 1996), as well as for the methodological merging of these two dimensions.

Conservation. Another dimension of public relations behavior that grew out of the concept of symmetrical communication—or, more specifically, asymmetrical communication—was articulated in recent scholarship as conservation (cf. Sha, 2004), meaning that organizations refuse to change; that is, they conserve their own fundamental agendas. Sha (2004) argued that symmetrical and asymmetrical behaviors should be viewed independently, rather than as opposite ends of a single continuum, and she applied Noether's theorem from mathematical physics to explicate this reasoning. In lay terms, Noether's theorem provides a mathematical formula that proves the existence of invariance of some aspect of any physical phenomenon in the face of simultaneous change produced by that phenomenon (Alekseevskii, 1988). Thus, Sha (2004) used Noether's theorem to argue that organizational changes wrought by symmetrical communication could only occur with the simultaneous maintenance (i.e., invariance or conservation) of some other aspect of the organization.

Sha (2004) related "conserving" public relations behaviors to what had previously been called asymmetrical behaviors or pure advocacy behaviors (cf. Dozier, L. A. Grunig, & J. E. Grunig, 1995). Empirical support for the conceptual separation of the original symmetrical/asymmetrical dimension had appeared as early as Rhee (1999), who found that symmetrical and asymmetrical behaviors were not mutually exclusive among practitioners in South Korea. Likewise, when they reexamined the Excellence data, L. A. Grunig, J. E. Grunig, and Dozier (2002) tested the reliabilities of both combined and separated scales for symmetrical and asymmetrical communication, finding that the separated scales yielded higher reliabilities.

Since the strategic dimension of conservation emerged from qualitative data collected by Sha (1999a), quantitative measures of this dimension were not found in the literature published to date, and, hence, no analysis of the construct's internal reliability is offered in this chapter's supplementary tables. However, an as-yet unpublished paper by Marks (2006) measured the dimension of conservation using the items proposed in this chapter and found an internal reliability of 0.60 for those items, which is an acceptable reliability standard for exploratory research (Nunnally, 1967).

Public relations tactics

Public relations tactics are about concrete ways in which organizations might execute or support their strategies (Cutlip, Center, & Broom, 2006; Wilson & Ogden, 2004). On the tactical side, three dimensions reflect actual ways in which public relations strategies might be executed; that is, through mediated communication, interpersonal communication, or social activities.

Mediated communication. The concept of mediated communication examines public relations behaviors that occur through some kind of mass media technology, such as the Internet or broadcast television. Because this dimension offers a concrete way of doing public relations activities, mediated communication is considered tactical in nature. Although J. E. Grunig (2001) placed mediated and interpersonal communication as two ends of the same dimension, other scholars (e.g., Huang, 1997; Rhee, 1999; Sha, 1999b) believed that separating the interpersonal/mediated dimension was more appropriate. Measuring mediated communication and interpersonal communication separately meant that organizations were no longer viewed as expressing one behavior at the expense of the other, which seems particularly relevant given the growth of new media technologies being used in public relations today.

The separation of the interpersonal/mediated dimension proved to be useful in explaining the communicative behaviors of the Democratic Progressive Party on Taiwan, which showed high levels of both interpersonal and mediated public relations behaviors (cf. Sha, 1999a, 1999b). These findings also underscored the importance of interpersonal relationships in public relations practice, perhaps first articulated by Sriramesh (1992) and reinforced by subsequent scholarship (e.g., Jo & Kim, 2004; Shin & Cameron, 2003). The use of separate scales to measure interpersonal communication and mediated communication was reinforced in L. A. Grunig, J. E. Grunig, and Dozier (2002).

Interpersonal communication. This tactical dimension of public relations centers on activities that take place either face-to-face or through one-on-one communication, such as via the telephone. Given the explosion in personal media technologies, however, much more research needs to be done on the tactical dimensions of public relations to determine whether some new activities, such as instant messaging, should be viewed as mediated or interpersonal communication—or perhaps as both!

Social activities. This tactical dimension of public relations looks at a special kind of interpersonal communication—activities that take place in a social context, but with a view toward accomplishing public relations objectives. Some examples of social activities might include banquets or gift-giving exchanges. The emphasis of this public relations dimension is on the opportunity to cultivate one-on-one, interpersonal relationships between organizational representatives and stakeholders; thus, large, impersonal special events, such as a rock concert, would not be considered reflective of this dimension.

The concept of social activities was offered by Huang (2004b) and grew out of her work in Asia showing the importance of social activities in the maintenance of organizational relationships. Specifically, Huang argued that social connections (called *guanxi* in Chinese), the cultivation of connections (*la guanxi*), and the strategic use of connections (*gao guanxi*) were essential elements of public relations practice that differed from simple behaviors or strategies of interpersonal communication. A second aspect of social activities lies in the Chinese term *xenqing*, which involves both maintaining and being supportive of one's social contacts. To measure this dimension, Huang (1997) extracted several of the items originally used to measure interpersonal communication.

The concept of the importance of interpersonal connections is not unique to Eastern cultures. In the West, this concept is reflected in American sayings, such as "Having friends in high places." Similarly, the French use the verb *pistonner* to refer to the use of *piston*—personal connections to important people—as a means of accomplishing something. Thus, the tactical dimension of social activities as a public relations behavior is relevant all around the world, and it is perhaps especially important to the development of theories to explain public relations practice in cultures where such personal networks are of particular import.

Measuring New Dimensions of Public Relations Practice

To measure these new dimensions, a meta-analysis of three studies (using four data sets) was conducted (see Appendices A and B for details). For a summary of reliability analyses of the dimensions across the three studies, see Table 1.

Directions for Research

This chapter concludes with a list of recommended items for each dimension based on the research results provided in Appendix B. These results offer clear directions for future scholarship on dimensions of public relations behavior as these relate to organizational strategies and program tactics. Specifically, the results reported here suggest that at least six of the seven dimensions may be reliably measured using specific items reviewed in this chapter. The suggested items for conservation have, to date, been used only in Marks (2006),

Table 1: Summary of Reliability Analyses of Dimensions of Public Relations Behavior

	Cronbach's Alphas			
Dimensions	Study 1	Study 2a	Study 2b	Study 3
1. Two-way	0.72	0.86	0.75	0.66
2. Mediated	0.78	0.79	0.76	0.71
3. Interpersonal	0.60	0.64	0.46	0.70
4. Social Activities	0.77	NA	NA	NA
5. Ethical	0.75*	0.83	0.79	0.63
6. Symmetrical	0.75*	0.78	0.69	0.71
7. Conservation	NA	NA	NA	NA

Note. The dimension of conservation emerged from qualitative data collected by Sha (1999b); quantitative measures of this dimension were not found in the literature published to date. NA denotes that the item was not applied in the study; * indicates that the alpha was for a combined construct of symmetrical/ethical communication. Study 1 was Huang (1997), portions of which were reported in Huang (2004b). Study 2a was Sha (1999b) for behaviors with internal publics. Study 2b was Sha (1999b) for behaviors with external publics. Study 3 was Rhee (1999), portions of which were reported in Rhee (2002).

as mentioned earlier. To provide public relations researchers with a clear articulation of these measures, this chapter concludes with recommended items for the measurement of two-way communication, symmetrical communication, ethical communication, conservation, mediated communication, interpersonal communication, and social activities.

Strategic Dimensions of Public Relations

Two-Way Communication:

- The organization listens to publics' opinions.
- Before conducting public relations activities, the organization researches and tries to understand publics' positions.
- After completion of public relations activities, the organization conducts an evaluation.

Symmetrical Communication:

- The organization not only tries to change the publics' attitudes and behaviors, but also tries to change the attitude and behavior of organizational management.
- The organization tries to change the organization's behaviors and policies after considering publics' opinions.
- The organization consults those affected by its policies during decision making.
- The organization plays an important role in mediating conflicts between the organization and its publics.

Ethical Communication:

- The organization considers how its activities might affect the public.
- The organization provides the public with accurate or factual information, even when this information may not portray the organization in a positive light.
- The organization considers the public interest more than individual or personal interests.
- The organization tells publics their motives and reasons for their actions.

Conservation:

- The organization subscribes to certain ideals or principles that the organization will never give up.
- When the organization communicates with an individual, the individual feels that the organization is more interested in accomplishing its own agenda than the individual's opinions.
- The organization's mission is unlikely to change in response to external pressures.
- The organization has a sense of purpose that remains unchanged since its founding.
- The organization exists primarily to accomplish its own goals.
- The organization often shifts from one vision to another (item reverse-scored).

Tactical Dimensions of Public Relations

Mediated Communication:

- The organization distributes news releases to generate media coverage.
- The organization uses mass media, such as television, radio, broadcasts, newspapers, or magazines.
- The organization distributes flyers, pamphlets, magazines, or other printed materials that represent the organization.
- The organization uses the Internet / World Wide Web to receive and send information.

Interpersonal Communication:

- The organization contacts members of the public in person.
- The organization contacts members of the public by phone.
- The organization primarily uses face-to-face methods of communication.
- The organization holds in-person meetings.

Social Activities:

- The organization holds dinners, banquets, or meal parties.
- The organization uses flattery.
- The organization gives gifts.
- The organization acknowledges events in the private/internal lives of its publics,

such as celebrating the birth of a child, expressing condolence at the passing of a loved one, or sending congratulations to a newly promoted leader.

Future research should measure the public relations behaviors of organizations—and of publics—using these items. Once all the items are administered in a single study, factor analysis may be conducted to further validate the dimensions articulated here.

Concluding Thoughts

These new dimensions change the conventional way of thinking about public relations practice as being one of four traditional models: press agentry, public information, two-way asymmetrical, and two-way symmetrical. Given the increasing complexity of public relations practice, compounded by new media technologies and global ramifications for organizational actions, thinking about public relations activities as strategic and tactical dimensions helps us to understand how a research-based practice of public relations can contribute to organizational effectiveness in the strategic planning process of our field. These new dimensions of public relations also offer a more complex way to measure organizational relationships with stakeholders across space, time, and new media, thus enabling more sophisticated organizational efforts in the research and evaluation phases of the four-step process.

For consideration

1 Which dimensions of public relations could be difficult for a practitioner to employ if he or she did not work for a management team that understood the importance of relationships in public relations?
2 Think about an organization that has been in the news, possibly because of its poor relationships with organizational stakeholders. How could this organization's public relations efforts be described, using the new dimensions explained in this chapter?
3 To what extent could each of the dimensions of public relations be carried out exclusively online by an organization?
4 Young people are often the earliest adopters of new media. Consider the measures for public relations dimensions proposed in this chapter. Would using these measurements cause organizations to miss important stakeholders from younger populations? How can these measurements be improved to better reflect communication practices among today's young people?
5 What other dimensions of public relations practice could evolve over the next twenty-five years?

For reading

Anderson, D. S. (1992). Identifying and responding to activist publics: A case study. *Journal of Public Relations Research, 4*, 151–165.

Grunig, J. E. (2001). Two-way symmetrical public relations: Past, present, and future. In R. L. Heath (Ed.), *Handbook of public relations* (pp. 11–30). Thousand Oaks, CA: Sage.

Huang, Y. H. (2004). Is symmetrical communication ethical and effective? *Journal of Business Ethics, 53*(4), 333–352.

Kim, Y., & Hon, L. C. (1998). Craft and professional models of public relations and their relation to job satisfaction among Korean public relations practitioners. *Journal of Public Relations Research, 10*(3), 155–175.

Sha, B.-L. (2004). Noether's theorem: The science of symmetry and the law of conservation. *Journal of Public Relations Research, 16*(4), 391–416.

References

Alekseevskii, D. V. (1988). Noether theorem. In M. Hazewinkel (Ed.), *Encyclopaedia of mathematics, Vol. 6: An updated and annotated translation of the Soviet "Mathematical Encyclopaedia"* (pp. 407–408). Dordrecht: Kluwer.

Anderson, D. S. (1992). Identifying and responding to activist publics: A case study. *Journal of Public Relations Research, 4*, 151–165.

Badran, B. A., Turk, J. V., & Walters, T. N. (2003). Sharing the transformation: Public relations and the UAE come of age. In K. Sriramesh & D. Vercic (Eds.), *The global public relations handbook: Theory, research, and practice* (pp. 46–67). Mahwah, NJ: Lawrence Erlbaum.

Bardhan, N., & Patwardhan, P. (2004). Multinational corporations and public relations in a historically resistant host culture. *Journal of Communication Management, 8*(3), 246–264.

Cancel, A. E., Cameron, G. T., Sallot, L. M., & Mitrook, M. A. (1997). It depends: A contingency theory of accommodation in public relations. *Journal of Public Relations Research, 9*, 31–63.

Chen, N. (1996). Public relations in China: The introduction and development of an occupational field. In H. M. Culbertson & N. Chen (Eds.), *International public relations: A comparative analysis* (pp. 121–153). Mahwah, NJ: Lawrence Erlbaum Associates.

Crable, R. E., & Vibbert, S. L. (1983). Mobil's epideictic advocacy: "Observations" of Prometheus-bound. *Communication Monographs, 50*, 380–394.

Cutlip, S. M. (1994). The unseen power: Public relations—a history. Hillsdale, NJ: Lawrence Erlbaum Associates.

Cutlip, S. M., & Center, A. H. (1978). *Effective public relations* (5th ed.). Englewood Cliffs, NJ: Prentice-Hall.

Cutlip, S. M., Center, A. H., & Broom, G. M. (2006). *Effective public relations* (9th ed.). Englewood Cliffs, NJ: Prentice Hall.

Dozier, D. M., Grunig, L. A., & Grunig, J. E. (1995). *Manager's guide to excellence in public relations and communication management*. Mahwah, NJ: Lawrence Erlbaum Associates.

Ekachai, D., & Komolsevin, R. (1996). Public relations in Thailand: Its functions and practitioners' roles. In H. M. Culbertson & N. Chen (Eds.), *International public relations: A comparative analysis* (pp. 155–170). Mahwah, NJ: Lawrence Erlbaum Associates.

Gonzalez, H., & Akel, D. (1996). Elections and earth matters: Public relations in Costa Rica. In H. M. Culbertson & N. Chen (Eds.), *International public relations: A comparative analysis* (pp. 257–272). Mahwah, NJ: Lawrence Erlbaum Associates.

Gregory, A. (2003). The ethics of engagement in the UK public sector: A case in point. *Journal of Communication Management*, *8*(1), 83–95.

Grunig, J. E. (Ed.). (1992). *Excellence in public relations and communication management*. Hillsdale, NJ: Lawrence Erlbaum Associates.

Grunig, J. E. (2000). Collectivism, collaboration, and societal corporatism as core professional values in public relations. *Journal of Public Relations Research*, *12*(1), 23–48.

Grunig, J. E. (2001). Two-way symmetrical public relations: Past, present, and future. In R. L. Heath (Ed.), *Handbook of public relations* (pp. 11–30). Thousand Oaks, CA: Sage.

Grunig, J. E., & Grunig, L. A. (1996, May). *Implications of symmetry for a theory of ethics and social responsibility in public relations*. Paper presented to the Public Relations Interest Group, International Communication Association, Chicago, IL.

Grunig, J. E., & Grunig, L. A. (1997, July). *Review of a program of research on activism: Incidence in four countries, activist publics, strategies of activist groups, and organizational responses to activism*. Paper presented to the Fourth Public Relations Research Symposium, Managing Environmental Issues, Lake Bled, Slovenia.

Grunig, J. E., Grunig, L. A., Sriramesh, K., Huang, Y. H., & Lyra, A. (1995). Models of public relations in an international setting. *Journal of Public Relations Research*, *7*(3), 163–186.

Grunig, J. E., & Hunt, T. (1984). *Managing public relations*. New York: Holt, Rinehart & Winston.

Grunig, L. A., Grunig, J. E., & Dozier, D. M. (2002). *Excellent public relations and effective organizations: A study of communication management in three countries*. Mahwah, NJ: Lawrence Erlbaum Associates.

Grunig, L. S., a.k.a., Grunig, L. A. (1986, August). *Activism and organizational response: Contemporary cases of collective behavior*. Paper presented to the Public Relations Division, Association for Education in Journalism and Mass Communication, Norman, OK.

Hallahan, K. (2001). Strategic media planning: Toward an integrated public relations media model. In R. L. Heath (Ed.), *Handbook of public relations* (pp. 461–470). Thousand Oaks, CA: Sage.

Hearit, K. M. (2001). Corporate apologia: When an organization speaks in defense of itself. In R. L. Heath (Ed.), *Handbook of public relations* (pp. 501–511). Thousand Oaks, CA: Sage.

Hellweg, S. A. (1989, May). *The application of Grunig's symmetry-asymmetry public relations models to internal communications systems*. Paper presented to the International Communication Association, San Francisco, CA.

Holtzhausen, D. R., Petersen, B. K., & Tindall, N. T. (2003). Exploding the myth of the symmetrical/asymmetrical dichotomy: Public relations models in the new South Africa. *Journal of Public Relations Research*, *15*(4), 305–341.

Huang, Y. H. (1990). *Risk communication, models of public relations and anti-nuclear activism: A case study of a nuclear power plant in Taiwan*. Unpublished master's thesis, University of Maryland, College Park.

Huang, Y. H. (1997). *Public relations strategies, relational outcomes, and conflict management strategies*. Unpublished doctoral dissertation, University of Maryland, College Park.

Huang, Y. H. (2004a). Is symmetrical communication ethical and effective? *Journal of Business Ethics*, *53*(4), 333–352.

Huang, Y. H. (2004b). PRSA: Scale development for exploring the cross-cultural impetus of public relations strategies. *Journalism and Mass Communication Quarterly, 81*(2): 307–326.

Jo, S., & Kim, Y. (2004). Media or personal relations? Exploring media relations dimensions in South Korea. *Journalism and Mass Communication Quarterly, 81*(2), 292–307.

Kim, Y. (2003). Professionalism and diversification: The evolution of public relations in South Korea. In K. Sriramesh & D. Vercic (Eds.), *The global public relations handbook: Theory, research, and practice* (pp. 106–120). Mahwah, NJ: Lawrence Erlbaum.

Kim, Y., & Hon, L. C. (1998). Craft and professional models of public relations and their relation to job satisfaction among Korean public relations practitioners. *Journal of Public Relations Research, 10*(3), 155–175.

Leichty, G., & Springston, J. (1993). Reconsidering public relations models. *Public Relations Review, 19*, 327–339.

Marks, K. S. (2006, May). *Actions speak louder than words: How Navy public affairs officers effectively communicate in support of public diplomacy goals*. Unpublished manuscript, San Diego State University, CA.

McElreath, M., Chen, N., Azarova, L., & Shadrova, V. (2001). The development of public relations in China, Russia, and the United States. In R. L. Heath (Ed.), *Handbook of public relations* (pp. 665–673). Thousand Oaks, CA: Sage.

Murphy, P. (1991). The limits of symmetry: A game theory approach to symmetric and asymmetric public relations. In L. A. Grunig & J. E. Grunig (Eds.), *Public relations research annual* (Vol. 3, pp. 115–132). Hillsdale, NJ: Lawrence Erlbaum Associates.

Nunnally, J. (1967). *Psychometric theory*. New York: McGraw Hill.

Penteado, R. (1996). *Effect of public relations roles and models on excellent Brazilian organizations*. Unpublished master's thesis, University of Florida, Gainesville.

Rhee, Y. (1999). *Public relations practices of top-ranked corporations: An exploratory study in South Korea*. Unpublished master's thesis, University of Maryland, College Park.

Rhee, Y. (2002). Global public relations: A cross-cultural study of the excellence theory in South Korea. *Journal of Public Relations Research, 14*(3), 159–184.

Rodino, V., & DeLuca, K. (1999, June). *Unruly relations: Not managing communication in the construction of the activist model of public relations*. Paper presented to the Public Relations Society of America Educators Academy Second Annual Research Conference, College Park, MD.

Sha, B.-L. (1995, October). *White schools and black student activists: Analyzing the power imbalance*. Paper presented to the Educators' Section, Public Relations Society of America, Seattle, WA.

Sha, B.-L. (1999a). *Cultural public relations: Identity, activism, globalization, and gender in the Democratic Progressive Party on Taiwan*. Unpublished doctoral dissertation, University of Maryland, College Park.

Sha, B.-L. (1999b, June). *Symmetry and conservation: Applying Noether's theorem to public relations*. Paper presented to the Second International, Interdisciplinary Public Relations Research Conference, College Park, MD.

Sha, B.-L. (2004). Noether's theorem: The science of symmetry and the law of conservation. *Journal of Public Relations Research, 16*(4), 391–416.

Sha, B.-L., & Huang, Y.-H. (2004). Public relations on Taiwan: Evolving with the infrastructure. In K. Sriramesh (Ed.), *Public Relations in Asia: An anthology* (pp. 162–185). Singapore: Thomson Learning.

Shin, J.-H., & Cameron, G. T. (2003) Informal relations: A look at personal influence in media relations. *Journal of Communication Management, 7*(3), 239–254.

Sriramesh, K. (1992). *The impact of societal culture on public relations: An ethnographic study of south Indian organizations*. Unpublished doctoral dissertation, University of Maryland, College Park.

VanSlyke Turk, J. (1996). Romania: From publicitate past to public relations future. In H. M. Culbertson & N. Chen (Eds.), *International public relations: A comparative analysis* (pp. 341–347). Mahwah, NJ: Lawrence Erlbaum Associates.

Wilson, L. J., & Ogden, J. D. (2004). *Strategic communications planning: For effective public relations and marketing* (4th ed.). Dubuque, IA: Kendall/Hunt.

Yarbrough, C. R., Cameron, G. T., Sallot, L. M., & McWilliams, A. (1998). Tough calls to make: Contingency theory and the Centennial Olympic Games. *Journal of Communication Management, 3*, 39–56.

Notes

1. An earlier version of this manuscript was presented as Sha, B.-L. (2005, August). *The death of the models: A meta-analysis of modern dimensions in public relations*. Paper presented to the Public Relations Division, Association for Education in Journalism and Mass Communication, San Antonio, TX.
2. Huang (2004b) refers to these dimensions as public relations strategies rather than behaviors. This chapter uses the term *behaviors* because strategies may be considered a type of behavior enacted to accomplish a specific objective. Thus, the term *behavior* is broader and, as a result, more likely to lend itself to theoretical extension. Also, the term *strategies* may be confused with an organization's rhetorical strategies (cf. Crable & Vibbert, 1983; Hearit, 2001). Use of the term *behaviors* thus clearly delineates this research as being in the vein of what Cutlip, Center, and Broom (2006) called organizational action strategies, rather than communication or message strategies.
3. Although L. A. Grunig, J. E. Grunig, and Dozier (2002) also published reliability data for scales measuring dimensions of public relations behavior, their study is excluded from this paper because the items used in their analysis were those worded to measure the original public relations models. The three studies examined in this paper used reworded items measuring public relations behaviors that had evolved from the items initially intended to measure the models of public relations.

Appendix A: Overview of Three Studies[3]

The first study examined in this meta-analysis was Huang's (1997) research on the public relations behaviors of members of the Legislative Assembly on Taiwan. The self-administered survey instrument was hand delivered to research participants, who were legislators and legislative assistants. Portions of Huang (1997) were published as one of three data sets reported in Huang (2004a) and Huang (2004b). The response rate reported in Huang (1997) was 45%, with 301 useable questionnaires completed, out of 671 delivered.

The second and third data sets discussed in this paper come from a study by Sha (1999b), which reported portions of findings detailed in Sha (1999a). The self-administered survey instrument was mailed to 577 members of the internally elected leadership of the Democratic Progressive Party on Taiwan. The response rate was 29%, with 166 useable questionnaires returned via mail. Because the survey asked participants about the party's internal and external public relations behaviors, those results are reported separately in this paper, just as they were in Sha (1999b). The findings with respect to the organization's external communication behaviors were published in Sha and Huang (2004).

The last study examined in the present meta-analysis was conducted by Rhee (1999) on public relations practitioners located in Seoul, South Korea, who worked in in-house public relations departments, public relations divisions of advertising agencies, and public relations agencies. Portions of the findings were reported in Rhee (2002), but nothing has yet been published dealing with the dimensions of public relations that Rhee (1999) measured. Both Rhee (1999) and Rhee (2002) reported a 61% response rate, with 212 of 344 mailed questionnaires returned, although the actual rate could be rounded up by one percentage point.

Appendix B: Research in Public Relations

Appendix B explains not only the actual items that make up each dimension of public relations, but also the reliability of the items across three studies. The higher the reliability score, the more appropriate it is for the items to be grouped together to measure a particular public relations dimension. This information is important for public relations scholars seeking measurements of the new public relations dimensions for theory-building purposes, as well as for public relations practitioners seeking better ways to measure organizational effectiveness in communicating with target publics.

Keeping in mind that no quantitative measures of conservation were available for analysis, Tables B1 through B6 offer a summary of the items that made up six of the seven dimensions discussed earlier in this chapter. Items constructing each of the dimensions were measured on a five-point Likert-type scale, with 1 being strongly disagree and 5 being strongly agree. As noted in the tables, several of the items were reverse scored because they were worded negatively. This is a common technique in social science research to make sure that respondents are not merely marking the same response to every question.

Strategic Dimensions of Public Relations

Two-way communication. For two-way communication, the four-item construct used by Huang (1997) and Sha (1999b) had greater internal reliability than the seven-item construct used by Rhee (1999), suggesting that the abbreviated measures are in fact a better operationalization of this dimension (see Table B1). The fourth item in this construct was dropped by Huang (1997) prior to computing Cronbach's alpha as a reliability measure; dropping this item also would have raised the alpha reported in Sha's (1999b) data set pertaining to internal communication behaviors.

Table B1: Items Constructing Two-Way Communication

Items (7 Items)	Alphas if Item Deleted			
	Study 1	Study 2a	Study 2b	Study 3
1. The organization listens to publics' opinions.	NR	0.82	0.68	0.56
2. Before starting public relations activities, the organization first researches and tries to understand publics' position with respect to a given issue or the organization itself.	NR	0.79	0.62	0.56
3. After completion of public relations activities, the organization conducts an evaluation.	NR	0.80	0.73	0.57
4. The organization talks and does not really listen to publics' opinions (R).	NR / D	0.87	0.72	0.60
5. Information flows out from the organization but not into it (R).	NA	NA	NA	0.61
6. The organization's public relations programs involve one-way communication from the organization to the publics (R).	NA	NA	NA	0.53
7. The organization's public relations programs involve two-way communication between the organization and publics.	NA	NA	NA	0.54
Cronbach's Alpha (Reliability Coefficients)	0.72	0.86	0.75	0.66

Note. Exact wording of items may have been modified to accommodate research participants in each study. The notation (R) indicates that the item was reverse scored. NR denotes that the data were not reported in the source. NA denotes that the item was not applied in the study. D indicates that the item was dropped from analysis prior to computing Cronbach's alpha for internal reliability. Study 1 was Huang (1997), portions of which were reported in Huang (2004b). Study 2a was Sha (1999b) for behaviors with internal publics. Study 2b was Sha (1999b) for behaviors with external publics, which was published in Sha and Huang (2004). Study 3 was Rhee (1999), portions of which were reported in Rhee (2002).

Symmetrical communication. With regards to symmetrical communication, explicated in Table B2, the internal reliabilities reported by Sha (1999b) and Rhee (1999) indicate that dropping the reverse-scored items (i.e., those items measuring asymmetry and thus possibly conservation) would make the remaining items a better measure of symmetry in public relations. The fourth item, originally a measure of asymmetry, should be dropped from the construct of symmetry, especially as doing so raises the alpha in both the second and third data sets (Sha, 1999b) reported in Table B2. Similarly, the second item might more appropriately serve as a measure of conservation, and the last item seems more appropriate to the construct of ethical communication.

Ethical communication. As indicated in Table B3, the five-item construct used by Sha (1999b) for ethical communication had higher internal reliability that the eight-item construct by Rhee (1999), suggesting again that the shorter list of items is preferable to the longer list. Thus, the reverse-scored items reported in Rhee (1999) were dropped, although the second item was modified to incorporate a broader aspect of disclosure. Both lobbying questions had been dropped by Huang (1997), and neither Sha (1999b) nor Rhee (1999)

Table B2: Items Constructing Symmetrical Communication

	Alphas if Item Deleted			
Items (7 Items)	Study 1	Study 2a	Study 2b	Study 3
1. The organization not only tries to change the individual's attitude and behavior, but also tries to change the attitude and behavior of organizational management.	NR / D	0.73	0.62	0.61
2. The organization tries to change the organization's behaviors and policies after considering publics' opinions.	NR / Y	0.71	0.53	0.62
3. The organization consults those affected by its policies during decision making.	NR / Y	0.69	0.62	0.64
4. The organization's main goal is getting the publics to do what it wants (R).	NR / D	0.82	0.77	0.28
5. The organization plays an important role in mediating conflicts between the organization and its publics.	NA	0.71	0.61	0.72
6. The organization tries to get publics to agree with its point of view (R).	NA	NA	NA	0.30
7. The organization provides the public only with information that portrays it in a good light (R).	NA	NA	NA	0.76
Cronbach's Alpha (Reliability Coefficients)	0.75*	0.78	0.69	0.71

Note. Exact wording of items may have been modified to accommodate research participants in each study. The notation (R) indicates that the item was reverse scored. NR denotes that the data were not reported in the source. NA denotes that the item was not applied in the study. D indicates that the item was dropped from analysis prior to computing Cronbach's alpha for internal reliability. Y indicates that the item was included in a combined symmetrical/ethical dimension. * indicates that the alpha was for a combined construct of symmetrical/ethical communication. Study 1 was Huang (1997), portions of which were reported in Huang (2004b). Study 2a was Sha (1999b) for behaviors with internal publics. Study 2b was Sha (1999b) for behaviors with external publics, which was published in Sha and Huang (2004). Study 3 was Rhee (1999), portions of which were reported in Rhee (2002).

used them. Although Huang (1997) combined the dimensions of ethical and symmetrical communication, the 0.75 reliability that she computed for the combined items was lower than three of four reliabilities that Sha (1999b) computed for the items separately, suggesting that the integration or nonintegration of symmetrical and ethical communication

Table B3: Items Constructing Ethical Communication

Items (11 Items)	Alphas if Item Deleted			
	Study 1	Study 2a	Study 2b	Study 3
1. The organization considers how its public relations activities might affect the public.	NR / Y	0.76	0.75	0.57
2. The organization provides the public with accurate or factual information.	NR / D	0.82	0.75	0.56
3. The organization considers the public interest more than individual or personal interests.	NR / Y	0.77	0.76	NA
4. The organization considers the public's interests more than its own organizational interests.	NR / D	0.83	0.75	0.56
5. The organization engages in open lobbying.	NR / D	NA	NA	NA
6. The organization engages in private lobbying.	NR / D	NA	NA	NA
7. The organization tells publics their motives and reasons for their actions.	NR / Y	0.77	0.73	0.62
8. The organization avoids disclosure of negative information about itself (R).	NA	NA	NA	0.55
9. Public relations' role is the promotion of organizational interests even at the expense of publics (R).	NA	NA	NA	0.55
10. The organization avoids dialogue with the public when it makes unpopular decisions (R).	NA	NA	NA	0.61
11. The organization disseminates only favorable information, while unfavorable information is kept from the public (R).	NA	NA	NA	0.56
Cronbach's Alpha (Reliability Coefficients)	0.75*	0.83	0.79	0.63

Note. Exact wording of items may have been modified to accommodate research participants in each study. The notation (R) indicates that the item was reverse scored. NR denotes that the data were not reported in the source. NA denotes that the item was not applied in the study. D indicates that the item was dropped from analysis prior to computing Cronbach's alpha for internal reliability. Y indicates that the item was included in a combined symmetrical/ethical dimension. * indicates that the alpha was for a combined construct of symmetrical/ethical communication. Study 1 was Huang (1997), portions of which were reported in Huang (2004b). Study 2a was Sha (1999b) for behaviors with internal publics. Study 2b was Sha (1999b) for behaviors with external publics. Study 3 was Rhee (1999), portions of which were reported in Rhee (2002).

remains—conceptually at least—an open question, depending on a particular scholar's research purpose and focus.

Table B4: Items Constructing Mediated Communication

	Alphas if Item Deleted			
Items (9 Items)	Study 1	Study 2a	Study 2b	Study 3
1. The organization distributes news releases.	NR	0.76	0.73	0.64
2. The organization uses advertisements.	NR	0.74	0.73	0.72
3. The organization holds news conferences.	NR / D	0.75	0.72	0.63
4. The organization uses mass media, such as television, radio, broadcasts, newspapers, or magazines.	NR	0.75	0.71	0.65
5. The organization offers information and news briefings.	NR / D	0.78	0.74	0.63
6. The organization gives speeches.	NA	NA	NA	0.67
7. The organization stages events, tours, or open houses.	NR	0.78	0.74	0.64
8. The organization distributes flyers, pamphlets, magazines, or other printed materials that represent the organization.	NR	0.76	0.73	0.65
9. The organization uses the Internet / World Wide Web to receive and send information.	NA	0.77	0.75	NA
Cronbach's Alpha (Reliability Coefficients)	0.78	0.79	0.76	0.71

Note. Exact wording of items may have been modified to accommodate research participants in each study. NR denotes that the data were not reported in the source. NA denotes that the item was not applied in the study. D indicates that the item was dropped from analysis prior to computing Cronbach's alpha for internal reliability. Study 1 was Huang (1997), portions of which were reported in Huang (2004b). Study 2a was Sha (1999b) for behaviors with internal publics. Study 2b was Sha (1999b) for behaviors with external publics, which was published in Sha and Huang (2004). Study 3 was Rhee (1999), portions of which were reported in Rhee (2002).

Tactical Dimensions of Public Relations

Mediated communication. For this dimension, the internal reliabilities were fairly consistent across the four data sets reported in Table B4, although the list of items varied somewhat. Huang (1997) dropped two items related to news conferences and briefings; Sha (1999b) added an item regarding use of new technologies, and Rhee (1999) added an item about speeches, although it was not clear how the latter counted as mediated communication. Overall, this list would be considered on the long side.

The use of advertisements could be dropped, as this item might be more a measure of publicity efforts than of mediated communication; dropping this item also raises the alpha reported in Rhee (1999). Similarly, the seventh item in the list could be dropped; although the item was intended to measure mediated communication in the sense of whether the listed activities resulted in mass media coverage, the actual activities represent face-to-face interactions. The inclusion of the ninth item—regarding the Internet—seems appropriate as we consider dimensions of public relations in an increasingly online age.

Table B5: Items Constructing Interpersonal Communication

	Alphas if Item Deleted			
Items (9 Items)	**Study 1**	**Study 2a**	**Study 2b**	**Study 3**
1. The organization contacts members of the public in person.	NR	0.57	0.44	0.67
2. The organization contacts government offices in person.	NA	NA	NA	0.65
3. The organization has informal contact with the public.	NA	NA	NA	0.64
4. The organization primarily uses face-to-face methods of communication.	NR	0.52	0.37	0.67
5. The organization holds meetings.	NA	NA	NA	0.64
6. The organization holds dinners, banquets, or meal parties.	NR / M	0.58	0.35	0.68
7. The organization uses flattery.	NR / M	NA	NA	NA
8. The organization gives token gifts of appreciation, such as party favors or memorabilia.	NR / M	0.57	0.39	0.67
9. The organization gives expensive gifts.	NR / M	0.64	0.45	0.71
Cronbach's Alpha (Reliability Coefficients)	0.60	0.64	0.46	0.70

Note. Exact wording of items may have been modified to accommodate research participants in each study. NR denotes that the data were not reported in the source. NA denotes that the item was not applied in the study. M denotes that the item was subsequently moved to the dimension of social activities. Study 1 was Huang (1997), portions of which were reported in Huang (2004b). Study 2a was Sha (1999b) for behaviors with internal publics. Study 2b was Sha (1999b) for behaviors with external publics, which was published in Sha and Huang (2004). Study 3 was Rhee (1999), portions of which were reported in Rhee (2002).

Interpersonal communication. Like the list for mediated communication, the items for interpersonal communication were numerous, as indicated in Table B5. For this construct, the third data set reporting behaviors with external publics (Sha, 1999b) was particularly problematic, with its very low alpha, so the present discussion excludes consideration of those data. The item by Rhee (1999) about personal contact with government offices reflected unique aspects of that study and therefore should not be included in general measures of this dimension.

The last four items shown in Table B5 were moved by Huang (1997) to the dimension of social activities, which makes sense given how that dimension has been conceptualized. However, this move would leave only two of the original items for the construct of interpersonal communication, although Huang (2004b) later added contact by phone, and Rhee (1999) added interpersonal communication via meetings. New items to measure this dimension might be deducted from Hallahan (2001).

Social activities. As for items constructing social activities, the internal reliability alpha of 0.77 shown in Table B6 as reported by Huang (1997) is acceptable, although these items need validation in additional studies. In particular, the item regarding the use of flattery requires additional explication. Furthermore, the giving of gifts might be collapsed into a single item. Another item that might get at the dimension of social activities could be one that explicitly explores whether communication between the organization and its publics goes beyond the business relationship to consider personal life events, such as celebrating the birth of a child or expressing condolence at the passing of a loved one.

Table B6: Items Constructing Social Activities

Items (4 Items)	Alphas if Item Deleted			
	Study 1	Study 2a	Study 2b	Study 3
1. The organization holds dinners, banquets, or meal parties.	NR	NA / IPC	NA / IPC	NA / IPC
2. The organization uses flattery.	NR	NA	NA	NA
3. The organization gives token gifts of appreciation, such as party favors or memorabilia.	NR	NA / IPC	NA / IPC	NA / IPC
4. The organization gives expensive gifts.	NR	NA / IPC	NA / IPC	NA / IPC
Cronbach's Alpha (Reliability Coefficients)	0.77	NA	NA	NA

Note. Exact wording of items may have been modified to accommodate research participants in each study. NR denotes that the data were not reported in the source. NA denotes that the item was not applied in the study. IPC denotes that the item was included with the construct for interpersonal communication. Study 1 was Huang (1997), portions of which were reported in Huang (2004b). Study 2a was Sha (1999b) for behaviors with internal publics. Study 2b was Sha (1999b) for behaviors with external publics. Study 3 was Rhee (1999), portions of which were reported in Rhee (2002).

Limitations of the Analysis

One limitation to this meta-analysis was that each of the studies examined here collected data in languages other than English. The Huang (1997) and Sha (1999a) studies were conducted in Chinese, and the Rhee (1999) study was conducted in Korean. Thus, some noise in the data reported by these studies may simply be the result of linguistic and cultural issues associated with translations and back-translations of the survey instruments. These threats to internal validity are an unfortunate fact of research life, and by no means do they invalidate the work of these scholars on dimensions of public relations behaviors.

Another limitation to this meta-analysis lies, of course, in the comparison of results from different populations, whose conceptions of "the organization" and "the publics" may have been quite different as they responded to the survey instruments. Yet, this limitation is inherent in all meta-analyses, which nevertheless remain useful for giving a broader perspective on theoretical phenomena, such as the shift in our field from public relations models to dimensions of public relations behaviors.

A Structurationist Approach to the Life Cycle of Internet Publics and Public Participation

Zoraida R. Cozier and Diane F. Witmer

Internet-based communication has altered publics' mobility, communication, and power relationships and, consequently, transformed identity movements. Although scholars and practitioners have embraced the move toward more interactive Web sites, discussion forums, and blogs to communicate with publics, further exploration of the ontological status of public participation is required to adequately segment, assess, and communicate with online publics. This chapter begins with a brief outline of the current research in public relations on an organization's Internet-based communication to its publics. Since the recommendations for organizations are based on traditional conceptions of publics, we outline these conceptions and their limitations. Three new developments based on a structurationist perspective of publics and public participation are presented in relation to the role of discourse in the formation and transformation of publics. An analytical strategy to link the three characteristics of a discourse public—institutions, identities, and ideologies—and its form of public participation is presented. This chapter concludes with implications for public relations communication with discourse publics.

Online communications have become ubiquitous in organizational life. Organizations are recognizing a "need to discover ways to move beyond thinking of public relations as a function of compliance experts and learn to think of it as an ongoing and inclusive process of discussion" (Hiebert, 2005, p. 3). The quest to increase organizational dialogue and public participation in public relations has led to a focus on Web site structure and ethics. Employers are hiring more communication graduates, particularly those with public relations emphases, to develop Web sites that foster what Kent and Taylor (1998) referred to as "dialogic communication" with various publics and conduct extensive research on electronic communications.

Internet-Based Research in Public Relations

Corporate Web sites. Corporations typically use Web sites to promote the organization's image and to increase their dialogue with publics. Esrock and Leichty (1998) conducted a content analysis of messages and structural features of the Web sites of one hundred Fortune 500 companies. The findings demonstrated that, while image building was common, Web sites focused on "disseminating corporate social responsibility information in much the same way as other traditional, one-way corporate communication vehicles" (p. 317). At the time of the study, "meaningful two-way interaction between organizations and their publics" (p. 317) was minimal.

Similarly, Taylor, Kent, and White (2001) later surveyed one hundred Web sites of activist publics to determine the extent to which these publics engage in dialogue with their publics. They applied five principles of dialogue to Web site design features: (a) dialogic loops, (b) ease of interface, (c) conservation of visitors, (d) generation of return visits, and (e) providing information. The findings revealed that these Web sites had not maximized their potential for dialogue. In an earlier work, Kent and Taylor (1998) suggested that organizations designate public relations staff members to serve as "Internet contacts" (p. 327).

Activists' roles. Public relations researchers have pointed out that activists could utilize the Internet to gain support from and mobilize individuals toward achieving their goals. Coombs's (1998) intriguing study of activist publics demonstrated that activists might gain power through the use of the Internet to connect with other stakeholders. Coombs concluded that activists with little power could (a) use the Internet for a low fee, (b) acquire access to online resources from other activists, and (c) pressure organizations to become more socially responsible. Hearit (1999) studied Intel's speech acts, such as online apologies in relation to a flawed chip. The study illustrated a newsgroup's ability to mobilize publics to an active stage quickly. Hearit recommended that companies with high-tech employees "use staff or hire firms to monitor all appearances online of a company's name, reputation, or products" (p. 304).

Situational publics. Studies have relied on traditional notions of publics and public participation that are based on situational theory (Grunig, 1977; Grunig & Hunt, 1984) and its incorporation into a theory of strategic management (Grunig & Repper, 1992). This approach distinguished stakeholders from publics. First, there is a stakeholder stage in which a group's actions have consequences on an organization or a public is affected by an organization's mission or procedures. These stakeholders become a public when they "recognize one or more of the consequences of a problem and organize to do something about it" (Grunig & Repper, 1992, p. 124). Other conceptualizations of publics that have been applied to public relations include the mass, rhetorical, and critical approaches, but situational publics have constituted the predominant model for public relations scholars. The extensive research on situational publics has contributed to public relations by operationalizing and segmenting an organization's stakeholders and presenting typologies of specialized publics based on their level of involvement or activity. Despite the significant contributions that the strategic management model of publics has provided to the field of public relations, researchers have identified the need for newer conceptions of publics and public participation.

Public relations professionals have enriched their relationships with publics (and markets) by developing online spaces to conduct research; sustaining blogs, listservs, and chat rooms; and implementing public relations campaigns.

But what constitutes public participation? Was it evident in the response of bloggers after Wal-Mart hired a public relations firm to post messages that reflected Wal-Mart's position against health insurance legislation (Barbaro, 2006)? Was it embedded in the voices of the female doctors and lawyers from Afghanistan who claimed to represent the disenfranchised women in Afghanistan on the Revolutionary Association of the Women of Afghanistan (RAWA) (2006) Web site? New technologies prompt divergent temporal and spatial relationships that, in turn, facilitate new participation frameworks for publics and public participation (Cozier & Witmer, 2001). The notion of public participation varies with the theoretical perspective.

Public Participation Defined

Publics are social organizations that enact stakeholder processes through the articulation of members' expertise. The participation status of publics evolves from members' attempts to penetrate alternative or resistant ideologies in an organization. Publics have the transformative capacity to make a difference, to enact change, or to reproduce their systems through the various degrees of ideological penetration. Public participation can be identified in the ways that alternative ideologies advanced by publics influence their or another organization's or public's (a) structuring, (b) public relations communication, and (c) potential or actual interactions with other organizations or publics.

Theoretical Perspectives of Public Participation

Much research of online public participation is based on stakeholder theories, which are drawn from systems perspectives. This research focuses on Web site features that generate responses from publics through feedback processes called "dialogic loops" (Kent & Taylor, 1998; Taylor, Kent, & White, 2001) and forms of activists' communication (Coombs, 1998; Hearit, 1999).

Situational theory tells us that publics engage in information-seeking behaviors and mobilizing efforts to support or oppose a problem or an issue. Public participation is considered ideal if Buber's concept of "genuine dialogue" prevails "as an intersubjective process" in which parties develop relationships based on mutuality and in which communication is the goal (as cited in Kent & Taylor, 1998, p. 321 and Taylor, Kent, & White, 2001, p. 280) or if individuals have the freedom to communicate about issues in a public forum (Heath, 1998). In addition, the role of online social networks that emerge through newsgroups and other online interactions is considered a critical component of public participation (Coombs, 1998; Van der Merwe, Pitt, & Abratt, 2005).

The critical and interpretive approaches to public participation idealize publics by reframing them as empowered agents and creators of their own identities (Cozier & Witmer, 2005; Leitch & Neilson, 2001). Duarte and Eiro-Gomez (2004) noted that "in cyberspace, publics can nowadays create or develop their own discussion groups via e-mail, web-logs, [and] mobile phones" (p. 7). Current research also develops alternative concepts of and approaches to publics by acknowledging the role of knowledge or discourse in their formation, in terms of heterogeneity (Asen, 2000; Cozier, 2001; Cozier & Witmer, 2001; Fraser, 1989; Motion, 2005); modernization, risk, identities, and subpolitics (Cozier, 2001; Cozier & Witmer, 2003; Jones, 2002); co-construction of meanings (Vasquez, 1996; Vasquez & Taylor, 2001); and identity construction of publics (Leitch & Neilson, 2001).

Public relations researchers have drawn on critical and postmodern works by Habermas, Giddens, and Foucault (cf. Chay-Nemeth, 2001; Durham, 2005; Leeper, 1996) to examine the relationships among discourse, identities, and ideologies of publics, particularly those that are marginalized. Public participation is based on the organization's role in stakeholder engagement processes. This chapter extends the conventional approaches to public participation by applying a structurationist perspective of public relations (Cozier & Witmer, 2005) to electronic public participation frameworks.

The Need: Alternative Model of Publics and Participation

The Internet challenges public relations scholars to think of alternative ways to communicate with and assess publics. In an effort to emancipate publics, scholars have proposed a move away from traditional conceptualizations that define publics and their participation in terms of level of activity, position on an issue or set of issues, or as merely respondents to an organization's public relations communication (Cozier & Witmer, 2001; Galloway, 2005; Hallahan, 2000; Leitch & Neilson, 2001; see also Monberg, 1998).

The mobility and diversity of online publics pose challenges to public relations professionals who want to target publics based on issues or actions. Practitioners need to recognize the role of experiences and meanings in the formation of publics (Cozier & Witmer, 2001; Galloway, 2005). Galloway (2005) argued that the mobile nature of electronic communications requires that practitioners move away from the concept of "delivering organizational messages to and from static reception equipment" (p. 572) and think in terms of the interactivity that generates and negotiates meaning systems. Nowhere is this more true than in the here-and-now environment of electronic discourse.

The Answer: A Structurationist Perspective of Public Participation

The structurationist perspective presents a discursive approach to publics and public participation that can be utilized online and offline and overcomes the limitations of situational publics. This perspective addresses the societal and technological changes that have simultaneously influenced the changes in the life cycle of public participation.

"Reflexive-self regulation" (Giddens, 1984) is an attribute that is absent in the biological metaphors of systems. Giddens (1979, 1984, 1993) proposed that actors, as knowledgeable agents, continually negotiate meaning (through structures of signification), enforce power (through structures of domination), and constitute a normative order (through structures of legitimation) as they interactively and institutionally create and sustain organizational environments. Moreover, the knowledgeability of actors is bounded by unacknowledged conditions and unintended consequences of action.

A structurationist perspective of public relations views the organization as the site of contestation of "ideological meaning systems" (Deetz & Mumby, 1990; Mumby, 1989). Giddens's (1979, 1984, 1990, 1991, 1993) structuration theory is based on societal changes in this late modern era. Giddens envisioned a reflexive and recursive relationship among societal systems, knowledge or expertise, and agents. First, late modernity is characterized by systems of expertise or knowledge that organize environments. Second, global and local structures are interconnected. Third, human agents and institutions are reflexively related. All three of these factors influence the formation of discourse publics and public participation across electronic environments.

Structuration theory (Giddens, 1993) explains that individuals draw on their experiences to make sense of realities and utilize these experiences in their daily interactions. As people communicate with one another, they create new meanings or reinforce common ones. Giddens called these *structures of signification*. These structures have intended and unintended consequences called *enablements* and *constraints*.

An additional characteristic of new communication technologies is the globalization that extends membership of publics (Monberg, 1998), which is difficult to project or estimate. Lurkers may not post messages but continually monitor electronic sites. They are part of the participation framework, although they are not always recognized as such. Goffman (1981) described a participation framework in which "others" come into the interaction and take on a particular participation status. Therefore, it is necessary to identify the key characteristics of online publics that attract members and influence their public communication and interactions.

Reciprocity of Communication: A Key Development of Public Participation

Three major developments that underlie the structurationist approach to public participation overcome these challenges. The first development entails the inclusion of the role of discourse in the formation of a collective identity and its relation to public participation. Drawing on the assumption that organizations are created through discourses, Cozier and Witmer (2001) characterized online publics as discourse publics that are discursive entities and comprised of multiple ideological stances. Discursive processes produce and transform shared realities and collective identities and facilitate the formation of identity movements (Cozier, 2001; Hardy, Lawrence, & Grant, 2005). An understanding of a group's set of identities can provide insight into the members' various motivations and intentions to act.

The second development entails adhering to the principles of genuine dialogue in which the concepts of interactivity and identities are made central to the analysis of discourse publics. Banks (2000) identified the need for the inclusion of identities in analyses of online communication and explained that genuine dialogue extends beyond feedback. This development discards the implicit rule of "organizational frameability" (Ehling, 1987) or the sense that public relations situations need to be viewed solely from the perspective of the focal organization. Of particular interest is the reciprocity of communications among organizations and publics.

A structurationist perspective focuses on the reciprocity of communication through the penetration of meanings and ideologies among online interlocutors and the affirmation of identities. The reciprocity of communication, a requirement for public participation, can be found in electronic "knowledge environments" (Cozier, 2001; Cozier & Witmer, 2005). This leads to the third development that permits the authors to address the concepts of publics and public participation as related, but distinct, formations or processes.

A social organization can act as a "disembedding mechanism" when organizational members incorporate global meaning systems into their local organizations (Witmer, 1997). Organizational environments, then, serve as *knowledge environments* in which members draw on ideologies and reappropriate them in other social systems. This means that organizations and publics are not separate, bounded entities, but rather they coexist in time-space relationships that extend beyond their copresence.

Prior to the presentation of the analysis of public participation embedded in the online knowledge environments, it is important to describe the role of discourse in the sustenance of the three main characteristics of discourse publics.

Discourse and Three I's: Institutions, Identities, and Ideologies

Discourse and institutions. The Internet has created an infinite number of possibilities for individuals to engage in discursive practices. The most entrenched practices are considered "institutions" (Giddens, 1984). Practices, according to Giddens (1993), entail situated doings of members. Discursive practices evolve as members articulate discourses on a regular basis.

Alvesson and Karreman (2000) refer to two types of discourses: the big and little *d*'s. Big *D* discourses provide the sources of signification (e.g., scientific expertise defines mental illnesses as biological). These sources of signification offer the meanings of texts and objects. The little *d* discourse deals with talk in interaction. This includes the discursive practices that members of publics engage in regularly. Online interlocutors employ various genres of discourse that form their discursive practices, including humor, Usenet groups, storytelling, and problem/solution talks in online social support groups, such as Lamplighters, which is purported to be the oldest and largest e-mail–based group of Alcoholics Anonymous.

Discourse and identities. Discourse creates organizations, such as online chat rooms and listservs. Individuals come together through ongoing conversations, and it is in these conversations that identities are shaped. Members engage in conversations, and it is in these conversations that realities of objects and ideas are formed. Hardy, Lawrence, and Grant (2005) view identities as discursive objects produced through conversations (pp. 61–62). These identities are always in a state of transformation, such that subsequent conversations have the ability to mold and transform them.

From this we can infer that, when individuals go online, their initial self-concept of identities becomes transformed through the seriality of conversations. An individual's self concept of identity and a group's conception of its collective identities are tied up with an individual or individuals' dominant ideologies.

Giddens (1991) characterizes two identity movements in terms of emancipatory politics and life politics. Emancipatory politics is marked by "a generic outlook concerned above all with liberating individuals and groups from constraints which adversely affect their life chances" (p. 210). RAWA (http://www. RAWA.org), for example, which aims to emancipate Afghan women from their plight, is an example of emancipatory politics. In contrast, life politics is an attempt to sustain traditions through a quest for self-actualization. Members ask moral and existential questions on Web sites, such as that for Burning Man fans (*Burning Man Project*, n.d.), or in discussion groups, such as those for Dead Heads (*Dead Net Central*, n.d.). These forms of identity politics implicate the dominant organizational ideology and the degree of permeability of the organization.

Discourse, meanings, and ideologies. Critical theorists consider organizations to be sites of contestation of meanings. Ideology both structures and is structured by an organization's practices to frame members' identities and ways of seeing as they serve vested interests. The meanings that are deeply embedded are considered ideological (Deetz & Mumby, 1990). To Hall (1985), ideological meaning systems, as systems of representation, "relay the way in which we represent ourselves to the world and others" (p. 103). Ideology is made available through discourse and "as involvements of beliefs within modes of lived existence" (Giddens, 1987, p. 183). MoveOn.org, for example, promotes civic and political action based on a deeply institutionalized ideology of democracy, freedom of expression, and a progressive political agenda.

An Analysis of the Life Cycle of Public Participation

A central theme of structuration theory is the time-space constitution of societal systems that are contingently accomplished by actors and collectivities. The life cycle of public participation entails the three *I*'s (institutions, identities, and ideologies) of a public that were

Figure 1: The Life Cycle of Public Participation

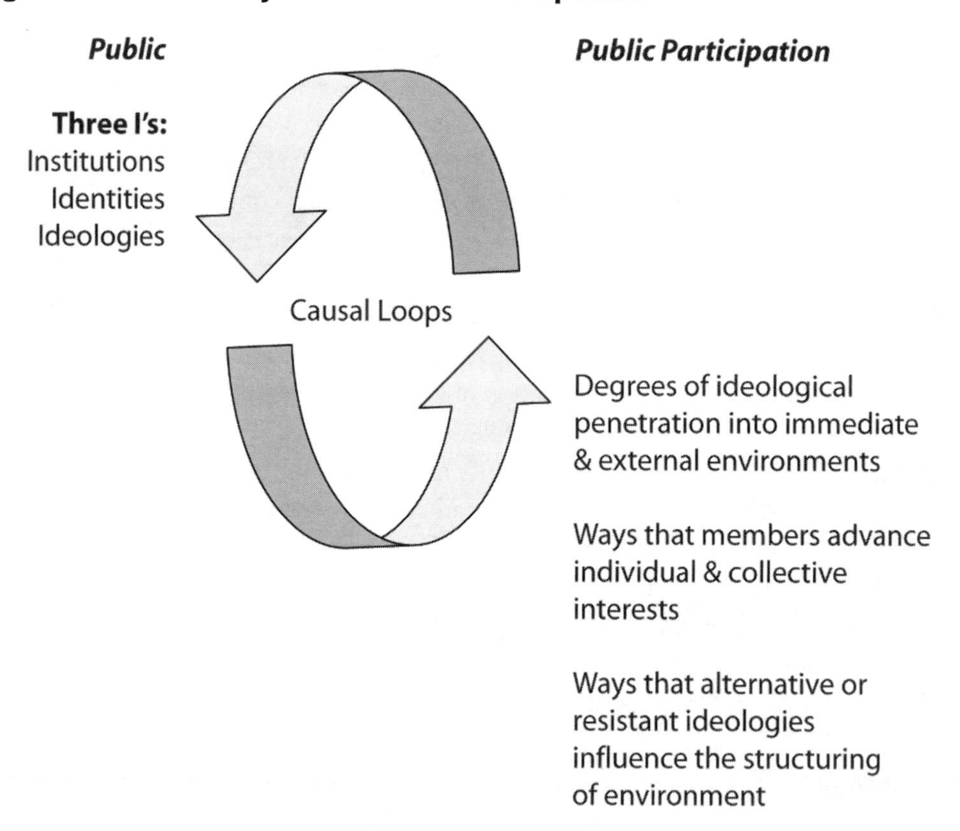

From *Restructuring public participation: A structurationist analysis to three alliances for the mentally ill*, by Z. R. Cozier and D. F. Witmer, 2005, November, paper presented to the Public Relations Division at the National Communication Association conference, Boston, MA. Reprinted with permission.

described previously, as well as the consequences, intended and unintended, of each characteristic. The communicative consequences become evident in degrees of ideological penetration or the extent to which members of a public align with or reject alternative or divergent ideologies. Figure 1 represents the life cycle of public participation.

Publics are not considered open or closed but rather have a low or high degree of permeability. Members act as "access points" (Giddens, 1990) and can filter out or penetrate ideologies from social systems. For instance, through a medical discourse, members of an online parental support organization can filter out ideologies associated with tough love or corporal punishment and penetrate behavioral modification treatments into the organization. The structures of signification associated with behavioral modification treatments have the potential to influence parents' interactions with educational and medical systems. Structures of legitimation or enforcement and structures of domination would be used by members to ensure that the behavioral modification approach becomes the dominant ideological meaning system.

Causal loops. In order to assess the public participation status of a public, research must assess the causal loops that sustain homeostasis, or balance, in organizations. Causal loops explicate the relationship between (a) each public's institution and communicative practices that perpetuate a dominant ideology and engender a collective identity and (b) the type and mode of public participation across knowledge environments.

An assessment of the public participation status of a public requires an analysis of its discursive strategies through the ways that publics advance individual or collective interests and the ways that alternative or resistant ideologies influence the structuring of its immediate or larger environments. The communicative consequences of these processes influence the three characteristics of a discourse public.

Causal loops represent the cyclical relationship among a public's dominant ideologies and its public presentation or choices to engage in interaction with other organizations or publics and vice versa. The expectations of public presentation or what occurs during public presentation influences the discourse publics' identities, ideologies, and institutions. It is important to note that members of publics do not have to physically interact with members of other publics or organizations for public-organizational interactions to occur.

Knowledge environments. Members have the capability to reconstruct their local environments through the rejection or alignment with ideologies. Knowledge environments exist at both the interactional and institutional levels. Knowledge environments are forms of public-public or public-institutional interactions.

Discursive strategies. Three discursive strategies can be used by members of a discourse public to reproduce or transform the public's ideologies. Discourses of resistance and domination can coexist. Members manage discourses of support and resistance by employing cooptation, bridging, and buffering strategies. Cooptation occurs when members extend their meaning systems or draw on meaning systems from other publics or organizations. This could become evident online through links to other sites and the public nature of blogs and newsgroups. Members become subjected to various meaning systems.

Bridging entails the development of alliances with other organizations and individuals that articulate competing and/or similar ideologies. This becomes evident online through online forums for political and social discussions.

The process of buffering entails filtering out access points to enact modes of domination over oppositional discourses and to justify a discourse public's dominant ideological stance. Buffering enables members to implement their identity politics.

In essence, a discursive analysis of discourse publics' online communication provides insight into the formation of a public via the three *I*'s and the attendant modes of public participation. An analysis of its public participation reveals the emergent causal loops that account for its public-public interactions, public-institutional interactions, discursive strategies, degree of permeability, type of knowledge environments, network of relationships, and calls to action. A cursory analysis of the National Alliance on Mental Illness (NAMI) Web site outlines these characteristics.

National Alliance on Mental Illness (NAMI): An Example of the Life Cycle of Online Public Participation

NAMI is an advocacy organization for individuals with mental illnesses. NAMI needs to manage the ideological influences across the three sectors of its movement: family members, medical professionals, and clients. Each sector employs its own institutions, ideologies, and identities, and each sector is mobilized to interact or refrain from interacting with other organizations. The NAMI Web site (National Alliance on Mental Illness, n.d.) at http://www.nami.org exhibits a collective identity of over 1,400 affiliates; it also exhibits the key characteristics of public participation.

Public characteristics. NAMI's Web site collapses the family members' institution of sharing, medical professionals' management practices, and the clients' rights to expression via discussion groups. Family members and clients engage in social support communication that entails the sharing of members' experiential realities. These discussions are sites of identity transformation. As members appeal to the scientific theories of mental illnesses, they become liberated from societal stigma and blame. Through therapeutic environments, members manage their existential questions and make sense of medical realities. This, in turn, implicates their public engagements.

The site also employs the institution of advocacy for the sectors that engage in political and social advocacy. The online political environment influences all sectors of NAMI's membership. Cozier's (2001) research shows that the family members who sustain the ritual of sharing refrain from engaging the public communication that advocacy requires, and, therefore, they limit their communication to social support meetings primarily with members. Some of the clients and medical professionals subscribe to ideologies of rehabilitation that propound views of independence. Consequently, these clients and medical professionals engage in political communication in public regions. The NAMI Web site's "About Public Policy" page attracts these members and calls on these members to act politically and socially.

Public participation. Public participation is evident in the ways that members advance their individual and collective interests and the ways that alternative or resistant ideologies influence the structuring of immediate or larger environments. NAMI's cooptation and bridging strategies exhibit its degree of ideological penetration. For instance, the NAMI Web site lists state and local organizations that represent its perspectives. But these organizations also display various ideological influences, so the list, in conjunction with the "actions" options, represents the degree of permeability or openness to alternative ideologies and the ways that NAMI can engage in public participation.

NAMI presents a report of each state's mental health systems. This report card in conjunction with its opposition to various politicians represents a set of buffering strategies. Therefore, NAMI members are called upon to respond.

An extensive study of NAMI's online discursive strategies would reveal how NAMI employs various discursive strategies and manages the multiple ideological influences. For instance, the *Advocate Magazine* represents various ideologies associated with the etiology of mental illnesses. The magazine includes articles that represent the neurobiological, genetic, and, at times, environmental theories of mental illnesses. The contradictory and supporting theories are not wholly supported by the NAMI affiliates, but NAMI has managed to express multiple ideologies of mental illnesses as it manages the tensions in its knowledge environments.

NAMI.org members and lurkers disembed meaning systems from the various theories presented and in turn, re-embed themes of these theories into their local environments (such as real-time sharing sessions). NAMI's discussion groups provide insight into the array of lurkers and/or visitors that are attracted to and respond to its calls for public participation. In sum, NAMI's Web site permits the sustenance of both life and emancipatory politics (see Giddens, 1991).

Implications for Public Relations Management

Identity politics. Central to the evolution of public participation of discourse publics are the identity politics that influence members' communicative decisions. The rise of online identity movements poses global implications. Publics are reconceptualized in terms of the three I's rather than solely on information-based and issue-driven features. This structuration analysis offers an insider's perspective to the ontological status of public participation as a cycle that includes both the process and product of discursive practices. Researchers can learn about membership rules, the members' management of supporting and resisting discourses, and global and local tensions among discourse publics (and lurkers).

Public participation. Public relations professionals and scholars typically consider a public's participation in terms of collective action. However, based on our characterization of a public, collective action would be difficult to discern without a clear understanding of the multitude of identities, ideologies, and institutions that are created and sustained across the interactions of publics. Moreover, public participation entails the members' management of environmental constraints and opportunities to sustain its dominant ideology.

Dialogue. Dialogic communication is more than an understanding of procedural rules or the development of feedback mechanisms. Dialogic communication entails full understanding and disclosures of identities. Discourse publics' Web sites and online spaces extend beyond linear channels for organizations to establish mediated dialogue with publics (see Kent & Taylor, 2002). The links on publics' Web sites provide insight into the acceptance and penetration of alternative and supportive ideologies. They also represent the discourse publics' choices for interactions or pursuit of goals. The links represent disembedding mechanisms in which public participation exists throughout the interactivity of public-institutional and public-public interactions.

For consideration

1. How could mediated communication be used to increase a person's level of comfort when he or she encounters ideologies different than his or her own online? For example, how could a political party make its ideologies more appealing to someone who didn't agree with its views but was willing to visit its Web site or participate in an e-mail–based discussion group?
2. What role could cognitive dissonance play when someone is faced with ideologies different than his or her own? How can mediated communication be used to offset the effects of cognitive dissonance?
3. To what extent do individuals change their identity online? To what extent do organizations change theirs?
4. In the context of an organization interacting with its stakeholders online, how does dialogue differ from feedback?
5. Consider the structures of domination, legitimation, and signification described in this chapter. How has each been demonstrated in organizations in which people live, work, and play?

For reading

Alvesson, M., & Karreman, D. (2000). Varieties of discourse: On the study of organizations through discourse analysis. *Human Relations*, *53*, 1125–1149.

Buber, M. (1970). *I and thou*. New York: Charles Scribner's Sons.

Cozier, Z. R., & Witmer, D. F. (2001). The development of a structuration analysis of new publics in an electronic environment. In R. Heath & G. Vasquez (Eds.), *Handbook of public relations* (pp. 615–623). Thousand Oaks, CA: Sage.

Hardy, C., Lawrence, T. B., & Grant, D. (2005). Discourse and collaboration: The role of conversations and collective identity. *Academy of Management Review, 30*, 58–77.

Scott, R. W. (1981). *Organizations: Rational, natural and open systems*. Englewood Cliffs, NJ: Prentice-Hall.

Taylor, M., Kent, M. L., & White, W. J. (2001). How activist organizations are using the Internet to build relationships. *Public Relations Review, 27*(3), 263–284.

Vasquez, G. V. (1996). Public relations as negotiation: An issue development perspective. *Journal of Public Relations Research*, 8(1), 57–77.

References

Alvesson, M., & Karreman, D. (2000). Varieties of discourse: On the study of organizations through discourse analysis. *Human Relations*, 53, 1125–1149.

Asen, R. (2000). Seeking the "counter" in counterpublics. *Communication Theory, 10*, 424–446.

Banks, S. P. (2000). *Multicultural public relations: A social-interpretive approach*. Ames, IA: Iowa State Press.

Barbaro, M. (2006, March 7). Wal-Mart enlists bloggers in its public relations campaign. *The New York Times*, pp. C1, C16.

Burning man project. (n.d.). Retrieved July 10, 2006, from http://www.burningman.com

Chay-Nemeth, C. (2001). Revisiting publics: A critical archeology of publics in the Thai HIV/AIDS issue. *Journal of Public Relations Research, 13*(2), 127–161.

Coombs, W. T. (1998). The Internet as potential equalizer: New leverage for confronting social irresponsibility. *Public Relations Review, 24*(3), 289–303.

Cozier, Z. R. (2001). An extension of systems public relations: A structurationist approach to an organization's public relations communication and communicative role (Doctoral dissertation, Purdue University, 2001). *Dissertation Abstracts International*, 63, 419.

Cozier, Z. R., & Witmer, D. F. (2001). The development of a structuration analysis of new publics in an electronic environment. In R. Heath & G. Vasquez (Eds.), *Handbook of public relations* (pp. 615–623). Thousand Oaks, CA: Sage.

Cozier, Z. R., & Witmer, D. F. (2003, November). *A structurationist perspective of public relations: A metatheoretical discussion of boundary spanning*. Paper presented to the Public Relations Division at the National Communication Association conference, Miami, FL.

Cozier, Z. R., & Witmer, D. F. (2005, November). *Restructuring public participation: A structurationist analysis to three alliances for the mentally ill*. Paper presented to the Public Relations Division at the National Communication Association conference, Boston, MA.

Dead net central. (n.d.). Retrieved July 10, 2006, from http://www.deadnetcentral.com

Deetz, S., & Mumby, D. K. (1990). Power, discourse, and the workplace: Reclaiming the critical tradition. In J. Anderson (Ed.), *Communication yearbook, 13* (pp. 18–47). Newbury Park, CA: Sage.

Duarte, J., & Eiro-Gomez, M. (2004, September). *Public relations and the public sphere: (New) theoretical approaches and empirical studies*. Paper presented at the 6th Annual Euprera/DG Puk Conference, Leipzig, Germany. Retrieved March 22, 2006, from http://www.ferpi.it/news_leggi.asp

Durham, F. (2005). Public relations as structuration: A prescriptive critique of the StarLink global food contamination case. *Journal of Public Relations Research, 17*(1), 29–47.

Ehling, W. P. (1987, May). *Public relations' function and adversarial environments*. Paper presented at the International Communication Association conference, Montreal, Canada.

Esrock, S. L., & Leichty, G. B. (1998). Social responsibility and corporate Web pages: Self-presentation or agenda-setting? *Public Relations Review, 24*(3), 305–319.

Fraser, N. (1989). *Unruly practices: Power, discourse, and gender in contemporary social theory*. Minneapolis, MN: University of Minnesota Press.

Galloway, C. (2005). Cyber-PR and "dynamic touch." *Public Relations Review, 31*(4), 572–577.

Giddens, A. (1979). *Central problems in social theory: Action, structure, and contradiction in social analysis*. Berkeley, CA: University of California.

Giddens, A. (1984). *The constitution of society*. Berkeley, CA: University of California Press.

Giddens, A. (1987). *Social theory and modern sociology*. Stanford, CA: Stanford University.

Giddens, A. (1990). *The consequences of modernity*. Stanford, CA: Stanford University.

Giddens, A. (1991). *Modernity and self-identity: Self and society in the late modern age*. Stanford, CA: Stanford University.

Giddens, A. (1993). *New rules of sociological methods: A positive critique of interpretive sociologies* (2nd ed.). Cambridge, UK: Polity Press.

Goffman, E. (1981). Replies and responses. In E. Goffman (Ed.), *Forms of talk* (pp. 5–77). Philadelphia: University of Pennsylvania Press.

Grunig, J. E. (1977). A situational theory of environmental issues, publics, and activists. In L. A. Grunig (Ed.), *Monographs in environmental education and environmental studies: Environmental activism revisited—The changing nature of communication through organizational public relations, special interest groups and the mass media*. Troy, OH: North American Association for Environmental Education.

Grunig, J. E., & Hunt, T. (1984). *Managing public relations*. New York: Holt, Rinehart & Winston.

Grunig, J. E., & Repper, F. C. (1992). Strategic management, publics, and issues. In J. E. Grunig (Ed.), *Excellence in public relations and communication management* (pp. 117–159). Hillsdale, NJ: Lawrence Erlbaum Associates.

Hall, S. (1985). Signification, representation, ideology: Althusser and the post-structuralist debates. *Critical Studies in Mass Communication, 2*(2), 91–114.

Hallahan, K. (2000). Inactive publics: The forgotten publics in public relations. *Public Relations Review, 26*(4), 499–515.

Hardy, C., Lawrence, T. B., and Grant, D. (2005). Discourse and collaboration: The role of conversations and collective identity. *Academy of Management Review, 30*, 58–77.

Hearit, K. M. (1999). Newsgroups, activist publics, and corporate apologia: The case of Intel and its Pentium chip. *Public Relations Review, 25*(3), 291–308.

Heath, R. L. (1998). New communications technologies: An issues management point of view. *Public Relations Review, 24*(3), 273–288.

Hiebert, R. E. (2005). Commentary: New technologies, public relations, and democracy. *Public Relations Review, 31*(1), 1–9.

Jones, R. (2002). Challenges to the notion of publics in public relations: Implications of the risk society for the discipline. *Public Relations Review, 28*(1), 49–62.

Kent, M. L., & Taylor, M. (1998). Building dialogic relationships through the World Wide Web. *Public Relations Review, 24*(3), 321–334.

Kent, M. L., & Taylor, M. (2002). Toward a dialogic theory of public relations. *Public Relations Review 28*(1), 21–37.

Leeper, R. V. (1996). Moral objectivity, Jurgen Habermas's discourse ethics, and public relations. *Public Relations Review, 22*(2), 133–150.

Leitch, S., & Neilson, D. (2001). Bringing publics into the public relations: New theoretical frameworks for practice. In R. Heath & G. Vasquez (Eds.), *Handbook of public relations* (pp. 127–138). Thousand Oaks, CA: Sage.

Monberg, J. (1998). Making the public count: A comparative case study of emergent information technology-based publics. *Communication Theory, 8*, 426–455.

Motion, J. (2005). Participative public relations: Power to the people or legitimacy for governmental discourse? *Public Relations Review, 31*(4), 505–512.

Mumby, D. K. (1989). Ideology and the social construction of meaning: A communication perspective. *Communication Quarterly, 37*, 291–304.

National Alliance on Mental Illness. (n.d.). Retrieved July 10, 2006, from http://www.nami.org

Revolutionary Association of the Women of Afghanistan. (2006). Retrieved June 22, 2006, from http://www.rawa.org

Taylor, M., Kent, M. L., & White, W. J. (2001). How activist organizations are using the Internet to build relationships. *Public Relations Review, 27*(3), 263–284.

Van der Merwe, R., Pitt, L. F., & Abratt, R. (2005, Spring). Stakeholder strength: PR survival strategies in the Internet age. *Public Relations Quarterly*, 39–48.

Vasquez, G. V. (1996). Public relations as negotiation: An issue development perspective. *Journal of Public Relations Research*, 8(1), 57–77.

Vasquez, G. M., & Taylor, M. (2001). Research perspectives on "the public." In R. L. Heath (Ed.), *Handbook of public relations* (pp. 139–154). Thousand Oaks, CA: Sage.

Witmer, D. F. (1997). Communication and recovery: Structuration as an ontological approach to organizational culture. *Communication Monographs, 64*, 324–349.

From Personal and Interpersonal Influence to Impersonal Influence

A Role for Media and Legitimacy in Relationship Management

Jennifer L. Bartlett

This chapter considers how the impersonal influence of mass media plays a central role in the symmetrical relationship maintenance strategy of legitimacy. In doing so, it challenges a number of existing conceptions of the role of media in public relations practice. Firstly, it conceptualises media as being in a symmetrical role in contributing to a mutual understanding of legitimacy between a range of publics and organizations. Secondly, it adds the impersonal influence of media to existing models of personal and interpersonal influence in relationship management. Thirdly, it conceptualises the role of media in terms of long-term effects not short-term outputs. By doing so, it suggests that media play an integral role in understanding the collective perceptions and expectations of others in the social environment within which organization-public relationships are developed.

This chapter considers the central role played by the impersonal influence of mass media in relationship maintenance. The media serve as an influential communication channel between organizations and publics. The common conceptualisation of the role of media within public relations is in terms of the asymmetric models of press agentry/publicity and public information. Existing relationship management models have considered personal and interpersonal influences in relationship management (Broom, Casey, & Ritchey, 1997; J. E. Grunig & Hon, 1999; Ledingham, 2003; Toth, 2000), or in media relations (Ledingham & Bruning, 1998, 1999). There has been little exploration of the specific influence of the media in relationship management for organizations.

The notion of the impersonal influence of media (Mutz, 1998) follows work in political communication about the role that media play in creating communities and their influence within technologically advanced societies with sophisticated media systems. It is in precisely these types of societies that public relations operates as an organizational function, seeking to create mutually beneficial relationships between organizations and publics because they are unable to be facilitated at the interpersonal level as a matter of course.

I suggest that the impersonal influence of media is inherent in the maintenance strategy of legitimacy (J. E. Grunig & Huang, 2000). Public relations is described as ecological; that is, the success or failure of organizations depends upon the relationship between the organization and its social environments (Cutlip, Center, & Broom, 2006). Legitimacy is a core concern of ecological studies (Hannan & Freeman, 1989), suggesting it is likewise inherent in this relationship. However, as expectations of legitimacy develop at a broad national or international level of the social environment, I argue that the legitimacy strategy of relationship management relies on media portrayals of collective perceptions and expectations. The mass media are in a unique position in modern economies. They can enable a common understanding of the views of the collective and of expectations of legitimacy.

This chapter begins with an overview of relationship management and how legitimacy is conceptualised within that literature. It then considers legitimacy and the impersonal effects of media. Finally, it discusses how impersonal effects of media can be incorporated into the influence of existing models in relationship management based on legitimacy and media theories.

Relationship Management and Legitimacy

Relationship management is now conceptualised as a central component of public relations, with explication as a general theory of the discipline (Ledingham, 2003). Ferguson (1984) called for the relationships—rather than the organizations and publics—to act as the unit of analysis, and subsequent relationship management research has evolved from that concept. Broom, Casey, and Ritchey (1997) explicated the concept of relationships, and the propositions surrounding the formation and impact of those relationships, where communication comprised a property of those relationships. Ledingham, Bruning, and Wilson (1999) have sought to further expand this theory through investigation of a series of relationship dimensions, such as perceptions of satisfaction, time in the relationship, and building and maintaining relationships.

J. E. Grunig and Huang (2000) sought to move public relations theory and evaluation from a focus on short-term campaign effects to an understanding of effects on long-term relationships. In doing so, they identified a number of maintenance strategies in the management of relationships. "Assurances of legitimacy" (p. 34) was one of the symmetrical dimensions identified and can be measured by expressions of legitimacy. Legitimacy has also been included in the considerations of Broom, Casey, and Ritchey (1997), who draw on organizational theory's perspectives of interorganizational relationships. Drawing on Oliver (1990), legitimacy here "refers to aspects of interorganizational relationships that lend justification and the appearance of agreement with prevailing norms, rules, beliefs, or expectations of external constituents" (p. 246).

A range of other authors also identify legitimacy as central to organization-public relationships and, more broadly, to the discipline of public relations (Boyd, 2000; Giaradelli, 2004; L. A. Grunig, J. E. Grunig, & Ehling, 1992; Jensen, 1997; Metzler, 2001). However, while legitimacy has been identified as a central component of relationship management and public relations, its specific role in the communication aspects of public relations, and to media in particular, has rarely been discussed. In addition, while legitimacy has been linked to expectations of parties within the organization-public relationship, existing discussions have not considered how a social-level phenomenon influences the interpersonal level of the relationships. The interpersonal level is the central focus of existing relationship management scholarship.

Legitimacy

The term *legitimacy* is widely used but rarely defined. It is even less frequently empirically tested (Suchman, 1995), not just in public relations, but across a range of management disciplines. Legitimacy has been described as a type of "cultural accounting system" (Meyer & Scott, 1992, p. 201) where actors in the social environment assess an organization's or individual's appropriateness within a socially constructed reality. Suchman describes legitimacy as "a generalised perception or assumption that the actions of an entity are desirable, proper or appropriate within some socially constructed system of norms, values, beliefs and definitions" (1995, p. 573).

Legitimacy is achieved when stakeholders with sufficient power and authority reach a form of cultural consensus about what is appropriate and acceptable (Meyer & Scott, 1992). These stakeholders can include regulators and courts that require organizations to meet their standards; educational and professional bodies that impose normative pressures on those practicing in an industry (Dowling & Pfeffer, 1975; Ruef & Scott, 1998); and public opinion, which sets and monitors acceptable social standards (Deephouse, 1996; Elsbach, 1994; Galaskiewicz, 1985; Meyer & Rowan, 1977; Meyer & Scott, 1992). Groups constantly vie for their standards to become dominant and acceptable, creating challenges for individual organizations that seek to build and maintain relationships with relevant publics. These broader conceptualisations of legitimacy are negotiated outside of the organizational domain.

There are a number of impacts arising from the stakeholder battle for consensus. First, stakeholders can impact a specific organization's legitimacy by destabilising perceptions of legitimacy. This could be through calling attention to shortfalls in expectations or demonstrating how an organization is inconsistent with cultural expectations (Elsbach & Sutton, 1992). Coombs (2000) noted that damage to relationships comes about because the parties in the relationship have different expectations of each other.

Second, organizations can require active, rather than passive, support for their activities. Organizations that require considerable support for their activities, say due to competitive pressures, have strong legitimacy demands from the social environment (DiMaggio & Powell, 1983; Suchman, 1995). These organizations demand a credible, collective account of what they are doing and why (Jepperson, 1991) to avoid claims that they are negligent or not needed (Meyer & Rowan, 1992). Hence, they can survive some challenges to legitimacy (Epstein & Votaw, 1978), but maintaining legitimacy is difficult if what is considered legitimate by the social system is conflicting, vague, and in flux (Meyer & Rowan, 1977). There can also be inconsistency about which organizations are deemed not legitimate (Massey, 2001).

Ecology

An emphasis on the mutual dependency of organizations and others in their social environment is reflected in conceptualisations of public relations as an ecological system (Cutlip, Center, & Broom, 2006). Ecological theory is based on two core organizational challenges that organizations need to address in order to garner resources and attain best fit with their social environment—legitimacy and competition (Hannan & Freeman, 1977, 1989). Initially, legitimacy is a prime concern, since, without the support of influential social actors in the social environment, an organization will not achieve legitimacy if it or its practices are not accepted as representing appropriate ways to conduct activity (Zucker, 1977, 1983). Neither will it be able to access resources required to operate, such as employees and finances, along with regulatory, customer, and community support (Pfeffer & Salancik, 1978).

Competition, on the other hand, arises when the activities of a number of organizations are taken for granted, resulting in competition among organizations for access to the best resources. The way that actors in the social environment decide which organization they will provide resources and support to then becomes a critical factor in an organization's success. Lomi (2000) further illuminates the ecological challenge organizations face by stating that competition is likely to be experienced at the organizational level. For example, competition is experienced as organizations, communities, competitors, and customers interact in the same market or as organizations within the same industry compete for the same employees.

The domain of competition, then, is the local level of experience (Lomi, 2000). Legitimacy, on the other hand, is linked to established patterns of cultural authority supported by nationwide and worldwide environments (Meyer & Rowan, 1977). Legitimacy is derived from a wider environment that can cross national, international, and cultural boundaries (Lomi, 2000; Meyer & Rowan, 1992) as standards of expected social behaviour are negotiated by a range of conferring social actors who tend to deal with so-called big picture issues rather than those of an individual organization or public.

This raises the question of how relationship management strategies involving legitimacy can be considered when the organization is unlikely to experience legitimacy at the interpersonal level. As legitimacy is negotiated amongst a range of social actors, the personal influence of an individual organization does not necessarily contribute to achieving legitimacy. Relationship management also fails to incorporate legitimacy as a symmetrical approach when standards of legitimacy are being determined outside the individual organization's sphere of influence. The challenge is that organizations need to understand both local level experiences—competition and interpersonal relationships—as well as broader level expectations—legitimacy.

Broom, Casey, and Ritchey (2000) may have identified part of one answer to this question when they noted that collective perceptions and expectations are one antecedent of relationships. The role of mass media in facilitating the creation of communities around issues, in the absence of interpersonal contact, has been long recognised (Park, 1938; de Tocqueville, 1956). One function the media can have through this perspective is to mirror opinions of others in the social system by indicating what others are thinking (Schoenbach & Becker, 1995). In doing so, media create impersonal influences on individual and organizational opinions and behaviours as perceptions about the state of the social world are influenced by media portrayals of the collective opinions and experiences (Mutz, 1998).

Media

Theories of impersonal influence of media operate when the information content of the media is outside the experience of individuals and others in their social networks (Mutz, 1998). The media has a significant impact on cognition and knowing about the social world, since situations not experienced personally may be made readily available through electronic media and technology. Even though they lack the trustworthiness of opinions on many more local matters, the media has an expertise as a reliable source of information about a global community (Mutz, 1998).

The other reason that media can play an important role in understanding the legitimacy strategy is that, over time, media coverage has shifted from an event to an issues base. There has been a move from episodic coverage to thematic coverage, and a greater proportion of coverage is now concerned with interpretation of events (Barnhurst & Mutz, 1997). This "long journalism" (Barnhurst, 1991, p. 110) presents an interpretation of events by reporting eyewitness accounts about which remote individuals would otherwise know nothing and by printing results of opinion polls and surveys that show what others are experiencing and thinking. The result is to create perceptions of the collective or generalised other and acceptable expectations of behaviour within a social environment that have been publicly discussed, shared, and agreed upon.

Impersonal influence theories incorporate aspects of a range of theories, including agenda setting, priming, cultivation, and risk perception. The matters with which legitimacy is concerned and their relationship to the expressions of legitimacy (J. E. Grunig & Huang, 2000) can be understood by drawing on central premises of agenda-setting theory. The main claim of agenda-setting theory is the relationship between media attention on a few public issues and the subsequent salience of those issues to the public and policy makers (McCombs, 2005; McCombs & Shaw, 1972). Media agendas can focus on issues, at the first level, as objects and, at the second level, as a series of attributes related to those issues (McCombs & Estrada, 1997). The way these issues and attributes are framed includes the use of both symbols and reasoning devices, such as catch phrases, exemplars, depictions, visual images, metaphors, and appeals to principles or moral claims (Gamson, 1992).

Media depictions and arguments surrounding the second level of the media agenda can be described as substantive and evaluative (Carroll & McCombs, 2003). Through the substantive function, organizations and publics come to know what legitimacy issues are of concern to the social environment. Through the evaluative function, inferences can be made about the appropriateness of the attributes, manifested as the practices associated with expectations about the legitimacy of organizational actions.

The media coverage primes evaluations of how organizational practices meet social expectations by displaying these attributes as an evaluative dimension (Deephouse, Carroll, & McCombs, 2001; Fombrun & Shanley, 1990; McCombs, Lopez-Escobar, & Llamas, 2000). One study evaluated and described the media's attribute agenda through a media favourability ranking (Deephouse, 2000), suggesting that media make and present assessments of organizational attributes through praising (favourability) or criticising (unfavourability) organizational actions.

This directional assessment is embedded in the perspective of the social actor presenting the view (Carroll & McCombs, 2003) and is usually based on impressions of issues of importance to the country or macro-social groups rather than to the individual. This has important implications for the proposed maintenance strategy of assurances of legitimacy, as the views of a range of social actors are presented in a public forum by evaluating an organization's compliance with agreed standards of legitimacy.

Legitimacy and the Media

Given that the media appear to be well placed as a communication channel for depicting expressions of legitimacy that are unlikely to be experienced personally, the discussion now turns to established relationships between legitimacy and media in the existing literature. Five different relationships have been identified in the literature—three in terms of the media content reflecting social expectations and two in which media are a tool for organizations to influence legitimacy. These relationships and the associated authors are depicted in Table 1.

Table 1: Role of Media in Relation to Legitimacy and Key Authors in Literature

Media Viewed from the Perspective of Publics	
Media confer legitimacy	Deephouse, 1996 Galaskiewicz, 1985 Hybels, Ryan, & Barley, 1994
Media measure legitimacy	Baum & Powell, 1995 Deephouse, 1996 Ruef & Scott, 1998
Media depictions aid evaluation of legitimacy	Deephouse (2000)
Media Viewed from the Perspective of the Organization	
Media act as tools for building cognitive legitimacy	Aldrich & Fiol, 1994 Baum & Powell, 1995
Media act as tools for managing legitimacy	Allen & Caillouet, 1994 Elsbach, 1994 Elsbach & Elofson, 2000 Elsbach & Kramer, 1996 Elsbach & Sutton, 1992 Massey, 2001

In the first three categories, media content is viewed from the perspective of publics. In the first, the media have been considered one of the influential social actors that confer legitimacy on an organization (Deephouse, 1996; Galaskiewicz, 1985) by endorsing the organization and its activities. The second role for media has also been conceptualised as providing a reflection and measurement of public support for an organization and, therefore, its legitimacy (Deephouse, 1996; Ruef & Scott, 1998); for example, as a proxy for public opinion. The third role is in priming evaluation of an organization and its practices in terms of shared expectations of legitimacy (Deephouse, 2000).

The second set of relationships between media and legitimacy are viewed from the perspective of organizations and how they use the media to manage perceptions of their legitimacy. For new organizations or innovations that an established organization is seeking to legitimise, the media are a tool to enable the provision of information to the social environment (Aldrich & Fiol, 1994; Baum & Powell, 1995). This asymmetrical approach would incorporate press agentry / publicity and public information communication models. In this way, media are used to build cognitive legitimacy or taken-for-grantedness for the organization or organizational activity. However, it is also important to remember that other organizations are also vying for their perspectives to be considered the appropriate standard of practice. Competing organizations can use the media and cognitive legitimacy to criticise others' practices and promote their own. The press agentry/publicity activities may come from a public and, in this example, media report attacks on organizational legitimacy (Dowling & Pfeffer, 1975; Pfeffer & Salancik, 1978).

The media can also be used as tools for the organization to rebuild legitimacy when it is being questioned. Through impression management (Allen & Caillouet, 1994; Elsbach, 1994; Elsbach & Elofson, 2000; Elsbach & Kramer, 1996; Elsbach & Sutton, 1992) and crisis management practices (Massey, 2001), organizations use the media to signal to the social environment that they are adopting legitimate activities (Elsbach & Sutton, 1992). In these cases, organizations seek to influence the perceptions of the social realm and the legitimacy of their actions.

There are long-term effects of these public attempts to question, build, and endorse expectations of legitimacy. As legitimacy is socially constructed, the legitimated practices provide blueprints for the appropriate way to conduct social interactions (Berger & Luckmann, 1967) in order to gain resources from the social environment (Hannan & Freeman, 1977). The practices of organizations that are most successful at achieving legitimacy are copied either voluntarily, through best practice, or by regulation (DiMaggio & Powell, 1983). The results are the routine and institutionalised behaviours that Broom, Casey, and Ritchey (2000) noted as a consequence of relationship management.

Role of Impersonal Influence in Relationship Management

The existing literature on relationships between media and legitimacy has not considered the impersonal influence of media content on perceptions of legitimacy. By doing so, a clearer contribution of the role of the media in developing an understanding of what constitutes assurances and expressions of legitimacy can be made. The media contributes to relationship management at the personal and interpersonal level by creating a shared understanding of expectations and perceptions about appropriate organizational and public activities within which the framework for relationships can be established. Through understanding the broader social context, relationship outcomes, such as trust, mutuality, satisfaction (J. E. Grunig & Huang, 2000), and the like, can be developed.

This perspective makes contributions at a number of levels:

- Conceptualises media in a symmetrical role as assisting in the development of a mutual understanding of legitimacy between a range of actors in a social environment,
- Includes the impersonal influence of media in understanding expectations of legitimacy that underpin interactions of personal and interpersonal influence, and
- Conceptualises media in terms of long-term effects not short-term outputs. Media play an integral role in understanding the collective perceptions and expectations of others in the social system that lead to institutional and routine behaviours.

The impersonal influence of media operates on both the organization and its publics at the personal level in their understanding of the social world. This, in turn, influences the interpersonal aspects of the relationship. This conceptualisation builds on the individual influence model (Toth, 2000) by adding the effect of media's impersonal influence on the personal understanding of the social world that organizations and publics bring to their relationship interactions. It also contributes to the symmetry of the relationship, making the assumptions upon which both organizations and publics operate transparently and accountably by incorporating the collective perspective rather than simply each individual party's views.

Implications for Relationship Management

The viewpoint proposed recasts the role of media in a symmetrical, rather than an asymmetrical, role. By the media bringing the views of a range of social actors into the public domain and forcing individuals and organizations to face experiences and opinions outside those that they personally experience, accountability and transparency of each of the parties is enhanced (Mutz, 1998). While organizations may seek for their perspectives to be presented in the media, by taking a longer-term perspective on the impact of that media coverage, and by incorporating it with the views of a range of publics, a view of the legitimate or taken-for-granted practices can be negotiated and evaluated in an open and transparent way. The resulting shared social expectations underpin both the beliefs of the individual parties within the relationship and also the way in which the relationship is conducted.

The changing nature of our social world has been facilitated and expanded by technology, including a new role for media. Accountability and transparency, rather than individual-level organizational efficiency, are paramount drivers for organizations (Argenti, 2003). New ways of viewing traditional media, such as an issues focus, combined with easy accessibility to new media have vastly changed the way organizations need to think about media. The media have moved from being seen as tools for creating influence to a means to share and build expectations amongst increasingly remote communities of individuals. New ways of thinking about all types of media therefore become imperative.

While an important shift has taken place in public relations scholarship to focus on relationship management as central to the discipline, the broader social and political frameworks within which relationships operate should not be ignored. This paper takes an early step toward considering how broader-level social expectations, facilitated by new conceptualisations of the media, can be incorporated into thinking about relationship management.

For consideration

1 Consider Suchman's (1995) definition of *legitimacy* as presented in this chapter and the idea of generalised social expectations. Now consider some of the current

issues in the media. What aspects of these issues could be related to what is *legitimate* at a national or international level? In what ways might organizations incorporate these expectations into managing their relationships with publics?

2. What role do the media play in the online management of relationships between organizations and their publics?
3. Commentators have been predicting the demise of traditional mass media with the increased use of new media. How can new media contribute to or inhibit a clearer understanding of collective perceptions and expectations in managing legitimacy in organizational-public relationships?
4. What are recent examples of legitimacy issues that have arisen outside an individual firm's control but still influenced relationships with its publics?
5. What tactics have publics used to influence societal expectations of appropriate organizational practices? What role have the media played in this process?

For reading

Broom, G. M., Casey, S., & Ritchey, J. (1997). Toward a concept and theory of organization-public relationships. *Journal of Public Relations Research*, 9(2), 83–98.

Deephouse, D. L. (2000). Media reputation as a strategic resource: An integration of mass communication and resource-based theories. *Journal of Management*, 26(6), 1091–1112.

Grunig, J. E., & Huang, Y. H. (2000). From organizational effectiveness to relationship indicators: Antecedents of relationships, public relations strategies, and relationship outcomes. In J. A. Ledingham & S. D. Bruning (Eds.), *Public relations as relationship management: A relational approach to the study and practice of public relations* (pp. 23–54). Mahwah, NJ: Lawrence Erlbaum Associates.

Mutz, D. C. (1998). *Impersonal influence: How perceptions of mass collectives affect political attitudes*. Cambridge, MA: Cambridge University Press.

Suchman, M. C. (1995). Managing legitimacy: Strategic and institutional approaches. *Academy of Management Review*, 20(3), 571.

References

Aldrich, H. E., & Fiol, M. (1994). Fools rush in? The institutional context of industry creation. *Academy of Management Review*, 19(4), 645–670.

Allen, M. W., & Caillouet, R. H. (1994). Legitimation endeavours: Impression management strategies used by an organization in crisis. *Communication Monographs*, 61, 44–62.

Argenti, P. (2003). *Corporate communication* (3rd ed.). Boston: McGraw Hill.

Barnhurst, K. (1991). The great American newspaper. *The American Scholar*, 60, 106–112.

Barnhurst, K., & Mutz, D. (1997). American journalism and the decline in event-centred reporting. *Journal of Communication*, 47(4), 27–53.

Baum, J. A. C., & Powell, W. (1995). Cultivating an institutional ecology of organizations: Comment on Hannan, Carroll, Dundon, and Torres. *American Sociological Review*, 60, 529–538.

Berger, P. L., & Luckmann, T. (1967). *The social construction of reality: A treatise in the sociology of knowledge*. New York: Anchor Books.

Boyd, J. (2000). Actional legitimation: No crisis necessary. *Journal of Public Relations Research, 12*(4), 341–353.

Broom, G. M., Casey, S., & Ritchey, J. (1997). Toward a concept and theory of organization-public relationships. *Journal of Public Relations Research*, 9(2), 83–98.

Broom, G. M., Casey, S., & Ritchey, J. (2000). Concept and theory of organization-public relationships. In J. A. Ledingham & S. D. Bruning (Eds.), *Public relations as relationship management* (pp. 3–22). Mahwah, NJ: Lawrence Erlbaum Associates.

Carroll, C. E., & McCombs, M. (2003). Agenda-setting effects of business news on the public's images and opinions about major corporations. *Corporate Reputation Review*, 6(1), 36–46.

Coombs, T. (2000). Crisis management: Advantages of a relational perspective. In J. A. Ledingham & S. D. Bruning (Eds.), *Public relations as relationship management: A relational approach to public relations* (pp. 73–94). Mahwah, NJ: Lawrence Erlbaum Associates.

Cutlip, S. M., Center, A. H., & Broom, G. M. (2006). *Effective public relations* (9th ed.). Upper Saddle River, NJ: Pearson Prentice Hall.

Deephouse, D. L. (1996). Does isomorphism legitimate? *Academy of Management Journal, 39*(4), 1024–1040.

Deephouse, D. L. (2000). Media reputation as a strategic resource: An integration of mass communication and resource-based theories. *Journal of Management, 26*(6), 1091–1112.

Deephouse, D. L., Carroll, C. E., & McCombs, M. (2001, May). *The role of the newsroom bias and corporate ownership on the coverage of commercial banks in the daily print media*. Paper presented at the 5th International Conference on Corporate Reputation, Identity, and Competitiveness, Paris, France.

de Tocqueville, A. (1956). *Democracy in America*. New York: Mentor. (Original work published 1835)

DiMaggio, P. J., & Powell, W. W. (1983). The iron cage revisited: Institutional isomorphism and collective rationality in organizational fields. *American Sociological Review, 48*, 147–160.

Dowling, J., & Pfeffer, J. (1975). Organizational legitimacy: Social values and organizational behavior. *Pacific Sociological Review, 18*, 122–136.

Elsbach, K. (1994). Managing organizational legitimacy in the California cattle industry. *Administrative Science Quarterly, 39*(1), 57–88.

Elsbach, K., & Elofson, G. (2000). How the packaging of decision explanations affects perceptions of trustworthiness. *Academy of Management Journal, 43*(1), 80–89.

Elsbach, K., & Kramer, R. M. (1996). Members' responses to organizational identity threats: Encountering and countering the *Business Week* ratings. *Administrative Science Quarterly, 41*(3), 442–476.

Elsbach, K., & Sutton, R. (1992). Acquiring organizational legitimacy through illegitimate actions: A marriage of institutional and impression management theories. *Academy of Management Journal, 35*(4), 699–738.

Epstein, E. M., & Votaw, D. (1978). Legitimacy. In E. M. Epstein & D. Votaw (Eds.), *Rationality, legitimacy, responsibility: Search for new directions in business and society* (pp. 69–82). Santa Monica, CA: Goodyear.

Ferguson, M. A. (1984, August). *Building theory in public relations: Interorganizational relationships*. Paper presented at the Association for Education in Journalism and Mass Communication conference, Gainesville, FL.

Fombrun, C. J., & Shanley, M. (1990). What's in a name? Reputation building and corporate strategy. *Academy of Management Journal, 33*, 233–258.

Galaskiewicz, J. (Ed.). (1985). *Interorganizational relations* (Vol. 11). Palo Alto, CA: Annual Reviews.

Gamson, W. A. (1992). *Talking politics*. New York: Cambridge University Press.

Giaradelli, D. (2004, May). *A schema-based conceptualisation of "image" and "reputation" in public relations*. Paper presented at the 55th Conference of the International Communication Association, New Orleans, LA.

Grunig, J. E., & Hon, L. (1999). Guidelines for measuring relationships in public relations. *The Institute for Public Relations Commission on PR Measurement and Evaluation*, 1–40. Retrieved March, 19, 2007 from www.instituteforpr.com

Grunig, J. E., & Huang, Y. H. (2000). From organizational effectiveness to relationship indicators: Antecedents of relationships, public relations strategies, and relationship outcomes. In J. A. Ledingham & S. D. Bruning (Eds.), *Public relations as relationship management: A relational approach to the study and practice of public relations* (pp. 23–54). Mahwah, NJ: Lawrence Erlbaum Associates.

Grunig, L. A., Grunig, J. E., & Ehling, W. P. (1992). What is an effective organization? In J. E. Grunig (Ed.), *Excellence in public relations and communication management* (pp. 65–90). Hillsdale, NJ: Lawrence Erlbaum Associates.

Hannan, M., & Freeman, J. (1977). The population ecology of organizations. *American Journal of Sociology*, *82*, 929–964.

Hannan, M., & Freeman, J. (1989). *Organizational ecology*. Cambridge, MA: Harvard University Press.

Hybels, R., Ryan, A., & Barley, S. (1994). *Alliances, legitimation, and founding rates in the U.S. biotechnology field, 1971–1989*. Paper presented at the Academy of Management Meetings, Dallas, TX.

Jensen, I. (1997). Legitimacy and strategy of different companies: A perspective of external and internal public relations. In D. Moss, T. MacManus, & D. Vercic (Eds.), *Public relations research: An international perspective* (pp. 225–246). London: International Thomson Business Press.

Jepperson, R. L. (1991). Institutions, institutional effects and institutionalism. In W. R. Powell & P. J. DiMaggio (Eds.), *The new institutionalism in organizational analysis* (pp. 143–163). Chicago: University of Chicago Press.

Ledingham, J. A. (2003). Explicating relationship management as a general theory of public relations. *Journal of Public Relations Research*, *15*(2), 181–198.

Ledingham, J. A., & Bruning, S. D. (1998). Relationship management and public relations: Dimensions of an organization-public relationship. *Public Relations Review*, *24*, 55–65.

Ledingham, J. A., & Bruning, S. D. (1999). Managing media relations: Extending the relational perspective of public relations. In J. Biberman & A. Alkhafaji (Eds.), *Business research yearbook* (Vol. 5, pp. 644–648). Saline, MI: McNaughton & Gunn.

Ledingham, J. A., Bruning, S. D., & Wilson, L. J. (1999). Time as an indicator of the perceptions and behaviour of members of a key public: Monitoring and predicting organization-public relationships. *Journal of Public Relations Research*, *11*(2), 167–184.

Lomi, A. (2000). Density dependence and spatial duality in organizational founding rates: Danish commercial banks, 1846–1989. *Organization Studies*, *March*, 433–461.

Massey, J. E. (2001). Managing organizational legitimacy: Communication strategies for organizations in crisis. *The Journal of Business Communication*, *38*(2), 153–182.

McCombs, M. (2005). The agenda-setting function of the press. In G. Overholser & K. H. Jamieson (Eds.), *The press* (pp. 156–168). New York: Oxford University Press.

McCombs, M., & Estrada, G. (1997). The news media and the pictures in our heads. In S. Iyengar & R. Reeves (Eds.), *Do the media govern? Politicians, voters and reporters in America* (pp. 237–247). Thousand Oaks, CA: Sage.

McCombs, M., Lopez-Escobar, E., & Llamas, J. P. (2000). Setting the agenda of attributes in the 1996 Spanish general election. *Journal of Communication*, *50*, 77–92.

McCombs, M., & Shaw, D. L. (1972). The agenda-setting function of mass media. *Public Opinion Quarterly*, *36*, 176–187.

Metzler, M. S. (2001). The centrality of organizational legitimacy to public relations practice. In R. L. Heath (Ed.), *The handbook of public relations* (pp. 321–334). Thousand Oaks, CA: Sage.

Meyer, J., & Rowan, B. (1977). Institutionalized organizations: Formal structure as myth and ceremony. *American Journal of Sociology*, *83*(2), 340–363.

Meyer, J., & Rowan, B. (1992). Institutionalized organizations: Formal structure as myth and ceremony. In J. Meyer & W. R. Scott (Eds.), *Organizational environments: Ritual and rationality* (pp. 21–44). Thousand Oaks, CA: Sage.

Meyer, J., & Scott, W. R. (1992). Centralization and the legitimacy problems of local government. In J. Meyer & W. R. Scott (Eds.), *Organizational environments: Ritual and rationality* (pp. 199–215). Thousand Oaks, CA: Sage.

Mutz, D. C. (1998). *Impersonal influence: How perceptions of mass collectives affect political attitudes*. Cambridge: Cambridge University Press.

Oliver, C. (1990). Determinants of interorganizational relationships: Integration and future directions. *Academy of Management Review*, *15*(2), 241–265.

Park, R. E. (1938). Reflections on communication and culture. *The American Journal of Sociology*, *44*(2), 187–205.

Pfeffer, J., & Salancik, G. R. (1978). *The external control of organizations*. New York: Harper & Row.

Ruef, M., & Scott, W. R. (1998). A multidimensional model of organizational legitimacy: Hospital survival in changing institutional environments. *Administrative Science Quarterly*, *43*(4), 877–905.

Schoenbach, K., & Becker, L. B. (1995). Origins and consequences of mediated public opinion. In T. L. Glasser & C. T. Salmon (Eds.), *Public opinion and the communication of consent* (pp. 323–347). New York: Guilford Press.

Suchman, M. C. (1995). Managing legitimacy: Strategic and institutional approaches. *Academy of Management Review*, *20*(3), 571–610.

Toth, E. L. (2000). From personal influence to interpersonal influence: A model for relationship. In J. A. Ledingham & S. A. Bruning (Eds.), *Public relations as relationship management: A relational approach to the study and practice of public relations* (pp. 205–220). Hillsdale, NJ: Lawrence Erlbaum Associates.

Zucker, L. G. (1977). The role of institutionalization in cultural persistence. *American Sociological Review*, *42*, 726–743.

Zucker, L. G. (1983). Organizations as institutions. In S. B. Bacharach (Ed.), *Research in the sociology of organizations* (Vol. 2, pp. 1–47). Greenwich, CT: JAI Press.

Public Relations and Complexity Thinking in the Age of Transparency

Sandra C. Duhé

This chapter combines the evolving science of complexity with the relatively recent notion of transparency to explore the implications of mediated relationship building for organizations. After a brief introduction to complexity science and complex adaptive systems, the chapter positions the Internet as a conduit of transparency and complexity for organizations and addresses the pursuant obstacles, uncertainties, and concepts (e.g., fitness landscapes, the edge of chaos, and self-organization) that complexity thinking can offer public relations practitioners. The organizational benefits derived from a more transparent approach to stakeholder interactivity are discussed, along with business-case rationale designed to assist practitioners seeking to apply these thought-provoking ideas, which often run counter to conventional management models.

In the not-too-distant past, a variety of factors, including a sense of legitimacy, financial success, inherent power, and related policies, enabled publicly held corporations to limit communication with certain stakeholders to the reporting of mandated data. Companies, for example, have been compelled by law to disclose executive salary information to shareholders and make public, via state agencies, air emissions data related to operating facilities. More obligatory than interactive, this corporate regurgitation of facts and figures has met legal requirements but has failed to facilitate any sort of exchange relationship between a company and its stakeholders.

The Internet has played a pivotal role in dissolving what were once considered impermeable information exchange barriers. Online sources of data, along with readily available opportunities for like-minded individuals to collaborate in cyberspace, have significantly increased the transparency of a company's operations, regardless of a company's desire or willingness to interact with its stakeholders. This elevated exposure of firm behavior advanced by new media can either give credence to or cast serious doubt upon corporate discourse within the public sphere. This lifting of the corporate veil brings both complexity and uncertainty to the practice of public relations.

Transparency is risky from a public relations perspective. As organizations become more transparent, they become more vulnerable to the judgments of their publics. While working in corporate public affairs, this chapter author recalls one analyst commenting that the closer one got to her company, the better it looked. Obviously, not all corporate entities share this attribute, but even companies that are lauded for outstanding social and shareholder performance can be rendered uneasy at the thought of having their every move examined at a microscopic level.

How stakeholders will react to a more open organization is unpredictable and beyond management's control. Of additional concern in a more open society is disclosure of information that is not intended for the public domain, such as trade secrets, confidential agreements, and internal bickering (Tapscott & Ticoll, 2003). With a few keystrokes, a disgruntled employee or customer can reveal information an organization would rather not broadcast.

To public relations practitioners and dominant coalitions, this heightened exposure can be unsettling. However, the boundary-spanning function of public relations is well positioned in its "unique perch" (Lauzen, 1995, p. 188) at the edges of organizations to guide executives through the uncertainties of transparency. Relationships are the hallmark of public relations practice (cf. Hon & Grunig, 1999; Ledingham, 2003; Ledingham & Bruning, 2000), and they are especially crucial for transparent organizations.

Organizations, whether for-profit or nonprofit, can choose to be more transparent by proactively sharing their operating practices and policies with stakeholders. Consent, however, is not a prerequisite for transparency. In an era when so much information is stored, transmitted, and accessed electronically, organizations are transparent whether or not they choose to be. The same holds true for elected officials, celebrities, political candidates, military commanders, and other public figures. Those in the public eye can expect to have their backgrounds investigated, decisions debated, and personal lives scrutinized online and in real time. Private citizens are also susceptible: Web sites, including Angie's List (http://www.angieslist.com) and RateMyProfessors.com (http://www.ratemyprofessors.com), impart scrutiny on service providers in business and academic environments. Such public comment sites appear to be growing in number and scope.

This chapter combines the evolving science of complexity with the relatively recent notion of transparency to explore the implications of mediated relationship building for organizations. After a brief introduction to complexity science and complex adaptive systems, the chapter positions the Internet as a conduit of transparency and complexity for organizations and addresses the pursuant obstacles, uncertainties, and concepts (e.g., fitness landscapes, the edge of chaos, and self-organization) that complexity thinking can offer public relations practitioners. Thereafter, the organizational benefits derived from a more transparent approach to stakeholder interactivity are discussed, along with business-case rationale designed to encourage application of these thought-provoking ideas. Although the chapter primarily focuses on complexity thinking in corporate environments, the ideas presented herein are adaptable for nonprofit and governmental organizations as well.

Murphy (2000) aptly notes that the tenets of complexity thinking are not radical breaks from the theories of conflict and change found throughout the public relations literature. Complexity concepts are actually *intuitive* for practitioners who have interacted with stakeholders on behalf of an organization. In practice, public relations practitioners are rarely the source of resistance to complexity and transparency; rather, the challenge practitioners more often face is to influence a change in *management*[1] thinking.

Organizations as Complex Adaptive Systems

Complexity science originated with the examination of how inherent instability—or chaos[2]—within complex physical systems, such as the weather, influenced the behavior of systems (Gaddis, 1997). Complexity in this sense does not mean complicated, yet its underlying concepts are paradoxical. That is, systems classified as complex are those in which randomness plays a large role, yet, upon investigation, emerging (and often surprising) behaviors are discovered to have quite simple rules, or choices, guiding them (cf. Bonabeau & Meyer, 2001; Durlauf, 1998; Kupers, 2001). In complex adaptive systems, initial conditions (i.e., history) matter, and very small changes can have quite significant, and unexpected, results (cf. Allen, 1982; Durlauf, 1998).

Complexity theory, referred to as "the science of sciences" (Clippinger, 1999, p. 1), has been applied to a broad range of fields, including economics (cf. Krugman, 1996), management (cf. Byrne, 1998; Richardson, 2005), e-commerce (cf. "Capitalist," 2000), health care (cf. McDaniel, 1997), and public relations (cf. Murphy, 2000).[3] Its appeal for use outside of the physical sciences stems from its holistic approach: The study of complex systems requires that the entire system be observed, not just one activity, function, or point in time (cf. Clippinger 1999; Newell & Meek, 2000; Stacey, 1992). The comprehensive and dynamic attributes of complexity science make it an appealing theoretical home for examining relationships between organizations and their publics, particularly in a new media environment.

Complex adaptive systems are not mechanical but rather emergent systems, in that properties resulting from the *interactions* of heterogeneous agents (e.g., people, organizations) are of a fundamentally different character than the agents themselves (cf. Durlauf, 1998; Sherman & Schultz, 1998). In other words, the whole is greater—and different—than the sum of its parts. For example, the behavior of a diverse *group* of people will produce different outcomes (e.g., a more powerful voice) than each individual could achieve acting on his or her own.

Building on the biological concept of an open system adapting to its environment, the complex adaptive systems perspective offers insight that resonates well with public relations practitioners: Organizations, like living matter, are interdependent with their environments. When interaction with outside forces (e.g., stakeholders) occurs, the potential for emergent behavior increases, and a profusion of unpredictable—though not entirely random—outcomes ensue (cf. Halal, 1998; Sherman & Schultz, 1998).

Organizations share the complex nature of biological systems described by Cambel (1993, as cited in Newell & Meek, 2000), in that they are (a) *nonlinear* with disproportionate cause and effects, (b) *dynamic* (the human dimension precludes organizations from remaining static), (c) *not fully predictable* (but not completely random), and (d) characterized by *permeable boundaries* with their environment. A combination of relevant literature explains that complex adaptive systems are characterized by inseparable components that can produce counterintuitive results, provide ambiguous and distant links between cause and effect, and are highly sensitive to some changes yet remarkably resistant to others (Kiel, 1994, 2000; Sherman & Schultz, 1998; Stacey, 1992).

So, how does this relate to public relations? Public relations practitioners interact with stakeholders across an organization's permeable boundaries, through which information exchange, via two-way communication, takes place. How open or closed those boundaries are depends on the issue or people involved. Complexity theory proposes that the results of these organization-stakeholder exchanges are neither fully predictable nor completely random. For example, a casual remark to a reporter, mistakenly believed to be "off the record," can trigger multiplying and unforeseen outcomes in numerous areas of a business. Likewise, a multimillion dollar public relations campaign can have little impact on influencing public opinion. In either case, the cause is disproportionate with the effect, which is characteristic of complex adaptive systems.

Complexity theory states that cause and effect can be difficult to trace and also separated by extended periods of time; these phenomena are also observed in the practice of public relations. For example, a root-cause investigation of an operational crisis that adversely impacted a community can reveal the culprit to be one minor lapse in maintenance judgment in years prior. A downturn in the economy can lead to massive layoffs (i.e., a highly sensitive response to change), while, at other times, organizations continue to thrive unscathed (i.e., a high resistance to change). Such varying response to change is found in complex adaptive systems.

Stakeholders, like organizations, act as agents in complex adaptive systems. Organizations that fail to adapt to change pressures from their stakeholder environments risk the likelihood of being selected out of a broader system of competing firms (Robertson, 2004). From their purview at the periphery of organizations, public relations managers are ideally positioned to address the challenge of seeking and catalyzing the *appropriate level of adaptation* to change pressures in order to prevent firm failure.

An organization will—and should—maintain an unchangeable core. That is, firms should not adapt to stakeholder demands so much as to lose their core values and purpose. Sha (2004) alludes to this point in her application of Noether's theorem[4] to public relations when she states that "the law of conservation implies that every organization, even as it engages in symmetrical communication, conserves some fundamental belief, principle, or purpose that will never be relinquished" (p. 398).

A firm's tolerance for adaptation to change pressures will differ based on the circumstances at hand; that is, firms may be less willing to change their established standards of business conduct (e.g., a prohibition against paying bribes to government officials) but more willing to yield to pressures that are in keeping with core ideologies (e.g., increased demand to include more women and minorities on boards of directors). Lauzen and Dozier (1994) note that "organizations with complex and/or turbulent environments ought to respond *dynamically* [italics added] to environmental challenges" (p. 163). Cancel, Cameron, Sallot, and Mitrook's (1997) proposal of a contingency theory of accommodation is particularly relevant for how best to relate to publics in an ever-changing environment:

> It depends upon the ethical implications in the situation. It depends on what is at stake. It depends upon how credible the public is. It depends upon a whole lot of things. (p. 33)

An "it depends" approach to public relations may resonate well with practitioners, but the idea can make managers who favor command-and-control tactics in stakeholder relations uneasy. Gaddis (1997) refers to the advent of complexity thinking among corporate agents as one of the forces that "threaten to cripple the [current] practice of strategic management with inhibiting doubt" (p. 38). A common premise in the complexity literature posits that the best way to determine the outcomes of modifying behavior within a complex adaptive system is to simply implement the change (e.g., establish a dialogue with an opposing nongovernmental organization), and then watch the results unfold. For managers not familiar with complexity principles, Bonabeau and Meyer (2001) warn that the idea of "group behavior that emerges—as if by magic—from the collective interactions of individuals can be a frightening concept for those unaccustomed to it" (p. 114). Such blatant uncertainty can lead corporate executives not to in-depth considerations of enabling an empowering, innovative, and adaptive environment but rather to vehement opposition to such a risky, foolhardy scheme!

At first glance, complexity concepts do not bode well in a boardroom comprised of individuals well scripted in the virtues of long-term planning, cause-and-effect linkages, and cost-benefit analyses. Anderla, Dunning, and Forge (1997) note the long-standing presence of these corporate doctrines, none of which are preserved in a complexity perspective:

> For the better part of a century business acumen has been viewed as resting on three pillars—an unshakeable belief in continuity and order, the idea of superior scientific methods, in particular scientific management, and reliance on quantitative data and our ability to analyse the "hard" facts. (p. 66)

At the core of complexity theory is the notion that adaptive systems are, for the most part, not externally controllable. Gayeski and Majka (1996) argue that linear models of strategy and persuasion developed in the formative years of management and communication theory fail to account for the undeniable presence of chaos and dynamism in an organization's operating environment. That is, stakeholders, whether peacefully latent or fully engaged in the emergence of an issue, continuously reconfigure patterns of meanings and exchanges. Social networks of relationships are constantly changing and evolving. Each of these fluid elements erodes the relevance of linear thinking in a nonlinear system. Gayeski and Majka further imply that successful adaptation requires public relations practitioners to realize that their role is one of influence and support—not creation—of stakeholder interpretations of the organization.

Both complexity theory and the practice of public relations reveal that corporate contact with stakeholders, even in its most rudimentary forms, is replete with unpredictability. One point of engagement can either conclude an exchange or trigger an unexpected and complex chain of events. Mediated stakeholder interaction across space and time adds another dimension of uncertainty to organizations. Nevertheless, competitive firms that fail to respond to the information-equalizing and complexity-enhancing capacity of the Internet and its inherent demands for adaptation are more likely to be criticized—and scrutinized—by their publics.

Transparency: The Corporation Has No Clothes

The Internet has leveled the playing field between corporate watchdogs and those who would prefer to keep information from them by providing activists with low-cost, expedient means to gather and act upon information previously not available outside of corporate circles (cf. Coombs, 1998). Hatcher (2003) asserts that "the Internet is creating a cyber citizenry which is fast eroding the power of political and business elites" (p. 33) and that the vocal, but not always rational, views of special interest groups are becoming a more significant factor in public policymaking.

Consequently, companies fall under the increasing scrutiny of stakeholders who are conscientiously mindful of lapses in financial, ethical, or social judgment and, through Internet channels, equally able to effectively organize and launch a response to perceived (and perhaps unfounded) malfeasance. Companies cannot see these so-called "brand assassins," and "accuracy is not necessarily the issue" ("Blog," 2006, para. 2) for groups that are loosely structured and much more difficult to identify and appease. This information-seeking process[5] can transpire in a manner and timeframe unbeknownst to firms, and it diminishes the relevancy of corporate command-and-control tactics.

Tapscott and Ticoll (2003) artfully describe the "the naked corporation" as one with decreasing ability to interrupt the free flow of information in the present age of transparency. The authors argue that transparency, defined as "the accessibility of information to stakeholders of institutions, regarding matters that affect their interests" (p. 22), is essential to successful enterprises and is driven largely by advances in communications technology. The firms most likely to prosper are those that are willing to openly engage in exchange relationships with their stakeholders. News releases, Web sites, blogs, e-mails, and annual reports each have a role in transparent communication and public relations. The Internet acts as both "the quintessential medium of transparency" and, in keeping with the principles of complexity thinking, a "technological challenge to traditional hierarchies" (p. 26).

The Global Environmental Management Initiative (GEMI) (2004), a nonprofit group of companies focused on corporate citizenship, similarly describes transparency as an organization's openness toward sharing information about how it operates. Of particular interest to public relations scholars is GEMI's recognition of the need for dialogic communication (see also, Kent & Taylor, 2002): "Transparency is enhanced by using a process of two-way, responsive dialogue" (Global Environmental Management Initiative, 2004, p. 1).

In today's business environment, transparency extends far beyond prudent, accurate financial reporting between corporate executives and regulating agencies (Tapscott & Ticoll, 2003). Demand for greater transparency increasingly flows across every stakeholder linkage to the organization, including corporate interest in the behavior and credibility of stakeholder groups. In a 24/7/365 Internet society, watchful observers of corporate behavior—whether internal or external, ally or opponent—no longer must solely rely on the company as a source of information. Rather, multiple alternative gatekeepers, with varying rates of response expediency, can be found. For nearly every report, complaint, media account, position paper, or blog entry filed in cyberspace, there is a search engine to find it and an electronic venue to share and expand upon it. Goodjik (2003) argues that, in an operating environment characterized by increasing focus on a company's relationship with its stakeholders, "transparency and accountability are urgent necessities" (p. 227).

Tapscott and Ticoll (2003) summarize the emerging movement of transparency into every aspect of corporate operations:

> Armed with new tools to find information about matters that affect their interests, stakeholders now scrutinize the firm as never before, inform others, and organize collective responses. The corporation is becoming naked. Customers can evaluate the worth of products and services at levels not possible before. Employees share formerly secret information about corporate strategy, management, and challenges. . . . in a world of instant communications, whistleblowers, inquisitive media, and googling, citizens and communities routinely put firms under the microscope. (p. xi)

The loss of control ushered in by "instant communications" brings little comfort to risk-averse managers seeking structure and certainty in a complex environment. The Internet gives stakeholders the freedom to both seek and respond to information in unpredictable ways; corporate and stakeholder agents communicate amidst the naturally organizing phenomena that typify complex systems (Gayeski & Majka, 1996). The literal mask provided by Internet-mediated communication allows many cyberspace participants to misrepresent themselves, violate social norms, and behave in ways (rude, crude, or otherwise) atypical of interpersonal interactions (Brignall & Van Valey, 2005; Ramirez, Walther, Burgoon, & Sunnafrank, 2002). Furthermore, mediated communication affords no opportunity to ascertain the nonverbal cues critical to interpersonal understanding (Brignall & Van Valey, 2005). Face-to-face contact, often impractical, is frequently needed to verify, debunk, or neutralize content.

The screen of relative anonymity offered by new media channels can provide courage, reach, and perceived power to those who would not otherwise be heard. However, the same mechanisms can also be used to conduct vicious smear campaigns and propagate false information about individuals, organizations, and industries. Computer-mediated communication is akin to broadcast communication and global commerce in that national borders cannot contain its reach (Holmes, 2002). Activist groups have increasingly utilized Internet resources to reach their constituents, build consensus, and expose corporate irresponsibility (Coombs, 1998). Burke (2005) adds that "no government, no organization, and no company is immune from the organizing potential of the Internet" (p. 23).

These sources of corporate anxiety are reinforced when the start-up costs for watchdogs are few. Rosenbloom (2004) emphasizes that "a blogger needs only a computer, Internet access, and an opinion" (p. 31) to bypass corporate gatekeepers and seek a share of voice. Granted, not every blog is credible, accurate, or worthy of corporate concern, but the ability to inexpensively, instantaneously, and broadly share one's perspective has clearly impacted the way information is presented and acted upon today by firms, individuals, and mass media alike. The marketplace of ideas has gone digital, and no organization is untouched by its effect on stakeholder relations.

The Need for Modern Mental Models

Mental models, or worldviews, set the course of corporate activities; they originate and propagate from the highest levels of organizations (Foster & Kaplan, 2001). Managers rely upon these well-honed presuppositions to solve organizational problems, yet unyielding entrenchment of outdated mental models can serve as a formidable obstacle to more interactive stakeholder communication via the Internet. Recognizing and challenging these models are an essential, and potentially cumbersome, first step for public relations practitioners attempting to move their organizations away from an exclusive reliance on conventional command-and-control thinking and toward more open and productive interactions with stakeholders in a complex environment.

Foster and Kaplan (2001) explain how mental models shape companies' views of their role in the economy and indicate their proclivity to seek more interactive relationships with stakeholders. When accurate, mental models can offer competitive advantages, but they must be based on information gathered and updated from sources other than the dominant coalition. Stakeholders have the potential to play a beneficial role in helping corporations ascertain the validity of such models, particularly when mediated, timely, and feedback-oriented communication venues are utilized.

Failure to overcome staid mental models that presume the certainty and value of management controls is a primary reason why many new insights, like proactive stakeholder engagement via the Internet, fail to realize their potential in organizational learning. Senge (1990) argues that mental models are powerful in affecting what organizations *do* because they, in large part, affect what corporate managers *see*. Outcomes within complex adaptive systems will vary greatly if all Internet-reliant activists opposed to company behavior are viewed as a nonsensical nuisance to be tolerated rather than as an alternative, and possibly beneficial, perspective to at least be heard, if not considered.

Instead of welcoming the noise, or environmental inputs, inherent to complex adaptive systems, Sherman and Schultz (1998) find that many conventional managers "in a misguided desire for order and certainty . . . would much rather stamp out ambiguity than see it as a necessary precondition to influence potentiality" (p. 4). Such thinking leads organizations to limit the interactivity of stakeholder communications. Burke (2005) urges companies to abandon a "command-and-control style of managing external affairs," noting that managers may tend to "rely on a style of relationships with external stakeholders that no longer work" (p. 57). These phenomena apply internally as well: Even within organizations, complexity and interdependencies have increased, while central control has decreased (Goodjik, 2003).

Mental models that perceive unexpected events as an annoyance rather than "as a desirable instrument of change and order" (Kupers, 2001, p. 15) prevent organizations from recognizing issues beyond boundaries that are too tightly defined (Sherman & Schultz, 1998). Faulty perceptions, created in an internal vacuum, hinder understanding of what actually defines a business's industry, environment, and stakeholders. Sherman and Schultz assert that such limited vision hampers an organization's ability to anticipate problems on the horizon. These distortions in awareness can have catastrophic public relations, economic, and/or societal effects.[6]

Effective stakeholder engagement involves seeking input from even those voices corporations may rather not, but need to, hear. Oftentimes, divergent ideas have the potential to create worthwhile opportunities when relationships—even mediated ones—are in place to capture those ideas. Pruzan (1998) calls for a fundamental shift in management training and practice away from efficiency and control to one of values-based management, noting that the former can be counterproductive for organizations. Referencing a variety of stakeholder groups, Pruzan warns, "It is unwise to attempt to plan and control what cannot be controlled without destroying vital qualities of those who are planned-for and controlled" (p. 1380).

Navigating the Fitness Landscape

The complexity literature refers to a field of action, or fitness landscape, upon which relationships, emerging consequences, and adaptive behaviors take place (cf. Kauffman, 1995). The Internet binds organizations and stakeholders within the landscape, providing not only enhanced transparency of information, but also the potential for two-way communication. Rules of engagement by which corporate and stakeholder agents maneuver the landscape are referred to as fitness criteria, which facilitate interaction and ultimately affect its outcomes.

In light of limited response resources, competing priorities, and source/content validity concerns, not all stakeholders can or should be actively engaged; this principle applies to both mediated and interpersonal environments. Although companies should expect to make some concessions as part of their stakeholder relationships, fitness criteria need to include suitable bounds guiding and limiting stakeholder involvement in firm decision making, as increased involvement restricts organizational autonomy to some degree. Proponents of complexity thinking recognize that organizations cannot be so transparent as to be indistinguishable from their environments; boundaries must be firm enough to contain the organization as an entity yet permeable enough to allow for beneficial exchange (Goldstein, 1994; Halal, 1998).

Feedback is an important component of the fitness landscape and a requirement for self-organizing systems (Allen, 1994; Durlauf, 1998; Kiel, 1994). Sherman and Schultz (1998) find that businesses operating in complex environments are subject to a continual process of feedback from multiple stakeholders. Mediated channels, however, are distinctive in that they permit these critical inputs and outputs to flow freely across time and distance, unlike relationships dependent on or limited to copresence. When corporations selectively facilitate the creation of exchange relationships with key stakeholders, cognitive capacities can be broadened for all agents involved.

Receiving and processing feedback enhances the learning process of an organization, a task for which new media are well suited. The ultimate goal is neither agreement among agents nor adaptation to all feedback exchanged; rather, the aim is to create a system able to cope with, and benefit from, multiple inputs. That is, the give-and-take nature of corporate-stakeholder engagement reveals new information to all participants. Through feedback, both organizations and their stakeholders give mental models an opportunity to be challenged and altered.

In complex systems, feedback takes two forms: negative, or damping, and positive, or amplifying (Newell & Meek, 2000; Stacey, 1992). Negative feedback serves to maintain an equilibrium state for organizations; positive feedback, on the contrary, escalates small changes and can become unmanageable if not limited in some way, such as through fitness criteria. In a corporate setting, mandated stakeholder communications, such as quarterly earnings announcements, are standardized and produced according to a relatively fixed plan. The role of negative feedback serves to dampen any deviation (e.g., doing more or less communicating) around stated goals and preserves predictability. Positive feedback, however, comes into play when stakeholder involvement in corporate decision making is allowed to escalate over time in unreasonable frequency and substance. If uncontrolled, too much positive feedback could allow stakeholders to overstep appropriate boundaries and adversely interfere with decisions requiring business expertise (e.g., price setting or management of trade secrets). The same can be said of the need to avoid improper encroachment of a corporation on stakeholders' interests.

A beneficial application of positive feedback is related to Durlauf's (1998) description of a conformity effect, in which the perceived benefit of a choice increases with the number of peers making the same choice. For instance, if a petrochemical plant implements an interactive online risk communication program, neighboring sites are likely to follow suit as public expectations for doing so escalate and benefits enjoyed by the originating site (e.g., enhanced credibility with stakeholders) are desired by others. Without any formal planning, the bar for public communication is considerably raised for industry in the region, highlighting the endemic nature of nonlinearities. Based on the proposition that "trend creates trend," Allen (1994) adds,

> [the] ability to produce innovation and change will drive the circumstances of others and drive evolution itself, favouring individuals [or organizations] capable of dealing with change, and eliminating those that are incapable. (p. 584)

Organizations that actively engage stakeholders in cyberspace are, in essence, functioning as open structures extracting energy (i.e., information) from their environments for maintenance and survival. Kiel, among others, refers to these structures as being *dissipative* (Prigogine, 1984, as cited in Kiel, 1994). Ongoing inputs, or fluctuations, test the stability of the structure. Sweet (2000) notes,

> In its application to organizations, energy is imported in the form of information. For that information to be utilized to assist in the organization's coevolution, it must be gathered from the environment at all levels of the system and dispersed throughout the system to support creative change. (p. 190)

The Internet is the global pipeline gathering, channeling, and altering energy in the form of information exchange between corporations and their stakeholders.

Communicating at the Edge of Chaos

Mental models often prevent organizational managers from embracing the uncertainties of Internet communications out of concern for inviting an excessive and possibly harmful level of positive feedback into the firm. Complexity theorists propose that organizations should strive to operate at the very fringe of chaos and order, finding the "sweet spot" between the perceived safe, yet stagnant, state of equilibrium in their landscapes (a point at which a system is at rest or not changing) and a level of chaos so unruly as to implode the organization (Clippinger, 1999; Stacey, 1992). This optimal space allows creativity and innovation to advance organizations beyond the restrictions of market conditions, technology, or institutional structure.

Adaptation occurs somewhat ambiguously on the part of both the corporation and its stakeholders via "bounded instability" at the edge of chaos. According to Stacey (1992),

> Together these intended and unintended feedback loops, of both the positive and negative kind, make it pointless to talk about a firm adapting to its environment. Managers and the people who constitute their firm's environment together create what happens through their interaction. It is not at all clear who adapts to whom. (p. 72)

Goldstein (1994) encourages organizations not to seek a state of equilibrium, where information is relatively limited, but rather to pursue far-from-equilibrium conditions whereby a business is connected to an environment from which it was previously isolated. Here, the corporation is primed to develop new ways of functioning and begins to move toward self-organization.

The Path to Self-Organization

A self-organizing system is one in which randomness and chaos suddenly evolve into unexpected order (Krugman, 1996). Coleman (1999) explains how self-organizing behavior emerges when organizations leverage the complexity, or interconnectedness, of stakeholder relationships and therefore enable information and ideas to be transformed into innovations. Ideally, boundaries between corporations and stakeholders, though certainly required, should not unnecessarily inhibit information exchange but rather enhance the likelihood of adaptive behaviors. The Internet can be a worthwhile partner in this two-way exchange. As a whole, self-organizing companies and stakeholder groups take on a form and function that are independent of their individual agents; subsequent behavior is not attributed to one particular subunit but rather to the interaction of various subunits (cf. Bonabeau & Meyer, 2001; Clippinger, 1999; Jones & Culliney, 2000; McDaniel, 1997). Management, then, becomes the art and science of influencing self-organization from below, as opposed to conventional models that dictate hierarchical control from above.

Goldstein (1994) notes, "The issue, then, is not how to pressure a system to change, but how to unleash the system's self-organizing potential to meet a challenge" (p. 9). Paradoxically, self-organization results not from *adding* structure, procedures, or controls to a business environment but rather from *removing* those mechanisms that inhibit self-organizing behavior.

Toward a Business Case for Complexity Thinking

The potential contribution of complexity theory to the area of mediated stakeholder relations is limited in organizations if only intangible benefits can be identified. In corporate settings, the ability to present bottom-line benefits of increasing interaction with stakeholders is essential to the advocate's success.

Halal (1998) suggests transition to a corporate community mindset that values information flow and is open to adaptive behaviors epitomizes democratic ideals, creating a private sector that is increasingly productive and socially beneficial. For example, he asserts that inflation could be better controlled given enhanced corporate productivity, hostile takeovers could be made more difficult because of increased interconnectedness within corporate environs, government regulations could be minimized as corporate stakeholders provide a more efficient checks-and-balances function, and social needs could be more effectively addressed through a combination of public and private resources and knowledge.

When stakeholder relationships are allowed to become more complex through Internet channels, and when public regard for a company improves, the increased robustness of corporate entities can better withstand potential shocks from exogenous events. Marion and Bacon (1999) propose that the ability for companies to steward relationships with multiple stakeholders is a condition of resilient infrastructure. Innovative thinking and previously untapped solutions emerge from stakeholder engagement but require a willingness on the part of all participants to experience some degree of vulnerability, uncertainty, and risk in the process. Trust, support, and mutual understanding develop over time but are not expected to dissolve the differences that uniquely define each agent.

Sherman and Schultz (1998) summarize the severe consequences that can result from corporate managers adhering to an illusion of safety and control in a closed system: a loss of organizational development and innovation. Marion and Bacon (1999) add that organizational extinction can result from the breakdown of networks and interactions among agents in a complex environment. Even sizable corporations able to survive with a relatively closed mentality face increasing demands for more specialized services, products, and stakeholder access. Anderla, Dunning, and Forge (1997) predict that these firms will eventually be forced to "adopt new attitudes and abandon earlier profiles of insular dominance" (p. 120). In the age of transparency, the risk of *not* engaging stakeholders could prove more daunting than the uncertainties encountered in complex environments.

Implications for Public Relations Management

Perceptions of the Internet and its impact on society vary widely across disciplines, cultures, and generations. Embraced by some organizations, it is avoided and even feared by others. Organizations that fall under both descriptions can find themselves interacting in a complex—and increasingly transparent—environment. It is, however, not so much what the Internet *is* as how it is *used* that affects an organization's relationships with its stakeholders.

Brignall and Van Valey (2005) highlight the neutrality of the Internet as a communication medium:

> The Internet itself is neither negative nor positive. It is inanimate, an object or a tool that can be used in various ways. To reify the Internet and suggest that it is somehow inherently liberating or enslaving is misleading. Nevertheless, it is important to look at how interaction on the Internet differs from other forms of interaction. These differences may play havoc with traditional social interaction rituals. (p. 340)

Kent and Taylor (1998) add, "Technology itself can neither create nor destroy relationships: rather, it is how the technology is used that influences organization-public relationships" (p. 324).

Profit-driven motives and related malfeasance have heightened public scrutiny of corporations. In today's business environment, transparency is a given; dialogic communication still remains a choice. Corporate watchdogs, with comparatively fewer resources at their disposal, have been arguably more effective and interactive online communicators compared to their corporate counterparts so often bound by outdated mental models. Those organizations willing to navigate uncertainty in order to facilitate learning have a greater likelihood of being not only a long-standing but also a welcomed member of the broader system.

Complexity theory offers an intriguing, though nascent, framework for the study of mediated corporate interaction with stakeholders and the resulting consequences of those emerging relationships. Overcoming mental models that oppose the shattering of conventional (and comfortable) corporate communication barriers is critical to establishing more transparent, and mutually beneficial, relationships with stakeholders, but there are challenges to doing so. For a variety of security, identity, and credibility reasons, online stakeholder engagement can be risky and unpredictable, but perhaps the greatest risk lies in corporations facing unfavorable events without stakeholder support. Corporate interaction with stakeholders, despite its unpredictable nature, increases connectedness and responsiveness to external environs and breeds institutional resilience. New media offer an increasing array of channels to establish and steward these connections.

The lesson of complexity science is for public relations managers to deliberately create environments conducive to learning. Stakeholders—including those with opposing views—can serve as valuable sources of otherwise untapped information and ideas. The principles of complex adaptive systems call for creation of more freeing structures that allow both organizational and stakeholder inputs to unfold, take shape, and influence the overall system. The Internet, as a conduit of transparency and complexity, channels and supplements the information gathering and feedback process needed for organizational evolution.

However, before an organization can realize the benefits of greater interactivity with its stakeholders, it must first be open to the risks and vulnerabilities involved. Public relations practitioners are ideally positioned to help organizations find the "sweet spot" between too little and too much stakeholder engagement and the appropriate bounds for organizational adaptation to change pressures.

The presence of stakeholders is a necessary and sufficient condition for complexity. Relatively few studies have examined the role of complexity in public relations, leaving much terrain yet to be explored. Future work can build on Lauzen's (1995) methodology of measuring environmental complexity through stakeholder relationships. Likewise, Murphy (2000) describes a number of complexity-based methods for public relations research. These authors, among numerous others in the social sciences, provide worthwhile points of entry into how complexity thinking can be incorporated into the study and practice of public relations.

Our charge as public relations professionals is to help guide organizations toward more transparent relationships with stakeholders while maintaining the core beliefs, values, and purpose that define those organizations. This balancing act becomes more challenging in a mediated environment but nevertheless remains a vital, enduring, and fulfilling part of the practice.

To be credible change agents, public relations practitioners must demonstrate an in-depth knowledge of their organizations' goals and objectives and, to the extent possible, the stakeholders involved in the issue at hand. Mediated and interpersonal channels assist both organizations and stakeholders in the research process. Research helps practitioners to select or respond to worthwhile opportunities for engagement and, more importantly, gain management support for delving into the unknowns of stakeholder relations in a complex environment.

Not to be underestimated is the courage required for practitioners to challenge outdated or inaccurate thinking in organizations, particularly when conventional command-and-control models and a *lack* of transparency appear to have served the bottom line of an organization well. The complex nature of relationships between practitioners and their internal and external stakeholders makes for a dynamic, sometimes frustrating, work environment. But the ability to facilitate change through mutually beneficial relationships makes for an extraordinarily rewarding career.

For consideration

1. To what extent is the outcome of interpersonal contact with stakeholders more or less predictable than online contact?
2. When should organizations use new media to gather stakeholder input about an issue—as it's emerging, once it's fully developed, or after the organization has responded to it? To what extent does it depend on the issue?
3. How could cross-disciplinary relationships *within* an organization help public relations practitioners build a business case for more transparent relationships with publics?
4. How can a public relations practitioner determine the proper bounds (or limits) for adaptation? To what extent can the practitioner make this determination on his or her own?
5. What role does expectation management play—both internally and externally—in the cultivation and stewardship of stakeholder relationships?

For reading

Global Environmental Management Initiative (GEMI). (2004). *Transparency: A path to public trust*. Retrieved June 9, 2006, from http://www.gemi.org/Transparency-PathtoPublicTrust.pdf

Robertson, D. A. (2004). The complexity of the corporation. *Human Systems Management, 23*, 71–78.

Sha, B-L. (2004). Noether's theorem: The science of symmetry and the law of conservation. *Journal of Public Relations Research, 16*(4), 391–416.

Sherman, H., & Schultz, R. (1998). *Open boundaries: Creating business innovation through complexity*. Reading, MA: Perseus Books.

Tapscott, D., & Ticoll, D. (2003). *The naked corporation: How the Age of Transparency will revolutionize business*. New York: Free Press.

Notes

1. Although public relations is defined as a management function (cf. Cutlip, Center, & Broom, 2006), and the practice of public relations is most effective when the head of the public relations function is a member of an organization's dominant coalition (Grunig, 1992), references to management thinking and management models, etc., throughout this chapter refer to dominant coalition executives *outside* of the public relations function with whom practitioners interact.
2. In a mathematical sense, chaos refers not to disorder and confusion but rather to stochastic, or random, behavior occurring within a deterministic system, that is, a system governed by exact and unbreakable law (Stewart, 1989).
3. See also Cottone (1993) and Murphy (1996) for chaos-related applications.
4. In mathematical physics, "Noether's theorem shows that symmetry—or change—can only exist simultaneously with conservation or invariance" (Sha, 2004, p. 391).
5. For further discussion, see Ramirez, Walther, Burgoon, and Sunnafrank (2002).
6. For a discussion of how distortions in awareness can impact policy making, see Brem (2000).

References

Allen, P. M. (1982). Evolution, modeling, and design in a complex world. *Environment and Planning*, *9*, 95–111.

Allen, P. M. (1994). Coherence, chaos and evolution in the social context. *Futures*, *26*(6), 583–597.

Anderla, G., Dunning, A., & Forge, S. (1997). *Chaotics: An agenda for business and society in the 21st century*. Westport, CT: Praeger.

The blog in the corporate machine. (2006, February 9). *The Economist*. Retrieved May 29, 2006, from http://www.economist.com/business/PrinterFriendly.cfm?story_id=5501039

Bonabeau, E., & Meyer, C. (2001, May). Swarm intelligence: A whole new way to think about business. *Harvard Business Review*, pp. 107–114.

Brem, R. J. (2000). The Cassandra complex: Complexity and systems collapse. In G. Morcol & L. F. Dennard (Eds.), *New sciences for public administration and policy: Connections and reflections* (pp. 125–150). Burke, VA: Chatelaine Press.

Brignall, T. W., III, & Van Valey, T. (2005). The impact of Internet communications on social interaction. *Sociological Spectrum*, *25*, 335–348.

Burke, E. M. (2005). *Managing a company in an activist world: The leadership challenge of corporate citizenship*. Westport, CT: Praeger.

Byrne, J. A. (1998, September 21). Virtual management. *Business Week*, pp. 80–82.

Cancel, A. E., Cameron, G. T., Sallot, L. M., & Mitrook, M. A. (1997). It depends: A contingency theory of accommodation in public relations. *Journal of Public Relations Research*, *9*(1), 31–63.

Capitalist e-construction. (2000, March). *Wired*, pp. 210–219.

Clippinger, J. H., III. (1999). Order from the bottom up: Complex adaptive systems and their management. In J. H. Clippinger III (Ed.), *The biology of business: Decoding the natural laws of enterprise* (pp. 1–30). San Francisco: Jossey-Bass.

Coleman, H. J., Jr. (1999). What enables self-organizing behavior in businesses. *Emergence*, *1*(1), 33–48.

Coombs, W. T. (1998). The Internet as potential equalizer: New leverage for confronting social irresponsibility. *Public Relations Review, 24*(3), 289–303.

Cottone, L. P. (1993). The perturbing worldview of chaos: Implications for public relations. *Public Relations Review, 19*(2), 167–176.

Cutlip, S. M., Center, A. H., & Broom, G. M. (2006). *Effective public relations* (9th ed.). Upper Saddle River, NJ: Pearson Education.

Durlauf, S. N. (1998, Winter). What should policymakers know about economic complexity? *The Washington Quarterly, 21*(1), 157–165.

Foster, R., & Kaplan, S. (2001). *Creative destruction: Why companies that are built to last underperform the market and how to successfully transform them*. New York: Currency.

Gaddis, P. O. (1997). Strategy under attack. *Long Range Planning, 30*(1), 38–45.

Gayeski, D. M., & Majka, J. (1996, September). Untangling communications chaos: A communicator's conundrum for coping with change in the coming century. *Communication World, 13*(7), 22–25.

Global Environmental Management Initiative (GEMI). (2004). *Transparency: A path to public trust.* Retrieved June 9, 2006, from http://www.gemi.org/Transparency-PathtoPublicTrust.pdf

Goldstein, J. (1994). *The unshackled organization: Facing the challenge of unpredictability through spontaneous reorganization*. Portland, OR: Productivity Press.

Goodijk, R. (2003). Partnership at corporate level: The meaning of the stakeholder model. *Journal of Change Management, 3*(3), 225–241.

Grunig, J. E. (Ed.). (1992). *Excellence in public relations and communication management*. Hillsdale, NJ: Lawrence Erlbaum Associates.

Halal, W. E. (1998). *The new management: Democracy and enterprise are transforming organizations*. San Francisco: Berrett-Koehler.

Hatcher, M. (2003). New corporate agendas. *Journal of Public Affairs, 3*(1), 32–38.

Holmes, D. (2002). Transformations in the mediation of publicness: Communicative interaction in the network society. *Journal of Computer-Mediated Communication*, 7(2). Retrieved August 26, 2006, from http://jcmc.indiana.edu/v017/issue2/holmes.html

Hon, L. C., & Grunig, J. E. (1999, November). *Guidelines for measuring relationships in public relations*. Gainesville, FL: The Institute for Public Relations Commission on PR Measurement and Evaluation. Retrieved August 26, 2006, from http://www.instituteforpr.org/index.php/IPR/research_single/guidelines_measuring_relationships

Jones, D., & Culliney, J. (2000). Rectifying the institution: Navigating the edge of chaos with complexity theory and Confucian maps. In G. Morcol & L. F. Dennard (Eds.), *New sciences for public administration and policy: Connections and reflections* (pp. 107–124). Burke, VA: Chatelaine Press.

Kauffman, S. (1995). *At home in the universe: The search for the laws of self-organization and complexity*. New York: Oxford University Press.

Kent, M. L., & Taylor, M. (1998). Building dialogic relationships through the World Wide Web. *Public Relations Review, 24*(3), 321–334.

Kent, M. L., & Taylor, M. (2002). Toward a dialogic theory of public relations. *Public Relations Review, 28*(1), 21–37.

Kiel, L. D. (1994). *Managing chaos and complexity in government: A new paradigm for managing change, innovation, and organizational renewal*. San Francisco: Jossey-Bass.

Kiel, L. D. (2000). The sciences of complexity and public administration: Complexity as challenge and goal. In G. Morcol & L. F. Dennard (Eds.), *New sciences for public administration and policy: Connections and reflections* (pp. 63–80). Burke, VA: Chatelaine Press.

Krugman, P. (1996). *The self-organizing economy*. Malden, MA: Blackwell.

Kupers, R. (2001). What organizational leaders should know about the new science of complexity. *Complexity, 6*, 14–19.

Lauzen, M. M. (1995). Toward a model of environmental scanning. *Journal of Public Relations Research, 7*(3), 187–203.

Lauzen, M. M., & Dozier, D. M. (1994). Issues management mediation of linkages between environmental complexity and management of the public relations function. *Journal of Public Relations Research*, 6(3), 163–184.

Ledingham, J. A. (2003). Explicating relationship management as a general theory of public relations. *Journal of Public Relations Research, 15*(2), 181–198.

Ledingham, J. A., & Bruning, S. D. (Eds.). (2000). *Public relations as relationship management: A relational approach to the study and practice of public relations*. Mahwah, NJ: Lawrence Erlbaum Associates.

Marion, R., & Bacon, J. (1999). Organizational extinction and complex systems. *Emergence, 1*(4), 71–96.

McDaniel, R. R., Jr. (1997, Winter). Strategic leadership: A view from quantum and chaos theories. *Health Care Management Review, 22*(1), 21–37.

Murphy, P. (1996). Chaos theory as a model for managing issues and crises. *Public Relations Review, 22*(2), 95–113.

Murphy, P. (2000). Symmetry, contingency, complexity: Accommodating uncertainty in public relations theory. *Public Relations Review, 26*(4), 447–462.

Newell, W. H., & Meek, J. W. (2000). What can public administration learn from complex systems theory? In G. Morcol & L. F. Dennard (Eds.), *New sciences for public administration and policy: Connections and reflections* (pp. 81–105). Burke, VA: Chatelaine Press.

Pruzan, P. (1998). From control to values-based management and accountability. *Journal of Business Ethics, 17*, 1379–1394.

Ramirez, A., Jr., Walther, J. B., Burgoon, J. K., & Sunnafrank, M. (2002). Information-seeking strategies, uncertainty, and computer-mediated communication: Toward a conceptual model. *Human Communication Research, 28*(2), 213–228.

Richardson, K. A. (Ed.). (2005). *Managing organizational complexity: Philosophy, theory, application*. Greenwich, CT: Information Age Publishing.

Robertson, D. A. (2004). The complexity of the corporation. *Human Systems Management, 23*, 71–78.

Rosenbloom, A. (2004, December). The blogosphere. *Communications of the ACM, 47*(12), 31–33.

Senge, P. M. (1990). *The fifth discipline: The art & practice of the learning organization*. New York: Currency Doubleday.

Sha, B-L. (2004). Noether's theorem: The science of symmetry and the law of conservation. *Journal of Public Relations Research, 16*(4), 391–416.

Sherman, H., & Schultz, R. (1998). *Open boundaries: Creating business innovation through complexity*. Reading, MA: Perseus Books.

Stacey, R. D. (1992). *Managing the unknowable: Strategic boundaries between order and chaos in organizations*. San Francisco: Jossey-Bass.

Stewart, I. (1989). *Does God play dice? The mathematics of chaos*. Malden, MA: Blackwell.

Sweet, V. K. (2000). Tolerating reorganizational anxiety: A nonlinear management theory of action for public administrators. In G. Morcol & L. F. Dennard (Eds.), *New sciences for public administration and policy: Connections and reflections* (pp. 177–194). Burke, VA: Chatelaine Press.

Tapscott, D., & Ticoll, D. (2003). *The naked corporation: How the Age of Transparency will revolutionize business*. New York: Free Press.

Minority Opinions Go Public

Implications for Online Issues Management and the Spiral of Silence

Damion Waymer

The Internet, with its ability to quickly mobilize a community of like-mindedness, is argued to lead to minority-opinioned publics forming and speaking out more readily. This manuscript argues that the spiral of silence hypothesis should be reconceptualized (in light of emergent technologies, such as the Internet) because individuals from marginalized groups are using technology to voice opinions that would have never been heard under the conventional understanding of the spiral of silence hypothesis. This chapter provides several examples of organizations that have used online technologies to speak out and challenge majority opinions. This chapter, through close textual analysis, examines Advocates for Animals and its Web site in greater detail. This chapter asserts that, in addition to conventional publics, issue managers must monitor and attend to online audiences who are willing to voice their concerns as well as conventional publics who are strategically using online technologies to augment their public relations strategies and campaigns.

> Never doubt that a small group of concerned citizens can change the world. Indeed, it's the only thing that ever has.
>
> *Margaret Mead*

Although this excerpt can be viewed as a call to arms for minority-opinioned publics, researchers have found that, when it comes to publicly voicing an opinion about controversial, unpopular public issues, "audiences whose opinions do not coincide with the majority opinion, as they perceive it, tend to maintain their silence" (Lin & Salwen, 1997, p. 129). Moreover, Shamir (1997) asserts that "people risk social isolation when openly expressing nonnormative or unpopular views, and the fear of isolation is a powerful motivator alerting them to constantly scan the social environment and assess the climate of opinion to avoid social sanctions" (p. 602). Noelle-Neumann (1991) states that "it is constant interactions among people, due to their social nature, that account for the transformation of the sum of individual opinions into public opinion" (p. 280). Thus, scholars (see Moy, Domke, & Stamm, 2001; Noelle-Neumann, 1974, 1991) have posited that the threatening of individuals by society and the individual-level fear of isolation work in concert to foster a single public opinion on controversial issues and influence the public expression of opinions. This, in part, is the spiral of silence hypothesis (Noelle-Neumann, 1974).

There are times, however, when audiences, despite their minority position and supposed fear of isolation, defy this spiral of silence, hold steadfast to their positions, and voice their opinions publicly. Issue salience, as well as people's perceived need to express their deeply rooted values, among other factors, can sometimes lead individuals to voice opinions other than the dominant public opinion (Lin & Salwen, 1997; Noelle-Neumann, 1991; Salmon & Kline, 1985; Shamir, 1997). Shamir asserts the following:

> We also found that people who perceive the climate being against them, and especially having turned against them, more often change their willingness to speak out, but these changes occur in both directions: some become silent and some become more vocal. Fear of isolation may be operating on some, but others may be driven by a need to express their deeply cherished values especially when in jeopardy, so as to define themselves and convince others. This motive may eventually override social pressures and encourage people's overt expression of opinion. (pp. 609–610)

I argue, however, that, with the emergence of new technologies, particularly the Internet (which is a public resource), individuals have the capability to go public with their minority opinions (allowances should be made, however, for the digital divide and the lack of universal access that in some ways reduces the Web's public access). This venue affords silenced minority opinions to be voiced, no matter the issue's salience, with minimal perceived threats of social isolation. Moreover, I argue that this audience, which was nonexistent or existed but found it quite difficult to come together, is an emerging audience to which political figures, government, media, and even corporate issues managers must now attend.

This chapter argues that, in light of emerging technologies, such as the Internet, the spiral of silence hypothesis must be reconceptualized, because it appears that individuals, whether or not they can be identified, are using technology to publicly voice minority opinions about controversial public issues; thus, opinions that may have never been heard under the conventional understanding of the spiral of silence are now being voiced. In addition, this chapter argues that all issues managers must add these emerging online audiences to their list of stakeholders (Hearit, 1999) as these managers continue to find additional ways to actively monitor their environments (Hearit, 1999; Thomsen, 1995).

In the following sections, I first provide a discussion about online identities and how they act as a means of safeguarding individuals from social isolation, followed by a discussion about online resources that enable individuals to access larger audiences in order to contribute to or respond to public opinion. Next, I explore relevant issues management literature and attempt to draw links among online minority-opinioned audiences, issue management, and the spiral of silence hypothesis. Finally, I discuss pragmatic and theoretical contributions and directions for future research and draw conclusions.

Online Identity: The Cloak of Anonymity

According to Turkle (1999), the "Internet links millions of people together in new spaces that are changing the way we think, the nature of our sexuality, the form of our communities, our very identities" (p. 643). For many individuals, online experiences challenge what they have traditionally called identity, because, in many instances, people can portray themselves however they choose in cyberspace. Turkle provides an illustration of this point in the following analysis of one of her participant's narratives:

In real life, I tend to be extremely diplomatic, nonconfrontational. I don't like to ram my ideas down anyone's throat. [Online,] I can be, "Take it or leave it." All of my Hepburn characters are that way. That's probably why I play them. Because they are smart-mouthed, they will not sugarcoat their words. (p. 645)

People's online identities are textual descriptions and representations of self; therefore, there is a perceived and/or real sense of anonymity online, because, without physically seeing them or hearing them speak, people must take at face value who other individuals say they are. This level of online anonymity, whether perceived or real, is simply not found in face-to-face interactions (Bowker & Tuffin, 2003).

The relatively anonymous aspect of online interactions gives people the chance to express themselves more freely when compared to the constraints that face-to-face encounters impose. Turkle (1999) argues that, "for some people, it [cyberspace] is a place to 'act out' unresolved conflicts . . . for others, it provides an opportunity to 'work through' significant personal issues, to use the new materials of cybersociality to reach for new solutions" (p. 644). Moreover, the new tools and materials that are readily available in cyberspace possibly enable individuals to reach new solutions more rapidly today as opposed to pre-cyberspace times. Although solutions can vary, from remedying individual mental, spiritual, and physical problems to addressing controversial public issues, the tools used to reach these ends are the same. In the following section, I explore, in greater detail, these new materials and their potential ability to empower minority-opinioned publics and to encourage them to go public with their opinions.

Online Resources: A Means of Granting Voice

According to Hurwitz (1999), "cyberspace was imagined alternatively as an 'electronic frontier' where free thought and egalitarian associations transcend political boundaries and an 'electronic commons' where netizens discuss issues and influence decision makers who are listening" (p. 655). In many instances, this imaginative vision has become a present reality. For example, Hearit (1999) used the case of the flawed Intel Pentium chip and management's (mis)handling of this issue for analysis. In this particular case, an online newsgroup challenged Intel's initial managerial decision; the decision was that Intel would only replace defective chips if individuals could demonstrate the need for replacement chips. The owners of computers with faulty chips had to display that their work involved "heavy duty scientific / floating point calculations" (p. 296) to warrant replacement chips. Through newsgroups, computer owners with the faulty chip were able to collectively voice their disdain toward Intel's decision. Ultimately, they were effective in getting Intel to replace all faulty chips, no questions asked, to those customers who wanted the chips replaced. Although this example arguably does not address a controversial public opinion issue, it does highlight ways in which individuals can use online resources to gain a voice and enact change.

In another example, Bickel (2003) analyzed the Revolutionary Association of the Women of Afghanistan (RAWA) Web site and found that these women, who were physically barred from public space in Pakistan due to Taliban and cultural dictates and faced with threats of violence, have used the Internet as a means of gaining a voice. RAWA was successful in getting petitions signed and monetary donations made, as well as garnering international support for their cause against fundamentalism and for women's rights, despite not having a physical office. All of their efforts and correspondence were conducted via the Internet.

Hyde and Rufo (2000) used a debate pertaining to euthanasia as a case for analysis. In this case, a disability civil rights group known as Not Dead Yet (NDY) conducted "what their opponents described as an 'invasion' of an electronic mailing list operated by the Euthanasia Research and Guidance Organization" (ERGO) (p. 1). NDY invaded ERGO's electronic mailing list for many reasons but none more important than NDY's fear that, if the public at large supported ERGO's position (that disabled persons should have the right to choose death and should act on this right if they feel that dealing with their disabilities is too much of a burden), this decision could serve to marginalize disabled persons who want to live and have a right to life. Others, including members of ERGO, may feel that these disabled persons are not fit to live, should die in peace, and/or should rid themselves of the torment of living with severe disabilities through euthanasia.

NDY was successful in raising the consciousness of ERGO members. Although it took five months for NDY's biggest opponents—those members of Right to Die (RTD) who subscribe to ERGO's mailing list—to offer open-minded and open-hearted responses to NDY's call for increased consciousness, members of NDY were able to get opponents to consider, engage in discussion about, and reflect on NDY's position (Hyde and Rufo, 2000). This example illustrates implicitly how individuals, through online resources, can contribute to, respond to, and/or refute public opinion.

Finally, Hurwitz (1999) showcases the point that netizens, via the Internet, can discuss issues and influence decision makers: "Advocacy and interest groups use it [the Internet] to organize their supporters for online lobbying of local, national, and foreign officials—who themselves need e-mail addresses to be credible in this Information Age" (p. 656). In addition, Hurwitz states the following:

> Ad hoc responses to major political events, like impeachment and massacres, can gain national attention in a few days, as network users redistribute petitions and sample letters to their personal distribution lists. Revolutionaries, like the Zapatistas in Chiapas, and dissidents, like the Serbian radio group B92, can webcast messages to audiences all over the globe and receive back moral and financial support. Survivalist groups can network by exchanging pointers to respective websites and thereby increase the possibilities of so-called "leadership resistance." (p. 656)

These examples illustrate that the Internet has become a new, useful tool and outlet that is at the disposal of various types of political and activist groups.

Hurwitz (1999) argues that "The Internet is an obvious and powerful tool for such democratic action because it can help create communities of interests that transcend space, time, and the need for formal introductions. Alerts can be spread quickly through supporters' pre-existing mail and distribution lists." (p. 660)

This rapid pace at which information can be transmitted, coupled with the speed at which spatially isolated individuals can mobilize, creates a situation where issue managers should be aware of these publics and be willing and prepared to address them (Hearit, 1999).

Issues Management of Emerging Publics

Crable and Vibbert (1985) state that "issues are not simply questions that exist. An issue is created when one or more human agents attaches significance to a situation or perceived problem" (p. 5). Thus, based on this definition, one could argue that the concerns of minority-opinioned publics can be considered legitimate issues. Effective issue managers should be aware of and prepared to address issues, including those that arise in cyberspace, that can possibly affect their organizations. This, however, is just one component of effective issues management. Heath (1997) defines issues management this way:

> Issues management entails efforts to achieve understanding and increase satisfaction between parties and to negotiate their exchange of stakes. It engages interlocking cultures that are in various states of compatibility and similarity. It fosters the interests of the stakeholders by helping an organization achieve its goals in a community or complementary competing interests. (p. 9)

Effective issue managers are proactive; they anticipate changes in their environments as well as attempt to influence "policies long before policy options are created by others" (Crable & Vibbert, 1985, p. 9). To further this point, Heath claims that "savvy organizations [or individuals] use issues management to monitor issues, sharpen their strategic business plans, improve their operations, and communicate in ways intended to build and strengthen relationships with key publics" (p. 4). Hence, it should be useful to issue managers to be cognizant of the possible minority opinions that have not reached national significance in addition to those issues that already have. To further support this point, I provide an example of a group that is voicing its opinion about a controversial public concern; this group has orchestrated a two-pronged attack against both a public policy and corporations. In addition, by using this organization's Web site as a text for analysis, I articulate that the spiral of silence hypothesis needs to be reconceptualized.

Advocates for Animals and Issues Management

Advocates for Animals, founded in 1912 as the Scottish Society for the Prevention of Vivisection, changed its name in 1989 to reflect that its interests and concerns for animals were no longer restricted to the use of animals in research. The organization campaigns and lobbies for legislation to protect animals throughout Scottish, United Kingdom, and European Parliaments, and it aims to ensure that animal welfare remains firmly on the agenda of all political parties (Advocates for Animals, 2001). Although the majority of its efforts have been directed in the United Kingdom, Advocates for Animals has also spoken adamantly against seal slaughtering in Canada as well as acts of animal cruelty in China.

Advocates for Animals, which takes what is arguably a minority position on the public issue of animal rights, is one of the leading animal protection organizations in the United Kingdom (Advocates for Animals, 2006). The organization opposes the use of animals in entertainment, experiments, and factory farming. Despite apparent obstacles the organization faces—only four million residents or 7% of adults and 12% of young people are vegetarians (*Vegetarianism*, 2006)—the organization has had several successful campaigns and touts its successes on its Web site and in its newsletters.

For example, in 2004, the organization's efforts, after several undercover investigations into Scotpigs Ltd. Farms, played a major role in having accreditation removed from Scotpigs Quality Meats of Scotland (Advocates for Animals, 2006). Essentially, losing the accreditation meant that Scotpigs could no longer supply meat to any major Scottish food outlets or food manufacturers; shortly after Scotpigs lost its accreditation, the organization began liquidating assets.

Advocates for Animals has launched a host of other strategic assaults against several entities, including the Scottish pet shop industry. Advocates for Animals has also helped to influence and shape animal rights legislation and policy in Scotland (Advocates for Animals, 2006). In 2002, the organization campaigned for a change in the existing law that would enable better protection for seals in UK waters; in 2003, the organization successfully battled pro-hunt supporters and helped to uphold the Scottish Hunt Ban, which outlaws mounted foxhunting, hare coursing, and fox bating in Scotland. In 2004, the organization attended several meetings with the Scottish Executive, which is the devolved government for Scotland that is responsible for "most of the issues of day-to-day concern to the people of Scotland, including health, education, justice, rural affairs, and transport" (Scottish Executive, 2006, para. 2); these meetings resulted in Scotland updating existing animal welfare legislation (Advocates for Animals, 2006).

This coalition has used the Internet as a tool to rally support for its causes. The organization has an e-protest link on its Web site. Advocates for Animals asks visitors to choose one or more e-protests and type in their contact information. Next, a campaign letter is sent to the appropriate legislators on behalf of the visitors.

This type of campaigning can be considered as nonprofit organizations' variant form of *Astroturf lobbying*. Astroturf lobbying refers "to apparently grassroots groups or coalitions which are actually fake, often created by corporations or public relations firms" (SourceWatch, 2006, para. 1); it has also been defined as a "grassroots program that involves the instant manufacturing of public support for a point of view in which either uninformed activists are recruited or means of deception are used to recruit them" (para. 2).

Although Advocates for Animals is not a manufactured organization, it, in some ways, uses a variant form of Astroturf campaigns to make its animal rights public relations campaigns appear as grassroots efforts that originate from general public opinion and not the organization itself, thus creating an illusion of broad-based public support.

Critics of this type of lobbying deem it unethical. U.S. Senate Democratic Leader Harry Reid has proposed new lobbying reform that addresses, in part, the ethicality of Astroturf lobbying and stealth coalitions (Senate Democratic Communications Center, 2006). However, the response of Advocates for Animals would likely be that the organization uses this variant form of Astroturf lobbying for good, not ill, will. It appears that Advocates for Animals' purpose for this tool is to make protesting easier: People with strong opinions and feelings about animal rights can more easily make their voices heard.

This example, in general, and Astroturf lobbying, in particular, highlight reasons why issues managers should stay attuned to public opinion, which is the communicated manifestation of public expectation. Thus, if publics' expectations are not met, then one could expect to see these publics' discontent reflected in public opinion. Simply put, whether orchestrated or not, publics have expectations of all organizations, no matter the size, and, when these expectations are not met, issues of legitimacy arise (Dowling & Pfeffer, 1975; Sethi, 1977).

Additionally, issues managers must be cognizant of Internet mobilization. Advocates for Animals uses the Internet as a vehicle of mobilizing audiences that would typically have a more difficult time coming together. This organization can unite animal activists across the globe and enable the marginalized voices of animal advocates to gain more power.

Through its Web site, Advocates for Animals allows members and visitors to receive the latest news, press releases, and timely updates; it also allows them to stay informed about upcoming events, campaigns, and important proposed legislature changes in the cue. Visitors can sign up for e-news releases; members and visitors can read backlogs of the organization's magazines for members.

The Web site also contains links and resources that provide members and visitors with a host of information, including, but not limited to, (a) reports about the organization's fight to free captive animals, animals in the entertainment industry, and farmed animals; (b) how to start local groups, recruit members, obtain free promotional materials, start campaigns, and stage effective protests; (c) ways members and visitors can change their lifestyles to be more animal friendly, including tips for diet, health, and clothing; and (d) how to donate, and bequeath money upon death, to the organization.

Although this organization existed before the advent of the World Wide Web, it has used the Web in recent years as a strategic public relations tool. Advocates for Animals uses the Web mostly to manage relationships with its existing member publics by keeping them informed and up-to-date with timely information. In addition, it uses the Internet to reach its nonmember publics; it aims to increase its membership and awareness about the organization's purpose.

It appears as if online technologies have enabled Advocates for Animals to be better champions for animal rights. Organizations that use these tools strategically pose new and serious challenges to issues managers; issues managers must be prepared to encounter these audiences.

Online technologies specifically have impacted both the way we practice and theorize about public relations. This chapter lays the foundation for inferring that the perceived fear of isolation from speaking out publicly about a controversial issue is removed when Internet users coalesce and petition over the Web. Moreover, it can be argued that the Internet possesses the capacity to empower individuals and make minority opinion holders feel less like a minority (when their views are validated by others online). Thus, the Internet impacts the spiral of silence framework because it emboldens minority opinion holders to take action, whether it is simply signing an electronic protest, joining an organization, and/or speaking out in conventional public forums.

Conclusions

If minority opinions are not expressed, issues managers may find it difficult or not worthwhile to take minority sentiment into account when making decisions; obviously, issues managers must be aware of or foresee issues in order to address them. Conversely, if minority opinions are being voiced, issues managers may be more inclined to address these voices because it becomes more difficult to ignore them. The rise of the Internet has made it easier for minority opinions to be voiced and more difficult for organizations to ignore these voices.

With the proliferation of online resources, such as chat rooms, blogs, Web sites, listservs, and bulletin boards, it is much easier for individuals who feel that their opinions are in the minority to locate one another. Hearit (1999) states it best in the following:

> Once these individuals who share a common problem recognize each other, it is much more likely that they will be able to organize and communicate to coordinate actions, whether it be to boycott a product or simply to damage an [organization's] reputation through repeated and continual charges that it is unresponsive and uncaring. (p. 303)

There are exceptions to the spiral of silence hypothesis, and these exceptions occur depending on whether a few die-hard advocates refuse to let go of an issue, the issue's general salience, and/or the value people place on the issue. In the animal rights issue, as well as the other cases discussed in this chapter, I did not attempt to determine whether the minority opinion holders that support the issues were die-hard advocates or found these issues to be very important to themselves. However, I did attempt to illustrate that, through the Internet, people can voice their minority opinions. Furthermore, these minority opinions expressed on the Internet, which would not likely otherwise be heard on a national or global stage, have the potential to and sometimes do influence public discussion, public policy, and enact change.

In addition, because of new media and emerging technologies that make various opinions readily and widely accessible, the spiral of silence hypothesis should be reconceptualized. Although people are still less likely to voice their minority opinions while in the physical presence of others, the Internet enables these minority opinions to bud, grow, and possibly flourish. The Internet's perceived cloak of anonymity provides minority-opinioned groups time to collaborate and form an organization in which they can later begin to make demands in the traditional public arena via public demonstrations, television, and radio, among other venues.

Although one of the major arguments of this chapter is that the spiral of silence hypothesis needs to be reconceptualized, it is beyond the scope of this work to provide a reconception of this theoretical framework. Future researchers should update the spiral of silence hypothesis in light of this analysis and emerging technologies.

One assumption that is made but not explored in this analysis is that the relative anonymity of the Internet reduces social pressures to conform and self-censor. Although it can be assumed that the Internet may override social pressures and sanctions that suggest silence and encourage people to publicly express their opinions on controversial public issues, this link has not been formally tested; thus, future research must be conducted to explore this relationship. By exploring the aforementioned relationship, empirical investigations can specifically assess how the emergence of the Internet has affected the spiral of silence hypothesis.

Researchers, such as Dahlberg (1998, 2001), have broadened their definition of the public sphere, and I argue that researchers should also employ this practice. More specifically, researchers should use a broader definition of the term *public;* rather than only including conventional speech acts as public communication, researchers should broaden this category to include online venues as acts of public communication. Studies using this broader definition of public communication, coupled with the spiral of silence hypothesis, should be conducted to see how these emerging technologies impact and/or help to shape public opinion. Finally, future work should assess the impact that the Internet has on the public policy process in general and pieces of legislation specifically.

In short, issues managers must recognize that the Internet has drastically changed the ways in which people can access information, opinions, and relationships. Via the World Wide Web, the world is more connected than ever before. For students, practitioners, and scholars, emergent technologies are a relatively new territory that is ripe for exploration; it is time for us to discover and uncover what's in store.

For consideration

1. In what instances could Astroturf lobbying be considered ethical?
2. What are some recent examples of how new media and online technologies have encouraged marginalized groups to speak out about issues of importance to them? How can the groups' effectiveness be measured?
3. How have new media and online technologies contributed to or detracted from the credibility of marginalized groups?
4. In what ways do online technologies create new challenges and pose potential risks to organizations and their issues managers? What new opportunities do online technologies offer to issues managers?
5. What are examples of issues that are more likely to fall prey to the spiral of silence but, if communicated more broadly, would serve the public interest?

For reading

Dahlberg, L. (1998). Cyberspace and the public sphere: Exploring the democratic potential of the Net. *Convergence*, 4(1), 70–84.

Lin, C.A., & Salwen, M. B. (1997). Predicting the spiral of silence on a controversial public issue. *The Howard Journal of Communications*, 8, 129–141.

Moy, P., Domke, D., & Stamm, K. (2001). The spiral of silence and public opinion on affirmative action. *Journal & Mass Communication Quarterly*, 78(1), 7–25.

Noelle-Neumann, E. (1991). The theory of public opinion: The concept of the spiral of silence. *Communication Yearbook*, 14, 256–287.

Thomsen, S. R. (1995). Using online databases in corporate issues management. *Public Relations Review*, 21, 103–122.

References

Advocates for Animals. (2001). Retrieved May 1, 2006, from http://web.archive.org/web/20010211185933/www.advocatesforanimals.org.uk/

Advocates for Animals. (2006). Retrieved May 1, 2006, from http://www.advocatesforanimals.org.uk/

Bickel, B. (2003). Weapons of magic: Afghan women asserting voice via the Net. *Journal of Computer-Mediated Communication*, 8(2). Retrieved May 31, 2007, from http://jcmc.indiana.edu/vol8/issue2/bickel.html

Bowker, N., & Tuffin, K. (2003). Dicing with deception: People with disabilities' strategies for managing safety and identity online. *Journal of Computer-Mediated Communication*, 8(2). Retrieved May 31, 2007, from http://jcmc.indiana.edu/vol8/issue2/bowker.html

Crable, R. E., & Vibbert, S. L. (1985). Managing issues and influencing public policy. *Public Relations Review, 11*(2), 3–16.

Dahlberg, L. (1998). Cyberspace and the public sphere: Exploring the democratic potential of the Net. *Convergence, 4*(1), 70–84.

Dahlberg, L. (2001). The Internet and democratic discourse: Exploring the prospects of online deliberative forums extending the public sphere. *Information Communication & Society, 4*(4), 615–633.

Dowling, J., & Pfeffer, J. (1975). Organizational legitimacy: Social values and organizational behavior. *Pacific Sociological Review, 18*, 122–136.

Hearit, K. M. (1999). Newsgroups, activist publics, and corporate apologia: The case of Intel and its Pentium chip. *Public Relations Review, 25*, 291–308.

Heath, R. L. (1997). *Strategic issues management: Organizations and public policy challenges*. Thousand Oaks, CA: Sage.

Hurwitz, R. (1999). Who needs politics? Who needs people? The ironies of democracy in cyberspace. *Contemporary Sociology, 28*(6), 655–661.

Hyde, M. J., & Rufo, K. (2000). The call of conscience, rhetorical interruptions, and the euthanasia controversy. *Journal of Applied Communication Research, 28*, 1–23.

Lin, C. A., & Salwen, M. B. (1997). Predicting the spiral of silence on a controversial public issue. *The Howard Journal of Communications*, 8, 129–141.

Moy, P., Domke, D., & Stamm, K. (2001). The spiral of silence and public opinion on affirmative action. *Journal & Mass Communication Quarterly, 78*(1), 7–25.

Noelle-Neumann, E. (1974). The spiral of silence: A theory of public opinion. *Journal of Communication, 24*(2), 43–51.

Noelle-Neumann, E. (1991). The theory of public opinion: The concept of the spiral of silence. In J. A. Anderson (Ed.), *Communication yearbook 14* (pp. 256–287). Newbury Park, CA: Sage.

Salmon, C. T., & Kline, F. G. (1985). The spiral of silence ten years later. In K. R. Sanders & L. L. Kaid (Eds.), *Political communication yearbook* (pp. 3–30). Carbondale, IL: Southern Illinois University Press.

Scottish Executive. (2006). *About Scottish Executive*. Retrieved June 24, 2006, from http://www.scotland.gov.uk/About/Intro

Senate Democratic Communications Center. (2006). *Reid statement on lobbying reform legislation*. Retrieved June 24, 2006, from http://democrats.senate.gov/~dpc/press/2006329C44.html

Sethi, S. P. (1977). *Advocacy advertising and large corporations: Social conflict, big business image, the news media, and public policy*. Lexington, MA: Lexington Books.

Shamir, J. (1997). Speaking up and silencing out in face of a changing climate opinion. *Journal & Mass Communication Quarterly, 74*(3), 602–614.

SourceWatch. (2006). *Astroturf*. Retrieved May 3, 2006, from http://www.sourcewatch.org/index.php?title=Astroturf

Thomsen, S. R. (1995). Using online databases in corporate issues management. *Public Relations Review, 21*, 103–122.

Turkle, S. (1999). Looking toward cyberspace: Beyond grounded sociology. *Contemporary Sociology, 28*(6), 643–648.

Vegetarianism. (2006). Retrieved May 1, 2006, from http://www.thesite.org/healthandwellbeing/fitnessanddiet/food/vegetarianism

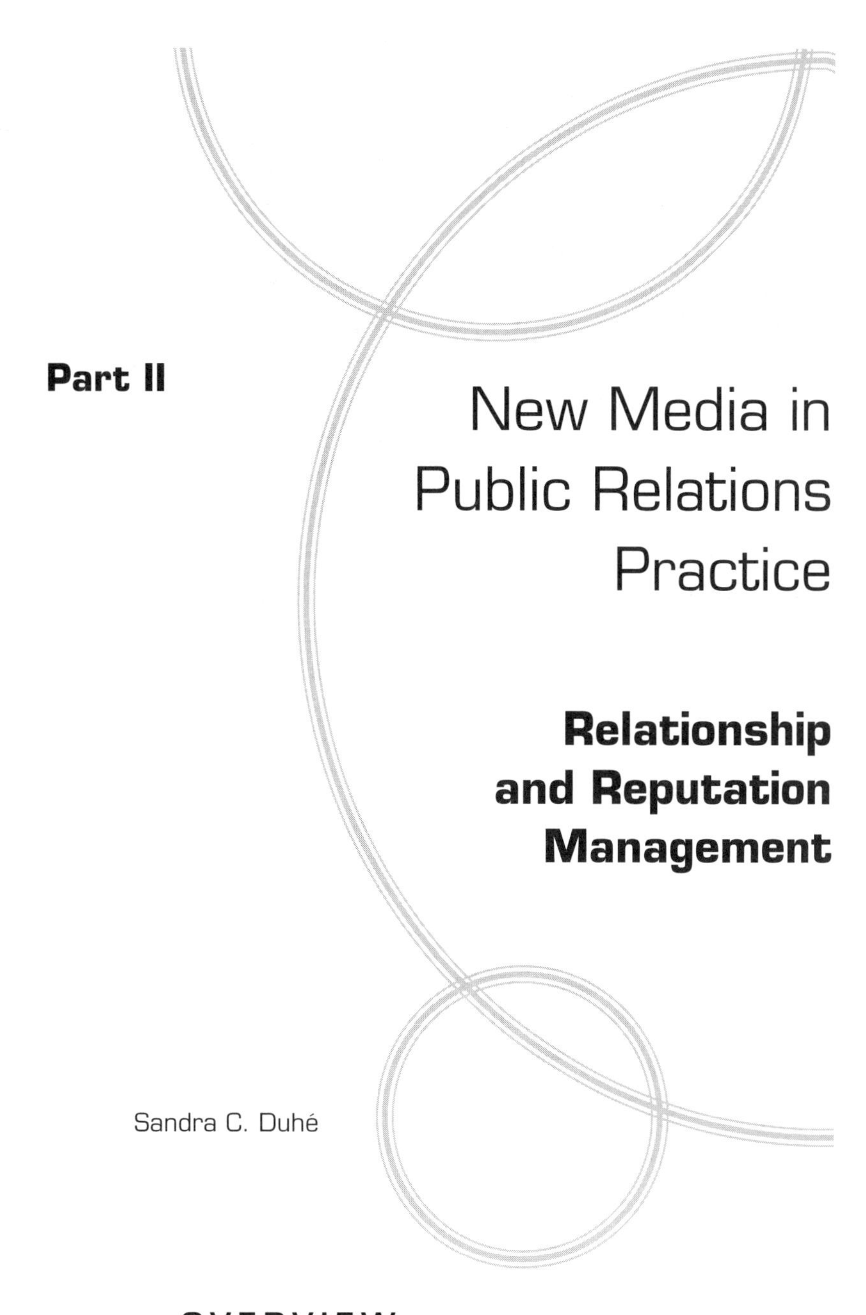

Part II

New Media in Public Relations Practice

Relationship and Reputation Management

Sandra C. Duhé

OVERVIEW

In Part II of the book, we transition from a primarily theoretical examination of how new media are changing the ways we think about public relations to an exploration of how new media are actually being used in contemporary practice. This first section focuses on relationship and reputation management: topics that are of broad interest to scholars and practitioners.

The core of public relations is *public relationships*. Relationships define an organization's connectedness to its operating environment. Whether an organization is a Fortune 100 multinational corporation, a regional network of volunteer-based literacy centers, or a local state-sponsored workforce development office, its public relations staff relies on stakeholder relationships for information exchange. Reputation is the outcome of multiple, interacting organizational behaviors; it is earned rather than created. Because the human dimension contributes to the dynamic nature of both relationships and reputations, it is presumptive for us to believe that we can manage relationships or reputations entirely.

Nevertheless, practitioners have an active role to play in both realms. Vincent Hazleton, Jill Harrison-Rexrode, and William Kennan expand on Hazleton and Kennan's previous work in social capital to examine how a practitioner's crucial ability to form and maintain relationships is affected by new developments in social media. The double-edged sword of blogs is explored by Kathleen Long, Peter Galarneau, Jr., Jeffrey Carlson, and Erin Bryan, not only from a content perspective, but also from an ethical and legal perspective, thereby raising some thought-provoking issues for practitioners to consider before employing this relatively new tool. Marcia Watson DiStaso, Marcus Messner, and Don Stacks reveal the results of their intriguing research into the framing of Fortune 500 companies in Wikipedia entries and discuss how this abolishment of "elitist gatekeeping" offers both challenges and opportunities in corporate reputation management. Idil Cakim provides practical, research-based advice for organizations eager to leverage what she views as the "underutilized communication channel" of the Internet to reach stakeholders, respond to crisis, and boost reputation. Debra Worley guides us through a comprehensive tour of how organizations are using the Web to communicate ethical and social responsibility attributes. This section concludes with Mihaela Vorvoreanu providing an informative description of (and accompanying questionnaires for) a relatively simple way to assess—and ultimately adjust—the quality of a visitor's interaction with an organization's Web site.

Rather than *replace* the interpersonal element of public relationship building, new media *supplement* the process in a compelling and unavoidable way. The following chapters illustrate how online communication is being used to connect with a variety of stakeholders.

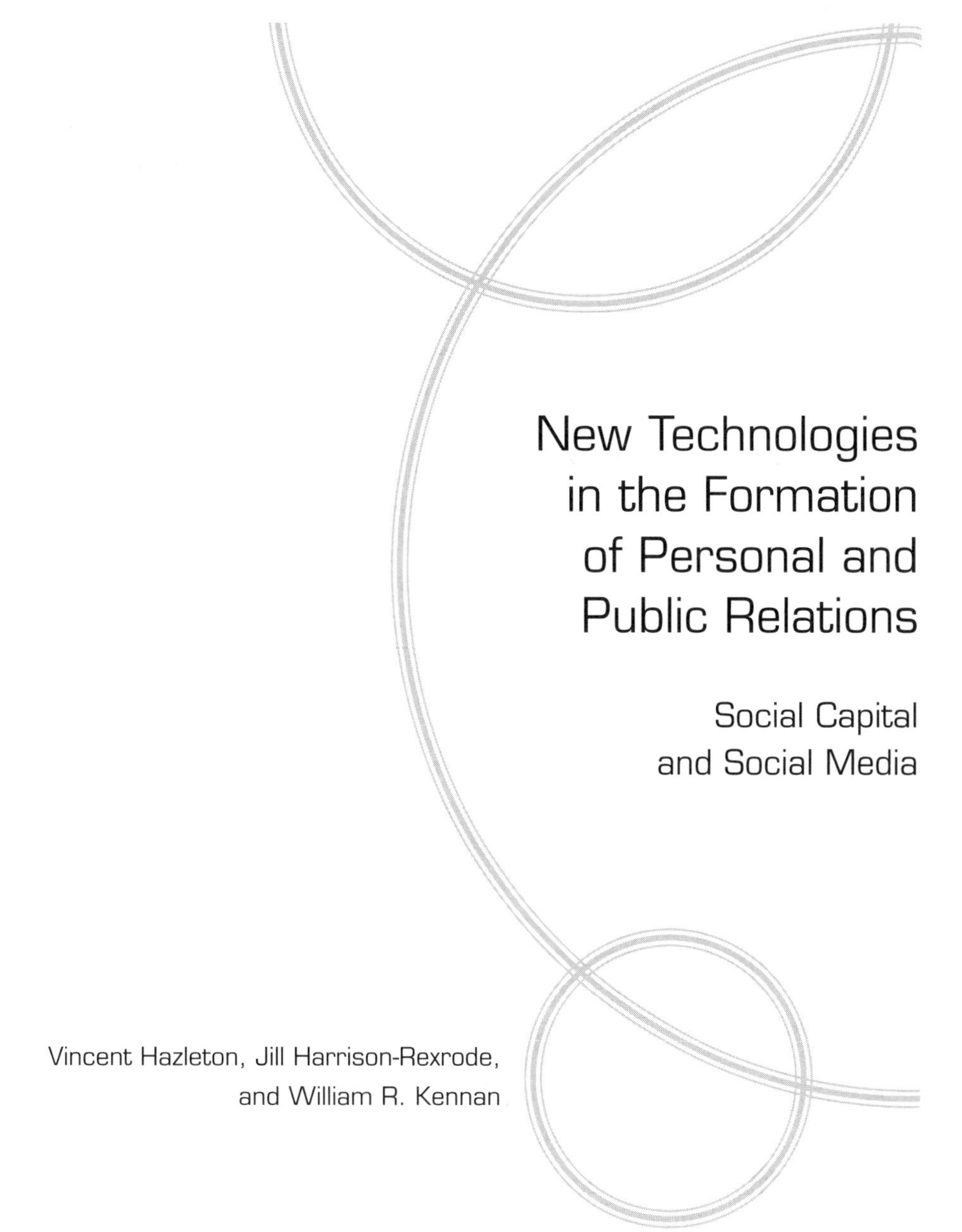

New Technologies in the Formation of Personal and Public Relations

Social Capital and Social Media

Vincent Hazleton, Jill Harrison-Rexrode, and William R. Kennan

How many times have you heard someone say something like the following? It's not what you know, but who you know, that is important. For public relations, both who you know and what you know are important. Who you know references the importance of relationships, both public and personal. What you know, particularly about communication and communication technologies, references methods and strategies for developing useful relationships. This chapter explains the impact of communication technologies on public relations practice, and it explores how the future might look by using the perspective of social capital theory.

Public relations practice is undergoing a revolution based on the emergence of communication technologies. The emergence of e-mail, instant messaging, Web pages, cell phones, etc., is changing the way public relations practitioners, individuals, and institutions communicate with each other, and it is changing how people get information about the world that surrounds them. Traditional methods of public relations (press releases and special events, for example) remain central to what public relations practitioners do on a daily basis. However, since individuals are changing the ways in which they communicate with each other, correspondingly, the methods used in public relations to create, maintain, and utilize relationships are changing as well.

This chapter explains the impact of communication technologies on public relations practice and it explores what the future might look like. In doing so, this chapter begins with a discussion of a useful theory called social capital theory that helps to clarify how and where communication technologies have their impact. Hazleton and Kennan (Hazleton & Kennan, 2000; Kennan & Hazleton, 2006) have proposed a social capital theory of public relations, which provides some useful ideas for understanding the role of new technologies and, particularly, new social media in public relations.

Second, we will examine how new technologies influence the development of social capital in public relations. Considerable differences of opinion exist as to whether new technologies increase the deposit of social capital available to actors. The main point of contention seems to be whether technological innovations, such as Web pages, e-mail, instant messaging, online gaming, and cell phones, increase isolation, thus reducing social capital, or whether new communication technologies actually allow relationships to be enhanced, thus increasing social capital. This chapter will continue by considering how changes in technology influence the way public relations professionals, individuals, and organizations communicate.

Social Capital

Social capital is defined as the ability to form and maintain relationships to facilitate goal and objective attainment (Fussell, Harrison-Rexrode, Kennan, & Hazleton, 2006; Hazleton & Kennan, 2000; Kennan & Hazleton, 2006). For public relations professionals, the ability to effectively and strategically use various communication strategies to develop personal and public relationships is essential for organizational success. What makes it important to organizations and public relations professionals is the fact that social capital can be transformed into other forms of capital, which directly and indirectly helps organizations to achieve their goals. People and organizations with social capital can draw on the resources from their relationships with others.

Relationships are embedded in social networks, and they emerge when connections are formed that link individuals or organizations to each other. This definition is of particular importance because it helps to clarify exactly what constitutes public relations practice. For our purposes, public relations is the communication process of creating, maintaining, and utilizing relational connections to assist clients in attaining their goals and objectives. As we will learn in what follows, the process of creating, maintaining, and utilizing relational connections is changing based on the emergence of various communication technologies and their adoption by broader segments of the population. Before continuing, however, we will discuss some characteristics of social capital that will help clarify this important perspective and why it is of such special importance to an understanding of the impact of technology on public relations.

First, the word *capital* is used as a metaphor. As such, it helps us to understand that, when public relations practitioners do their work, they are creating network connections with other people and organizations and forming relationships. Conceptually, if we were to create a bank account of social capital, one could better visualize social capital as a resource. Imagine that you are doing media relations for your organization, and you make a cell phone call to a reporter for your local newspaper. You chat for a few minutes, and then hang up. Now there is a network connection between you, and while you aren't necessarily friends, it will be easier for you to call and ask that person to place a press release for you when the need arises. Imagine also that you make many contacts with many people in television stations, newspapers, newsletters, and radio stations. All of these various connections are really like a bank deposit. Those relational connections are in your account, waiting for you to withdraw and use to achieve various goals and objectives.

Second, the more social capital you have deposited into your account, the more likely it is that you and your organization will succeed. The more social capital available to public relations professionals, the more likely they will be able to effectively represent the interests of their organization. Public relations professionals who have no social capital lack the social connections required to have an impact on the marketplace of ideas regarding their industries. In other words, social capital confers a competitive advantage that allows individuals and organizations to achieve their goals and objectives.

Imagine that you are working as a public relations professional for a large organization. The organization has recently decided to become actively involved in community development. They want you to develop a community development campaign. First, you will need to identify the needs of the community. You may do this by e-mailing the city planner to get more information. This communicative exchange creates a social connection between you and the city planner. He or she has helped you by providing you with needed information. This creates an expectation for him and an obligation for you. He expects you to provide him with information in the future, should he need it, and you are obligated to repay your debt.

Third, social capital is embedded in relationships in networks, and this involves obligations and expectations. A reporter calls you for a quote from your chief executive officer (CEO) for a story she is writing. You oblige and provide the requested quotes. What has emerged is an obligation on the part of the reporter, who knows she owes you something, and an expectation on your part that you can call on the reporter when you have a situation that requires the assistance of the reporter. This is now a relationship, in a network, that involves obligations and expectations, which is a deposit in your social capital account.

It is important to remember that you are never guaranteed profits from investments. Some individuals who invest their financial capital in stocks or businesses lose their investment when companies perform poorly. Similarly, investments in social capital do not always yield positive returns to individuals or organizations. In some cases, people do not return favors or meet their social obligations. Some of the factors that influence the potential for social capital to produce a positive return on investment are discussed in the following section.

Fourth, social capital (like money) can be wasted. You have, at some point, impulsively purchased something you didn't need or perhaps did not necessarily want. We have all done that, and, in retrospect, we wonder why we wasted our money. Social capital is like economic capital in that it can be spent in ways that do not lead to positive outcomes. Sometimes, public relations practitioners are not good communicators, despite the fact that this is their job. Perhaps a public relations professional sends out a press release with typos or inaccurate information. The reporter who reads this may grow suspicious of the credibility of the source and might be more reluctant in the future to accept material from that source.

Here is another quick example of how social capital can be expended unwisely. Cell phones are ubiquitous. It is hard to find someone who does not have one these days. Public relations people carry them so they can be in constant contact. These people often excuse themselves to answer their phones to make sure that relationships and networks are carefully maintained. But, what happens when someone fails to respond promptly? With cell phone technology, there is now a norm that there is no excuse for not immediately returning calls. Failing to do so can actually become a dysfunctional expenditure of social capital.

Fifth, social capital is not like money in that it can't be given to someone else easily. You can give me $20, and the value of the money is instantly transferred to me with full value, and I can immediately spend it. However, it is hard to give me your social capital. Your relationships and your networks are, in many ways, yours alone. Certainly, you can say, "Call my friend, Sarah, who can help you with this event." But it is not really the same, is it? Sarah doesn't know me, and we both have to work through the formation of a relationship before we can really proceed. So, the value of the social capital does not necessarily transfer in the same way that economic capital would.

What makes social capital of particular importance from a public relations perspective is twofold. First, it makes relational connections of various kinds of central importance, and this constitutes a unique perspective on public relations. Second, it helps us to understand that communication is the essence of social capital development. Third, this perspective makes it clear that social capital can serve as an asset that can make the organization successful.

Dimensions of Social Capital

In developing the connection between communication and social capital, Hazleton and Kennan (2000) identified three dimensions of social capital: structural, relational, and communication.

First, the structural dimension focuses on the connectedness of social actors within a system. A connection has to exist before a relationship can be established. One way of understanding social capital is in terms of the number and organization of connections that a person or organization has.

The arrangement and proximity of actors within a network facilitate the development and use of social capital. Three structural aspects of networks that influence social capital are access, referral, and timing (Burt, 1992). Access indicates the degree to which individuals believe that they have a usable connection to individuals within a network that can produce effective action. It is the "who you know" factor in social capital. The more useful people or organizations you have relationships with, the more social capital you have.

Referral indexes the degree to which people can find information they need through existing network connections currently available to them. If you do not know someone who has the information that you need, it is likely that someone that you know does know someone else who has the information and can introduce you to them. This is best understood in terms of the popular idea that everyone is separated by only six degrees (connections).

Timing refers to the degree to which individuals can get information connected to the issue at hand in an appropriate time frame. That is, how easy or difficult is it to use the network connections that are available? One of the authors of this chapter was in Asia while we were writing, and it took days, rather than our usual minutes, to collaborate. At the university, our offices are next to each other, and we interact through talk and e-mail many times a day. During his travels, Bill did not have regular Internet access or the use of his cell phone. An eighteen-hour time difference also made a difference. We were not necessarily awake at the same times. Obviously, the length of time it takes to get and use information influences its value. If you only have hours to solve a problem, and it takes days to get to the people that can help you, then you are not likely to successfully solve your problem.

Second, the relational dimension focuses on the necessary conditions needed to maintain relationships within a network. It is in the relational dimension that norms of reciprocity are formed between social actors. The concepts of "a deed done now, will be repaid in the future" and "you help me now, and I'll help you later" serve as an unspoken rule in the network.

There are two relational conditions that affect this dimension. They are trust and the strength of social connections, also known as ties. Coleman (1988) identifies two types of trust in relationships: fragile trust and resilient trust. Fragile trust tends to rely on formal, contractual, and, often, written agreements that predicate how and when the expectations and obligations within a relationship will be fulfilled. Coleman implies that fragile trust is indicative of weak ties. The impact of having access to weak ties has been shown to provide access to information, aid in job attainment, create economic ramifications, and impact diffusion of innovations (Granovetter, 1973, 1983).

Resilient trust is indicative of strong ties and does not require contracts indicating how the expectations and obligations of the relationship will be fulfilled. Social actors within networks with resilient trust are more likely to engage in the spontaneous give-and-take associated with social capital (Coleman, 1988). "I'll help you now because I trust you would help me if I needed it and will help me in the future" is the attitude that fuels the reciprocal behavior. There is no need to have formal contracts or lengthy agreements boasting contingencies for the fulfillment of expectations and obligations between social actors.

Imagine you are out to lunch with a friend. When the waiter brings the check, you pay the tab knowing that, at some point in the future, your friend will offer to pay for lunch. There is no need to discuss when this will occur, because your friendship has resilient trust.

Strong ties are often essential to survival. The time and interaction required to create strong ties makes them limited in number. Family and friends are often considered to be strong ties because of the time and interaction invested in the relationship. If you needed to borrow $100, you are more likely to ask a strong, rather than a weak, tie for money.

Unlike strong ties, weak ties are easier to cultivate. Weak ties require very little time and interaction and are often used for information exchange. In their own way, weak ties are important and useful as well: They are bridges to the external world and environment. Strong ties with employees, suppliers, and financial supporters are probably necessary to organizations. Social networks without weak ties lack the coherence and ability to recruit members beyond a small sector of the network; weak ties provide the necessary connections for individuals to integrate into society (Granovetter, 1983). Weak ties are especially useful in job advancement and attainment when the weak tie has a connection to persons high in the organizational hierarchy (Lin, Ensen, & Vaughn, 1981). As a public relations professional, it is important to develop and maintain both strong and weak ties.

Identifying the function and strength of social ties or connections within a social network, and understanding the environment in which those relationships are most likely to create and maintain social capital, enables public relations professionals to effectively and strategically manage relationships to facilitate organizational success. It may appear that having strong ties is better than having weak ties. This, however, is not necessarily true. Both types of ties are useful and necessary.

Third, the communication dimension, identified by Hazleton and Kennan (2000), focuses on human communication as a foundation for the creation of social capital and as a tool for accessing social capital resources. Communication is a necessary condition for relationship development. Communication serves as a symbolic process for managing the capital created through relationships within a network. An expenditure of social capital by one person creates an obligation on the part of the receiver. To maintain the norms of reciprocity associated with social capital, a communicative process must occur to move the pendulum in the direction of repaying a debt. Communication alone, however, does not predict social capital. Social capital cannot exist without communication. Social capital cannot exist in any one of these dimensions alone. Structural, relational, and communication variables must be combined to create social capital.

Social Media and Public Relations

From a public relations perspective, if social capital is the creation of networks or relationships that have value and can be expended to achieve goals and objectives, then the central question is how is technology influencing this process? We will begin by considering the impact of technology on the general public, which sometimes forms an important audience for messages. There are two conflicting perspectives on this issue, which are briefly discussed below.

Robert Putnam (1995) penned a famous article titled "Bowling Alone: America's Declining Social Capital," which he later developed into a book (2000). In that article, he blamed, among other things, increased television viewing for the decline in civic memberships, such as Parent-Teacher Associations (PTAs), Kiwanis Clubs, etc. His argument was that people were watching more TV and hence spending less time in direct contact with other human beings. Hence, his argument was that media were causing a decline in social capital.

His central metaphor is that, while more people than ever before are bowling, there are fewer people bowling in leagues. Hence, there is a decline in social capital (civic engagement) as people spend more time alone.

A bit later, a Stanford Research survey (as cited in Nie & Erbring, 2000) revealed that people are spending more and more time with Internet activities and, in fact, spending less time watching TV. The conclusion is that face-to-face social interaction is in decline, and, therefore, people are becoming more isolated. All of this conspires to reduce the social capital available to individuals and communities.

There is another argument championed by scholars, such as Lin (2001), who argue that new technologies include the opportunity for increased, continual connection with other individuals. So often, the authors of this chapter see their students immediately reach for their cell phones as they leave class to connect with friends and family. Instant messaging and Web sites such as Facebook (http://www.facebook.com) offer instantaneous connection with friends twenty-four hours a day. Lin and others would argue that, in fact, these new technologies are increasing social capital by increasing connectivity.

There are several important issues here. First, Lin's argument presumes access to these technologies. Take Malaysia, for example, where 80% of Malaysians have cell phones (Malaysian Communications and Multimedia Commission, n.d.), a statistic that supports Lin's argument. However, if you travel to the northeast part of Malaysia, toward Kota Bahru through rural central Malaysia, the penetration of cell phone technology is far less pronounced. So, if you don't have access to these technologies, what happens to your deposit of social capital?

Second, in the United States, there is a generation gap. Older Americans are less familiar with these technologies and use them less often, although there is evidence that older Americans are adopting technology in an increasingly rapid fashion. Here is an example that demonstrates the social capital divide that can exist where technology impacts the preferences for communication channels: One of the author's close friends, who is in his mid-fifties, reports trying to make a cell phone call for twenty minutes with no success. His wife wandered by and gently explained to him that the call might go through if he used his cell phone instead of the TV remote to make his call. So, there is something of a social capital divide based on age.

The penetration of cell phone communication technology among college students is quite different. The results of a Student Monitor survey ("Statistics," 2006) are revealing:

- 92% of undergraduates own a cell phone
- 85% used their cell phone to send or receive text messages, and the average number of text messages per month was 115.
- 60% of students who have phones are in a family plan.

This survey highlights the differences. Younger individuals are much more immersed in new technologies than older individuals, and this undoubtedly has an impact on the nature of social capital formation, maintenance, and use. As one can imagine, there is potentially a bit of a divide between younger individuals who feel more comfortable handling relationships via cell phones and older people, like the person previously mentioned, who are less comfortable and prefer more traditional channels of communication. Another kind of isolation is now possible: Older individuals who have not involved themselves in new communication technologies may be isolated from younger individuals who see these technologies as normal, regular, and essential.

In an international context, there are also examples of social capital divides based on technology. Bluetooth technology is changing how young men meet women in Saudi Arabia ("Saudi Youth," 2006). Saudi Arabians tend to be members of the very conservative Wahabbi Islam sect. Women are veiled with only the eyes showing, and young men and women are not allowed to meet in public or, really, anywhere else. So, Bluetooth technology allows young people to exchange profiles and phone numbers, initiating a network link, without having to speak or even acknowledge the other's presence. In the past five years, cell phone use has increased from 1.7 million to 14.5 million in Saudi Arabia. This situation is causing considerable concern among older Saudis who are concerned about how connections are made between young men and women. Cell phones create a level of freedom that is difficult to monitor and control.

Third, there are also gender differences. Lin (2001) argues that, while men and women have the same amounts of social capital in the workplace, men get more benefit from theirs. That is, given similar deposits of social capital, men reap more rewards from their deposits than do women. The literature increasingly points to basic differences in technology use, which probably accounts for some of the differences in social capital effectiveness. Here are three examples from a survey conducted by Websense ("Men More Likely Than Women," 2006):

- 65% of men who access the Internet at work access non-work-related Web sites during work hours, compared to 58% of women.
- 41% of men who access non-work-related Internet sites at work said they would rather give up coffee than Internet access to non-work-related sites, compared with 47% of women.
- 20% of men use instant messaging as compared to 14% of women who use instant messaging.

These results reflect gender differences in technology use. Consequently, these gender differences are likely to influence the degree and kind of investment and return on social capital.

So, one has to take Lin's argument with a grain of salt. New technologies are being adopted quite quickly among younger people (especially college students), although we have met young people who are not connected. So, the social capital effect is probably not exactly uniform, meaning that, for some, social capital grows exponentially, while, for others, there is not much of an effect.

Public relations as a profession was founded on the use of mass media to achieve organizational goals through developing relationships with publics. The communication environment of the late nineteenth century and twentieth century was dominated by the emergence and growth of a limited number of mass media with large audiences. Public relations in this one-way environment primarily focused upon mass media and media relations.

Previous media forms used by public relations professionals were not conducive to the formation of social capital. Reading a newspaper, for example, can be done in isolation and does not involve human interaction. The same may be said for television and radio. The developments of satellite and computer technologies have, however, permanently changed the media environment for public relations in two significant ways. First, today there are a large number of media choices available to publics. The mass audiences of the last century have become fragmented, pursuing individual needs and interests because they can. Second, the new media environment promotes interactive two-way communication rather than passive one-way communication, which was characteristic of print and broadcast mass media. Interaction is central to the creation of relationships. The interactive nature of social media makes it an important variable of social capital study.

The fragmentation of mass media has had several important consequences for public relations. First, the power and importance of traditional advertising has diminished with the fragmentation of mass audiences. This increases the cost effectiveness of public relations, which has experienced significant growth as the new media environment has emerged. Second, users of new media tend to be active rather than passive participants in the communication process. They are more likely to already have knowledge, opinions, and objectives that influence how they respond to messages. They also have goals and purposes in mind when they use these media. Third, the knowledge and skills necessary to use new media are different from the knowledge and skills required to use traditional media. These differences are identified within our discussions of social media that follow.

Over the last couple of years, the term *social media* has been used to describe emerging new technologies that are important to public relations. While the term was initially used to describe the interactive nature of blogging, we feel that it can be broadly applied to all interactive technologies.

The word *social* implies relationships. Social media can include, but is not limited to, e-mail, instant messaging, online groups, blogs, Internet social networks, and cell phones. These new technologies allow public relations professionals to choose from a variety of communication tools to create connections with others. These new technologies have a direct impact on the structural, relational, and communication dimensions of social capital.

Social media use can impact the structural dimension of social capital. Social media, like e-mail, instant messaging, social networks, and cell phones, allow for greater frequency of communicative exchanges or interaction between social actors. Through the use of social media, information can be easily accessed, timing can be immediate, and referrals can be shared. Just as social media impact the structural dimension, they also impact the relational dimension of social capital.

In the relational dimension, social media impact the two important predictive variables of social capital: trust and social ties. Do you have a blog on MySpace (http://www.myspace.com) or Facebook (http://www.facebook.com)? Do you instant message or e-mail on a daily basis? These new social communication tools directly affect trust and networks. The rise in social media use has changed the structure of social networks. E-mail, instant messaging, and blogging have emerged as new ways to initiate weak ties with others; these weak ties serve as key links between social actors. The frequency of interaction allowed by social media creates the potential for strong ties to be developed from weak ties.

Finally, the use of social media impacts the communication dimension of social capital. Communication and interaction are necessary to form relationships. Social media are often used to communicate and to interact with others. Take a moment and think about your communication and interactions with others using social media. Do you use e-mail or instant messaging to communicate? Interactive technological media, like e-mail and instant messaging, have replaced traditional face-to-face communication in certain contexts. Like all forms of communication, people using social media have stylistic expectations.

All communication tools have different expectations in different situational and receiver contexts. Public relations students may use instant messaging, blogging, e-mail, and cell phones as preferred and frequent communication tools in college, but their eventual workplaces may not share that same preference. As a generation, college students are accustomed to using various types of social media to interact with others.

Take a moment to think of the last e-mail or instant message that you sent. Did you type complete sentences, and use correct punctuation and spelling? Probably not. The expectation for interactions via social media is different in a college environment than that found in the workplace. Instant messaging may be a preferred communication tool in an organization, but the stylistic expectations for these exchanges are likely to be different for instant messages exchanged between colleagues than instant messages exchanged between peers. For those accustomed to using social media, it is easy to assume that everyone communicates the way they do. This, however, is not the case, as there are generational differences in the use of social media. Next, we will explore the various types of social media and their implications for the creation, maintenance, and expenditure of social capital.

Cell Phones

Although cell phones were originally used by business professionals or purchased for safety concerns, they have evolved over time to become a popular communication device for social interaction. Anyone who has attended a meeting of public relations professionals has seen firsthand the use of cell phones during meeting breaks. College students have been known to use cell phones to text message during class. Many people feel such a strong need to remain connected that they turn their cell phones to vibrate rather than turning them off entirely. This obviously has implications for the structural dimension of social capital, especially with respect to timing.

Igarashi, Takai, and Yoshida (2005) saw that the intimacy and connection between individuals who communicate both face-to-face and through cell phone messages were higher than between those individuals who only communicated face-to-face. Social networks formed through cell phones might be slower to form than those that include face-to-face communication, but they will eventually reach comparative levels of intimacy and connectedness. Smith and Williams (2004) found that individuals who sent and received messages on their cell phones reported feeling more included and involved with others around them. Those who do not use cell phones for messaging purposes reported an increase in feeling ostracized from social communities. The increase in feeling ostracized led Smith and Williams to conclude that using technology in productive ways can increase connections and have a positive effect on an individual's social capital.

E-mail

E-mail has become one of the most commonly used communication tools. As a technological medium, e-mail allows for less time-consuming and less intrusive communication when compared to face-to-face interaction and traditional letter writing. Professionals use e-mail to work collaboratively. Multiple communicative exchanges can occur in a matter of minutes. Because of its ease of use, e-mail can be effectively and efficiently used to maintain existing weak and strong ties. E-mail alone, however, may not be an efficient means of establishing strong ties. The ease of sending common messages to multiple receivers also facilitates the efficiency of e-mail compared to traditional mail.

Instant Messaging

A technology that is not as popular in business but is likely to become more important in the future is instant messaging. This is a technology prevalent among college students.

There is a variety of research that discusses the importance of instant messaging among the college student population. Hu, Wood, Smith, and Westbrook (2004) found that instant messaging served as a supplement to face-to-face communication. They also observed that instant messaging dialogue created a desire to communicate face-to-face. Ruppel and Fagan (2002) suggest that college students prefer instant messaging their peers rather than telephoning or visiting them in person. Lee and Perry (2004) found instant messaging to be the primary medium for communication: The students they studied reported using instant messaging more often than the phone, e-mail, and face-to-face communication. Technologies, such as instant messaging, can be used to increase social capital with those with whom it would be physically impossible to interact face-to-face.

Flanagin (2005) found that individuals established new relationships through instant messaging. It can be a viable medium for developing new relationships among geographically distant groups and individuals. Read (2004; see also Wellman, Hasse, Witte, & Hampton, 2001) stated that most people have a social network that extends beyond their local community and that technology enables them to maintain those connections without excessive cost or personal strain.

Instant messaging is often used as a means of informal communication to update people about daily activities. Although instant messaging is mostly a social communication tool, people also use instant messaging to coordinate on real-time tasks. One of the authors uses instant messaging to communicate with students during office hours. Students have told him that they prefer instant messaging to phone calls.

Blogs

Blogs, which originally developed as online personal journals, have quickly been adopted by public relations and marketing as organizational tools. Having employees or company experts write regularly about products is becoming common. Another common tactic involves providing resources and information to private individuals who write blogs about a company or its products. The advantage, or disadvantage, of blogs is their interactive character. Blogs routinely provide opportunities for readers to comment on the content that they have read. Comments appear to have a characteristic similar to dialogue in conversation. It is not uncommon for a blogger to respond to comments, creating a thread of related communications. Such exchanges may also lead to offline relationships and communication, creating the possibilities of stronger ties and enhanced social capital.

Social Networking

As the Internet has developed, so, too, have new resources for online communication and information exchange. One recent resource is the development of social networking communities, such as MySpace and Facebook. The primary use of these two network communities has been social interaction, although both allow the formation of interest groups around careers and personal interests. The social network concept is rapidly being expanded to include business interests. One of the chapter authors belongs to LinkedIn (http://www.linkedin.com), a social networking community with a focus on professions that also lists employment opportunities. Craig Carroll, a friend and network connection of one of the authors, described (via e-mail) his experience with LinkedIn in the following manner:

I have used it for staying in contact with friends and colleagues from college and previous jobs, finding jobs for my students now (have placed two students through linked in), and extending my own network within the PR professional community. It provides a segue, a reason to reestablish contact with old colleagues and to e-mail them when it might a bit weird to e-mail them out of the blue when too much time has passed by. "I just found this new gadget on the Web, saw you were listed there, and thought I'd say hi. By the way, how are you?" And it provides a way to "integrate" the various roles that you play and to take control of your network when it seems a bit out-of-hand and overwhelming. (One of the consequences of having an extensive network is that it can become overloaded and then interfere with productivity.) (personal communication, June 2, 2006)

The links to social capital formation and use are obvious in Carroll's description of his use of LinkedIn. This technology, like many others discussed in this chapter, takes advantage of the speed and interactive characteristics of new social media, which are useful for both enhancing and maintaining social capital.

Conclusions

Public relations professionals' use of social media can have a direct impact on their personal and professional stock of social capital. Take a second and think of the number of social ties you have. Are you a member of the Public Relations Student Society of America (PRSSA), Facebook, MySpace, a blog, or instant messaging? New public relations professionals will serve as important inputs to society and organizations. Social capital developed while preparing for a career can be transferred to one's professional life and organization.

You may find that your current social ties provide friendship but very few resources that can be capitalized for mutual benefit. Social ties are likely to evolve and change over time. A college friendship with another student who is a member of PRSSA may not, at the time, seem like an important social tie. In ten years, however, that same person may be a critical resource in finding a job.

Social media are not a substitute for face-to-face interaction. They are a complement. Through social media, public relations professionals may develop personal relationships that they then can utilize to provide benefits for their employers. They may also build public relationships between the organization they represent and those with whom they communicate.

For consideration

1 How does the importance of face-to-face communication compare with that of social media in the development of weak and strong ties?
2 How can a university use new technologies to develop and enhance its social capital with students, faculty, and alumni? To what extent would research play a role in making that determination?

3 What are the possible consequences of generational differences in uses of new technologies for developing public relations programs?
4 How much social capital do you have?
5 Join the debate. Does the Internet allow for increased connectivity among individuals or increased isolation? What situational and contextual factors affect individual outcomes of new media use?

For reading

Fukuyama, F. (1995). *Trust: The social virtues and the creation of prosperity*. New York: The Free Press.

Fussell, H., Harrison-Rexrode, J., Kennan, W., & Hazleton, V. (2006). The relationship between social capital and transaction costs: A case study. *Corporate Communication: An International Journal*, *11*, 148–161. Retrieved March 20, 2007, from http://www.emerald-library.com/info/journals/ccij/ccij.jsp

Hazleton, V., & Kennan, W. (2000). Social capital: Reconceptualizing the bottom line. *Corporate Communications*, *5*, 81–86.

Igarashi, T., Takai, J., & Yoshida, T. (2005). Gender differences in social network development via mobile phone text messages: A longitudinal study. *Journal of Social and Personal Relationships*, *22*(5), 691–713.

Putnam, R. D. (2000). *Bowling alone*. New York: Simon and Schuster.

References

Burt, R. S. (1992). *Structural holes: The social structure of competition*. Cambridge, MA: Harvard University Press.

Coleman, J. S. (1988). Social capital in the creation of human capital. *American Journal of Sociology*, *94*, 95–120.

Flanagin, A. J. (2005). IM online: Instant messaging use among college students. *Communication Research Reports*, *22*, 175–187.

Fussell, H., Harrison-Rexrode, J., Kennan, W., & Hazleton, V. (2006). The relationship between social capital and transaction costs: A case study. *Corporate Communication: An International Journal*, *11*, 148–161. Retrieved March 20, 2007, from http://www.emerald-library.com/info/journals/ccij/ccij.jsp

Granovetter, M. (1973). The strength of weak ties. *American Journal of Sociology*, *78*, 1360–1380.

Granovetter, M. (1983). The strength of weak ties: A network theory revisited. *Sociological Theory*, *1*, 201–233.

Hazleton, V., & Kennan, W. (2000). Social capital: Reconceptualizing the bottom line. *Corporate Communications*, *5*, 81–86.

Hu, Y., Wood, J. F., Smith, V., & Westbrook, N. (2004). Friendships through IM: Examining the relationship between instant messaging and intimacy. *Journal of Computer-Mediated Communication*, *10*(1). Retrieved February 28, 2006, from http://jcmc.indiana.edu

Igarashi, T., Takai, J., & Yoshida, T. (2005). Gender differences in social network development via mobile phone text messages: A longitudinal study. *Journal of Social and Personal Relationships*, *22*(5), 691–713.

Kennan, W. R., & Hazleton, V. (2006). Internal public relations, social capital, and the role of effective organizational communication. In C. Botan & V. Hazleton (Eds.), *PR theory II* (pp. 311–338). Mahwah, NJ: Lawrence Erlbaum Associates.

Lee, K. C., & Perry, S. D. (2004). Student instant message use in an ubiquitous computing environment: Effects of deficient self-regulation. *Journal of Broadcasting and Electronic Media*, *48*(3), 399–421.

Lin, N. (2001). *Social capital: A theory of social structure and action*. New York: Cambridge University Press.

Lin, N., Ensen, W., & Vaughn, J. (1981). Social resources, strength of ties and occupational status attainment. *American Sociological Review*, *46*, 393–405.

Malaysian Communications and Multimedia Commission (MCMC). (n.d.). Retrieved August 24, 2006, from http://www.cmc.gov.my/index.asp

Men more likely than women to engage in personal Web surfing at work. (2006, May 16). *PR Newswire*. Retrieved August 24, 2006, from LexisNexis Academic database.

Nie, N. H., & Erbring, L. (2000). *Internet and society: A preliminary report* (Stanford Institute for the Quantitative Study of Society, Stanford University). Retrieved May 24, 2001, from http://www.stanford.edu/group/siqss/Press_Release/Preliminary_Report-4–21.pdf

Putnam, R. D. (1995). Bowling alone: America's declining social capital. *Journal of Democracy*, *6*, 65–78.

Putnam, R. D. (2000). *Bowling alone: The collapse and revival of American community*. New York: Simon & Schuster.

Read, B. (2004, May 28). Have you "facebooked" him? *The Chronicle of Higher Education*, pp. 29–31.

Ruppel, M., & Fagan, J. C. (2002). Instant messaging reference: Users' evaluation of library chat. *Reference Services Review*, *30*, 183–197.

Saudi youth use cell phone savvy to outwit the sentries of romance. (2006, August 6). *The Washington Post*, p. A01.

Smith, A., & Williams, K. D. (2004). R u there? Ostracism by cell phone text messages. *Group Dynamics: Theory, Research, and Practice*, *8*, 291–301.

Statistics on college student cell-phone use (article insert). (2006, May 19). *eSchool News*. Retrieved August 24, 2006, from http://www.eschoolnews.com/news/showStory.cfm?ArticleID=6310

Wellman, B., Hasse, A. Q., Witte, J., & Hampton, K. (2001). Does the Internet increase, decrease, or supplement social capital? Social networks, participation, and community commitment. *The American Behavioral Scientist*, *45*(3), 436–459.

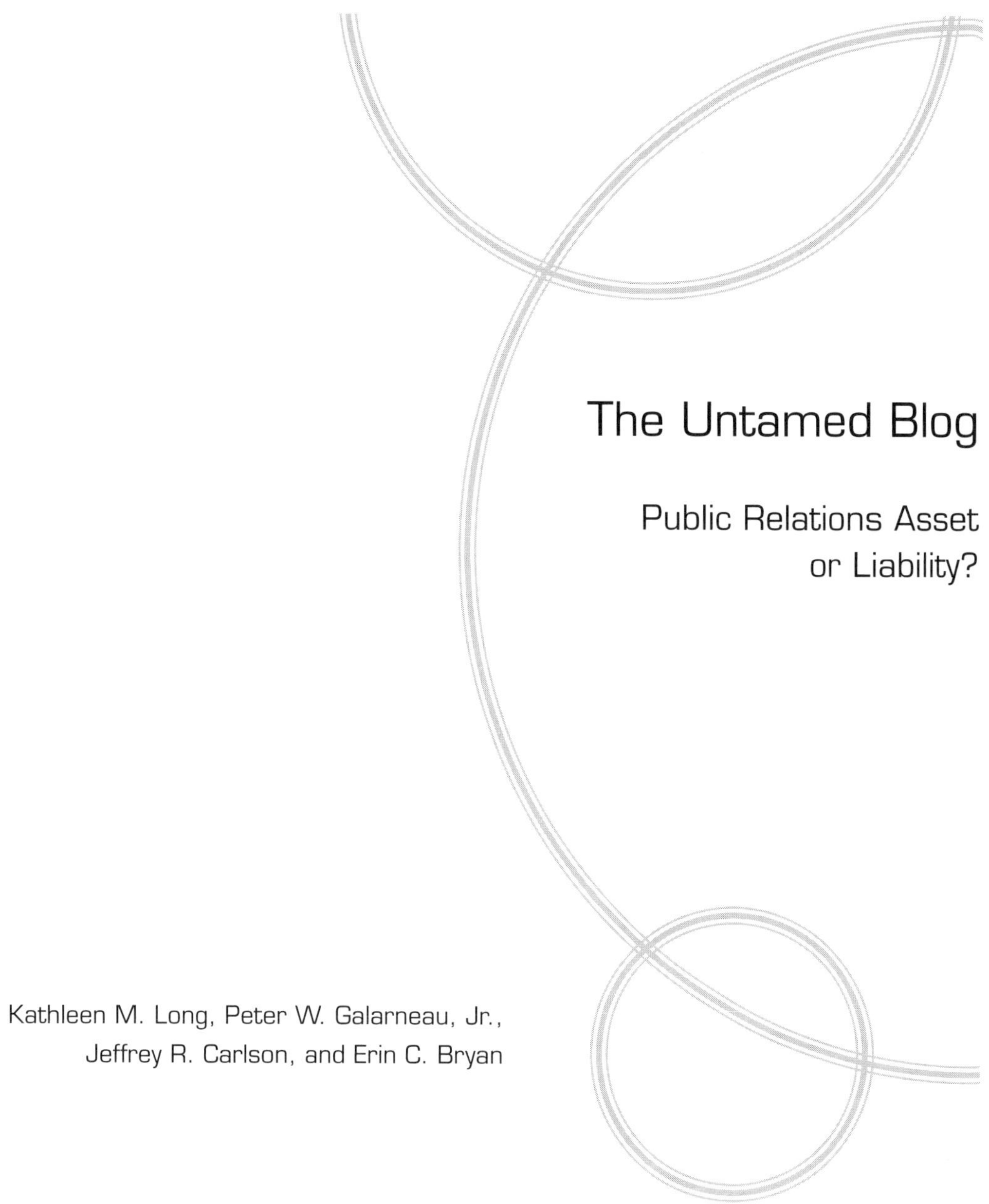

The Untamed Blog

Public Relations Asset or Liability?

Kathleen M. Long, Peter W. Galarneau, Jr., Jeffrey R. Carlson, and Erin C. Bryan

The public relations industry is being challenged by a new information paradigm that has shifted editorial control into the hands of the many, not just the few. Yet, public relations practitioners view this new media outlet as a method through which they can build and maintain relationships with their publics. Web logs, better known as blogs, instill individuals with the power once awarded by editors and publishers to select writers. Because blogging bypasses traditional gatekeepers of information, practitioners can directly communicate with both internal and external Internet-using audiences. This chapter first discusses the history of blogging and follows with a discussion of trends in blogs as a public relations tool, including examples of organizations that have used them. Blogging is then examined in the context of organizational structures and the traditional communication theories of organizational structure and gatekeeping. A discussion of the similarities and differences between blogs and other tools of public relations in terms of style and form, cost, time, potential audience reach, and the ethical and growing legal considerations of blogging is also addressed.

Public relations practitioners have recently been presented with a new media outlet through which they can build and maintain relationships with their publics. Web logs, better known as *blogs*, imbue individuals with the power once awarded by editors, publishers, and other business managers to select writers. Because of their ability to undermine hierarchical power controls, blogs provide an outlet for the purest form of U.S. First Amendment rights, granting expression to all on the social ladder.

Before examining blogs as a new public relations tool, we first must understand a blog's influence on today's society. In doing so, it is beneficial to remember that human behavior is guided by several desires, one of the greatest being the need for social contact. Before the Internet, social contact was concentrated. Either you knew someone who engaged your psychological desires and communicated with him or her through faster channels, such as the telephone, in a mostly one-to-one relationship, or you interacted face-to-face with a group of socially acceptable individuals, many of whom you might not have known but with whom you shared common interests. The latter social group usually entailed the transportation of oneself to a common location at a predetermined time. To socialize with those who shared your spiritual interests, you might drive to your local church at ten o'clock in the morning. Romance might have required a weekend trip to the local nightclub. A group could be found for just about any social need, but, to participate, you had to find a way to get there.

The Brief History of Blogging

And then came the new road paved not with asphalt but with electricity. The Internet provided a way to transcend time and space. Its first use and reason for existence was for social, common-interest networking. Researchers from universities needed a way to share data. Traveling was too prohibitive, given the time sensitivities associated with experimentation. A bunch of computers wired together and communicating via a common language was the answer.

Once the Internet took hold of the larger population some twenty years later, it took very little time before social networks engaged the speed and ease of the new electrical road. These new digital communities used technologies such as Usenet, e-mail lists, bulletin board systems, and forums. Each brought together people who shared affinities. No longer did a person need to drive to church at ten o'clock in the morning to engage in conversation about his or her spiritual beliefs. All one had to do was click on a power switch and surf.

Blogs have become one of the premier online social networks of today, surpassing all digital community technologies of the past. The Pew Internet and American Life Project found that, in 2004, around eight million people in the United States had created blogs (Marken, 2005). Today, the popular online blog search engine Technorati tracks more than fifty million blogs worldwide, and these are only the blogs about which it knows.

Why the popularity? It is because a blog is easy to create and easy to maintain. All you need is a connection to the Internet; libraries provide this service for free. There exist several Web sites that provide free blog space. And, for the most part, no one controls what you say in your blog, so the posting of commentary is immediate and uncensored. You have a completely free and easy way to express your freedom of speech. You have the means to supplant media controls on what is published. You can potentially influence elections, reverse a corporation's swing on the stock exchange, and uncover the lies of politicians, chief executive officers (CEOs), and presidents. You can erode institutional power.

Blogs Evolve into a Public Relations Tool

In the early evolution of blogs, this medium was viewed as only in the purview of individuals wishing to share their ideas and opinions with the world. However, after observing the power to connect with targeted publics, organizations and their public relations teams began to formally develop blogging as an additional strategic tool to manage relationships with both internal and external target publics. The discussion of blogs, as a public relations strategy, is limited, in this chapter, to diaries or journal postings whereby a member of an organization communicates purposefully and directly with Internet-using audiences via a posting. Yet these public relations–controlled blogs are not the only ones operating within an organization. For instance, in 2005, Hewlett-Packard (HP) had more than 2,500 blogs, and IBM had more than 4,500 blogs written by employees (Marken, 2005). Moreover, these internal blogs are only a small portion of the estimated fifty million blogs worldwide.

Stakeholders use blogs to communicate among themselves, and public relations practitioners have discovered that their relationships with customer or client publics can be greatly and quickly impacted by postings on the blogosphere. As van der Merwe, Pitt, and Abratt (2005) state, "public relations becomes more than the management of effective communication with stakeholders—it evolves to become the management of communication between stakeholders as well" (p. 40). For example, the personal computing company Dell has strategically used blogs as two-way communication to garner feedback to meet customer needs and to explain company policies. Both positive and negative customer experiences are used as part of the research function, and responses or changes in everything from policies to product line have grown Dell to #1 in the personal computing market (Marken, 2005). But let us now examine two cases in which public relations practitioners conceived and designed a blog as a strategy or tactic to build, continue, or repair a relationship with a target public.

The soft drink company Dr. Pepper / 7-Up initiated a blog strategy in March 2003 to increase product awareness of a new flavored milk called Raging Cow. The blog was started by inviting a group of young bloggers to be briefed on Raging Cow and encouraging them to blog about the new product (Bruner, 2004). Raging Cow's blog / public relations initiative was one of the first efforts by a mainstream company to use a blog to promote product awareness. The main blog told a story about a cow's life and the shift the cow made from an unexciting, dairy life to starting a crusade against boring milk. According to Bruner, the initiative fell short because the younger bloggers were asked not to mention that they had been briefed on the new product, creating the illusion that their positive remarks were purely their own ideas. Furthermore, the company offered financial awards to the young bloggers for blogging about the actual Raging Cow blog being cool (Ochman, 2006). Although Dr. Pepper / 7-Up maintains that the blog was successful for its target audience, there was a community of bloggers that accused the company of being deceitful (Bruner, 2004). To this day, the Raging Cow blog is commonly referred to as what not to do when using a blog as a public relations strategy (Ochman, 2006).

The cable channel FX initiated a blog strategy in spring 2006 to increase viewers for the season finale of the television show *Nip/Tuck*. Initially, FX used the social network MySpace to stir interest in *Nip/Tuck* characters, especially The Carver. Blogs were used to leave hints about The Carver's identity, and the show's Web site also had both written and video blogs promoting the suspense. This tactic created hype about the finale show, which resulted in the highest rating for a final show for niche cable channels last spring (Safo, 2006). Clearly, the blogs developed a relationship between viewers and the characters in *Nip/Tuck*, and FX's goal was achieved.

Generally, public relations strategies and their associated tactics are designed to influence, maintain, or change relationships with stakeholders or publics important to an organization. Usually, tactics are selected and messages are designed as part of a public relations plan within the processes of research and evaluation. Tactics are often classified by type of message format or channel (e.g., special event, newsletter, media kits, annual reports, Web sites, press conferences, press release, bill inserts, and speeches). These channels are also categorized on a continuum of controlled to uncontrolled media, meaning the degree to which the message is composed and distributed directly to a target audience by the practitioner and the organization versus when someone else serves as the distributor and determines the form the message takes, as well as the timing of the distribution.

How does one classify blogs as a public relations tactic? Is a blog a controlled or an uncontrolled medium; that is, is the message and its presentation on the blog managed and constrained by the organization? Is a blog a public relations tactic distributed through a public or mass medium? Can entries on the blog be posted by everyone with access or is posting restricted to targeted bloggers? From the previous examples and discussion, the answer is, it depends.

If the blog is accessible only through the intranet of an organization, and the blog's intended audience is employees, then this blog is restricted to a targeted audience. However, this type of blog could either be controlled or uncontrolled. If management used the blog to explain various positions on upcoming changes in benefit plans or to discuss ideas about new product or market development and limited access to posting, then this blog would be a controlled medium. If, however, the blog was open for postings by all employees on whatever topic was of interest or importance to them, then the blog would be uncontrolled but still nonpublic. If the blog is initiated by the organization—for example, NBC's blog The Daily Nightly, where Brian Williams and NBC News correspondents, producers, and staff share their perspectives on how *NBC Nightly News* is edited—then the blog is public in that anyone with an interest in news editing and access to the Internet can access the blog, but it is controlled in that NBC is managing the timing, presentation of content, and tone of this blog. But what if the organization opens a blog on its Web site and invites both internal and external publics to post messages about whatever topic they choose? This would be uncontrolled and public, much like the rogue Web sites that have popped up that usually criticize the product or the management of a particular organization, such as Starbucks, Dell, and Chase Bank.

What is different about the use of blogs as an uncontrolled medium is that the distribution is not in the hands of the traditional media or news organizations, but in the hands of individuals, and, thus, the usual advantages may not apply. With other public relations tactics, uncontrolled media were evaluated as more credible, as there was an implicit third-party authorization. In addition, uncontrolled media are usually less expensive in terms of both staff time and energy and production costs. These advantages may not transfer to blogs, as the credibility of the message will depend on the credibility of the individual bloggers, and the monitoring of the postings may involve considerable staff time and effort. With controlled blogs, the traditional advantage of the form, timing, and frequency of the distribution of the message being in the power of the organization remains, but the credibility of the message may be lessened. Thus, blogs require a new look at the structure of organizations and the dissemination of public relations strategies both within the organization and between the organizations and its publics.

Blogging and Organizational Communication Theory

In the context of organizational structure, a blog is a double-edged sword for public relations practitioners. Although structures vary among organizations, one common thread among structure type is the controlled flow of information within and outside of organizations. Traditionally speaking, a gatekeeper is an individual who is able to manage and control messages flowing within and outside of an organization (Richmond & McCroskey, 2001). This control allows the gatekeeper to screen, edit, and filter information as it flows through the organization.

The concept of a gatekeeper was first introduced by Lewin (1947) and defined as an individual or group that has the power or control over information flowing in, out, and through the organization. Relying on Lewin's initial research, White (1950) analyzed a local newspaper and the gatekeeper that held the power to make the initial judgment as to whether a story was important or not. At that newspaper, it was the gatekeeper's job to select from the wire copy provided by the Associated Press, United Press, and International News Service what copy would be read on the front page of the morning newspaper by the thirty thousand families to which it was distributed. White found that many of the reasons the gatekeeper provided for the rejection of the stories fell into the category of highly subjective judgments. However, the most important point is the fact that the gatekeeper held the power to make a judgment, not that the judgment was subjective.

Since White's (1950) study, the conceptualization of a gatekeeper has been continuously refined. A gatekeeper is no longer seen as an individual free to make his or her own decision, but rather as a person who is shaped by internal (e.g., employees, marketing, and public relations) and external (e.g., customers, interest groups, and government) forces during the decision making process. Furthermore, a message may go through not just one, but several gatekeepers to reach the targeted audience (Messner, 2005; Shoemaker, 1991). For example, a news release will not reach the public until it travels through several internal and external gatekeepers (e.g., public relations practitioner, manager, and editor). Although the news release eventually reaches the public, each gatekeeper has the same power and opportunity: the ability to edit and filter.

Within a formal structure, a public relations practitioner is a gatekeeper for any given organization who manages information released to specific audiences. For example, in the formal structure, it may be explicit that a public relations practitioner prepares and edits publications for internal and external audiences. In this case, the practitioner is clearly controlling information within and outside of the organization. But what if a practitioner created a blog that included input from all aspects of a business, such as marketing, sales, design, research, etc.? Tom Murphy, a public relations manager for Cape Clear Software, hosted several blogs that included contributions from marketing, product management, and engineers (Karpinski, 2003). In this case, the public relations practitioner may not be able to control the flow of information within the organization, and the practitioner's role as a gatekeeper is more limited.

Moreover, blogs represent an opportunity and temptation for any given employee to become an informal gatekeeper and bypass the formal structure of an organization. This changes our conception of the gatekeeper's role, because, with the emergence of a blog, unedited information can reach the public on a continuous basis (Messner, 2005). Employees with Internet access can bypass an organization's public relations practitioner to directly communicate with both internal and external Internet-using audiences. From a public relations point of view, this can become problematic if employees release information contrary to organizational and public relations goals. McGregor (2004) notes that allowing employees to speak directly to customers requires a large amount of trust by the organization, because a loose cannon might reveal corporate secrets, send the wrong message, or open the company up to legal trouble. Given an employee's ability to blog about an organization's affairs, Karpinski (2003) proposes the following for consideration:

- How do you keep company blogs on-message with corporate public relations strategy?
- Should organizations direct or influence the information about which an employee writes?
- Who should run the company blog?
- Do organizations need a corporate blogging policy?
- Should a company embrace or control employees who blog on their own time?

The employee's ability to blog not only changes the conceptualization of a gatekeeper, but also impacts the formal organizational structure. According to Barzilai-Nahon (2005), a major claim of the traditional concept of gatekeeping is that the opportunity to circumvent the gatekeeper is minimal. Previously, the only opportunity to circumvent the gatekeeper was to go to another gatekeeper within the organization. Thus, a gatekeeper maintained control not only because of the formality of the organizational structure, but also because of the one-way flow of communication from the individual to the gatekeeper. Blogs limit the amount of control held by an organization's traditional gatekeepers by creating opportunities for individuals to bypass gatekeepers.

From an organizational viewpoint, this not only allows an individual to bypass the formal structure (i.e., the explicit gatekeeper) of that organization, but also creates an informal structure for conversing with others. An informal network is any type of communication that "does not follow the hierarchical path or chain of command" (Richmond & McCroskey, 2001, p. 27). Informal networks can grow among employees and management and have no correlation with the formal organizational structure (Richmond & McCroskey). These networks are not new with blogs; grapevines exist in all organizations. What is new is that the conversation—who is talking to whom about what—is neither limited to the individual's social network nor the informal networks that exist in the organization. Employees who decide to blog on their own time about organizational events may have an audience larger than their social acquaintances and friends. Furthermore, the blog reaches beyond the internal organization to any external individual who has Internet access. Before the use of the Internet and blogs, this was not a feasible possibility and fell outside of the theoretical realm of a traditional gatekeeper.

The only current theory that encapsulates how the Internet and blogs change gatekeeping is network gatekeeping theory (NGT) (Barzilai-Nahon, 2005). NGT shows that, through networks, an individual can circumvent gatekeepers. For example, through publishing a blog, an employee can respond to and comment on events without the intervention of a gatekeeper. In a sense, each employee becomes an individual gatekeeper who is capable of filtering organizational information to the external environment.

Blogging as a Public Relations Tool

Similar to employee bypass of traditional gatekeepers, public relations practitioners can also utilize a blog to circumvent traditional gatekeepers of information (e.g., newspapers, television, and Web sites) to communicate directly with specific audiences. This allows practitioners much more freedom than that allowed by many of the traditional mediated channels. Messner (2005) noted that, for the first time, it is not solely the traditional media that determines which news is allowed to reach the public. Companies and public relations practitioners can choose to publish blogs without the media editing and filtering that information. For example, Macromedia, the developer of Flash, Dreamweaver, and other creative tools, created several blogs pertaining to its products (Carroll, 2002). Staff members update blogs that feature a running commentary with news, tips on product fixes, hints, and links to sites that have been built using the product, as well as customer feedback. Under the scope of organizational structure and gatekeeping theory, the blog can be viewed as a powerful tool with both positive and negative impacts.

The Legal and Ethical Aspects of Public Relations Blogging

The legal and ethical aspects of blogging that the public relations practitioner needs to consider differentiate themselves into three categories: (a) blogs that could be described as internal, that is, owned but only sometimes controlled by the organization; (b) blogs that are internally controlled but externally directed; and (c) blogs that exist externally to the organizational boundaries and thus are not controlled by the organization.

Externally uncontrolled blogs. Beginning with the latter category, the uncontrolled blogs that exist outside of the organization in cyberspace are the epitome of our use of the phrase "double-edged blog." Just like practitioners monitor the traditional media, monitoring the content of new media, like blogs, is imperative. Blogs allow users the opportunity to share their beliefs, opinions, known facts, or even blatant untruths with potentially millions of people; thus, the rationale for monitoring is clear. Therefore, Web logs offer an organization equal opportunity to receive positive and negative attention. What can a public relations manager do about content that is harmful to his or her organization?

Simply put, the same legal considerations apply to blogs as to any other form of speech about which a public relations practitioner would be concerned. In the United States, freedom of speech certainly applies to bloggers, but so does libel law, which, as any public relations practitioner knows, is a complex body of legal knowledge with a difficult burden of proof. In an October 2005 judicial ruling, the burden of proof was made more difficult in court cases involving charges against anonymous bloggers (Merriweather, 2006).

In a small-town political showdown, a Delaware town councilman filed a defamation lawsuit against several John Does—bloggers who had posted degrading comments about the councilman. Originally, the Delaware Supreme Court allowed an order that directed Comcast Corporation, the local cable television and Internet provider, to provide personally identifiable information about the blogging subscriber or to appear in court to explain why the customer in question could not be identified. However, in October 2005, the order was overturned, allowing the identity of the blogger to remain unknown (Merriweather, 2006).

Thus, with the complex nature of libel law and the ever-changing judicial precedent and legislation of intellectual property and freedom of speech, the most accurate answer to the question, What can you do? would have to come from a qualified attorney.

Blogging employees. With the number of existing blogs reaching into the millions, it is probable that some employees of any given organization are blogging on personal time through means they provide themselves (e.g., Internet access, computer). Not surprisingly, then, personal blog postings by employees that focus on or mention the employing organization are not uncommon. Therefore, it is important that public relations practitioners are aware of the potential results of negative content coming from a disgruntled employee's personal blog. There is also the possibility that the content of an employee's personal blog could contain information that is inappropriate to be discussed online or is circumventing the purpose of having an organizational gatekeeper of information. In addition, there are employees who blog anonymously, also on personal time, without specifically identifying their employers, colleagues, or even their work. Yet, at times, blog readers are able to deduce the author's and or the organization's identity, and, subsequently, corporate secrets or other information organizational leaders would rather keep quiet is shared.

So, what can be done about an employee who is harming the organization through his or her personal blog? Despite the lack of a concrete answer to the prior question about rogue bloggers from the general public, there are steps a practitioner can take to help prevent employees from knowingly or unknowingly publishing negative content on their own time in personal blogs.

These steps begin with the question of how much control an employer should have over an employee's private life and actions. This question has been discussed more recently in both the traditional and new media, as this issue was brought to greater attention when Catherine Sanderson was dismissed from the French accounting firm Dixon Wilson in 2006 upon the firm's discovery of her personal blog site, Petite Anglaise: Slices of My Life in Paris. The firm believed that her blog's content was disgraceful to the firm—even though Sanderson never mentioned herself or the firm by name, though her picture was clearly shown ("Bridget Jones," 2006; Sanderson, 2006).

When a blog is conceptualized as nothing more than a diary, which happens to be public, it is not surprising that details of a blogger's work life would be part of the content. Questions for public relations practitioners and their organizations to consider is whether it is ethical to fire an employee whose personal blog contains information that casts the employing organization in a negative light. In comparison, would an employee be fired for sharing face-to-face the latest water cooler gossip with an organizational outsider? Or would he or she be terminated if caught ranting about a horrible boss or administrator via the home telephone or, perhaps, in the local newspaper's letters to the editor? What makes publishing through a blog similar or different to the above situations? How much influence does the potential mass audience of a blog have on the answers to the above questions?

Although not necessarily relevant to the content in Sanderson's blog, a rhetorical question was posed by one of the Petite Anglaise blog readers, as reported by CNN: "Say you worked for a large corporation, and in your spare time you wrote an anonymous 'insider's view' column for the Financial Times. Would you expect anything less than termination upon discovery?" ("Bridget Jones," 2006, para. 21). On the other hand, there were other commentators who believed that Sanderson had committed no wrong and that employers should have little to no say about an employee's personal life (Sanderson, 2006).

In the United States, the answer to the question about ranting or disgruntled blogger-employees is certainly impacted by the First Amendment. But, as previously mentioned, not all employees are necessarily posting in anger. Although the details of the Petite Anglaise story do not accuse Sanderson of the dissemination of corporate secrets or the disclosure of clients' personal information, such a situation could happen, and it could be due to simple naivety of an employee-blogger believing that he or she is posting in anonymity. Thus, as a precautionary measure, wise organizations should draft an organizational policy on employee's participation in cyberspace, even when that participation occurs on personal time. These policy suggestions should not limit themselves to blogging but should expand to address employees' construction of personal Web sites, participation in other public dialogue, and images in varying publication forms via cyberspace. Such policies not only establish clear guidelines about the employer's expectations for the employee, but also educate otherwise happy employees who may naively include too many organizational details in their blogs, despite their assumed anonymity.

With such policies, though, comes the ethical question previously mentioned: How far will the organization go in attempting to control its employees' private lives? The answer to this is not an easy one. It is possible that such policies could lead to a question of legality and constitutionality; nonetheless, such policies would make the organization's preferences about employee behavior known. Therefore, if a practitioner determines such a policy is wise for his or her organization, it would not be beyond the scope of the practitioner's job to initiate the creation of such policy, as the implications of having or not having such a policy fall within the concern of the public relations department.

Internally uncontrolled blogs. The discussion of the previous policy matters segues into another category of legal and ethical blogging considerations: blogs that are created internally but intended to be read by both internal and external audiences. Many large companies today, like Hill and Knowlton and Sun Microsystems, are encouraging their employees to blog on company time, using the company's computing hardware and software. Blogging by employees can potentially foster innovation, cut down on problem-solving time, improve employee morale, and attract new employees and, perhaps, even clients. The real legalistic consideration for these blogs is, again, policy. Business Blogwire, a Web log focused on discussing corporate and business blogging and offering tips and guidelines, highlighted Hill and Knowlton's and, later, Sun Microsystems' blogging policy in an October 2005 post (Ellsworth, 2005a, 2005b). Included in the praise of Sun Microsystems are key points of Sun Microsystems' policy: Encourage your employees to blog, warn against divulging company secrets, and, as a blogging employee, do not assume that including simple disclaimer language releases you from liability for your content (Ellsworth, 2005b; Sun Microsystems, n.d.). In May 2005, IBM published its own blogging policy and guidelines, which left the decision of whether "to blog or not to blog" to employees (IBM Corporation, 2005). Although the decision was left to the individual, IBM clearly proclaimed blogging and other types of online dialogue to be opportunities for learning and contributing to the company and to the world. In its policy, IBM clearly described eleven specific guidelines for those who engage in the practice of blogging, including attention to confidentiality, copyright, fair use, financial disclosure laws, specifications on identifying oneself when writing, and the assertion that, "In general, what you do on your own time is your affair. However, activities in or outside of work that affect your IBM job performance, the performance of others, or IBM's business interests are a proper focus for company policy" (p. 3).

Clearly, blogging is becoming a tactic used by organizations for multiple purposes, and policy seems to be the key in preventing blogs from backfiring for the organization. However, there is a third, final type of blog used by organizations that merits specific and careful discussion—the blog as a specific public relations or marketing tool.

Internally controlled, externally directed blogs. As previously mentioned, public relations professionals, like those for the television show *Nip/Tuck*, have employed the use of blogs to help relate or market to their publics. Many questions arise from this practice. Certainly, the residual external impact of blogs that are used principally for internal purposes, like those at Hill and Knowlton or IBM, is legitimate and ethically allowable. However, what about blogs that are written, edited, and displayed by an organization with the sole and specific intention of influencing readers to take an action that is favorable to the organization?

For example, West Virginia University (WVU) (2006) began an admissions blog in March 2006. Content included postings that described alumni connections across the country, unique faculty-student programs, the transportation system that was voted better than Disney World's, outdoor adventure orientation classes, popular music groups that headline campus events, and nationally ranked athletic teams.

Clearly, the blog was used as a promotional tool to build a relationship with prospective students. However, the blog clearly identified its purpose as keeping "students, parents, and high school guidance counselors up-to-date on deadlines, events, the admissions process, and to answer frequently asked questions about WVU" (West Virginia University, 2006, Welcome section, para. 1). Given its stated purpose as an information-providing tool and its apparent purpose as a promotional tool, is the use of the term *blog* appropriate? For example, do these types of online publications categorically fall under the definition of *blogging*? If the WVU blog was an information sharing tool, as the university stated, why not use traditional online bulletin boards or frequently asked questions pages, which are common outlets for online information seekers?

Because these traditional online tools were not used, one wonders whether the use of the term *blog* for publications similar to the WVU example mislead readers into believing the content is shared *without* bias or a persuasive intent. On the other hand, were the blogs for *Nip/Tuck* misleading, given the fictitious nature of the show? If any of these uses of blogs is deemed misleading, the use of the term *blog* then becomes a major ethical concern.

Examples like the WVU site are not uncommon, particularly in higher education recruiting efforts. Regardless of the popularity, however, a public relations manager needs to consider what the target audience believes the use of the term *blog* indicates. If the audience believes the blog content is not controlled by a gatekeeper and is not a strategy that is part of an overall public relations or marketing plan, then the potential for misleading readers exists, and ethical concerns are likely.

The Future of Blogging as a Public Relations Tool

Blogging is clearly a popular tactic in public relations and marketing efforts at this point in time. Will it last? It is possible that it will, but it is also possible that people will catch on to the persuasive intent and find that there are much more straightforward and less biased avenues of obtaining information they wish to know. On the other hand, it would not be surprising if true blogs stick around and retain their potential for both positive and negative public relations interactions with multiple publics.

For consideration

1. What are the functional characteristics of a true blog, that is, one designed for the free and open expression of ideas? To what extent should organizations use uncontrolled blogs in building relationships with publics?
2. Ultimately, are organizations better served by attempting to control employee expression of ideas (on and off the job)?
3. What could guidelines for using blogs in public relations strategy look like?
4. How can public relations practitioners establish themselves as credible and active contributors to company policy making?
5. To what extent are uncontrolled customer blogs beneficial to an organization? Under what circumstances would it be appropriate and ethical for a public relations practitioner to post a message on a customer blog aimed at his or her company?

For reading

Business blog consulting. (n.d.). Retrieved August 30, 2006, from http://www.businessblogconsulting.com/

Business blogwire. (n.d.). Retrieved August 30, 2006, from http://www.businessblogwire.com/

Marken, G. A. (2005). To blog or not to blog, that is the question? *Public Relations Quarterly*, 50(3), 31–33.

Richmond, V. P., McCroskey, J. C., & McCroskey, L. L. (2005). *Organizational communication for survival: Making work, work* (3rd ed.). Boston: Pearson.

van der Merwe, R., Pitt, L., & Abratt, R. (2005). Stakeholder strength: Public relations survival strategies in the Internet age. *Public Relations Quarterly*, 50(1), 39–48.

References

Barzilai-Nahon, K. (2005). Network gatekeeping theory. In S. Erdelez, L. McKechnie, & K. Fisher (Eds.), *Theories to information behavior: A researcher's guide* (pp. 247–254). Medford, NJ: Information Today.

"Bridget Jones" blogger fire fury. (2006, July 19). *CNN.com*. Retrieved July 19, 2006, from http://www.cnn.com/2006/WORLD/europe/07/19/france.blog/index.html

Bruner, R. E. (2004, June 30). Raging cow: The interview. *Business blog consulting*. Retrieved September 6, 2006, from http://www.businessblogconsulting.com/2004/06/raging_cow_the_.html

Carroll, J. (2002). *Jumping on the corporate blog wagon*. Retrieved May 5, 2006, from http://www.jimcarroll.com/articles/mktg22.htm

Ellsworth, E. (2005a, October 25). Corporate blog review: Hill and Knowlton. Message posted to Business Blogwire Web log, archived at http://www.businessblogwire.com/2005/10/corporate_blog_review_hill_and.html

Ellsworth, E. (2005b, October 25). Great corporate blogging policy: Sun Microsystems. Message posted to Business Blogwire Web log, archived at http://www.businessblogwire.com/2005/10/ great_corporate_blogging_polic.html

IBM Corporation. (2005, May 16). *IBM blogging policy and guidelines*. Retrieved August 15, 2006, from http://www.snellspace.com/IBM_Blogging_Policy_and_Guidelines.pdf

Karpinski, R. (2003, April 4). Corporate blogs make personal connections. *B to B*, *88*(4), 1–2.

Lewin, K. (1947). Frontiers in group dynamics: II. Channels of group life; social planning and action research. *Human Relations*, *1*, 143–153.

Marken, G. A. (2005). To blog or not to blog, that is the question? *Public Relations Quarterly*, *50*(3), 31–33.

McGregor, J. (2004, April). It's a blog world after all. *Fast Company*, *81*, 84–86.

Merriweather, J. (2006, February 3). Stepdaughter was blogger, mayor says. *The News Journal*. Retrieved February 3, 2006, from http://www.delawareonline.com

Messner, M. (2005). Open gates everywhere: How Web logs open opportunities for public relations practitioners. In M. L. Watson (Ed.), *The impact of public relations in creating a more ethical world: Why can't we all get along?* (International Public Relations Research Conference proceedings, March 10–13, pp. 309–320). Miami: Institute for Public Relations.

Ochman, B. L. (2006). What is a business blog? *The Sideroad*. Retrieved September 6, 2006, from http://www.sideroad.com/Business_Communication/business_blog.html

Richmond, V. P., & McCroskey, J. C. (2001). *Organizational communication for survival: Marking work, work* (2nd ed.). Boston: Allyn and Bacon.

Safo, N. (2006, September 5). Cable networks gain viewers with edgy shows. *NPR*. Retrieved September 6, 2006, from http://npr.org/templates/story/story/php?storyId=5767759

Sanderson, C. (2006, July 20). Suspendered. Message posted to Petite Anglaise Web log, archived at http://www.petiteanglaise.com/archives/2006/07/20/suspendered

Shoemaker, P. (1991). *Gatekeeping*. Newbury Park, CA: Sage.

Sun Microsystems, Inc. (n.d.). *Sun news—Sun blogs*. Retrieved August 15, 2006, from http://www.sun.com/aboutsun/media/blogs/policy.html

van der Merwe, R., Pitt, L., & Abratt, R. (2005). Stakeholder strength: public relations survival strategies in the Internet age. *Public Relations Quarterly*, *50*(1), 39–48.

West Virginia University. (2006, March 20). This blog's for you! Message posted to Admissions Counselor Web log on WVU Blog Network, archived at http://blogs.wvu.edu/admissionscounselor/articles/2006/03

White, D. M. (1950). The "gate keeper": A case study in the selection of news. *Journalism Quarterly*, *27*, 383–390.

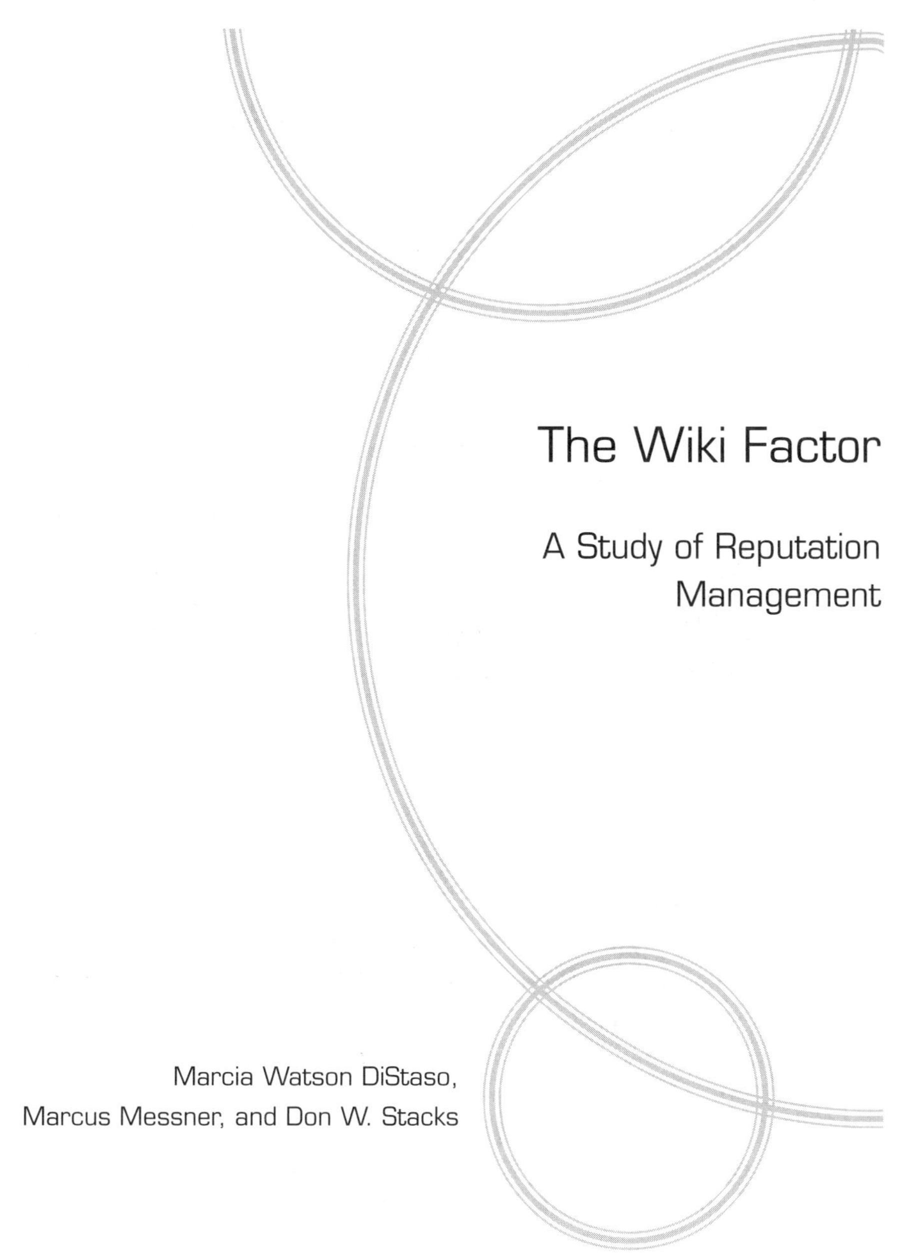

The Wiki Factor

A Study of Reputation Management

Marcia Watson DiStaso,
Marcus Messner, and Don W. Stacks

The study presented in this chapter examined Wikipedia's implications for corporate relationship management. A content analysis of ten Fortune 500 companies was conducted and found that Wikipedia was largely neutral, but almost 40% of content was either positive or negative, including content on issues such as scandals and social responsibility. This is significant for public relations practitioners, as Wikipedia often ranks high on Internet search engines. In addition, Wikipedia's editing process provides ways to identify those passionately for or against a company—essentially providing a list of an active public that may form public opinion about a company.

November 30, 2005, was a wake-up call for all public relations practitioners in the United States when journalist John Seigenthaler disclosed his very personal case of "Internet character assassination" (Seigenthaler, 2005, p. 11) by putting the latest online revolution under the spotlight. The former editorial page editor of *USA Today* and assistant to Robert F. Kennedy disclosed that, for four months, a false biography had been published under his name in the free online encyclopedia Wikipedia, which is collectively written by the public and is read by millions around the globe ("Battle of Britannica," 2006; Lamb, 2006).

"At age 78, I thought I was beyond surprise or hurt at anything negative said about me. I was wrong," Seigenthaler (2005) wrote in *USA Today* that day. He was surprised to learn from the Wikipedia entry that he "was thought to have been directly involved in the Kennedy assassinations of both John and his brother Bobby" (Seigenthaler, 2005) Although, "nothing was ever proven," this was what it said in the Wikipedia entry (Seeyle, 2005, p. 1). In addition, Seigenthaler was accused of having temporarily moved to the Soviet Union.

One week after Seigenthaler published his accusations against "volunteer vandals with poison-pen intellects" (Seigenthaler, 2005) in *USA Today*, his Wikipedia biographer was revealed. Brian Chase of Nashville, Tennessee, where Seigenthaler once had been editor of *The Tennessean*, saw his additions as a joke on a co-worker, with whom he had simply discussed the well-known Seigenthaler family (Kirtley, 2006; Page, 2005).

While *The Economist* ("Battle of Britannica," 2006) categorized this first highly publicized incident involving Wikipedia as a "scholarly footnote in media history" (p. 65), the implications of the collaborative wiki model are far-reaching for public relations practitioners. How do agencies react when their clients' Wikipedia biographies are altered by unknown contributors? What do public relations practitioners do if an old company scandal occurs in the Wikipedia profile?

Hill (2006) sees "opportunities as well as threats" (p. 15) in Wikipedia. While it can be a tool for public relations practitioners to communicate with online constituents, it can also include negative information on scandals or subpar annual performances. *Wired News* sees even broader implications of Wikipedia as the "new paradigm in human discourse" (Sjöberg, 2006), which raises even more questions. If any individual can contribute to what is ultimately considered the truth about a person or a company, then what influence remains for public relations practitioners? Holtz (2006) warns that "most businesses have not yet awakened to just how powerful that voice will be" (p.24). Berrisford (2006) argues that, with the explosion of Web information, communicators inevitably

> need to be flexible and adaptive; able to continually construct meaning from disparate and fragmented information flows, and by the same token to continually deconstruct the distributive media which constitute the flow. We may need to radically reconsider our concept of the organization as a whole. (pp. 27–28)

This chapter explores the implications of the wiki model used by Wikipedia. Wikipedia is different from traditional encyclopedias, such as *Encyclopaedia Britannica*—a leader in reference and education publishing based on an expert-centered model of knowledge production (Encyclopaedia Britannica, 2006). *Encyclopaedia Britannica* (1768 inception) established its credibility by involving some of the greatest minds in history, such as Albert Einstein, Sigmund Freud, and Marie Curie.

Wikipedia, on the other hand, is a user-contributed online encyclopedia that is collaboratively edited and utilizes the wiki concept—the idea that any Internet user can change content on any page within the Web site (Chawner & Lewis, 2006; Ebersbach & Glaser, 2004; King, 2006). Remy (2002) describes the goal of the Wikipedia project as being "to create a truly free universal encyclopedia" (p. 434). Paradoxically, this process has created an Internet reference that is gaining credibility. Stemming from the word *wiki*, the Hawaiian word for quick, Wikipedia employs the fast, up-to-date, and information-based communication advantage often sought in society today (Hill, 2006). This is why Wikipedia continues to grow in popularity.

This chapter furthers theory-driven research by analyzing the framing of *Fortune 500* companies in Wikipedia. How these companies are presented is an important subject of study, because they are among the most powerful in the United States. The contestation of their public image is open and unbridled in the wiki process. For those who wish to influence contestation, including public relations practitioners, stockholders, and activists, a better understanding of this process is essential.

Literature Review

Wikipedia was formally launched on January 15, 2001, as a single English-language edition at http://www.wikipedia.com. Today, there are about one hundred active language editions of Wikipedia, and the English version hit one million articles on March 1, 2006. Overall, Wikipedia claims a total of 2.5 million articles with 600,000 users (Wikipedia, 2006).

According to Alexa (2006), a Web traffic monitor, Wikipedia reaches a daily average of thirty million users and is the sixteenth most popular Web site in the world. For comparison, Yahoo, Google, and MSN are currently the top three Web sites; CNN is the twenty-third most popular Web site, AOL is the twenty-fifth, and *The New York Times* is fifty-fourth. Wikipedia has also been among the top five rising searches on Google in 2005 ("2005's Top Rising," 2006). Scholars see Wikipedia as a part of the emergence and practice of citizen journalism, which also includes the increasing popularity of Web logs (Bowman & Willis, 2005; Dorroh, 2005).

Wikipedia's benefits allow *anyone* to edit *anything*, which has led to its popularity and expansion, while simultaneously opening it to criticism and misuse. In February 2006, congressional staffers were reported to be "playing Wikipolitics" by editing profiles of congressmen and senators on Wikipedia on various issues ranging from personal attributes to political views. For instance, Senator Robert C. Byrd was listed to be 180 years old (McCaffrey, 2006). Senator Tom Coburn had an erroneous entry claming that he "was voted the most annoying senator by his peers in Congress" (Noguchi, 2006, p. 1). The offices of Senators Joe Biden and Dianne Feinstein, on the other hand, deleted from their senators' Wikipedia entries embarrassing references to a plagiarism scandal and a conflict-of-interest issue. Politicians are not alone in altering their biographies. Former MTV veejay and podcasting entrepreneur Adam Curry was caught anonymously changing Wikipedia's podcasting entry to remove credits of other people and inflate his role in podcasting's creation (Nuttall, 2006).

During the 2004 presidential campaign, supporters of President George W. Bush and challenger John Kerry started a Wikipedia war by editing and re-editing candidate entries minute by minute. At one time, the President's picture was replaced by one of Adolf Hitler. Consequently, the entries were made unavailable for editing until after the election (Boxer, 2004).

Although at times abused, Wikipedia does have rules, according to its Web site:

> As anyone can *edit* any article, it is of course possible for biased, out of date, or incorrect information to be posted. However, because there are so many other people reading the articles and monitoring contributions using the Recent Changes page, incorrect information is usually corrected quickly. Thus, the overall accuracy of the encyclopedia is improving all the time as it attracts more and more contributors. You are encouraged to help by correcting articles, validating content, and providing useful references. (Wikipedia, 2006)

Additionally, Wikipedia encourages people to create a login and requires this registration to create a new article (an entry or Web page is called an *article*). To guard against abuse, when edits to existing pages are made without a login, the user's Internet Protocol (IP) address is recorded.

Although the motives and expertise of Wikipedia contributors are generally not known, users are encouraged to maintain a neutral point of view by presenting the facts in a manner upon which supporters and opponents can agree (Denning, Horning, Parnas, & Weinstein, 2005; Lih, 2004). Contentious topics may be debated and can sometimes lead to "editing wars" (Lipczynska, 2005). A benefit of this format is that current events and news are kept up-to-date. Breaking information from news stories is added and updated as events unfold. As one supporter commented, "The criticisms that have been leveled at Wikipedia are, in fact, the very sources of its strength" (Lipczynska, 2005, p. 7).

Wikipedia is the most popular wiki-model project; however, several other wikis have developed and are gaining popularity. Wiktionary, for instance, is an online dictionary, Wikitravel is a travel guide, and Wikinews is a compilation of news stories (Ebersbach, Glaser, & Heigl, 2006).

In contrast to Wikipedia, *Encyclopaedia Britannica* claims to be "one of the world's most-trusted sources of information on every topic imaginable," and it aspires to "take all human knowledge, organize it, summarize it, and publish it in a form that people find useful" through assistance from "the very best minds" (Encyclopaedia Britannica, 2006, n.p.). *Encyclopaedia Britannica* was recently compared to Wikipedia in the release of "an expert-led investigation carried out by *Nature*—the first to use peer review to compare [the encyclopedia's] coverage of science" (Giles, 2005, p. 900). The study revealed numerous errors in both encyclopedias, but, among the forty-two entries tested, the difference in accuracy was not particularly great: The average science entry in Wikipedia contained approximately four inaccuracies, whereas *Encyclopaedia Britannica* contained approximately three. (*Encyclopaedia Britannica* has since disputed the accuracy of the methodology and subsequent findings of this study [Panelas, 2006; "Wiki Principle," 2006].)

Essentially, the difference between Wikipedia and *Encyclopaedia Britannica* is that Wikipedia circumvents the gatekeeping process. Wikipedia's creation is bottom up, where users contribute and edit information, thereby adapting and changing to meet the needs of readers. In effect, the differences between the two encyclopedias can be explained in that each contains the same information, only *framed* differently.

How an article is framed can influence a reader's perception of that topic. The framing of an issue often determines whether it climbs or falls from agendas. The selection of certain aspects of an issue can make it important, because issues are the basis around which publics are organized and public opinion is formed (Grunig & Hunt, 1984).

Communication by and between publics and organizations is a form of public relations activity. The relationship management dimensions of trust, satisfaction, commitment, and control mutuality can be explored through an analysis of how companies are described in the media. Essentially, a company's framing provides insight into how writers and readers feel about that company.

Framing theory, as defined by Gamson (1989), offers insight into the concept of accentuating certain attributes of information while subordinating other content. Framing becomes important when multiple ways to interpret and present a set of facts exist. Through the selection and emphasis of certain facts, interpretations may shape and develop public opinion (Mahon & Wartick, 2003). In journalism, choosing a frame for a story is the most consequential decision that can be made (McCombs & Shaw, 1993).

The way an object is framed can have measurable behavioral consequences (McCombs & Shaw, 1993). Issues gain relevance when they are made prominent (Dearing & Rogers, 1996). According to Wanta, Golan, and Lee (2004), the public learns through salience the importance of issues based on the amount of coverage those issues receive, thereby setting the public's agenda.

Framing can be studied through the analysis of tonality of media content. According to Michaelson and Griffin (2005), "tonality is an analysis that uses a subjective assessment to determine if content is either favorable or unfavorable to the person, company, organization or product discussed in the text" (p. 4). Tonality therefore describes the positive, neutral, or negative framing of an issue, an approach previously applied in business contexts. Smith (1996), for instance, studied framing in consumer product advertising and the effect of positive and negative frames on viewers. Kweon (2000) examined biases in the coverage of company mergers by analyzing the favorable, neutral, and unfavorable tones in news magazine coverage. These studies established that tonality of media content is an important field of study, as it impacts how an audience perceives issues or images. Research on tonality is important in public relations because it provides a means of studying company image portrayals.

Purpose of the study. Although Wikipedia is increasingly mentioned in the press, there is currently little, if any, scholarly research on the subject. As its popularity grows, the need for a better understanding of Wikipedia increases. Content related to corporate pages is especially important to exploring issues concerning organization-public relations. It is important to understand how and what attributes are selected and presented to the public, because they can influence the public's perception of a company. The purpose of this study was to examine Wikipedia's implications for corporate relationship management.

Methodology

To analyze trends in Wikipedia articles about U.S. companies, a content analysis of the top ten company reputations identified by the 2005 *Fortune 500* list (Wal-Mart Stores, ExxonMobil, General Motors [GM], Ford Motor, General Electric, Chevron Texaco, ConocoPhillips, Citigroup, American International Group [AIG], and International Business Machines [IBM]) was conducted. This sample was chosen because it includes companies that, by virtue of their membership on this list, are the largest and most widely followed companies in the country.

The sample included ten articles from Wikipedia (one for each company). The unit of analysis was the sentence, and each of the ten articles was reviewed sentence by sentence for tonality and topic. The researchers also analyzed the companies for search engine hits, supplemental content, and editing. Overall, a total of 884 sentences was analyzed with an average of eighty-eight sentences per article (with a range from 6 to 168).

Procedure. The content was analyzed for the framing of each company, which was operationalized through a combination of tonality and the topics covered in the entry on the company. To obtain tonality, each sentence was coded as positive, negative, both positive and negative, or neutral, as established in previous research (Kweon, 2000; Michaelson & Griffin, 2005; Smith, 1996). For instance, a positive statement would be, "In 2004, XYZ Company was named the number one company for employers and employees on the Forbes 500 Global Player list." On the other hand, an example of negative tone would be "XYZ Company has been involved in many scandals." An example of a sentence that would be coded as both positive and negative is "XYZ Company has committed itself to pay ongoing wages to nonworking employees displaced through automation." A neutral statement would be "XYZ Company headquarters are in Irving, a suburb of Dallas, Texas."

Along with tone, the topic of each sentence was recorded by indicating the presence or absence of the following predetermined list:

- *Corporate* (i.e., sentences about the company or management, such as "In the United States, XYZ Company is the largest underwriter of commercial and industrial insurance");
- *Historical* (i.e., sentences about historical events defined as pre-2000, such as "In 1908, the XYZ Company released the the first car");
- *Performance* (i.e., sentences such as "For the first time ever, in 2004, the total number of cars produced by all makers in Ontario exceeded those produced in Michigan");
- *Employees* (i.e., sentences such as "XYZ Company refers to its employees as 'associates,' and encourages managers to think of themselves as 'servant leaders'");
- *Financial* (i.e., sentences about company finances or stock, such as "XYZ Company has revenues of $91 billion annually");
- *Competition* (i.e., sentences such as "XYZ Company has often had one of the highest brand loyalty rates for gasoline in America");

- *Community efforts/social responsibility* (i.e., sentences such as "In 2005, XYZ Company launched its Ecomagination initiative in an attempt to position itself as a 'green company'");
- *Legal concerns/scandals* (i.e., sentences such as "In May 2004, XYZ Company agreed to pay $2.65 billion, to settle a class action lawsuit");
- and *Other*.

These topics were chosen because they represented broad categories into which the sentences could fit and are often used in investment strategy decisions (Investopedia, 2006). Because the sentences were coded for the presence or absence of each topic, it was possible for any sentence to contain more than one topic.

The coding manual was pretested and revised before coding began to increase intercoder and intracoder reliability. Two trained coders each analyzed approximately half of the sample. In addition, 60% was coded by both to calculate intercoder reliability (six Wikipedia articles). Overall, the data reflected an intercoder reliability of 97% using Holsti's (1969) formula. Reliability estimates were as follows: company, 100%; type, 100%; tone, 92%; and topic, 94%. Scott's *pi* (1955) was calculated for an overall agreement at 92%.

Results

Before examining the content in the Wikipedia articles, the researchers explored its prominence in search engines. As Table 1 shows, Wikipedia is a top Web site that comes up in using the most common search engines when looking for corporate information. At the time of the study, the researchers found, on average, that, 80% of the time, Wikipedia was listed on the first page of search results (within the top ten Web sites listed) (n = 24). Fifty percent of the time, the Wikipedia article was in the top five search results (n = 15), with the IBM Wikipedia article link consistently in the top three results.

Table 1: Fortune 10 Web Hits

Fortune Companies	Number of Web Hits		
	Yahoo	MSN	Google
Wal-Mart Stores	9	5	13
ExxonMobil	3	9	7
General Motors	3	3	7
Ford Motor	12	11	8
General Electric	10	4	12
Chevron Texaco	7	5	6
ConocoPhillips	3	3	4
Citigroup	3	5	9
AIG	4	17	4
IBM	3	2	2

Framing. The analysis of tone revealed that Wikipedia's entries not only offered readers neutral information about the largest companies in the United States, but also contained negative and positive information. Overall, Wikipedia had significantly more neutral sentences (n = 556, 63%), followed in frequency by negative sentences (n = 155 sentences, 18%), positive sentences (n = 109 sentences, 12%), and sentences that were both positive and negative (n = 64 sentences, 7%).

Table 2 provides a breakdown of each by topic used in the Wikipedia articles. Overall, they largely pertained to company information (n = 334; 28.7%) and/or historical information (n = 309; 26.6%). The topic of community efforts / social responsibility (n = 71; 6.1%) was balanced with legal issues / scandals (n = 69; 5.9%).

Table 2: Sentence Topic Breakdowns in Wikipedia Articles

Sentence Topic	Number of Sentences
Corporate	334 (28.7%)
Historical	309 (26.6%)
Financial	119 (10.2%)
Employees	105 (9.0%)
Community Efforts / Social Responsibility	71 (6.1%)
Legal Concerns / Scandals	69 (5.9%)
Competition	36 (3.1%)
Performance	9 (0.8%)
Other	4 (0.3%)

Note. Each sentence was coded for the absence or presence of each topic, so many sentences contained multiple topics, and percentages may not total 100% due to rounding.

Supplemental content. Pictures were included in seven out of the ten corporate articles, tables or charts were in nine articles, and references and/or additional links were in all ten articles. The findings for the identified sources and links referred to as "See Also," "References," and "External Links" are presented in Table 3. There was an average of twenty-one links per company (N = 207). Links to the sections or divisions of the corporate Web site were the most common (n = 65; 31%), followed by press articles (n = 48; 23%) and then activist organizations (n = 35; 17%).

Edits. Each of the company Wikipedia articles contained edits. At the time of the study, the total number of edits for the ten companies was 6,165, with a high of 2,567 for Wal-Mart and a low of forty-eight for ConocoPhillips. Overall, there was an average of seven edits per sentence based on the total of 884 sentences for all the companies, with a high of twenty-eight for Wal-Mart and a low of two for Citigroup. There was a total of 2,385 edits made by anonymous users, for an average of 239 per company, with a high of 920 for Wal-Mart and a low of eighteen for ConocoPhillips. The combined total of registered users was 1,474, with a high of 418 for Wal-Mart and a low of twenty-five for ConocoPhillips. Table 4 provides breakdowns by company.

Table 3: Link Type Breakdowns in Wikipedia Articles

Link Type	Number of Links
Corporate Web site	65 (31.4%)
Press	48 (23.2%)
Activist organizations	35 (16.9%)
Books	19 (9.2%)
Financial	19 (9.2%)
Research	7 (3.4%)
Customer sites	6 (2.9%)
Other	8 (3.9%)

Content is available on most editors who register with Wikipedia. This allows practitioners to see who is writing what. For example, an edit on the Wal-Mart article was made by a self-proclaimed "retired commercial beekeeper that specialized in pollination service for fruit and vegetable growers." He lives in South Carolina and is "occupied with writing and photography, particularly about nature and local interest themes" and has written a couple of books and magazine articles, along with editing a small town newspaper (Pollinator, 2006).

Information like this on active publics or people actively participating in the creation of a company's image in Wikipedia can be useful for companies in determining how to communicate about a problem or possibly prevent one before it occurs.

Table 4: Edits to Wikipedia Articles

Company	Total Number of Edits	Average Edits per Sentence	Total Number of Edits by Anonymous Users	Total Number of Registered Users
Wal-Mart Stores	2567	28	920	418
Exxon Mobil	397	4	162	128
General Motors	503	4	229	153
Ford Motor	1071	6	456	258
General Electric	300	5	98	98
Chevron Texaco	109	3	31	46
ConocoPhillips	48	8	18	25
Citigroup	368	2	106	90
AIG	118	3	61	37
IBM	684	6	304	221

Implications for Relationship Management

This study examined Wikipedia's implications for relationship management by exploring a new communication technology for corporate content. With corporate logos, pictures, and links directly to the company Web site, Wikipedia may appear to have company involvement and/or sponsorship. Therefore, does the company have a right to edit or add any information it likes or dislikes?

Although the Wikipedia articles remained largely neutral, the positive and negative extremes should be of concern. In considering the companies' framing, not only is tonality important, but knowing the topics of discussion also becomes critical. This should provide companies with a better understanding of easily available information about themselves. Although unavoidable, negative content, such as legal issues and scandals, is one area that companies may not want to highlight but on which the public focuses. The Wikipedia articles contained 12% positive content and 18% negative content, with 6% of content focusing on legal issues or scandals specifically.

The findings of this study indicate a call to action for companies: Monitoring Wikipedia or similar technologies should become a common practice for public relations practitioners. On the World Wide Web, Wikipedia draws readers to its Web site, as demonstrated by high rankings on common search engines. Wikipedia often ranks at the same level as the official company Web sites, which is generally the first search result.

Wikipedia has abolished the elitist gatekeeping of traditional encyclopedias by opening the defining of issues to the public, thereby including or erasing controversies and biased viewpoints from corporate images. By applying this approach to the framing of portrayals of major corporations, the consumers and the public as a whole gain the opportunity to help shape companies' images. The public does not have to influence a gatekeeper to have an impact on the issues anymore.

The breakdown of the gatekeeping process should alarm public relations practitioners. While, in the past, practitioners could rely on a neutral framing of companies in encyclopedias, practitioners must now monitor Web sites, such as Wikipedia, which have a greater focus on controversies and the potential of changing every second, as users change the entries frequently.

The editing process, however, allows public relations practitioners to identify not only those who change and frame entries, but also topics of concern and targeted company messages. An advantage for companies is the list of active users making edits on a company's article. This list can be reviewed to identify these people, providing a way to identify an Internet community and a critical public relations public. Public relations practitioners must quickly adapt to this new environment to regain control over companies' image portrayals.

For consideration

1 How, if at all, should a company respond to frequent editors of its Wikipedia entry?

2 To what extent can a company control its image portrayals in a communications environment increasingly populated by new media technologies?
3 Should public relations practitioners alter the Wikipedia entries for their employers or clients? Why or why not?
4 How has the abolishment of elitist gatekeeping affected the credibility of openly editable sources?
5 How could organizations, whether for-profit or nonprofit, imitate the Wikipedia model on their own Web sites? That is, what type of content could actually benefit from collaborative editing?

For reading

Berrisford, S. (2006). How will you respond to the information crisis? *Strategic Communication Management, 10*(1), 26–29.

Denning, P., Horning, J., Parnas, D., & Weinstein, L. (2005). Wikipedia risks. *Communications of the ACM, 48*(12), 152.

Ebersbach, A., Glaser, M., & Heigl, R. (2006). *Wiki: Web collaboration*. Berlin: Springer.

Holtz, S. (2005). The impact of new technologies on internal communication. *Strategic Communication Management, 10*(1), 22–25.

Porter, L. V., Sallot, L. M., Cameron, G. T., & Shamp, S. (2001). New technologies and public relations: Exploring practitioners' use of online resources to earn a seat at the management table. *Journal of Mass Communication Quarterly, 78*(1), 172–190.

References

Alexa. (2006). *Top sites*. Retrieved May 14, 2006, from http://www.alexa.com

Battle of Britannica. (2006, April 1). *The Economist, 379*(8471), 65–66.

Berrisford, S. (2006). How will you respond to the information crisis? *Strategic Communication Management, 10*(1), 26–29.

Bowman, S., & Willis, C. (2005). The future is here, but do news media companies see it? *Nieman Reports, 59*(4), 6–10.

Boxer, S. (2004, November 10). Mudslinging weasels into online history. *The New York Times*, p. E1.

Chawner, B., & Lewis, P. H. (2006). WikiWikiWebs: New ways to communicate in a Web environment. *Information Technology and Libraries, 25*(1), 33–43.

Dearing, J., & Rogers, E. (1996). *Agenda-setting*. Thousand Oaks, CA: Sage Publications.

Denning, P., Horning, J., Parnas, D., & Weinstein, L. (2005). Wikipedia risks. *Communications of the ACM, 48*(12), 152.

Dorroh, J. (2005). Wiki: Don't lose that number. *American Journalism Review, 17*(4), 50–51.

Ebersbach, A., & Glaser, M. (2004). Towards emancipatory use of a medium: The wiki. *International Journal of Media Ethics, 2*, 1–9.

Ebersbach, A., Glaser, M., & Heigl, R. (2006). *Wiki: Web collaboration*. Berlin: Springer.

Encyclopaedia Britannica. (2006). *History of Encyclopedia Britannica and Britannica Online*. Retrieved March 2, 2006, from http://corporate.britannica.com

Gamson, W. (1989). News as framing. *American Behavioral Scientist, 33*(2), 157–161.

Giles, J. (2005, December 14). Internet encyclopaedias go head to head. *Nature, 438*. Retrieved March 4, 2006, from http://www.nature.com/news/2005/051212/full/438900a.html

Grunig, J. E., & Hunt, T. (1984). *Managing public relations*. New York: Holt, Rinehart and Winston.

Hill, A. (2006, February, 24). Wikipedia—friend or foe on the Net? *PRWeek*, p. 15.

Holsti, O.R. (1969). *Content analysis for the social sciences and humanities*. Reading: Addison-Wesley.

Holtz, S. (2006). The impact of new technologies on internal communication. *Strategic Communication Management, 10*(1), 22–25.

Investopedia. (2006). *Stock-picking strategies: Qualitative analysis*. Retrieved March 2, 2006, from http://www.investopedia.com/university/stockpicking/stockpicking2.asp

King, W. R. (2006). The collaborative Web. *Information Systems Management, 23*(2), 88.

Kirtley, J. (2006). Web of lies. *American Journalism Review, 28*(1), 66.

Kweon, S. (2000). A framing analysis: How did three U.S. news magazines frame about mergers or acquisitions? *The International Journal on Media Management, 2*(3/4), 165–177.

Lamb, G. M. (2006, January 5). Online Wikipedia is not Britannica—but it's close. *Christian Science Monitor*, p. 13.

Lih, A. (2004). *Wikipedia as participatory journalism: Reliable sources? Metrics for evaluating collaborative media as a news resource*. Paper presented at the 2004 International Symposium on Online Journalism, Austin, TX. Retrieved March 2, 2006 from http://journalism.utexas.edu/onlinejournalism/2004/papers/wikipedia.pdf

Lipczynska, S. (2005). Power to the people: The case for Wikipedia. *Reference Reviews, 19*(2), 6–7.

Mahon, J. F., & Wartick, S. L. (2003). Dealing with stakeholders: How reputation, credibility and framing influence the game. *Corporate Reputation Review, 6*(1), 19–35.

McCaffrey, S. (2006, April 30). Political revisions disrupt Wikipedia encyclopedia. *Miami Herald*. Retrieved May 1, 2006, from http://www.miami.com

McCombs, M., & Shaw, D. (1993). The evolution of agenda-setting research: Twenty-five years in the marketplace of ideas. *Journal of Communication, 43*, 58–67.

Michaelson, D., & Griffin, T. L. (2005). *A new model for media content analysis*. Gainesville, FL: Institute for Public Relations.

Noguchi, Y. (2006, February 9). On Capitol Hill, playing Wikipolitics. *The Washington Post*. Retrieved March 4, 2006, from http://www.washingtonpost.com

Nuttall, C. (2006, February 8). Wikipedia users expose flattery by political staff. *Financial Times*. Retrieved March 4, 2006, from http://www.ft.com/cms/s/37859d9c-98df-11da-aa99–0000779e2340.html

Page, S. (2005, December 12). Author of false Wikipedia biography apologizes. *USA Today*, p. 4A.

Panelas, T. (2006, March 24). *Britannica rips* Nature *magazine on accuracy study*. Retrieved April 24, 2006, from http://corporate.britannica.com/press/releases/nature.html

Pollinator. (2006). In *Wikipedia, The Free Encyclopedia*. Retreived March 6, 2006, from http://en.wikipedia.org/wiki/User:Pollinator

Remy, M. (2002). Wikipedia: The free encyclopedia. *Online Information Review, 26*(6), 434–435.

Scott, W. (1955). Reliability of content analysis: The case of nominal scale coding. *Public Opinion Quarterly, 17*, 321–325.

Seeyle, K. Q. (2005, December 4). Snared in the Web of a Wikipedia liar. *The New York Times*, p. 1.

Seigenthaler, J. (2005, November 30). A false Wikipedia biography. *USA Today*, p. 11A.

Sjöberg, L. (2006, April 19). The Wikipedia FAQK. *Wired News*. Retrieved April 20, 2006, from http://www.wired.com/news/columns/0,70670–0.html

Smith, G. E. (1996). Framing in advertising and the moderating impact of consumer education. *Journal of Advertising Research, 36*(5), 49–64.

2005's top rising searches on Google. (2006, January 9). *Financial Times*, p. 18.

Wanta, W., Golan, G., & Lee, C. (2004). Agenda setting and international news: Media influence on public perceptions of foreign nations. *Journalism & Mass Communication Quarterly, 81*(2), 364–377.

Wikipedia. (2006). *Wikipedia:About*. Retrieved March 2, 2006, from http://en.wikipedia.org/wiki/Wikipedia:About

The wiki principle. (2006, April 22). *The Economist, 379*(8474), 14–15.

Digital Public Relations

Online Reputation Management

Idil Cakim

Corporations are accustomed to containing crises by communicating with journalists, publishing newspaper advertisements, or answering calls from concerned customers. These means of crisis control remain essential, but online communication is of growing import. The widespread use of Web-based technologies among key business stakeholders—the media, investors, employees, customers, and vendors—has made online communication an indispensable part of crisis management plans. Companies can no longer forgo the opportunity to communicate with their online stakeholders who retrieve information from company Web sites and buzz about executives, products, and services on social networking Web sites, blogs, and discussion boards.

With the advent of Internet communications, such as online social networks, news, and company Web sites, the practice of public relations can no longer be circumscribed by tactics, such as press releases, events, and media interviews. Key business audiences form their perceptions of companies, brands, products, and services by reviewing online news and information. Upon reading an article, hearing a conversation between colleagues, or observing the interaction between consumers and a brand, opinion leaders turn to company Web sites to verify the facts and get in-depth information.

These Web-based technologies drive business by boosting communication between companies and their audiences. Today, knowledge is at our fingertips—whether it is an instant message (IM) conversation with colleagues, an e-newsletter update, or a document sent through a wireless modem at an airport to a home office in another city. Receiving e-mail news alerts, commenting on blogs, and tracking discussion board messages are typical activities for connected stakeholders. Indeed, audiences can learn about the dealings of a company almost instantaneously. They can use search engines or directly visit the company's online news center and carry the information to their networks through online and offline conversations.

For instance, breaking news from *The Wall Street Journal* and CNN can be communicated via e-mail or Blackberry the moment coverage is reported. Text messages can be sent to cell phones with daily headlines. Apple's iPOD can be synched to an Internet connection to download recent live media coverage. The digital age makes information accessible to business leaders and consumers worldwide. As a result, communications practitioners face the challenge of following what journalists publish and broadcast, as well as what critical bloggers, consumers, and shareholders view and post on the Internet.

In an age when news immediacy is as important as accuracy, and audiences have access to a wealth of sources, corporations must take the lead in providing clear and sufficient information. Company Web sites offer a terrain where corporations have the chance to tell their side of the story, respond to the media, squelch rumors, and appease stakeholder worries. Therefore, these online properties play pivotal roles in communicating with online audiences during a crisis.

Crises in the Internet Era

Using its Web site audit tool PRePARE, Burson-Marsteller conducts ongoing research on online crisis communications. The major global business newspapers (e.g., *The New York Times*, *Financial Times*, *The Globe and Mail*, and *The Wall Street Journal*) are reviewed daily to spot emerging corporate crises and issues. Each company Web site is evaluated for sixty-two features, including the following:

- Mention of the crisis on the homepage
- Press release regarding the crisis
- Special Web section devoted to the crisis
- Response from top management
- Crisis-related links (e.g., links to government offices)
- Legal information for employees, customers, and investors

Between January 2002 and March 2006, Burson-Marsteller identified close to seven hundred ($N = 668$) corporate crises and issues. The majority (76%) of the reviewed companies are based in North America. Some 18% of the sample consisted of European companies. Among the crises tracked in the research, the most frequently reviewed types were SEC (U.S. Securities and Exchange Commission) investigations (18%), pending lawsuits (18%), and chief executive officer (CEO) / high-level executive turnover events (18%).

There were noteworthy differences between the types of crises North American and European companies faced. For instance, SEC investigations were more common among North American companies (26%) than among European companies (11%). Meanwhile, European companies were more likely than their North American counterparts to face leadership issues (13% versus 5%).

Companies' Online Crisis Response

Burson-Marsteller's Web audit revealed that, across all regions, the reviewed companies showed similar tendencies in their online responses to crises. The following are key trends identified across all audited companies:

Companies neglect to address crises online. A substantial segment (43%) of crisis-ridden companies did not address the crisis on their Web sites.

Online crisis responses are buried deep in company Web sites. Regardless of the length and detail of online responses, corporate crisis communications were typically buried deep within company Web sites. Less than one-quarter (18%) of companies had a crisis-related statement on their homepages—their initial point of contact with online stakeholders. For example, when faced with bankruptcy, Delta Airlines began its communications with stakeholders with a message posted on the homepage, detailing its restructuring plan.

Press releases are the most common form of online crisis response. One-half (52%) of companies addressed the crisis with a press release posted on their Web sites.

Crisis contacts are overlooked Web site features. Most companies did not inform their audiences about additional crisis communication channels available to them beyond the company Web site. Among the reviewed crisis-stricken companies, approximately one-third (31%) listed special contacts on their Web sites. After Hurricane Katrina wreaked havoc on New Orleans, Tulane University communicated with concerned audiences through its Web site and posted a crisis contact.

Some companies discuss crises in special online sections. Approximately one-tenth (12%) of companies facing crises provided responses in special Web site sections. Graco, a children's product company, addressed a product recall in detail in a Web site section dedicated to the matter.

Few CEOs and high-level executives take the helm in online crisis communications. Some 11% of companies dealing with crises posted letters from their CEOs and high-level executives on their Web sites. When faced with the Oracle takeover, PeopleSoft posted a letter on its Web site from its CEO to address potential customer and partner concerns.

Few companies link to third-party Web sites. Only 9% of reviewed companies provided links to crisis-related online sources, such as government, law firm, or newspaper Web sites. Following the Columbia Shuttle accident, the National Aeronautics and Space Administration (NASA) linked to reports from congressional hearings in special Web sections devoted to the incident.

Other straightforward tactics to help demystify a crisis and appease stakeholders' worries include creating a glossary of legal terms and a list of contacts for investors in a special section of the company Web site.

The rarity of detailed and accessible online crisis response points to the Internet as an underutilized communication channel. Corporate Web sites provide spokespeople with the opportunity to disclose information in real time. Companies can respond to multiple regions simultaneously, update and elaborate as events unfold, begin dialogues with stakeholders through Contact Us Web site sections, and protect their reputational assets.

Creating a Plan

The following are strategic recommendations for companies building an online crisis-communication plan.

Before a crisis occurs

Plan for multiple scenarios. The amount of information appropriate for each type of crisis response varies. Knowing what to include before the crisis occurs will save time.

Build a Web team. Designate communications officers to collaborate with Webmasters in publishing online responses.

Prepare a dark Web site. Create response templates for each type of crisis that can later be modified to include case-specific information, ensuring both a timely and thorough response.

When a crisis hits

Have an online press release. Address the crisis in a press release and have the statement available online. Also, maintain past press releases in an archive system, ensuring disclosure in chronological order.

Provide easy and visible access. Link to critical information from the company homepage.

Be accessible and offer customized assistance. Ensure that each audience group has a contact and knows where to direct further questions or concerns. Provide basic information in multiple languages for international audiences.

Consider a special crisis section. Having a Web site section devoted to the crisis keeps crisis information and updates organized and reader friendly.

Boost credibility with objective sources. Consider providing links to other Web sites related to the crisis and/or posting additional information, such as legal documents or research from nonprofit organizations.

Synchronize. Match offline and online crisis responses. Carry offline advertising messages online.

And always . . .

Demonstrate leadership. Consider including a message from the CEO to various stakeholders.

Be transparent. Even when facing legal restrictions on disclosure, acknowledge the crisis. Indicate that you have information about the crisis, are collaborating with respective agencies, and will respond to those affected in a timely fashion.

Integrating Online Communications with Proven Public Relations Practices

The Internet is an established part of both corporate and consumer spheres. Organizations are thus required to expand their relationship management efforts beyond classic communication approaches. To manage the flow of information from their Web sites to public and business circles, companies now have to be even more vigilant about keeping the content on their sites up-to-date and accurate. They must also respond to their Web site visitors' queries as promptly and as thoroughly as possible. On the one hand, the company Web site can be a terrific promotional medium to underscore the organization's mission and vision, introduce its management team, and provide product and service snapshots. Breaking news such as partnerships, quarterly profit announcements, and new product launches can bolster a positive corporate image, strengthening the company's brand and presenting it as a valuable asset to investors. On the other hand, having a Web site that does not address publicly known crises can be a reputational liability.

The Birth of Digital Public Relations and Online Crisis Management

The Internet, as the always-on medium where information can be available for years, has given birth to a new branch of communications called digital public relations. Consumer and business audiences hold companies accountable for publicly released information. Having an online presence now requires companies to have online brand and stakeholder management mechanisms in place.

Public relations specialists need to plan for both upbeat news and crisis situations, such as lawsuits, unexpected executive turnover, bankruptcy, and even natural disasters. When a company is faced with a critical issue, a lack of appropriate response or hard-to-access information on the Web site can add to online stakeholders' confusion and skepticism, ultimately generating negative word of mouth and deteriorating valuable business-to-consumer or business-to-business relationships. Inability to manage communications on the company Web site can deem the organization vulnerable and leave a void for speculation.

Online Communication Tools for Practitioners and Audiences

Corporate communications practitioners are increasingly using Web-based tools, such as Flash, multimedia streams, blogs, Webcasts, podcasts, and RSS feeds to tell and spread their organizations' stories through the Internet. Communications professionals' work not only delivers current information to online audiences globally, but also builds a news base that can be accessed for years to come. For instance, Web sites like Waybackmachine.com[1] allow visitors to sift through archived pictures of Web sites, enabling curious consumers to dig deeper into a company's history with just a few clicks.

The same easy-to-use online publishing tools that allow public relations practitioners to quickly update Web sites and deliver news also help vocal online stakeholders to distribute information to their peers. When online influencers visit company Web sites and cannot find credible information to put the news they have heard or read into context, they can take to discussion boards, Usenet newsgroups, and blogs to get an insider's story and comment on the company's position. Subsequently, these online public opinion leaders create achievable trails of user-generated media (also referred to as UGM) on the Internet, paving yet another platform for communication specialists to watch closely and respond if need be.

User-Generated Media: Latest Alternative to Corporate Press Releases

For those audiences who are inundated with information from traditional media sources and have lost their confidence in professionally crafted news, user-generated media is increasingly becoming a trusted source of information. Alternative online news is especially persuasive when fueled by social, economic, and political factors (e.g., rising oil prices, rampant corporate scandals, and fear of terrorist attacks) that shake trust in public institutions.

With the addition of user-generated media to the pile of publicly available content about companies, brands, products, and services, communication specialists must constantly measure the sentiments of online stakeholders, note simmering issues, and prevent them from turning into crises. Today, businesses must manage both online and offline relationships in a manner that will maintain a proactive stance on the corporation's mission. Amidst information overload and rampant public skepticism about the business world, corporate Web sites are critical communications vehicles that give companies the opportunity to respond directly to rumors or issues stemming from user-generated media. These Web areas can provide a reliable, credible, candid information alternative to discussion boards, blogs, and newsgroups where online stakeholders may be reading counterviews or unfounded stories about the company.

New Public Relations Skills

With the dual nature of Internet as both a formal and an informal communication channel, public relations professionals need to broaden their skill sets and job functions while being savvy about the following concepts:

Online audience statistics. Communication specialists need to be proficient in online audience metrics to assess the value of a story placed online. Basic questions they should be able to answer include the following: How many visitors does the Web site attract per day? When does the site traffic reach its peak? Which content areas are the most popular? Who are the online audience members, and which ones are more influential? What content keeps audiences on the site longest, and which stories are forwarded most frequently?

Search engine optimization. Online search engines, such as Google, Yahoo, and MSN, are often the primary destinations of lean-forward types of audiences who want to find out about current events and news from the best-of-breed sources. When end users enter their key words into search engines, company Web sites compete with online news Web sites, blogs, and competitor sites for online audiences' attention. To reach their target audiences online, company Web sites need to rise above the clutter and rank among the top listings of the most popular search engines. As a result, communication specialists need to think through the following issues when developing content for the company Web site: How long does the story stay accessible through search engines? Where does it rank when audiences run searches using key words about the issue or the concept? Who else refers or links to the story? What are the competing stories that come up on search engines when searching for the story or the company Web site in general?

Online media relations. Although digital public relations can start with the company Web site, it extends much beyond the proprietary content area. Besides consumers, investors, customers, and business partners, journalists are frequent visitors to company Web sites, digging for additional information on a story or to pick up a lead. As many publications enter 24/7 news cycles and keep feeding news to their readers offline and online, media relations experts need to familiarize themselves with journalists who publish on the Internet and consider the following questions: Who are the key online journalists? Do online journalists have blogs where they can place additional information that does not make it into their articles? Can these online stories be syndicated in other online or print publications? How does the value of an online placement compare to placing a story in a traditional news outlet?

Online crisis communications. Communication specialists can choose from a plethora of innovative Web site features and tools to help tell their stories online and spread their news to the widest possible audiences. However, when faced with a crisis situation, it is best to put the bells and whistles aside and focus on delivering clear, concise, and relevant information to all key stakeholder groups. The principal questions that should be addressed when devising an online crisis communication strategy include the following: How much information is appropriate to post? How frequently should the content be updated? How should the crisis response be placed on the Web site? What sorts of tools should be used to make it easier for business audiences to access the pertinent information?

Digital public relations tools. Online communications can help keep a company and its Web properties ahead of the never-ending content flow generated by third parties. They enable a corporation to get its information out first and respond effectively to emerging issues. Public relations practitioners who want to stay abreast of new developments and expand their skill sets should be thinking about the following questions: What are the latest technologies that can help distribute online information to the right audiences? Which ones are most effective in achieving communication goals at hand?

The fusion of online and offline information. Organizing information and audiences by media channel is an efficient way of addressing business, marketing, and communication problems. However, publics often use multiple information sources simultaneously—traditional and alternative, offline and online—as they try to get the news and verify the facts. They talk about breaking news they received via e-mail alerts; they visit company Web sites upon noticing a newspaper or magazine blurb. Online information and offline information meld together in the public conscience and equally contribute to forming opinions.

Thus, communication strategists of the future need to think through the following questions: How does online information impact offline conversations among stakeholders? How does offline knowledge about an issue drive audiences to online content? Do audiences distinguish between online and offline information? Do they value offline information more than online content?

Internet-related technologies are permeating through ever-widening circles and across borders of social, economic, and political differences. The public relations and marketing communications industries are going through phenomenal structural changes while simultaneously adapting to new ways of delivering news and managing stakeholder relations. Information technologies will continue to evolve, shortening the news cycle, extending stories' lifetimes, and forcing communication specialists to stay abreast of such developments to remain effective and succeed in managing corporate reputation.

For consideration

1. What are possible reasons why so many companies either deeply bury or do not address critical information on their Web sites?
2. What are some possible situations in which dark Web sites could be useful?
3. In addition to providing a letter from the CEO during times of struggle, what are some other steps companies could take to demonstrate leadership through new media channels?
4. To what extent can user-generated media (UGM) negatively or positively affect a company's brand image?
5. To what extent can online crisis response impact a company's bottom line, if at all?

For reading

Bishop, T. A. (2006, May 28). Regaining consumers' trust: Bausch & Lomb comes clean on concerns about ReNu. *Baltimore Sun Times*, p. 1C.

Fearn-Banks, K. (2001). *Crisis communications: A casebook approach. Student workbook.* (LEA's Communication Series, 2nd ed.). Mahwah, NJ: Lawrence Erlbaum Associates.

Gaines-Ross, L., Cakim, I., & Dietz, S. (2003, November 17). Using the Web to communicate in a crisis. *IR Web Report*. http://www.irwebreport.com

Gottschalk, J. A. (Ed.). (1993). *Crisis response: Inside stories on managing image under siege*. Canton, MI: Visible Ink Press.

Harvard Business Review on Crisis Management (A Harvard Business Review Paperback). (2000). Boston: Harvard Business School Publishing.

Lukaszewski, J. E. (2005). *Crisis communication plan components and models*. White Plains, NY: The Lukaszewski Group, Inc.

Mitroff, I. I., & Alpaslan, M. C. (2003, April 1). *Preparing for evil (HBR OnPoint enhanced edition)*. Boston: Harvard Business Review.

Swann, P. A. (2006, May 1). Got web? Investing in a district Web site: An effective site that can help you reach your organizational goals. *School Administrator, 63*(5), 24–30.

Note

1. Retrieved July 16, 2006, from http://waybackmachine.com/

Relationship Building in an Internet Age

How Organizations Use Web Sites to Communicate Ethics, Image, and Social Responsibility

Debra A. Worley

At the heart and soul of public relations practice is the fundamental premise that effective and ethical public relations create and maintain mutually beneficial relationships between an organization and the stakeholders upon which its survival depends. This chapter discusses how organizations, using the Web, can communicate in ways that develop dialogic, interpersonally effective relationships with stakeholders. Implications for the ethical practice of public relations and the ethical development of mutually beneficial relationships with stakeholders are discussed.

As many of the chapter authors in this book point out, effective public relations practice creates and maintains mutually beneficial relationships between an organization and the numerous stakeholders upon which the organization's survival depends. Effective public relations requires public relations practitioners to know who those stakeholders are, to understand the needs and expectations that exist among members of the stakeholder group, and to communicate and act in ways that match those needs and expectations. In other words, effective and ethical public relations means doing the right thing, at the right time, for the right reasons, for stakeholders.

Heath and Coombs (2006) believe:

> Relationships are strongest when they are mutually beneficial and characterized by "win-win" outcomes. Relationships are best when people share information that is accurate and relevant. Relationships require a commitment to open and trustworthy dialogue, a spirit of cooperation, a desire to align interests, a willingness to adopt compatible views/opinions, and a commitment to make a positive difference in the lives of everyone affected by your organization. (p. 5)

Today's relationship building in public relations occurs in as many different communication contexts as the imagination and technology can develop, but, more and more, today's public relations relies on the Internet and online technology to create and maintain important relationships. Globalization has made communication and relationship development both more efficient and more complex in our diverse and fast-paced society. Organizations have the ability to communicate more quickly to larger numbers of stakeholders and can take advantage of advances in technology to make this communication more focused toward the needs and expectations of those stakeholders.

The Pervasiveness of Mediated Communication

Think over the past week. How many times did you have the radio or television on? Did you watch cable TV? How often did you access your email, enter a chat room, or use instant messaging? Are you a blogger? How often did you access your organization's intranet? How often did you to play games, gather information, or shop online? Did you use a digital subscriber line (DSL) Internet service provider (ISP) or a dial-up ISP? Did you listen to a CD? How many billboards have you passed as you drove your car? Have you read an e-magazine or e-book? When did you last see a film in a theatre or rent a video? Do you have an iPOD?

In all likelihood, these questions make sense to you, even though they have quite a bit of jargon. The fact, alone, that you understand this specialized language points out the pervasiveness of mediated communication in our culture, to say nothing of the frequency with which we use mediated communication.

This pervasiveness of mediated communication in our culture and throughout the world provides us opportunities to see, hear, think, and do things of which people prior to the 1950s never dreamed. We can travel to far away places through the Internet or the Discovery channel. We can chat with people from every corner of the globe with a click of a mouse or the touch of a button. The media at our disposal provide opportunities to expand our knowledge and understanding of the world faster and more comprehensively than ever before. Understanding the unique characteristics of mediated communication helps us reflect upon its impact on our lives and learn about its influence so that we can be more astute consumers and users of these media. Understanding the pervasiveness of mediated communication and the opportunities for relationship building in public relations through these media expands the practice of public relations and builds credibility within the profession.

Mediated Communication as Interpersonal Communication

Coombs (2001) suggests that, despite globalization and the ability to reach larger numbers of stakeholders, contemporary public relations practice must engage in dialogue that maximizes understanding and a more *interpersonal* approach to communication. Interpersonal communication occurs when the parties believe in mutual influence and interdependence. Interpersonal communication is different from impersonal communication in that there is attention to the relationship, a *mindfulness* about the other that requires investment of time and energy and attentiveness. When an organization engages in relationship building based upon interpersonal dimensions, such as trust, involvement, investment of time and effort, and commitment to continue the relationship, both the organization and the stakeholders benefit.

Researchers studying the effectiveness of public relations strategies have found that the Internet has great potential for following the interpersonal, two-way symmetrical model of public relations far more than any other medium (Elliot, 1997; Marken, 1998). New communication technologies, particularly the Web, provide practitioners with the potential to establish much richer dialogues with stakeholders, and "because of the low costs of information and opinion delivery, companies, governmental agencies, and activists are more on par" (Heath, 1998, p. 273). The Internet, then, allows many different types of companies, whether large or small, similar opportunities to communicate with their customers, investors, communities, employees, and various other stakeholders. The Internet may be the great equalizer in terms of relationship building.

Use of the Internet and World Wide Web abound in today's organizations. While the Internet is a complex system of telecommunications linkages among major computer facilities worldwide, the World Wide Web is the sophisticated application of the Internet that allows widespread access to graphics and information, other Web sites, and interactivity. The World Wide Web is used to gather information, check on the activities of other organizations, access news almost immediately, and shop for virtually anything twenty-four hours a day. Organizations use Web sites as external communication tools whose objectives are information dissemination, reputation and issues management, employee recruitment, sales and customer service, philanthropy and civic engagement, and many others.

The World Wide Web has become an important tool in building dialogic relationships with numerous stakeholders. Weber (1996) suggests, "Smart companies take advantage of the one-to-one communication opportunity the Web offers by allowing reporters, analysts and opinion leaders to register on site and specify the breadth and depth of information that interests them" (p. 20). A December 2000 issue of *Marketing Week* ("Digital Revolution") went so far as to suggest that the Internet has caused a "fundamental shift" in information distribution and that the public relations industry has the most to gain from this trend.

Increasing use of the Web as a tool to disseminate organizational discourse, coupled with knowledge that critical stakeholders to any organization functioning today demand clear, consistent, and comprehensive discourse on the social responsibility of the organization, suggests that we need to understand how organizations utilize the Web to strategically communicate ethical values, practices, and policies.

The Internet and Ethical Dialogue in Relationship Building

To what extent do multinational corporations communicate about their social responsibility, and what is the nature of that communication on corporate Web sites? If the Internet (in this case, corporate Web sites) is becoming one of the most important communication channels available today, contemporary public relations professionals must understand how successful companies are utilizing this channel to communicate about corporate ethics.

Web site purpose and target audiences. As far back as 1995, researchers began to focus on the importance of the Internet, or the "information superhighway," as a tool for use by public relations professionals. Bobbitt (1995) identified the features of the Internet that "are the most beneficial to public relations professionals," including electronic mail, the World Wide Web, Usenet, CompuServe, and Prodigy (p. 27). In the more than ten years since Bobbitt's prediction, advances in technology have made some of these features already obsolete. Bobbitt suggested that professionals could use the Web to provide news releases for the media, "but it's not as effective as using e-mail because the media will have to know where to look for them" (p. 31). He was wrong in that prediction. Every one of the top ten Fortune 500 Web sites in 2006 has not only included news releases but also has a virtual news room where you can get news releases, speeches, reports, or other communications with the click of a mouse.

An *Investor Relations Business* article in 2001 ("Corporate Websites") reported that journalists were not satisfied by the content in corporate Web sites. Journalists found fault in the Web sites for lack of comprehensive information and for no available process for fact checking. Just five years later, corporate Web sites have comprehensive information for anyone interested in finding out more about a company. Today's journalists can find background about virtually every aspect of publicly held organizations.

Organizations use Web sites for recruiting, creating organizational image, creating a brand, communicating corporate citizenship and values, and increasing media coverage. For example, General Motors (GM) (n.d.) posted the following on the Corporate Responsibility link of its Web site:

> GMability is designed to demonstrate corporate responsibility by transparently presenting company information, and GM's ability to make a difference in the world. Read our 2004/05 Corporate Responsibility Report. (n.p.)

In the Corporate Responsibility Report, visitors could read about GM's overall commitment to social responsibility and safe products, highlights of philanthropic and environmental responsibility, and its "social performance," which discussed a commitment to employees, suppliers, and educational enhancement.

Another example of effective communication of social responsibility could be seen in General Electric's Web site. General Electric (n.d.) said this about its commitment to responsible actions:

> Naturally, we measure our Company's performance through financial results and stock price. But we also view how that performance is achieved in a broader context: the health, safety and opportunities for workers, the impact of our operations on the environment and communities, our interaction with governments and regulatory agencies around the world, and our compliance with legal and accounting rules. Our goal is to grow responsibly while engaging stakeholders. (n.p.)

In addition to this values statement, General Electric communicates through fact sheets, annual reports, proxy statements, letters to stakeholders, citizenship reports, executive bios and photos, and position papers, among many other documents available on the Web site. Every item is available in a portable document format (PDF) or Word document and can be easily downloaded.

There is no doubt that competitive organizations today incorporate Web sites into their communication mix. The body of research on the Internet as a communications tool ultimately suggests that relationships can be created and maintained through this channel. As organizations become more sophisticated in developing sites—and in communicating to a broader range of stakeholder groups—the function, purpose, and significance of sites increase. And as stakeholders become much more sophisticated, cynical, and demanding in their desire for more information, organizations that include discourse on ethics and social responsibility may more successfully enhance relationships with those stakeholders.

Ethics and social responsibility communication on Web sites. What does the word ethics bring to mind? News headlines that describe yet another example of misconduct by a government official or of another faulty product sold to unsuspecting consumers? If so, you may have a misleading picture of ethics. Although examples of unethical behavior are seen every day in newspapers, television, and other media, most of us never make the headlines. Yet, in our daily lives, we are faced with decisions that have ethical dimensions. In any situation in which one thinks about a decision in terms of right and wrong or good and bad, or when one considers what the consequences of a decision might be to one's self, friends, family, co-workers, or a significant other, that person has begun to touch on the ethical dimension of communication. Ethical responsibility as public relations professionals begins with how we define ethical responsibility as individuals in our everyday lives.

Ethical decisions surround us in the workplace. According to the Ethics Resource Center (2005) annual survey of businesses in 2005,

> More than half of American workers have observed at least one type of ethical misconduct in the workplace, a slight increase from 2003, despite an increase in worker's awareness of formal ethics programs, according to the 2005 National Business Ethics Survey (NBES) released today by the Ethics Resource Center. Employee reporting of misconduct they observe is also down by 10 percentage points. Despite the decrease in ethical conduct, according to the NBES report, "Ethics and

> compliance programs can and do make a difference. However, their impact is related to the culture in which they are situated." (Home Page section, para. 1)

Public relations professionals are responsible for their individual ethical behavior outside of the workplace, their ethical behavior as employees of an organization, and the ethical communication between the organization and external audiences.

The importance of ethical corporate behavior and communication about that behavior has come to the forefront of our discussion of responsible public relations. An article in *The Economist* published in April 2000 ("Business Ethics") suggested,

> Indeed, companies face more ethical quandaries than ever before. Technological change brings new debates, on issues ranging from genetically modified organisms to privacy on the Internet. Globalization brings companies into contact with other countries that do business by different rules. Competitive pressures force firms to treat their staff in ways that depart from past practice. Add unprecedented scrutiny from outside, led by non-governmental organizations (NGOs), and it is not surprising that dealing with ethical issues has become part of every manager's job. (p. 66)

A variety of resources is available today to assist organizations in understanding social responsibility expectations and developing rhetorical strategies to meet those expectations. For example, the purpose of DePaul's Institute for Business and Professional Ethics, as noted on its Web site (DePaul College of Commerce, n.d.), was "ethical deliberation in decision-makers by stirring the moral conscience, encouraging moral imagination, and developing models for moral decision-making" (para. 1). Today's organizations have numerous options in how to communicate about ethical values and practices, but early efforts by corporations to communicate about their ethical behavior took the form of printed corporate credos, ethics codes, or mission and value statements (Murphy, 1995). Now all of this can be found on organizational Web sites.

Codes of ethics are usually more detailed discussions of policies and the nature of relationships within and outside of the organization. Research in the early 1990s suggested that more than 90% of large corporations had ethics codes; most of the ethics codes appeared in the form of pamphlets and booklets. The availability of materials, such as pamphlets and booklets, limits opportunities for discourse, however, to a narrow selection of audiences. Making the information available on the organizational Web site expands the number of stakeholders with access to the information and expands the possibilities for relationship development.

The nature of the communication about conduct or ethics codes varies on organizational Web sites, but virtually all of them discuss the mission, vision, and expectations for employees, suppliers, and others who impact their business. For example, Starbucks (n.d.) discussed its six guiding principles for decision making, as well as its environmental mission statement, in the About Us / Corporate Social Responsibility / Mission Statement section of its Web site. In addition, Starbucks created a Supplier Code of Conduct, which began with this statement of expectations:

Expectations of Our Suppliers:

> Starbucks is committed to treating all individuals with respect and dignity, and protecting the environment. As part of Starbucks corporate social responsibility, we believe these principles should

> be reflected throughout our supply chain, and embraced by Starbucks suppliers. This was the motivation for creating Starbucks Supplier Code of Conduct. The Code was introduced in September 2003. Suppliers are required to have an officer or owner of the company sign an acknowledgement that they agree to comply with our Code and standards. New suppliers are required to comply as a condition of doing business with Starbucks. (About Us, Corporate Social Responsibility, Supplier Code of Conduct section, para. 1)

Weaver, Trevino, and Cochran (1999) found that practices associated with corporate ethics programs have grown to encompass a much wider array of strategies, including (a) ethics-oriented policy statements, (b) formalization of management responsibilities for ethics, (c) free-standing ethics offices, (d) ethics and compliance telephone reporting/advice systems, (e) top management and departmental involvement in ethics activities, (f) use of ethics training and other ethics awareness activities, (g) investigatory functions, and (h) evaluation of ethics program activities. Their study suggested that, while a wide array of strategies exists in corporations today, there is also wide variability in the extent to which the policies are implemented.

Internet Communication, E-Commerce, and Brand Image

Socially responsible communication by an organization helps to build and maintain many types of relationships. Obviously, companies use their Web sites for e-marketing and e-commerce communication. Therefore, understanding appropriate and ethical e-marketing and e-commerce communication is as critical to effective relationship-building as is the communication of mission, values, and other more obvious ethical communication. Today, a person can live as a virtual recluse in his or her home and be assured that everything needed to survive can be purchased on the Web and shipped.

In a survey of marketing executives regarding their perceptions of Internet regulation, ethical issues in Internet marketing, and the role of ethics and Internet marketing in their organizations, researchers found that there are significant concerns about marketing using this new medium (Bush, Venable, & Bush, 2000). Businesses are unsure if and how it should be regulated. There are pervasive concerns about issues of privacy and security. Over 80% of the respondents suggested that organizations should develop codes of ethics specifically for the Internet and for Internet marketing.

Legislation has been proposed to cover issues of privacy in online environments, with mixed success, including the Consumer Privacy Protection Act, the Privacy and Identity Protection Act of 2000, the Notice of Electronic Monitoring Act, the Consumer Internet Privacy Enhancement Act, the Secure Online Communication Enforcement Act of 2000, and the Freedom from Behavioral Profiling Act of 2000.

Although e-marketing and e-commerce make up a large percentage of what is communicated by companies through Web sites, other important information is beginning to appear with more consistency. Wonnacott (2000) suggests that if you "visit today's best sites . . .

you'll get much more than a marketing pitch" (p. 64). A Web site, she suggests, is a powerful tool for strengthening brand and building consumer loyalty. Starbucks (n.d.) created its brand and reputation as one of *Fortune's* 100 Best Places to Work by stating:

> The bottom line: We always figured that putting people before products just made good common sense. So far, it's been working out for us. Our relationships with farmers yield the highest quality coffees. The connections we make in communities create a loyal following. And the support we provide our baristas pays off everyday. (About Us section, para. 1)

Today's effective organizations communicate much more than marketing and e-commerce, and they create and maintain more of their relationships with their Web sites. Virtually every stakeholder group can be targeted through this communication tool. The development of organizational discourse about ethical philosophies and practices began in the middle to late 1980s and is still growing strong today. Social responsibility communication is becoming an increasingly important part of the message mix. But what types of information should an organization communicate through its Web site in order to meet stakeholders' expectations for ethical communication?

Categories of Social Responsibility Discourse

In order for an organization to develop and maintain legitimacy—the perceptions among stakeholders about how organizational activities relate to general norms and standards for appropriate conduct—the company must communicate publicly. "Specifically, organizations communicate about their social value, seeking to persuade stakeholders and the general public that their activities are consistent with appropriate social norms and values" (Seeger, 1997, p. 111). Norms often include economic viability and rational operation or organization. It is stakeholders who, through their understanding of and perceptions about an organization, confer legitimacy. That is, an organization's image or reputation is always dependent upon the perceptions of stakeholders: their understanding of, feelings about, and behavior toward an organization. Organizations must therefore develop clear communication about their values, policies, and actions that match stakeholder expectations.

Organizational legitimacy is closely associated with organizational responsibility. "Organizational responsibility entails a set of organizational obligations to specific communities and to society in general" (Seeger, 1997, p. 19). Responsibility is part of an organization's obligation to support important social goals and values and to help solve social problems that affect society and the organization. Responsibility "refers to meeting social expectations before they become part of the organization's legal or regulatory framework" (p. 121). Seeger identifies four predominant clusters of responsibilities: (a) philanthropy, (b) products and services, (c) the environment, and (d) workers.

Philanthropy responsibility. Philanthropy was one of the first, and remains the most widespread, forms of organizational responsibility. Philanthropy includes "supporting worthwhile causes in a wide variety of areas including charities, education, programs for the poor and homeless, and the arts" (Seeger, 1997, p. 124). Organizational philanthropy first led organizations to become involved in alleviating social ills, such as poverty, hunger, and disease. These efforts have expanded significantly to include support for social service organizations, arts and cultural groups, educational issues, sports, medical and community health efforts, and a broad range of nonprofit groups.

As an example of commitment to philanthropy, IBM developed its Global Innovation Outlook initiative, which is "an examination of three areas that affect broad swaths of society and are ripe for innovation: the future of healthcare; the relationship between government and its citizens; and the intersection of work and life" (IBM, n.d., Our Company, Relationships, Introduction section, para. 3). IBM's Web site spoke explicitly to the types of relationships that are important to the company, as in the following statement about philanthropy:

> IBM's innovative, consultative partnership approach to our relationships with schools, community and nongovernmental organizations is exemplified in our philanthropic programs. We form multiyear relationships with schools and community partners who serve as equals in the development of software and services to support schools and communities in need. We rely on organizations like SeniorNet, MentorNet, the Tomás Rivera Policy Institute, the New York Hall of Science, and the Egyptian Museum—as well as on school districts around the world—to beta-test new software and services in order to better serve the needs of these organizations' constituencies. Through these unique partnerships we develop cutting-edge software and solutions that bring together the best technology and know-how our company has to offer, with the expertise, insight and on-the-ground experience of community organizations. (Our Company, Relationships, Communities section, para. 3)

Products and services responsibility. The second responsibility category concerns products and services the organization provides. "Product and service responsibility concerns the organization's market relationship to society" (Seeger, 1997, p. 126). Responsible products and services are those that do not pose a risk of harming consumers. An organization may be seen as negligent when it fails to exercise reasonable care in providing the product or service, when the product or service causes harm or injury to consumers, or when an organization fails to provide consumers with a warranty for products and services. Consumers look for explicit or implicit claims regarding the warranty. "The consumer has a reasonable expectation that the product or service will be consistent with these claims. When the product is inconsistent with the warranty, thereby causing an injury, the organization may be held liable" (p. 126).

FedEx, for example, communicated online about a variety of relationships in its mission statement, including products and services (Federal Express, n.d.):

> FedEx will produce superior financial returns for shareowners by providing high value-added supply chain, transportation, business and related information services through focused operating companies. Customer requirements will be met in the highest quality manner appropriate to each market segment served. FedEx will strive to develop mutually rewarding relationships with its employees, partners and suppliers. Safety will be the first consideration in all operations. Corporate activities

> will be conducted to the highest ethical and professional standards. (About FedEx, Mission section, para. 1)

Environmental responsibility. A third category is environmental responsibility. "The idea that organizations have a fundamental responsibility to the environment has taken hold as a major organizational ethic since the 1970's and has captured the imagination of a number of groups, agencies, and organizations" (Seeger, 1997, p. 127). Environmental responsibility includes those practices that an organization might use to show itself as a green organization. These may include product labels or information on recycling or recyclability, use of nonpolluting manufacturing techniques, and other aspects of general environmental friendliness. Today, a new environmentalism has developed within and among organizations and communities. "This form of environmental ethics is characterized by strong public concern for environmental issues, the growth of green marketing and consumerism, broad diffusion and social acceptance of environmental values, and increased governmental regulation concerning environmental issues" (p. 129).

ExxonMobil (n.d.) made this statement about its environmental responsibility:

> We pledge to be a good corporate citizen in all the places we operate worldwide. We will maintain the highest ethical standards, obey all applicable laws and regulations, and respect local and national cultures. Above all other objectives, we are dedicated to running safe and environmentally responsible operations. (Guiding Principles, Communities section, para. 1)

Responsibility to workers. The final category of responsibility focuses on worker's rights and the organization's relationship to employees. Worker's rights have been a significant issue for organizations for one hundred years. But responsibility to workers encompasses more than rights; it includes safe working conditions, privacy and confidentiality, lack of worker exploitation, and a whole host of labor issues. Within this category are issues that not only affect U.S. workers, but also focus on questions of how and where workers are employed in developing countries. As an example, Nordstrom (n.d.), a Fortune 500 company and a Hall of Fame member of *Fortune*'s 100 Best Companies to Work for, speaks to responsibility to workers in this way:

> One of the best ways we can provide excellent customer service is by working to reflect the communities we serve. Having a workplace that embraces and encourages diversity benefits both our customers and our employees. We believe each of us should have the opportunity to realize our potential and contribute to the success of our company.
>
> Nordstrom is an equal opportunity employer, committed to recruiting, hiring and promoting qualified people of all backgrounds, regardless of sex; race; color; creed; national origin; religion; age; marital status; pregnancy; physical, mental or sensory disability; sexual orientation; gender identity or any other basis protected by federal, state or local law. (Nordstrom Careers, Diversity section, para. 1 and 2)

When organizations act in ways that match the values and expectations of important stakeholders, they are taking the first step toward creating and maintaining mutually beneficial relationships with those stakeholders. When organizations consciously and continuously communicate their values and actions to those stakeholders, they are taking a second step in relationship building. The final step is to ensure that this communication is dialogic. In other words, communication should be a negotiation of meaning and understanding, not one-way dissemination of information. Dialogue in relationships is an exchange of ideas and a process of negotiating meaning. Ethical organizations today create communication that is dialogic and grounded in values of truth telling, non-maleficence (do no harm), beneficence (do good), confidentiality, and fairness. These values are fundamental to both the Public Relations Society of America (n.d.) Member Code of Ethics and the International Association of Business Communicators (n.d.) Code of Ethics for Professional Communicators.

Conclusion

Increasing use of the Web and the Internet as tools to create and maintain relationships with stakeholders and to communicate in ways that match stakeholder values and expectations has become a major factor in ethical public relations practice. Communication and relationship building through organizational Web sites must be created in a dialogic, interpersonally effective manner based upon trust, involvement, investment of time and effort, and commitment to continue the relationship. Organizational values and actions must be created and communicated so that stakeholders understand the organization's commitment to philanthropy, products and services, the environment, and workers.

The Internet as a new medium not only expands the opportunities for communication between and among an organization and its stakeholders, but also contracts the world into a smaller, more interpersonal set of relationships. There is no doubt that stakeholder expectations for truthful, timely, and ethical communication guide the expanding use of organizational Web sites as a tool for relationship building in today's public relations. When organizations develop Web sites that are grounded in an interpersonal, dialogic, and mutually beneficial philosophy of communication, everyone benefits.

For consideration

1. How has the emergence of new media influenced the ethical concerns of public relations practitioners?
2. Do most people want to know the truth? Or do they want to be left in the dark if the truth would be harmful?
3. What, if any, are the limits to social responsibility? What are the proper bounds of responsibility for a corporation versus a charity versus a government agency?
4. Which is more important for a public relations professional to be effective—ethics or expertise?
5. What indicators differentiate between an organization that *communicates* about all of the wonderful and ethical things in which it believes and one that actually *engages* in ethical and effective behavior?

For reading

Diggs-Brown, B. (2007). *The PR style guide: Formats for public relations practice* (2nd ed.). Belmont, CA: Thomson Learning.

Halbert, T. & Ingulli, E. (2002). *Cyber ethics*. Mason, OH: West Legal Studies Division of Thompson Learning.

Marlow, E. (1996). *Electronic public relations*. Belmont, CA: Wadsworth.

Radford, M. L., Barnes, S. B., & Barr, L. R. (2002). *Web research*. Boston: Allyn & Bacon.

Ringle, W. J. (1998). *TechEdge*. Boston: Allyn & Bacon.

References

Bobbitt, R. (1995). An Internet primer for public relations. [Electronic version]. *Public Relations Quarterly, 40*(3), 27–33.

Bush, V. D., Venable, B. T., & Bush, A. J. (2000, February). Ethics and marketing on the Internet: Practitioners' perceptions of societal, industry and company concerns. *Journal of Business Ethics, 23*(3), 237–248.

Business ethics: Doing well by doing good. (2000, April 22). *The Economist*, pp. 65–67.

Coombs, W. T. (2001). Interpersonal communication and public relations. In R. L. Heath (Ed.), *The handbook of public relations* (pp. 105–114). Thousand Oaks, CA: Sage.

Corporate Websites fail basic needs of reporters. (2001, April 2). *Investor Relations Business*, pp. 4–5.

DePaul College of Commerce Institute for Business and Professional Ethics. (n.d.). Retrieved May 1, 2006, from http://www.depaul.edu/ethics/

Digital revolution heralds a golden age for PR. (2000, December 14). *Marketing Week*, p. 41.

Elliot, C. M. (1997). *Activism on the Internet and its ramifications for public relations*. Unpublished master's thesis, University of Maryland, College Park.

Ethics Resource Center. (2005). *National business ethics survey*. Retrieved May 24, 2006, from http://www.ethics.org/research/nbes-2005.asp

ExxonMobil. (n.d.). *Guiding principles*. Retrieved May 1, 2006, from http://exxonmobil.com/corporate/About/ViewPoints/Corp_V_GuidingPrinciples.asp

Federal Express. (n.d.). *Mission*. Retrieved May 2, 2006, from http://www.fedex.com/us/about/today/mission.html

General Electric. (n.d.). *Citizenship*. Retrieved May 24, 2006, from http://www.ge.com/en/citizenship/

General Motors. (n.d.). *Corporate Responsibility*. Retrieved May 24, 2006, from http://www.gm.com/company/gmability/

Heath, R. L. (1998). New communication technologies: An issues management point of view. *Public Relations Review, 24*(3), 273–288.

Heath, R. L., & Coombs, W. T. (2006). *Today's public relations: An introduction*. Thousand Oaks, CA: Sage.

IBM. (n.d). *Relationships*. Retrieved May 1, 2006, from http://www.ibm.com/ibm/responsibility/company/relationships/

International Association of Business Communicators. (n.d.). *Code of ethics*. Retrieved May 1, 2006, from http://www.iabc.com/about/code.htm

Marken, G. A. (1998). The Internet and the Web: The two-way public relations highway. *Public Relations Quarterly*, *43*(1), 31–33.

Murphy, P. E. (1995). Corporate ethics statements: Current status and future prospects. *Journal of Business Ethics*, *14*(9), 727–748.

Nordstrom. (n.d.). *Nordstrom careers: Diversity*. Retrieved May 1, 2006, from http://www.recruitingsite.com/csbsites/nordstrom/company/culture/diversity.asp

Public Relations Society of America. (n.d.). *PRSA member code of ethics*. Retrieved May 2, 2006, from http://www.prsa.org/aboutUs/ethics/index.html

Seeger, M. W. (1997). *Ethics and organizational communication*. Cresskill, NJ: Hampton Press.

Starbucks. (n.d.). Retrieved May 24, 2006, from http://www.starbucks.com

Weaver, G. R., Trevino, L. K., & Cochran, P. L. (1999). Corporate ethics practices in the mid-1990's: An empirical study of the Fortune 100. *Journal of Business Ethics*, *18*(3), 283–294.

Weber, L. (1996, November). Internet rewrites rules of public relations game. *Public Relations Tactics*, *3*(11), 20.

Wonnacott, L. (2000, December 4). Strengthen your brand online with a gift in kind: Charitable giving helps brand your site. *InfoWorld*, p. 64.

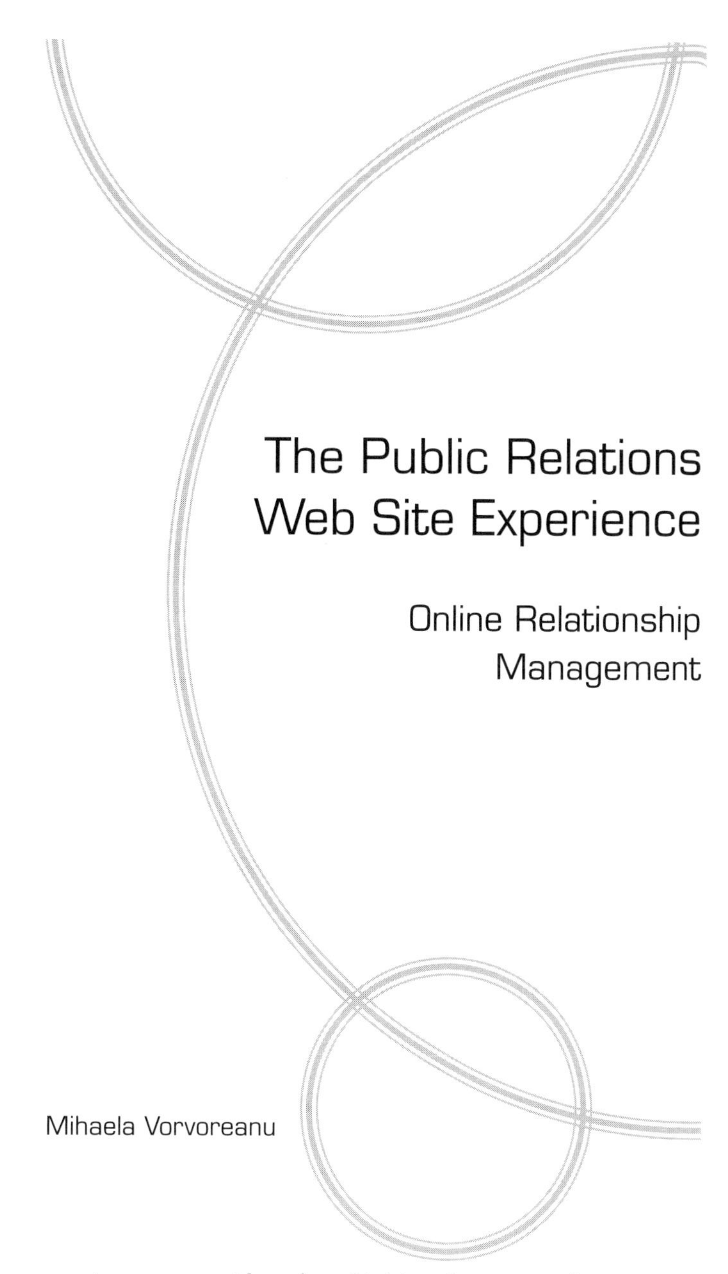

The Public Relations Web Site Experience

Online Relationship Management

Mihaela Vorvoreanu

This chapter presents three important issues to consider when thinking about using the organizational Web site for relationship management. The first issue concerns the importance and the pros and cons of online relationship management. The second issue is that of designing a good public relations Web site. The third issue is that of considering what kind of public relations experience the Web site offers to the organization's publics. The chapter introduces a research method used to understand and analyze a public's online experience.

"The Internet is like a store front for world business. If you're not there, you're not even close to doing business this day and age," said a public relations practitioner quoted in a research study (White & Raman, 2000, p. 413). A World Wide Web presence has become as common and necessary for an organization as being listed in the telephone book. Building a Web site is relatively easy, but building a Web site that builds relationships is a more complex problem. This chapter addresses the issue of organization-public relationship building and maintenance (that is, management) on an organization's Web site. It starts by explaining why the organizational Web site is an appropriate avenue for relationship management, provides some guidelines for online relationship management, and explains a research protocol, Website Experience Analysis, that can help public relations practitioners assess how publics perceive relationship management efforts on an organization's Web site.

Why Use the Organization's Web Site for Relationship Management?

Perhaps a more appropriate question is, Is it possible *not* to use the organization's Web site for relationship management? Once an organization has a Web site, and that Web site is seen by members of the organization's publics, it plays a role in that organization's relationship with its publics whether or not the organization realizes it. Consider, for example, a banking institution's Web site. Even if no one at this bank ever intended to use the Web site for public relations or relationship management, it does not make a difference to its publics. The Web site will still influence the publics' perceptions of their relationship with the bank. They might be impressed with the easy options to make electronic payments, they might be annoyed at the slow loading speed, and so on. Even the layout and the color scheme of the site might have an influence on the relationship between the bank and its publics.

Therefore, it is safe to assume that, if an organization has a Web site, this Web site *will* play a role in the organization's relationships with its publics. In fact, there is wide agreement in the public relations field that Web sites are very important relationship management tools. Both public relations scholars and practitioners feel that organizational Web sites play an important role in building and maintaining dialogue and relationships with publics (Coombs, 1998; Hallahan, 2003; Heath, 1998; Kent & Taylor, 1998), increasing positive perceptions of an organization, and maintaining loyalty (Goldie, 2003; Hurst & Gellady, 1999; Newland Hill & White, 2000; Nielsen, 1997; Nielsen & Norman, 2000; White & Raman, 2000).

In the professional world, there is also agreement that corporate public relations Web sites influence a public's perceptions of an organization and loyalty toward it. Goldie (2003), Hurst and Gellady (1999), Nielsen (1997), and Nielsen and Norman (2000) state that providing a positive experience on the organization's Web site is crucial to maintaining a good relationship with publics, enhancing reputation and customer loyalty, and, ultimately, surviving as a business. Interviews with public relations practitioners have shown that they too believe that Web sites contribute to an organization's reputation, and so it is important to have an up-to-date, competitive site (Johnson, 1997; Newland Hill & White, 2000; White & Raman, 2000).

These arguments about a Web site's important role in organization-public relationship management are supported by experimental studies that show organizational Web sites influence public perceptions of relationships with an organization, corporate credibility and goodwill, and identification with the corporation (Jo & Kim, 2003; Len-Rios, 2003).

Web sites are very important for public relations. This does not mean, however, that one should embrace Web sites without considering the pros and cons of using them for relationship management.

Web sites have many characteristics that make them attractive for public relations (Esrock & Leichty, 1998). One of the main advantages is that the gatekeeping function of mass media does not operate on the World Wide Web. That is, organizations are free to make public any information they desire, without having to pass through the selective process of the mass media. Another advantage of Web sites is that they can have built-in interactive features that facilitate two-way communication. Through the use of feedback features, an organization can collect information, monitor public opinion, and get input from its publics. Furthermore, Web sites tend to serve active audiences who seek information and are already motivated to listen to what the organization has to say.

All of these are obvious reasons why organizational Web sites are appropriate media for relationship management. Less obvious, however, might be the dangers, or drawbacks, of engaging in online relationship management. Consider, for example, the fact that a Web site is available to any and all people with Internet access. This level of openness has several implications, many of which an organization cannot control. To address all publics, the Web site might be fairly generic, which might not satisfy any of the publics. To address a narrow public, the Web site (or a part of it) might be very specific and narrowly targeted. However, this information would also be available to other, unintended audiences. Moreover, some publics might not have Internet access at all, and it is important not to leave them out. The issues of audience targeting, unintended audiences, and Internet access are some important things to consider when thinking about online relationship management.

Guidelines for Online Relationship Management

This section introduces some guidelines for online relationship management, and it attempts to answer the following question: What should a good public relations Web site be like?

First, it should be a clear, easy-to-use site. Guidelines for effective and easy-to-use Web site design abound. Amazon.com offers over 2,500 titles about Web design. Although it is virtually impossible and hardly necessary to compile a comprehensive review of the entire Web design literature, a few major ideas will be reviewed here.

Without exception, all the Web style guides reviewed for the purpose of writing this chapter (Brinck, Gergle, & Wood, 2002; IBM, 2003; Lynch & Horton, 2002; Nielsen, 2000; Van Duyne, Landay, & Hong, 2003) share the perspective of user-centered design. The main principle behind their recommendations is usability, defined as ease-of-use of the Web site. They each emphasize the requirements of simplicity, clarity, predictability, speed, and consistency in the design of sites. It is essential for public relations practitioners to become familiar with principles for good Web site design. Even if the practitioner herself will not be the one designing the site, it is just as important to understand this communication medium as it is to understand any other medium used in public relations. The two Web design books in the list of recommended readings at the end of this chapter are an excellent place to start.

Besides being well designed, a good public relations Web site should be able to create and maintain relationships with publics. Kent and Taylor (1998) propose five principles for public relations Web sites, which are briefly reviewed here.

Kent and Taylor's (1998) first principle is that of having a dialogic loop. They state that it is important for Web sites to take advantage of two-way communication features and to offer site visitors the possibility of communicating with the organization through e-mail and interactive forms. Of course, the e-mail account should be checked regularly, and a trained professional should respond to public inquiries.

The second principle states that information on a good public relations site should be useful to all publics. The Web site should contain information of general value, such as the organization's background and history, along with specialized information for specific publics. The hierarchy and structure of the information are also important aspects to consider, as they facilitate access to content. In other words, avoid burying useful content in a poor structure—this will make it difficult to find and possibly render it useless.

The third principle is the generation of return visits. A relationship, of course, is not a one-time event. For a Web site to be successful in relationship management, it needs to attract return visits. Updated information, changing issues, and interactive forums and discussion sessions are features likely to motivate public members to return to the Web site.

Kent and Taylor's (1998) fourth principle states that the Web site's interface should be intuitive and easy to use. Again, the structure and organization of the site, clear lists of links, and a site map contribute to ease of use. Also important is that the Web site download quickly and not keep visitors waiting for the page to display. A good Web site will avoid unnecessary, distracting bells and whistles and will emphasize solid, useful content.

The fifth and final principle is the rule of conservation of visitors. The principle refers to avoiding losing visitors by leading them to other Web sites through external links and distracting advertising. External links should be carefully chosen and strategically placed so as not to lead people away from your Web site. Once people leave the site, it is unlikely that they will return.

Kent and Taylor's (1998) five guidelines for building relationships on an organization's Web site do a nice job of integrating Web design and public relations principles. And they are good to keep in mind when creating or examining a public relations Web site.

Web Site Experience Analysis

This section of the chapter provides some tools to ensure that the public relations Web site is offering users a positive experience. So far, this chapter has discussed Web sites as if they were objects, artifacts, or texts. Now, we are going to change the point of view and focus on people's experiences on Web sites. It is important to differentiate between the Web site itself and the Web site experience. On one Web site, each visitor will have his or her own experience—and these experiences can be very different. Of course, there cannot be a Web site experience without a Web site. But the Web site is not the only thing that influences a person's Web site experience. Factors, such as personal goals and characteristics (age, gender, education, etc.) and personal history, will also influence a person's Web site experience. Because no two people are identical, no two Web site experiences are identical either.

Public relations practitioners want to offer Web site visitors an experience that builds and maintains relationships. To do that, it is necessary to implement principles of both Web site design and public relations. Yet, it takes more than that to ensure that publics will have a positive relationship-building and -maintenance experience on the site. In order to understand and assess the quality of a public's Web site experience, one needs to analyze their experience on the Web site, learn from it, and modify the Web site accordingly. A relatively simple research protocol, Website Experience Analysis (WEA), was designed and tested by the author for this purpose (Vorvoreanu 2004, 2006a, 2006b, 2006c).

WEA requires selecting a small number of people (five to ten people are enough) who are representative of the organization's public. If the Web site targets multiple publics, it is advisable to conduct WEA with five to ten representatives of each public. The WEA research protocol requires participants to review the Web site and then answer a number of questions about it. The questions tap directly into five important aspects of organization-public relationships: trust, commitment, involvement, openness, and dialogue. The remainder of this chapter explains the WEA procedure in some detail.

The environment and supplies necessary to conduct WEA are a quiet room to conduct the research and one or two computers. One computer is used for the research participant to look at the Web site and another for him or her to input answers into a computer-based questionnaire.

WEA uses two series of questions—that is, two questionnaires. The first questionnaire includes background and demographic questions about the research participant. A sample background questionnaire is included in Appendix A at the end of this chapter. The second questionnaire (i.e., the WEA questionnaire) includes questions about the Web site experience (see Appendix B). The questions are divided into three parts, mirroring the three parts of the experience: first impression, exploration, and exit.

Questions 1 through 6 in the WEA questionnaire are about the first phase of the Web site experience: first impression. Questions 7 through 21 tap into the main phase: exploration. Questions 22 through 27 ask about opinions Web site visitors would have formed by the time they engage in the last phase: exit. Question 28 is a check to assess any difficulty research participants might have had understanding the questions.

The questions in the WEA questionnaire are grouped in pairs. Each pair contains a closed-ended and an open-ended question on the same topic. Both questions are equally important. The answers to the closed-ended questions (the ratings) provide a numerical assessment of each aspect of the experience. The answers to the open-ended questions provide information about what specific aspects of the Web site influenced the research participants' opinions.

We will now take a closer look at the questions that tap into the main phase of the Web site experience, or exploration. After forming a first impression of the Web site, visitors will engage in the exploration phase. The first step of the exploration phase involves orienting oneself and figuring out a rudimentary mental map of the site; this step is called orientation. Questions 7 through 11 in the questionnaire address the orientation step. Questions 12 through 21 address the other step of the exploration phase: engagement. At this point in the Web site experience, users engage the content of the site; therefore, this step offers the most opportunities for relationship management. This is why the questions corresponding to this step, questions 12 through 21, address five core components of organization-public relationships: trust (questions 12 and 13), involvement (questions 14 and 15), investment (questions 16 and 17), openness (questions 18 and 19), and dialogue (questions 20 and 21). These five components of organization-public relationships have been isolated from an extensive review of public relations research articles on organization-public relationships (Broom, Casey, & Ritchey, 1997, 2000; Bruning & Ledingham, 1998, 1999, 2000; Grunig, 2002; Grunig & Huang, 2000; Huang, 2001; Ledingham, 2003; Ledingham & Bruning, 1998, 1999).

It is essential to understand the questionnaires before starting the research. The questionnaires may be modified to suit a specific situation, public, or organization. The questions can be uploaded to an online survey management service, such as SurveyMonkey.com (2006), to enable computer-based data collection.

Once the environment, the supplies, and the questionnaires are ready, the researcher needs a protocol to follow with each research participant. The protocol's steps are as follows: First, the researcher welcomes the research participant into the room, explains the nature of the research and of the task (to examine and evaluate a Web site), and obtains his or her informed consent to participate. Then, the researcher invites the participant to fill out the background questionnaire. Once that is completed, the participant is directed to the organization's Web site and instructed to examine the Web site and form an opinion of it.

As the participant looks at the site, at some point, she will click a link off the home page. At this moment, the researcher interrupts her and asks her to fill out the first six questions in the WEA questionnaire related to her first impression. After the participant has answered these questions, she is free to browse the site as much as she wants, until she feels ready to answer more questions about the site.

People generally take about ten minutes to examine the site and another ten to fifteen minutes to answer the questions. When the participant is done answering the questions, the researcher thanks her for her help, offers to answer any questions she might have, and, if applicable, hands her a small payment or reward for participating in the research.

After collecting data from five to ten members of the organization's public, the researcher examines their answers to each question. The answers will provide insightful information about the participants' Web site experience and will enable the public relations practitioner to optimize that experience by improving the Web site. For example, when analyzing the Web sites of several corporations, this chapter's author found that a student public was not satisfied with the mere presence of a Contact Us link (Vorvoreanu, 2004, 2006a). In terms of the organization's interest to engage in dialogue with its publics, the research participants felt that a link was not sufficient, and, because there was no explicit, emphasized invitation to provide feedback and engage in dialogue, they did not perceive that the organizations were interested in two-way communication. This was an unexpected finding with direct implications for changing the Web site's design.

Conclusion

This chapter has addressed the complex issue of public relations Web sites—that is, Web sites that play a role in the management of relationships between an organization and its publics. The purpose of this chapter was to consider important issues about online relationship management: pros and cons of using Web sites for relationship management, the elements of good public relations Web site design, and the public's experience on an organization's Web site. The chapter switched points of view, considering first the Web site itself, then the experience it provides to Web site visitors. The WEA research methodology for assessing whether a Web site provides a positive relationship management experience to an organization's publics was also presented.

Although this chapter does not, by far, exhaust the complex topic of online relationship management, it provides a starting point for thinking about public relations Web sites. The issues discussed in this chapter are key questions to consider before engaging in online organization-public relationship management.

The Internet is changing the field of public relations just like it has changed so many aspects of contemporary society. Web sites make dialogue (essential to any relationship) between organizations and publics technically possible and relatively easy. The traditional paradigm of public relations thinking will have to accommodate the public's increasing demand for one-on-one, meaningful dialogue.

The public relations profession has always been subjected to time pressures. The use of online technologies will only increase this pressure, making instant updates and communication the norm. New technologies have gotten publics accustomed to getting the news as it happens or as it is about to happen. Take, for example, the surreal story of passengers watching on TV a story about the possible crash of the aircraft in which they were flying! (Associated Press, 2005). Crisis communication plans and the ability to engage in rapid communication are becoming paramount. The race to be the first to frame the issue is getting tighter by the (split) second.

Good understanding of new and emerging communication media and a grasp of how their use is changing the field of public relations are essential for successful public relations practice. Practitioners and scholars alike face the new responsibility of keeping up with the times and staying informed. At the pace at which things are moving, by the time this chapter is read, it may be a piece of history. As frightening as that might seem, one must keep in mind that solid knowledge of communication and public relations principles can provide a much-needed grounding in this exciting, ever-changing world of new technologies.

For consideration

1. To what extent can Web site characteristics influence a person's relationship with an organization? In an increasingly mediated society, is interpersonal contact needed for a relationship to be established?
2. What are specific risks, pitfalls, or disadvantages of using the organizational site for relationship management?
3. Consider Kent and Taylor's (1998) Web design principle number three: the generation of return visits. What features of current Web sites exemplify this principle?
4. Is relationship management via new media tools more important for corporations, nonprofits, or government agencies? Why?
5. How could a public relations practitioner establish a business case for conducting WEA research on behalf of an organization?

For reading

Brinck, T., Gergle, D., & Wood, S. D. (2002). *Usability for the Web: Designing Web sites that work* (1st ed.). San Francisco: Morgan Kaufmann Publishers.

Holtz, S. (2002). *Public relations on the Net: Winning strategies to inform and influence the media, the investment community, the government, the public, and more!* (2nd ed.). New York: AMACOM.

Kent, M. L., & Taylor, M. (1998). Building dialogic relationships through the World Wide Web. *Public Relations Review, 24*(3), 321–334.

Nielsen, J. (2000). *Designing web usability*. Indianapolis, IN: New Riders.

Vorvoreanu, M. (2006). Web site experience analysis. In R. Reynolds, R. Woods, & J. D. Baker (Eds.), *Handbook of research on electronic surveys and measurements* (pp. 281–284). Hershey, PA: Idea Group.

References

Associated Press. (2005, September 22). Emergency landing televised on JetBlue flight. *MSN*. Retrieved August 6, 2006, from http://www.msnbc.msn.com/id/9430871/

Brinck, T., Gergle, D., & Wood, S. D. (2002). *Usability for the Web: Designing Web sites that work* (1st ed.). San Francisco: Morgan Kaufmann Publishers.

Broom, G. M., Casey, S., & Ritchey, J. (1997). Toward a concept and theory of organization-public relationships. *Journal of Public Relations Research*, 9(2), 83–98.

Broom, G. M., Casey, S., & Ritchey, J. (2000). Concept and theory of organization-public relationships. In J. A. Ledingham & S. D. Bruning (Eds.), *Public relations as relationship management. A relational approach to the study and practice of public relations* (pp. 3–22). Mahwah, NJ: Lawrence Erlbaum Associates.

Bruning, S. D., & Ledingham, J. A. (1998). Organization-public relationships and consumer satisfaction: The role of relationships in the satisfaction mix. *Communication Research Reports, 15*(2), 198–208.

Bruning, S. D., & Ledingham, J. A. (1999). Relationships between organizations and publics: Development of a multi-dimensional organization-public relationship scale. *Public Relations Review, 25*(2), 157–170.

Bruning, S. D., & Ledingham, J. A. (2000). Perceptions of relationships and evaluations of satisfaction: An exploration of interaction. *Public Relations Review, 26*(1), 85–95.

Coombs, W. T. (1998). The Internet as potential equalizer: New leverage for confronting social irresponsibility. *Public Relations Review, 24*(3), 289–303.

Esrock, S. L., & Leichty, G. B. (1998). Social responsibility and corporate Web pages: Self-presentation or agenda setting? *Public Relations Review, 24*(3), 305–319.

Goldie, P. (2003). *Experience matters—more now than ever.* Retrieved March 25, 2003, from http://www.macromedia.com/newsletters/edge/march2003/section0.html

Grunig, J. E. (2002). *Qualitative methods for assessing relationships between organizations and publics.* Gainesville, FL: Institute for Public Relations.

Grunig, J. E., & Huang, Y.-H. (2000). From organizational effectiveness to relationship indicators: Antecedents of relationships, public relations strategies, and relationship outcomes. In J. A. Ledingham & S. D. Bruning (Eds.), *Public relations as relationship management: A relational approach to the study and practice of public relations* (pp. 23–53). Mahwah, NJ: Lawrence Erlbaum Associates.

Hallahan, K. (2003, May). *A model for assessing Web sites as tools in building organizational-public relationships.* Paper presented at the International Communication Association Convention, San Diego, CA.

Heath, R. L. (1998). New communication technologies: An issues management point of view. *Public Relations Review, 24*(3), 273–288.

Huang, Y.-H. (2001). OPRA: A cross-cultural, multiple item scale for measuring organization-public relationships. *Journal of Public Relations Research, 13*(1), 61–90.

Hurst, M., & Gellady, E. (1999). *Building a great customer experience to develop brand, increase loyalty, and grow revenues (Creative Good White Paper One).* Retrieved March 10, 2003, from http://www.creativegood.com/creativegood-whitepaper.pdf

IBM. (2003). *IBM ease of use Web site: Web design guidelines.* Retrieved March 20, 2003, from http://www-3.ibm.com/ibm/easy/eou_ext.nsf/Publish/572

Jo, S., & Kim, Y. (2003). The effect of Web characteristics on relationship building. *Journal of Public Relations Research, 15*(3), 199–233.

Johnson, M. A. (1997). Public relations and technology: Practitioner perspectives. *Journal of Public Relations Research, 9*(3), 213–236.

Kent, M. L., & Taylor, M. (1998). Building dialogic relationships through the World Wide Web. *Public Relations Review, 24*(3), 321–334.

Ledingham, J. A. (2003). Explicating relationship management as a general theory of public relations. *Journal of Public Relations Research, 15*(2), 181–198.

Ledingham, J. A., & Bruning, S. D. (1998). Relationship management in public relations: Dimensions of an organization-public relationship. *Public Relations Review, 24*(1), 55–65.

Ledingham, J. A., & Bruning, S. D. (1999). Time as an indicator of the perceptions and behavior of members of a key public: Monitoring and predicting organization-public relationships. *Journal of Public Relations Research, 11*(2), 167–183.

Len-Rios, M. L. (2003, May). *Communication rules and expectations in consumer use of information and e-commerce Web sites.* Paper presented at the International Communication Association Conference, San Diego, CA.

Lynch, P. J., & Horton, S. (2002). *Web style guide: Basic design principles for creating Web sites* (2nd ed.). New Haven, CT: Yale University Press.

Newland Hill, L., & White, C. (2000). Public relations practitioners' perceptions of the World Wide Web as a communications tool. *Public Relations Review, 26*(1), 31–51.

Nielsen, J. (1997, August 1). *Loyalty on the Web*. Retrieved March 20, 2003, from http://www.useit.com/alertbox/9708a.html

Nielsen, J. (2000). *Designing web usability*. Indianapolis, IN: New Riders.

Nielsen, J., & Norman, D. A. (2000, January 14). Web-site usability: Usability on the Web isn't a luxury. *InformationWeek*. Retrieved March 24, 2003, from http://www.informationweek.com/773/web.htm

SurveyMonkey.com. (2006). Retrieved August 6, 2006, from http://www.surveymonkey.com/

Van Duyne, D. K., Landay, J. A., & Hong, J. I. (2003). *The design of sites: Patterns, principles, and processes for crafting a customer-centered Web experience*. Reading, MA: Addison-Wesley.

Vorvoreanu, M. (2004). *Building and maintaining relationships online: A framework for analyzing the public relations Web site experience*. Unpublished doctoral dissertation, Purdue University, West Lafayette, IN.

Vorvoreanu, M. (2006a, November). *Studying relationship building on corporate Web sites: A new research protocol*. Paper presented at the National Communication Association Annual Convention, San Antonio, TX.

Vorvoreanu, M. (2006b). Web site experience analysis. In R. Reynolds, R. Woods, & J. D. Baker (Eds.), *Handbook of research on electronic surveys and measurements* (pp. 281–284). Hershey, PA: Idea Group.

Vorvoreanu, M. (2006c). Online organization-public relationships: An experience-centered approach. *Public Relations Review 32*(4), 395–401.

White, C., & Raman, N. (2000). The World Wide Web as a public relations medium: The use of research, planning, and evaluation in Web site development. *Public Relations Review, 25*(4), 405–419.

Appendix A: Web Site Experience Analysis: Sample Background Questionnaire

1. Age: ________

2. Sex:
   a) Male
   b) Female

3. Education (please check the highest level completed):
   a) High school
   b) College—currently enrolled
   c) College
   d) Graduate school—master's degree
   e) Graduate school—PhD

4. Marital status:
   a) Single

b) Married
c) Divorced
d) Separated
e) Other

5. How many children do you have?
a) 0
b) 1
c) 2
d) 3
e) 4
f) 5 or more

6. Overall annual family income:
a) Under $15,000
b) $16,000—$25,000
c) $26,000—$35,000
d) $36,000—$45,000
e) $46,000—$55,000
f) $56,000—$65,000
g) $66,000—$75,000
h) $75,000—$100,000
i) Over $100,000

7. How long have you been using a computer?
_______ Years _______ Months

8. How often do you use a computer?
a) Several times a day
b) Every day
c) 2 to 3 times a week
d) Once a week
e) 2 to 3 times a month

9. Where do you use a computer?
a) At home
b) At work
c) At the library
d) At a friend's or relative's house
e) Other:_______________

10. How long have you been using the Internet?
_______ Years _______ Months

11. How frequently do you use the Internet?
    a) Several times a day
    b) Every day
    c) 2 to 3 times a week
    d) Once a week
    e) 2 to 3 times a month

12. Where do you use the Internet?
    a) At home
    b) At work
    c) At the library
    d) At a friend's or relative's house
    e) Other:______________

13. Do you enjoy using computers?
    a) Very much
    b) Somewhat
    c) Neutral
    d) Not really
    e) Not at all

14. Are you familiar with [organization name]?
    a) Very familiar
    b) I've heard the name, but I don't know much about it
    c) Not familiar at all

Appendix B: Web Site Experience Analysis Questionnaire

Questions 1 through 6 to be completed after the first click off the homepage.

1. My first impression of this Web site is:
   (very bad) 1 2 3 4 5 6 7 8 9 10 (very good)

2. Please describe your first impressions of the Web site. In your description, point out those Web site aspects upon which your first impressions are based.

3. I expect to find good quality content on this Web site.
   (not at all) 1 2 3 4 5 6 7 8 9 10 (very much)

4. What aspects of the Web site make you feel the way you do?

5. I expect this Web site to be interesting.
   (not at all) 1 2 3 4 5 6 7 8 9 10 (very much)

6. What aspects of the Web site make you feel the way you do?

Please continue browsing the Web site. When you have formed an opinion of it, return to this questionnaire.

7. The organization of this Web site is:
   (very confusing) 1 2 3 4 5 6 7 8 9 10 (very clear)
8. Navigating the links on this Web site is:
   (very frustrating) 1 2 3 4 5 6 7 8 9 10 (very easy)
9. Please list the aspects of the Web site that help you find your way around the site:
10. Please list the aspects of the Web site that you find confusing and/or disorienting:
11. What on this Web site maintained your interest while browsing the Web site?
12. Do you feel you can trust this organization?
    (not at all) 1 2 3 4 5 6 7 8 9 10 (very much)
13. What on this Web site makes you feel this way?
14. Do you feel this organization is interested in maintaining a relationship with you?
    (not at all) 1 2 3 4 5 6 7 8 9 10 (very much)
15. What on this Web site makes you feel this way?
16. Do you think this organization enjoys helping others?
    (not at all) 1 2 3 4 5 6 7 8 9 10 (very much)
17. What on this Web site makes you feel this way?
18. Do you think this organization is open about sharing information?
    (not at all) 1 2 3 4 5 6 7 8 9 10 (very much)
19. What on this Web site makes you feel this way?
20. Do you feel that this organization is interested in listening to what people like you have to say?
    (not at all) 1 2 3 4 5 6 7 8 9 10 (very much)
21. What on this Web site makes you feel this way?
22. Overall, how do you evaluate this Web site?
    (very bad) 1 2 3 4 5 6 7 8 9 10 (very good)

23. Please describe your overall opinion of the Web site. In your description, please point out those Web site aspects upon which your opinion is based.

24. Overall, how do you rate your opinion of this organization?
(very negative) 1 2 3 4 5 6 7 8 9 10 (very positive)

25. Please describe your overall opinion of [organization name]. In your description, please point out those Web site aspects upon which your opinion is based.

26. Would you visit this Web site again?
(definitely no) 1 2 3 4 5 6 7 8 9 10 (definitely yes)
Why, or why not?

27. The questions I had to answer about the Web site were:
(very confusing) 1 2 3 4 5 6 7 8 9 10 (very clear)

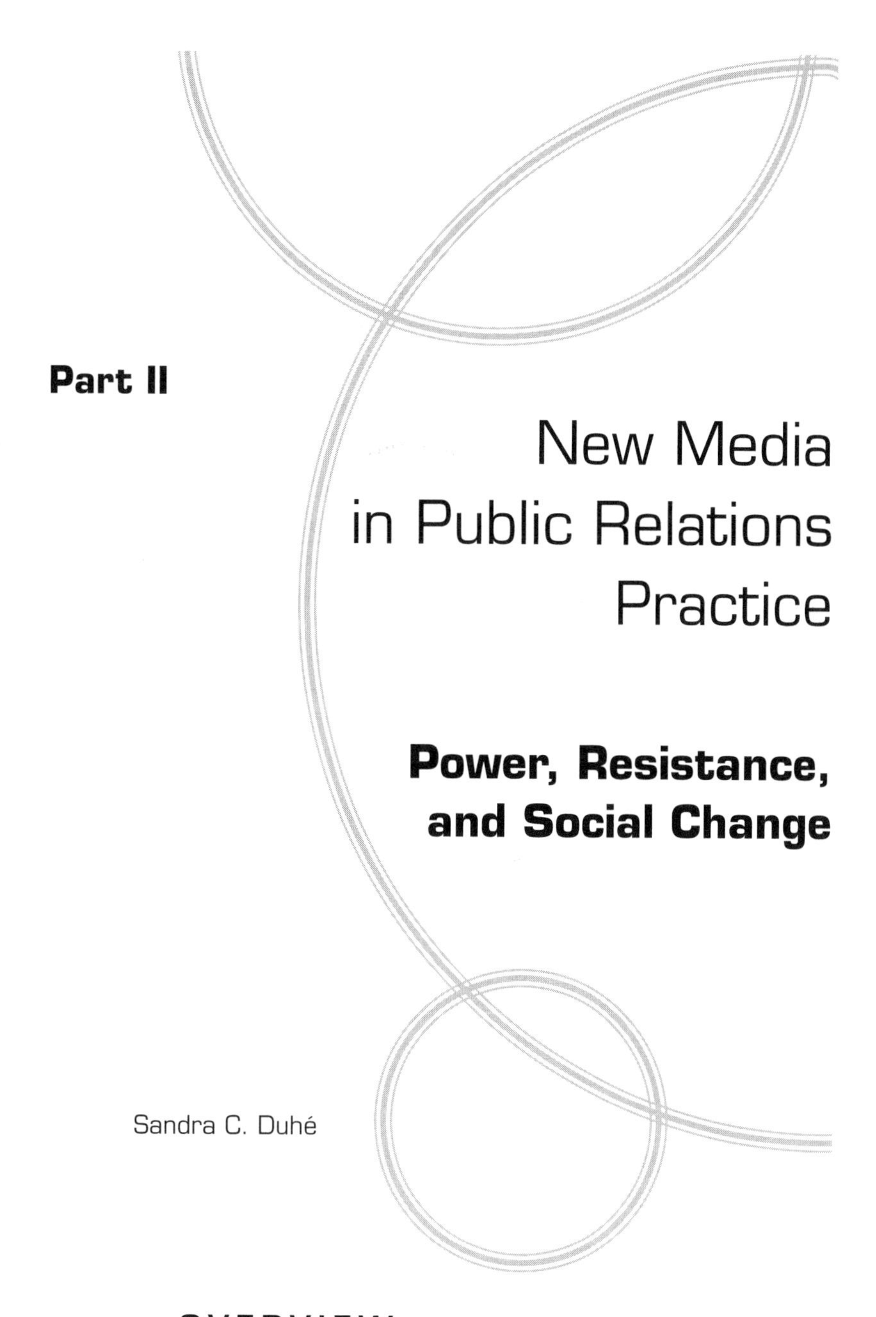

Part II

New Media in Public Relations Practice

Power, Resistance, and Social Change

Sandra C. Duhé

OVERVIEW

Aside from its global reach, low-cost accessibility, and information-equalizing capability, perhaps the most notable impact of the Internet has been the increasingly audible presence of minority voices. New media have effectively facilitated a shift in power and voice from the elite few to the underrepresented many in political, economic, and humanitarian spheres.

In essence, the ability for like-minded individuals to interact in cyberspace has eroded any organization's monopoly on information. Dominant regimes still remain, but they are no longer immune to change pressures exerted by those who disagree with their principles and practices.

With shifts in power come resistance. More pronounced voices accustomed to being heeded resist change to established ways of doing things, while previously disenfranchised voices, given a newfound ability to be heard, resist being silenced ever again. This dynamic interplay of wills within the global marketplace of ideas should ideally lead to more informed, and transparent, social change.

Each of the six authors in this section explains how the desire to be heard has the potential to catalyze change at a societal level. W. Timothy Coombs and Sherry Holladay begin with their description of how the uncontrollable nature of Internet "contagions" enables resistors to influence corporate activities. A case study of an historic, and successful, Web-based shareholder rebellion against the Walt Disney Company is then chronicled by Sarah Bonewits Feldner and Rebecca Meisenbach. Thereafter, Mohan Dutta and Mahuya Pal analyze how a local struggle for environmental and social justice in western India has become a global model for online activism. Continuing in an activist vein, Romy Fröhlich and Jeffrey Wimmer examine how a counterculture organization in the German political sphere communicates effectively despite its decentralized network structure. Maria de Fátima Oliveira explains how a Brazilian advocacy group successfully uses its Web site to frame and redefine dialogue about crime and social inequities. Finally, Mahmoud Eid provides a compelling portrayal of how the political economy of Arab nations has contributed to a digital divide and inhibited global interaction.

New media promote social change from the bottom up by providing marginalized groups with opportunities for collaboration, regardless of participants' geographical location or socioeconomic status. This virtual landscape, though riddled with resistance, advances the free flow of information and the democratic ideals of public relations.

Consumer Empowerment through the Web

How Internet Contagions Can Increase Stakeholder Power

W. Timothy Coombs and Sherry J. Holladay

Negative word of mouth has always been a concern for corporations because of its negative impact on profits and reputations. Conversely, it also is a way for consumers to exercise power and have their concerns heard by corporations. This potential of the Internet to empower resistors is reflected in organizations' concerns for attack Web sites and critical comments on blogs and discussion groups. This chapter uses the idea of Internet contagions to explain the power that consumers/resistors are finding by organizing and managing their issues/concerns via the Internet. Resistors can implement principles of buzz marketing to accrue power and influence. We explore how and why specific features of Internet communication can empower consumers and increase their ability to have their concerns heard by organizations.

While the Internet has provided organizations with opportunities to promote products and services, the Internet also has emerged as a valuable tool for consumers who seek to challenge organizations. Individuals and groups intent on promoting consumer resistance, especially when utilizing the more active resistance strategies of complaining and boycotting, can utilize the Internet to widen the range of people receiving their messages and, ultimately, to increase the power they wield when confronting organizations (Fournier, 1998). Well-known examples of the use of the Internet to challenge organizations include Greenpeace's challenge of Shell's planned dumping of the Brent Spar, People for the Ethical Treatment of Animals' (PETA's) protests of Burger King's and Wendy's chicken purchasing practices, and the Flaming Ford owners versus Ford Motor Company. We refer to the power associated with the spread of messages in cyberspace as an *Internet contagion*. The idea of Internet contagion parallels principles of buzz or viral marketing. This chapter explores how resistors can implement principles of buzz marketing and negative word-of-mouth (WOM) communication in order to accrue power and influence corporations. An examination of Internet contagion provides a key to understanding the influence process. What tools can be used to create an Internet contagion? How do we know when the Internet contagion is building power for resistors? Both organizations and resistors can examine specific channel and content indicators to gauge the effectiveness of Internet contagions.

Why Stakeholders Become Resistors

The growing emphasis on corporate social performance has opened organizations to greater stakeholder scrutiny. The public expects more from corporations because of their increasing prominence as sociopolitical forces, as well as economic forces. Stakeholders are no longer limited to stockholders, employees, or community members. The growing importance of organizations as institutions has led stakeholders to feel they have legitimate interests in the operations of corporations. Additionally, globalization has led stakeholders to become concerned about how corporations operate in other countries.

Activist stakeholders (resistors) can have a variety of concerns about organizational operations stemming from their products (e.g., value, safety, and health concerns), business operations (e.g., business strategies, including the use of sweatshop labor, fair trade products, outsourcing, discrimination, and advertising), and environmental effects (e.g., impact on air and water quality). In addition to challenging corporations on a variety of operations, activists can challenge them on more macro, societal issues. Some activist stakeholders are concerned with macro issues, such as antiglobalization, corporate support of oppressive governments, or environmental concerns, such as global warming or destruction of rain forests.

Various stakeholder groups develop expectations for how corporations should behave, and these expectations are often at odds with one another. Because the activist stakeholder groups may hold different, often contradictory, expectations grounded in disparate values and beliefs, classifying stakeholders and the threats they may pose is problematic. In spite of the complexity of monitoring the range of stakeholder activities that may impact corporations, corporations should be highly motivated to do so because of their potential to threaten their reputations and operations. Additionally, corporations should be motivated to monitor stakeholders because of the information they can provide concerning the extent to which the corporation is meeting expectations. Stakeholders are a significant barometer of public sentiment concerning a wide range of issues pertaining to corporations.

Regardless of the basis for resistance, activist stakeholders are becoming increasingly aware of the power of the Internet in assisting their causes. While, traditionally, we may have visualized activists engaging in public protests, we must also recognize the power of virtual activism to influence others and corporations. Several Web-based avenues exist for resistors, including Web pages, blogs, e-mail lists, and discussion groups. The Internet has emerged as an important tool for stakeholders who want to demand attention for their claims and influence operations (e.g., van de Donk, Loader, Nixon, & Rucht, 2004). Virtual activists can use the Internet in numerous ways to affect sentiments, disseminate information, connect like-minded people, and mobilize actions.

The Internet-based activities of activists can even become a news source for the traditional media and thereby garner increased attention and credibility. When Web sites are cited in news reports by mainstream media organizations, the sites and the causes gain legitimacy and extend awareness of the issues. For example, the Flaming Ford Web site gained attention when CNN, NPR (National Public Radio), and major news networks cited the Web site in news reports. In spite of federal investigations, Ford had resisted a recall of some of its vehicles for several years. The Web site is credited for sparking attention and forcing Ford to recall the defective cars and trucks (Coombs, 1998). In this way, the Internet can work in conjunction with traditional media to challenge the activities of corporations (Bennett, 2004).

From Buzz Marketing to Internet Contagion

Activist stakeholders can learn a great deal from corporate marketing strategies. The marketing concept known as *buzz* or *viral marketing* is especially instructive. The basic idea is to create a message that is then spread by consumers. The point of this type of promotion is to have a central message picked up and distributed by e-fluentials (online opinion leaders) and/or average consumers. The benefits are twofold: (a) Consumers do the marketing work for the organization, and (b) the organization benefits from the credibility of WOM. Like a virus, the message spreads through the Internet, infecting others. A buzz marketing effort involves a mix of orthodox Internet communication avenues, such as Web sites, e-mail, blogs, and discussion boards, in the hope that the message will penetrate various regions of cyberspace (Holt, 2004).

An example of buzz marketing was Lee Jeans' use of a free video game to boost sales of its product Lee Dungarees. Through a marketing firm, e-mail messages were sent to one thousand young males with heavy Internet use. The messages contained a Web site and brief message. The e-mail was forwarded to an average of six people. At the Web site, people played a video game, but to play at higher levels required codes found on Lee's products. If customers wrote Web logs about the game or posted messages to discussion boards, the buzz was even stronger.

The danger is that buzz cannot be controlled. Once a message is injected into cyberspace, it can mutate in unpredictable ways. Buzz marketing is a variety of publicity and WOM communication. In publicity, an organization sends a message to a media outlet hoping that the message then becomes a story. However, publicity lacks control; the organization does not know if a story will be run or how the story will be framed. WOM is viewed as a credible source when consumers want information. The Internet substitutes for and/or reinforces the telephone or face-to-face communication. Buzz is electronic WOM. Corporations take buzz seriously as a marketing tool that can produce results (Middleberg, 2001).

Resistors can use the same conventional Internet communication avenues, including buzz marketing, plus unorthodox avenues. The publicity parallel informs this point. Corporations can attempt to curry publicity through orthodox tactics, such as news releases, press conferences, and pitch letters. Activists can go beyond the orthodox to include protests and boycotts. The danger is that the activists' tactics become the message or the focus of attention or that the group is further marginalized. For example, the news media may cover the protest and not the reason for it. We see people chained to fences but have no idea why they have chained themselves or are told the people are extremists. Unorthodox Internet communication avenues include attack Web sites and complaint portals.

Corporations are expressing concern over resistors using either orthodox or unorthodox avenues. Firms offer specialty services in coping with attacks and in monitoring mentions of corporations or products in blogs and discussion groups (Middleberg, 2001). The owners of complaint portals repackage and sell the information to the objects of scorn (the companies). People are making money from corporate fears of resistors using buzz. Why? A negative message spreading through cyberspace is a potential threat to an organization's reputation and how it is perceived by stakeholders. Reputation is recognized as an extremely valuable, intangible asset for an organization. A positive reputation has been linked to improved sales, stronger stock prices, and higher-quality job applications (Dowling, 2001). Negative media coverage and WOM are real threats to a reputation. An Internet contagion, unfavorable messages spreading on the Internet, is recognized as a serious threat as well.

How the Internet Can Build Power through Contagions

The Internet has increased the power potential of stakeholders who wish to become activists. Stakeholders' Internet actions can make it difficult for corporations to resist the claims of activist groups. The following outlines the potential threats that can be posed by Internet activists through their strategic use of the Internet to manage issues.

Coombs (2002) proposed the issue contagion perspective to explain and prescribe what issue managers should do in maximizing the potential of the Internet. "Issue contagions are issues that spread rapidly through the Internet" (p. 216). When an issue gains momentum and spreads among people, it becomes a contagion. Issue contagions can be the province of the issue managers, those who assume responsibility for helping an issue gain salience to corporations. This discussion draws upon Coombs's (2002) work on the issue contagion perspective to highlight how the Internet can be used to effectively manage issues and to empower activists.

Issues are points of contention between organizations and stakeholders (Coombs, 1992). Organization-stakeholder relationships are complex, as there are myriad relationships. The relationship exists when the stakeholder or the organization recognizes an interdependency. This sets the stage for potential conflicts of interests. Interests can be values, beliefs, and expectations, among other things, that lead people or corporations to act. When resistors perceive conflicts of interest, they may be motivated to act in order to influence the corporation.

The primary corporate interest is survival. Organizational success is largely dependent on reputations, a valued resource. Organizations also depend on others to not interfere with their operations. On a daily basis, an organization may expect delivery of raw materials and access to technology, among other things, in order for it to operate. Disruptions to the routine standard operating procedures create problems. Organizations are more or less dependent on stakeholders to not interfere with their business and to not contest their reputation. Challenges—or the potential for challenges—to either reputation or operations are problematic and may necessitate a response.

This dependency provides the backdrop for the development of stakeholder power. A stakeholder may perceive that an organization poses a threat to his/her interest(s) and decides to challenge the organization. A disgruntled stakeholder in isolation may not be perceived to pose much threat. However, if several stakeholders share their concerns and mobilize, they can gain power. "Issues emerge only when people share problems with others through communication" (Hallahan, 2001, p. 28). Stakeholders can create awareness of a problem through their activities on the Internet. A stakeholder or stakeholder group can move issues onto a public agenda and become salient to the organization.

Prior to the Internet age, activist stakeholders had to rely on media advocacy for media coverage. This media publicity might allow them to attract other like-minded people. However, media advocacy is a form of uncontrolled publicity that cannot guarantee story placement or how the story will be framed in the media (Treadwell & Treadwell, 2000). Extreme actions (e.g., publicity stunts) designed to attract attention to an issue often leave activists looking more foolish than wise and often overlook their message in favor of highlighting the extreme action (activists chaining themselves to trees, walking naked in public, etc.).

In contrast to traditional media advocacy and publicity, the Internet empowers stakeholders to become more significant players in the issues management game. The Internet enables a greater degree of control and presents more options for activists to publicize a problem with the organization-stakeholder relationship. By enabling high-involvement individuals to communicate with others, the Internet has altered how issues emerge (Hearit, 1999; Holtz, 1999). The Internet can facilitate movement from an individual problem to a public issue and demand attention from a corporation. It can become a contagion when it spreads rapidly among stakeholders. Activists can use the Internet to demonstrate that their concerns pose a real threat and to increase the pressure on corporations to take their demands seriously.

The following discussion draws upon three areas of research to further explore how the Internet contagion perspective helps us understand how stakeholder activists / issue managers can help an issue gain traction and thereby gain power. Insights from stakeholder theory, network analysis, and reputation management are reviewed briefly and synthesized with the issue management literature to inform this analysis.

Stakeholder theory. Stakeholder theory offers an explanation of how stakeholders come to be viewed as powerful and thereby important to management. Mitchell, Agle, and Wood (1997) propose three dimensions for evaluating stakeholder salience: (a) power (the ability to get an actor to do something she or he would not do otherwise), (b) legitimacy (the actions are perceived as appropriate, desirable, or proper within the context of some belief system), and (c) urgency (the extent to which time frame is important; the call for immediate action due to the importance of the claim or the relationships to stakeholders). Stakeholder salience is a function of the ability to demonstrate these attributes. The more attributes the stakeholders are perceived to possess, the greater their salience and the greater the pressure on management to deal with them and their issues. This is complicated by the fact that, sometimes, the demands of different stakeholder groups are contradictory or mutually exclusive. The importance of various stakeholders can change depending on the situation. The issues advocated by competing groups must be prioritized according to their threat level as determined by their ability to damage the organization and their probability of developing momentum. However, resistors will be perceived to have greater salience when the attributes of power, legitimacy, and urgency are strong (Coombs, 2002).

Stakeholders can increase salience by managing the three attributes through their Internet activities. The idea is that corporations will monitor the Internet-based activities of resistors in order to assess their threat potential. Savvy stakeholders can engage in coalition building and social construction of reality (management of interpretations) to enhance the perception that they can exercise power and force the organization to pay attention to their demands (Mitchell, Agle, & Wood, 1997). For example, by using Internet communication, the resistors can demonstrate that they are a powerful group that has legitimate claims and can act quickly to disrupt operations. Resistors can organize an e-mailing campaign to corporate officials stating that they will boycott products if a health benefits policy covering same-sex partners is implemented. The orchestrated action signals the power of the group to organize and affect the corporation's financial status. The group's legitimacy claims (their reasons for implementing the boycott) may be grounded in references to biblical teachings that they say represent their foundational values. The e-mail campaign is designed to create a sense of urgency and persuade the corporation that their actions will be swift. If the management perceives the attributes in the way desired by the resistors, they will be motivated to address the issue.

For example, at http://www.killercoke.org (*Campaign to Stop Killer Coke*, 2006), visitors could learn about Coca-Cola's alleged support of paramilitary groups in Columbia who threatened union leaders and workers who attempted to unionize through the National

Union of Food Industry Workers (Sindicato Nacional de Trabajadores de la Industria de Alimentos [SINALTRAINAL]). Coca-Cola was accused of not intervening in the mistreatment of union workers at a Colombian bottling facility. Numerous links to related sites and a variety of news stories were provided to lend legitimacy to the charges. Links to accounts of university students' protests, along with successful attempts to end the sale of Coke on campuses, were outlined to demonstrate the power of resistance. SunTrust Banks, Coca-Cola's principal financial ally, also was targeted for protests. Resistors could download a letter to Coca-Cola board members to protest worker treatment in Colombia. The letter included board members' pictures and affiliations, along with their current employers and other boards of which they are members. Another letter encouraged readers to participate by making a financial contribution and/or volunteering to distribute leaflets and protest both Coca-Cola and SunTrust. The site informed and attempted to mobilize resistance by creating a sense of urgency through horrific stories supported by news media accounts, showcasing successful resistance, demonstrating connections to other groups, and making it easy to protest to corporate officials.

In sum, indicators of stakeholder salience include power, legitimacy, and urgency. All three can be demonstrated via the Internet to persuade the corporation that stakeholders' concerns should be considered. Power can be shown by claiming growing numbers of supporters as demonstrated through active participation in discussion groups and e-mailing lists. Web sites can offer automated complaint letters to be sent to corporations. This activity can signal both power and urgency. Urgency can be shown via reference to corporate plans (e.g., indicating a timetable for blocking corporate decision making). The legitimacy of the resistors' concerns can be demonstrated by using the information storage and interactive features of the Internet. Through the posting of scientific reports, expert testimony, and other documents, as well as links to well-recognized organizations, evidence for the legitimacy of stakeholder demands can be established. The interactive features of the Internet also can be used to support the legitimacy of the claims. For example, discussion threads and TrackBacks can signal that Web site visitors share perceptions of the corporation's misdeeds. Discussions may signal that interpretations are shared and that emotions run high surrounding the issue.

Network perspective. Rowley (1997) argues for the importance of using a network perspective to understand the power and potential influence of stakeholders. Network analysis offers a method for examining structural relationships as a system by examining interconnections among various stakeholders and the organizations that comprise the environment. Because organizations have relationships with many different stakeholders, they must consider the complex web of relationships. Relationships with some stakeholders gain prominence when those stakeholders can exercise power and are difficult for an organization to ignore (Rowley, 1997).

Network structure represents opportunities and constraints for actors within the network (Wasserman & Galaskiewicz, 1994). The corporation is an actor, as are the other stakeholders, and they operate within a larger environment consisting of similar corporations, suppliers, and other relevant stakeholders (Rowley, 1997). For our purposes, the network boundaries are limited to those who share an interest in an issue (Knoke, 1994).

In network terms, power is a function of the issue manager's / stakeholder's position within the larger network of stakeholders. *Centrality* is a critical variable because it enables communication and is an indicator of an actor's power and importance within the network (Coombs, 2002; Rowley, 1997). The Internet offers a wide range of opportunities for resistors to develop relationships with others and with other organizations that may support the issue.

Coombs (2002) applied the concept of centrality to the analysis of Internet contagions. Centrality is a function of a resistor's closeness and degree (Rowley, 1997). The *closeness* aspect of centrality refers to the ability of the issue manager to independently access others across the network. A resistor who demonstrates closeness can spread information quickly and directly without intermediaries. *Degree centrality* indicates the extent to which the issue manager is well connected. *Degree* refers to the number of ways a stakeholder can be linked to others in the network. It signals the potential for access to many different sources of information (Rowley, 1997). A stakeholder who demonstrates the degree dimension of centrality can use a variety of Internet-based communication tools to facilitate access to others, including issue Web sites, discussion groups, e-mail alerts, Web logs, and posts to complaint portals (Coombs, 2002).

In sum, centrality can create power through direct Internet communication with other stakeholders. The power of the issue manager is important to the ability of an issue to become a contagion. The structure of the stakeholder network contributes to this power. The closeness aspect of centrality means the issue manager can independently access others through the network. The degree aspect of centrality highlights the importance of using different avenues of Internet communication (Coombs, 2002).

Corporations that monitor the Web activity of activist groups will be able to observe this centrality and make inferences about power. Indicators of strong issue manager power through the centrality measure include Web site traffic, links to the Web site search engine results placement (it should be close to the organization being challenged and within the top ten placements), number of subscribers to e-mail alert systems, discussion group traffic, and complaint portal traffic (Coombs, 2002). The indicators signal the ability of the issue manager to connect with others through a variety of methods.

Reputation management. Protecting reputations is an important motivation in corporate behavior. Favorable reputations are associated with positive outcomes, including financial success, ability to recruit talent, community support, and admiration. Reputational threats carry weight. When issues are evaluated as having the potential to go public, corporations are more motivated to respond.

As part of their own issues management function, savvy corporations monitor the activities of groups that might influence their activities and instigate an issue contagion. They hope to anticipate threats through risk assessments and to plan strategies to counter stakeholder claims that they perceive are not in their best interests. Corporations are aware of the potential of the Internet to grant legitimacy to stakeholder claims. Corporations are worried about attack sites, blogs, discussion groups, e-mailing lists, and other forms of e-communication that can gain traction and generate attention.

Using various Internet channels can alter a network dynamic, increase resistors' power, and make them more powerful in the eyes of an organization. Table 1 reviews some basic markers of success in using Internet channels to spread a contagion. The focus is on the number of people reached, number of connections, and variety of connections.

Table 1: Evidence of Issue Contagion Success (Power through Connections)

Markers of Issue Contagion Success

- Traffic to the Web sites
- Links to the issue Web sites
- Search engine results placement of the issue Web sites
- Number of subscribers to the e-mail alert systems
- Traffic to discussion groups
- Traffic to complaint portals
- Popularity of blog site
- Number of TrackBacks to a blog
- Search engine results placement (it should be close to the organization being challenged and on the first page)
- Number of subscribers to e-mail alert systems
- The variety of Internet-based channels used (e.g., Web sites, blogs, discussion groups, e-mail lists, etc.)

Legitimacy: Content Concerns

When discussing contagions or buzz, too often the focus is simply channel based. An example is the current preoccupation with blogs in public relations. The talk centers on the need to *use* the new channel, not the *content* of the blogs. What a message says is also important. Because legitimacy is critical to a contagion, the perceived *legitimacy* of the issue must be a preeminent concern for activist stakeholders. While, for some, the legitimacy of the issue may seem self-evident, creating a contagion that is difficult for a corporation to ignore requires more than the involvement of a few committed, so-called true believers. The creation of legitimacy for an issue and the resistors themselves requires effective *framing*. Resistors must be concerned with how the issue is described, how the involvement of relevant corporations is depicted, and how recommendations for managing the issue are presented. The framing process requires skill in managing perceptions and meanings.

Overall, issue framing, including the types of support provided and the language used, should be viewed as a *social construction process* (van de Donk, Loader, Nixon, & Rucht,

2004). Communicators contribute to the definition and development of the issue. This social construction process holds significant implications for the perceived legitimacy of the issue. Legitimacy resources should be used to demonstrate that an issue is worthy of public concern. The incorporation of legitimacy resources aids the cause by demonstrating that the issue is not merely an obsession of the lunatic fringe. For this reason, careful attention should be devoted to the process through which legitimacy is socially constructed in cyberspace. Table 2 describes the two common strategies for developing legitimacy: *endorsement* and *self-evidence*.

Attempts to build issue legitimacy via the Internet will rely most heavily on issue Web sites and discussion groups (Coombs, 2002) (see Table 3). Given the increasing attention given to blogging, this method of communication also should be added. Blogs can contain personal stories or provide acceptable evidence to support a claim.

Issue Web sites can incorporate a variety of endorsement and self-evidence legitimacy resources to support the issue. The key is to attract other stakeholders who accept the legitimacy claims. The need to create legitimacy has an obvious, concomitant persuasive dimension. The hope is to attract and win over additional stakeholders who can benefit the cause and contribute to the contagion (Coombs, 2002). For example, the Web site may include links to news media stories related to the issue; links to scientific reports appearing in established research journals; outlines, time lines, or maps of the history of the development of the issue (e.g., a time line demonstrating destruction of rain forests or a time line of regulatory actions or litigation); testimony from scientists who are working on research related to the issue; comments from government officials in countries where the issue has gained prominence; and stories from individuals who have personally experienced adverse effects from the failure of corporations to address the issue (e.g., survivor stories).

High-quality Web sites are more likely to be seen as legitimate. Witmer (2000) con-

Table 2: Legitimacy Strategies

Legitimacy Strategies
Endorsement: a person who is perceived to possess legitimacy supports the issue. Endorsers' legitimacy may derive from their position (e.g., a member of a regulatory board), credibility (perceived trustworthiness and expertise; e.g., a noted biologist offers testimony), and/or charisma (they possess extraordinary characteristics that attract others; e.g., popular politicians and celebrities).
Self-evidence: draws upon features of the issue itself.
Tradition: uses precedents that demonstrate that things have been done this way in the past and there is no reason to change.
Rationality: provides logical reasoning and evidence to support the legitimacy of an issue.
Emotionality: creates strong feelings and reactions to effectuate persuasion.

Note. Based on information from "The failure of the task force on food assistance: A case study of the role of legitimacy in issue management," by W. T. Coombs, 1992, Journal of Public Relations Research, 4, pp. 101–122. Adapted with permission.

Table 3: Indicators of Attempts to Build Legitimacy

Legitimacy Markers

- Issue Web sites (quality of Web site, utilization of legitimacy resources)
- Discussion groups (quality of the post, utilization of legitimacy resources)
- Blogs (quality of blog writing, utilization of legitimacy resources)

tends that a quality Web site includes qualified creators, presents objective information, offers evidence to support claims, and is up-to-date. Although stakeholders who are already strongly committed to the issue may enjoy reading diatribes by other like-minded individuals, Web sites that are composed only of highly biased information or opinions may fail to attract and keep new stakeholders. Corporations can more easily ignore the rants of zealots than the well-reasoned, fact-based arguments of scientists or respected humanitarians.

Markers for the success of legitimacy reflect that others are accepting an issue and suggest an issue is building momentum. Markers of legitimacy success include (a) the number of different people posting or commenting on a blog, (b) the valence of the posts or comments (supports or refutes the issue), (c) the length and valence of threads (exchanges of messages between people), (d) the valence of crossover stories, and (e) TrackBacks to blogs. The number of different markers indicates growing interest rather than a limited debate between a few people. Negative valence indicates people are not seeing the issue as legitimate, while positive valence indicates acceptance of the issue's legitimacy. Crossover stories are when the news media report on the Web site. Media stories with a positive valence signal that the media are endorsing the issue's legitimacy (Coombs, 2002). TrackBacks indicate other people believe a blog is worthy of attention.

Conclusion

Internet contagions are important to resistors because they enable them to gain the power they need to influence corporate activities. As issue managers, resistors must be concerned with the channels and contents of Internet activity related to the issues. Corporations should actively monitor the Web-based activities of resistors in order to collect information about stakeholder concerns, identify potential contagions that could negatively impact their operations and reputations, and consider engaging resistors online.

For consideration

1. Which is more effective: resistors appealing to emotions or facts?
2. How have organizations targeted by resistors used new media to refute or to accommodate the challenges?
3. How should corporations respond to resistors when the resistors have no interest

in dialogue? How can concepts from stakeholder theory, network analysis, and reputation management inform that response decision?

4 How important is it for Internet activists to gain the attention of mainstream media?
5 Are there limits to the appeal of Internet activism and the creation of contagions? Would resistors experience the same degree of involvement and satisfaction with virtual and in-person activism?

For reading

Bennett, W. L. (2004). Communicating global activism: Strengths and vulnerabilities of networked politics. In W. van de Donk, B. D. Loader, P. G. Nixon, & D. Rucht (Eds.), *Cyberprotest: New media, citizens and social movements* (pp. 123–146). New York: Routledge.

Coombs, W. T. (2002). Assessing online issue threats: Issue contagions and their effect on issue prioritisation. *Journal of Public Affairs*, 2(4), 215–229.

Holtz, S. (1999). *Public relations on the Net: Winning strategies to inform and influence the media, the investment community, the government, the public, and more!* New York: AMACOM.

Middleberg, D. (2001). *Winning PR in the wired world: Powerful communications strategies for the noisy digital space*. New York: McGraw-Hill.

Mitchell, R. K., Agle, R. A., & Wood, S. J. (1997). Toward a theory of stakeholder identification and salience: Defining the principle of who and what really counts. *Academy of Management Review*, 22(4), 853–886.

References

Bennett, W. L. (2004). Communicating global activism: Strengths and vulnerabilities of networked politics. In W. van de Donk, B. D. Loader, P. G. Nixon, & D. Rucht (Eds.), *Cyberprotest: New media, citizens and social movements* (pp. 123–146). New York: Routledge.

Campaign to stop killer Coke. (2006). Retrieved April 14, 2006, from http://www.killercoke.org

Coombs, W. T. (1992). The failure of the task force on food assistance: A case study of the role of legitimacy in issue management. *Journal of Public Relations Research*, 4, 101–122.

Coombs, W. T. (1998). The Internet as potential equalizer: New leverage for confronting social irresponsibility. *Public Relations Review*, 24(3), 289–303.

Coombs, W. T. (2002). Assessing online issue threats: Issue contagions and their effect on issue prioritisation. *Journal of Public Affairs*, 2(4), 215–229.

Dowling, G. (2001). *Creating corporate reputations: Identity, image, and performance*. New York: Oxford University Press.

Fournier, S. (1998). Consumer resistance: Societal motivations, consumer manifestations, and implications in the marketing domain. *Advances in Consumer Research*, 25, 88–90.

Hallahan, K. (2001). The dynamics of issues activation and response: An issues process model. *Journal of Public Relations Research*, 13, 27–59.

Hearit, K. M. (1999). Newsgroups, activist publics, and corporate apologia: The case of Intel and its Pentium chip. *Public Relations Review*, 25(3), 291–308.

Holt, D. (2004). *How brands become icons: The principles of cultural branding*. Cambridge, MA: Harvard Business

School Press.

Holtz, S. (1999). *Public relations on the Net: Winning strategies to inform and influence the media, the investment community, the government, the public, and more!* New York: AMACOM.

Knoke, D. (1994). Networks of elite structure and decision making. In S. Wasserman & J. Galaskiewicz (Eds.), *Advances in social network analysis: Research in the social and behavioral sciences* (pp. 3–25). Thousand Oaks, CA: Sage.

Middleberg, D. (2001). *Winning PR in the wired world: Powerful communications strategies for the noisy digital space*. New York: McGraw-Hill.

Mitchell, R. K., Agle, R. A., & Wood, S. J. (1997). Toward a theory of stakeholder identification and salience: Defining the principle of who and what really counts. *Academy of Management Review, 22*(4), 853–886.

Rowley, T. J. (1997). Moving beyond dyadic ties: A network theory of stakeholder influence. *Academy of Management Review, 22*(4), 887–910.

Treadwell, D., & Treadwell, J. B. (2000). *Public relations writing: Principles in practice*. Boston: Allyn and Bacon.

van de Donk, W., Loader, B. D., Nixon, P. G., & Rucht, D. (2004). *Cyberprotest: New media, citizens and social movements*. New York: Routledge.

Wasserman, S., & Galaskiewicz, J. (1994). *Advances in social network analysis: Research in the social and behavioral sciences*. Thousand Oaks, CA: Sage.

Witmer, D. (2000). *Spinning the Web: A handbook for public relations on the Internet*. New York: Longman.

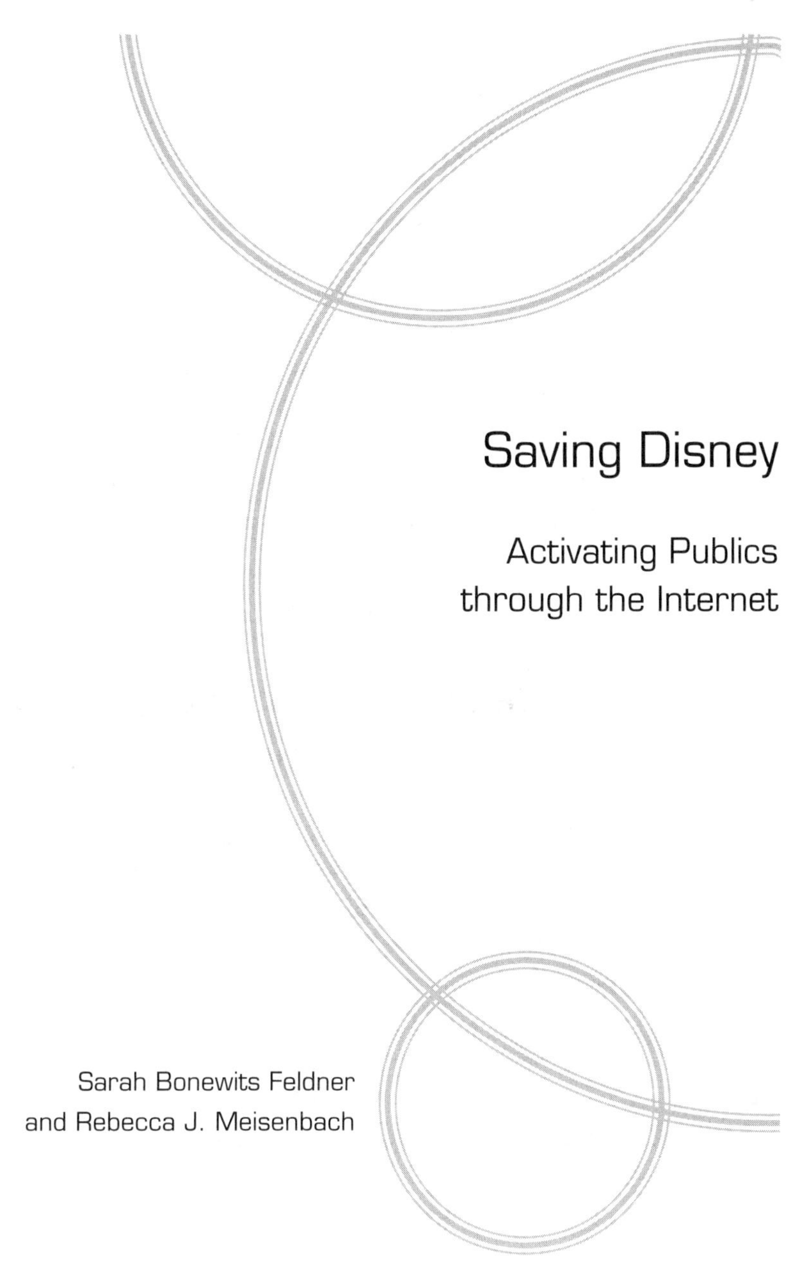

Saving Disney

Activating Publics through the Internet

Sarah Bonewits Feldner
and Rebecca J. Meisenbach

In December 2003, Roy Disney and Stanley Gold launched the SaveDisney.com Web site, marking the beginning of an innovative Internet-based public relations campaign. While public relations scholarship focuses on the need to listen and respond to publics, publics often do not have the means to participate actively in a meaningful way in this process. The emergence of the Internet has created an inexpensive and easily accessible forum for public organizing, which may, in turn, dramatically increase the presence, voice, and power of publics. This case represents the use of the Internet by dissatisfied shareholders to advocate for changes in corporate governance. The SaveDisney campaign demonstrates how the strategic use of technology, such as the Internet, changes the public relations landscape by shifting the balance of public relations practice in ways that give activist publics a significant and much needed voice.

The Walt Disney Company theme parks are known throughout the world as "the happiest places on Earth." Yet, in the boardroom, Disney is not unlike other large corporations.

Problems and disagreements arise. At the start of the twenty-first century, Roy Disney, the only remaining Disney family member working for the company, quit his job and helped launch a campaign challenging and attacking the legitimacy of Walt Disney Company policies and practices. He and another former board member used the Internet, particularly the World Wide Web, as the primary medium through which to voice their concerns and to activate shareholders, employees, and other stakeholders to demand change. Although many of the predictions about the Internet's potential to facilitate democratic and equal voice in the public sphere have not come to fruition, the SaveDisney campaign is a fascinating example of how Web sites can be used by organizations to accomplish change. As such, it behooves both corporate and activist publics and public relations experts to consider how change in the Walt Disney Company was spurred by the use of a Web site as a public relations medium.

The World Wide Web and the Changing Face of Public Relations

There are few aspects of organizational life that remain untouched by the virtual explosion of new technology and media. Public relations practice is no different, as scholars and practitioners alike have been grappling with the challenges and opportunities presented by new technologies for the past decade. Among the technologies that are changing current thinking about public relations, the Internet and World Wide Web are perhaps the most prevalent and fastest growing (Howard, Rainie, & Jones, 2001). Many have cited the role of the Internet in providing greater access to both obtaining and presenting information (e.g., Badaracco, 1998; Coombs, 1998; Springston, 2001).

In looking at the role of the Internet in reshaping public relations, there are two primary lines of thinking (Gregory, 2004). First, there are those who consider the Internet as an extension of current one-way public relations tactics (e.g., e-mail is the new form of the memo). However, a second perspective suggests that this technology provides for two-way communication or a more dialogic process (Gregory, 2004; Springston, 2001; Taylor, Kent, & White, 2001).

Intuitively, the notion that the Internet allows for a dialogic process makes a great deal of sense. Both in providing greater access to information in a usable format and in providing avenues for feedback through chat rooms, discussion boards, and e-mail responses, the Internet can be the key to realizing the dialogic or relational model of public relations that has been the focus of a great deal of commentary and discussion among public relations scholars (e.g., Botan, 1997; Kent & Taylor, 2004). Yet, despite this theorized potential, surveys of Web sites over the past several years demonstrate that few organizations have taken full advantage of the Internet's dialogic potential (Kang & Norton, 2004; Kent, Taylor, & White, 2003; Taylor, Kent, & White, 2001).

To date, most studies exploring the dialogic potential of the Internet have focused on activist and nonprofit organizations. Having an Internet presence gives activist organizations a status that is commensurate with corporate organizations (Kent, Taylor, & White, 2003). Despite this potential, most activist and nonprofit organizations focus simply on the Web content (i.e., sharing information via the Web) versus having a Web presence (i.e., using the technology for relationship building) (Curtin & Gaither, 2004). This failure on the part of activist organizations affirms Kent, Taylor, and White's (2003) contention that most activist organizations, although radically different in functioning than for-profit organizations, often use the same public relations tactics.

In a discussion of adapting public relations theory for activist groups, J. Grunig (2001) suggested that interpersonal communication would be the most successful method for organizing unempowered publics, since these publics "seldom pay attention to mass-mediated messages" (p. 19). However, we argue that this contention should be revisited in light of the opportunity presented by mediated messages disseminated via the Web that adopt a relational perspective. Taylor, Kent, and White (2001) suggested that public relations dialogue does not have to be that different from interpersonal dialogue. Both can provide for attraction, interactivity, trust, maintenance, and cost and rewards, which are factors associated with any relationship. Taking these two arguments together solidifies our belief that effective Internet use provides the perfect opportunity for activist organizations to build relationships with their publics. As few, if any, exemplars of relational use of the Internet are available, we are left to question what a public relations campaign that integrates the opportunity provided by Internet use—while embracing the relational perspective—would look like. The SaveDisney campaign offers an enticing intersection of technology, activism, and dialogue, providing a partial answer to these questions. Thus, we examine the SaveDisney campaign in an effort to determine in what ways the campaign managers and involved stakeholders took advantage of the dialogic potential afforded by the Internet.

The Online Campaign to Save Disney

On November 30, 2003, Roy E. Disney, Walt Disney's nephew, resigned from his dual posts as chair of the feature animation division and as vice-chair of the board of directors at the Walt Disney Company. The following day, Stanley P. Gold, who had served on the board for over fifteen years, announced his resignation from the Walt Disney Company Board of Directors. Their resignation letters called for the resignation of Michael Eisner, then chief executive officer (CEO) and chair of the board of directors. The news of Disney's resignation spread quickly, and, within a month, *The Wall Street Journal* had a special feature available on its Web site that chronicled the turmoil at Disney ("Battle for Disney," 2004). However, media coverage of events at the company was not the only version of the story that was being told. R. Disney and Gold turned to the Internet to voice their side of the story and to encourage others to join them.

Creating a voice, starting a campaign. Well-designed public relations campaigns begin with careful research about the organization, the opportunity at hand, and the relevant publics. Traditionally, the practitioner does research on his or her own organization or client. But, in this case, the activist nature of SaveDisney led to an unconventional process of defining the issues and identifying the organizational target. Even though the campaign began with only two people with a cause, according to L. Grunig (1992), this is all that is required to be considered an activist organization. She defined activist organizations as "two or more individuals who organize in order to influence another public or publics through action that may include education, compromise, persuasion tactics, or force" (p. 504).

Organization. Whereas SaveDisney was the organization that launched the campaign, the major organizational player under scrutiny was the Walt Disney Company. From their various posts within the company and board positions, R. Disney and Gold were privy to the day-to-day practices, finances, and overall strategies guiding the Walt Disney Company and felt uniquely qualified to judge what was in the company's best interests. Their resignation letters outlined the specifics of what they saw as "failed initiatives," "flawed plans," and "unfulfilled promises" (SaveDisney.com, n.d., letters/spg_resign_letter, para. 4; accessed February 4, 2004). They saw problems with "Disney's poor financial performance, a loss of creative leadership, and board accountability" (Stewart, 2005, p. 496).

Problems/opportunities. R. Disney and Gold argued that these problems were all tied to Eisner, suggesting that Eisner's leadership was weakening the company. Therefore, they focused on Eisner's control of the company and board as the major problem. In their letters of resignation, both R. Disney and Gold claimed that they had advocated for change for years, both in meetings and in letters, to no avail. They resigned from the board in the hope that they might effect change from the outside. The yearly shareholder meeting provided one of the only opportunities available to those outside the boardroom to voice concerns, and they felt they could use the meeting to generate a significant vote of no confidence in Eisner and other board members. To accomplish their goal, they needed to reach the individuals who owned shares either directly or through their mutual funds, which led to consideration of publics.

Publics. In terms of research on publics, R. Disney and Gold were aware of employee unrest and dissatisfaction. They counted themselves as a public audience of the Walt Disney Company and wanted to find others who shared their vision. Thus, their initial Web site,[1] which appeared originally in December of 2003, threw out a wide net, appealing to anyone who identified himself or herself as "concerned about the welfare of The Walt Disney Company and its future direction" (SaveDisney.com, n.d., main section, para. 1; accessed February 4, 2004). As the Web site developed, and they collected more information about who was accessing the site, they began targeting their audiences. In this case, the shareholder meeting meant that they gave high priority to Walt Disney Company shareholders as an audience and also to the proxy advisory services that controlled a majority of the stock votes. However, their campaign did not focus exclusively on the shareholders. The campaign also targeted employees, known as cast members, and Disney consumers.

What will save Disney? Within a week of their resignations, R. Disney and Gold started their SaveDisney campaign, launching the Web site that initially outlined one specific goal. The fine print at the bottom of the December 2003 version of their Web page indicated that "This website has been established to provide a forum for discussing, analyzing and critiquing the performance, direction and management of The Walt Disney Company" (SaveDisney.com, n.d., footnote; accessed February 4, 2004). However, the creators and managers of the campaign had a larger goal than just creating a quality Web site: They wanted "to restore Disney to its position as the preeminent entertainment company in the world." For R. Disney and Gold, the return to greatness for the Walt Disney Company was going to be defined, in part, by the removal of Eisner from his posts as chair of the board and CEO. As stated on the Web page in a posting dated January 27, 2004, the campaign wanted shareholders to vote "NO on the re-election of Michael Eisner, George Mitchell, Judith Estrin, and John Bryson as directors," because "they symbolize, respectively, the poor management, poor governance, poor compensation practices, and lack of board independence that are impeding the development of long-term shareholder value at The Walt Disney Company" (SaveDisney.com, n.d., Letters, Just Say No section, para. 5; accessed July 21, 2005). The concrete objective was to generate a critical number (10% to 20%) of no votes for Eisner in order to spark the board to make changes (Stewart, 2005).

Making it happen. R. Disney and Gold used both Internet-based and in-person strategies to achieve their objectives. The primary vehicle for communicating with interested stakeholders was through the SaveDisney.com Web site. The innovative Web site was modeled on 2004 presidential-hopeful Howard Dean's Internet campaign for the Democratic nomination (Stewart, 2005).

The first modest version of the Web page contained only text with a short message from R. Disney. They introduced the site as being "devoted to those concerned about the welfare of The Walt Disney Company and its future direction" (*SaveDisney.com*, n.d.). Along with this introduction, site visitors could find links to the resignation letters and contact information for R. Disney, Gold, and members of the Walt Disney Board of Directors, but little else. This first version was framed primarily as shareholders communicating with shareholders.

In January 2004, R. Disney and Gold launched a more stylized and strategic version of the Web site that included graphics mimicking those used by the Disney company Web site. The contrast between these two versions was striking, as the updated version included graphics, color, streaming audio and video, and multiple links. The site also included poll questions, an invitation to join the SaveDisney.com mailing list, a statement from R. Disney, and postings of news and commentaries relating to the company from around the world. The updated page had distinct sections that addressed families, consumers, and employees who associated themselves with Disney.

R. Disney and Gold attracted people to the Web site largely by word of mouth. In the weeks following their resignations, they received thousands of e-mails supporting their efforts (Holson, 2003). Visitors were invited to "Join the Fight" and had the option of joining an e-mail list that would keep them updated throughout the campaign. Registered visitors received a free SaveDisney bumper sticker and regular graphics-enhanced e-mail announcements about the progress of the campaign. E-mails contained a message from R. Disney, along with links to SaveDisney.com pages and relevant news stories and announcements.

The e-mail updates sent by the SaveDisney campaign managers included specific requests to inform others about the campaign. A March 26, 2004, e-mail[2] stated:

> We are asking for your help in this effort by contacting your friends and family and getting them to join the SaveDisney Nation. The link below will allow you to forward a message to 20 of your contacts at a time and invite them to become involved in this historic grassroots campaign to preserve Disney for future generations. (personal communication, para. 3)

The e-mails consistently directed recipients back to the Web site to "check out what has been happening at SaveDisney." As of January 2004, 15,607 e-mails were sent to supporters by the SaveDisney campaign as a result of this opt-in process. These e-mails resulted in 5,435 forwards, and 3,865 of those forwards were opened ("SaveDisney Rewrites the Rules," 2004). The Web site had four thousand registered activists in mid-January, and that number grew to 35,000 by March 5, 2004 (Magill, 2004).

Although the Web page was the center of their efforts, SaveDisney did use other, more traditional, strategies. Throughout February 2004, R. Disney and Gold promoted the Web site and their cause in interviews and on television (Stewart, 2005). Specifically, they made presentations to several of the larger proxy firms to convince them to support SaveDisney's objectives. For example, on February 2, 2004, they visited the Institutional Shareholder Services (ISS) to argue their case. Just over a week later, ISS recommended that shareholders withhold on their vote for Eisner while still supporting the other members of the board. Such visits and their successes were promoted in press releases on the Web site and were highlighted on the SaveDisney time line.

R. Disney and Gold also used the Web site and the e-mails to promote their own rally, held the day before the annual shareholder's meeting. The SaveDisney rally in Philadelphia was open to anyone who saw herself or himself as a Disney supporter. The campaign even helped people get to Philadelphia, securing and providing links to negotiated airline discounts. More than eight hundred people attended the SaveDisney rally in Philadelphia.

Have we saved Disney? The next day at the shareholder meeting, the campaign appeared to be successful. R. Disney was allowed to speak and received a standing ovation from the three thousand shareholders in attendance (Orwall, Steinberg, & Lublin, 2004). According to final official numbers released in April 2004, Michael Eisner received a no-confidence vote from 45.37% of shareholders, and board member George Mitchell received a 25.69% no-confidence vote ("Walt Disney Co.," 2004). Clearly, the campaign had met and exceeded its goals of generating up to 20% no-confidence votes for Eisner. Furthermore, the Disney board members met immediately after the shareholder meeting and decided that George Mitchell would replace Eisner as chair of the board and that the roles of CEO and chair of the board would be separated.

Being a good steward: SaveDisney round two. Public relations campaigns do not start and end in a vacuum; rather, they stem from and contribute to ongoing public dialogue.

Kelly (2001) argued that using the evaluation of one campaign to determine the direction of future campaigns and analyzing how well participants in the campaign are thanked for their participation and contributions are essential to understanding how public relations engages in long-term relationship maintenance. Indeed, the SaveDisney campaign not only spent time thanking those who voted and participated, but also used its first success to advocate for further change. The Web site soon posted a note:

> *Welcome to Round Two in the fight for Disney!* Stanley and I are grateful to all of you for your support and encouragement in what we are now calling 'Round One' of the battle to remove Michael Eisner from his position. . . .
>
> Your votes—each and every one of them—were vitally important to the effort, and now we want to encourage you to STAY WITH US at SaveDisney.com in the coming weeks and months. We promise you that we will continue the fight . . . Thanks for your vote, and let's stick together to bring back the magic! (Letter from Roy Disney section, para. 1 and 2 accessed March 31, 2004)

Thus, while the SaveDisney campaign had clearly impacted the shareholder meeting, Gold and R. Disney declared their intent to continue. After congratulating readers for their efforts, R. Disney suggested that the board's response of promoting Mitchell, who also received a large vote of no confidence, was outrageous and sent a signal that further change was needed at Disney. The Web site began to focus not only on the platform of corporate governance, but also on nostalgic reminiscences about the glory days of the Walt Disney Company and current problems in animation and the parks. In September 2004, when Eisner announced his resignation as CEO of the Walt Disney Company effective September 2006, SaveDisney welcomed the news. However, R. Disney and Gold responded that two more years with Eisner was not acceptable and called for his immediate replacement.

In March 2005, the announcement that Robert Iger would succeed Eisner as CEO was not well received by the managers of the SaveDisney campaign. The SaveDisney.com postings began to actively address the selection process. Specifically, R. Disney and Gold questioned whether the board of directors had conducted a thorough and open search. In a March letter, R. Disney and Gold questioned the presence of Eisner at all interviews and suggested that the actions had "eroded whatever faith" (*SaveDisney.com*, n.d., Open Letter to the Walt Disney Company Board, March 10, 2005, para. 2; accessed July 21, 2005) they had in the search process.

Finally, after several meetings with Iger in July 2005, the Walt Disney Company, R. Disney, and Gold issued a joint statement that they had come to a resolution and the SaveDisney campaign would come to an end (Gentile, 2005). The last few updates to the SaveDisney.com Web site cited positive changes at Disney. In addition, R. Disney was given an emeritus director position on the board and was re-hired as a consultant. With a statement thanking the SaveDisney supporters for their faith, trust, and support, R. Disney announced that the Web site would be taken down as "we here at SaveDisney have come to a mutual agreement with the new management at the Walt Disney Company regarding our mutual relationship" (*SaveDisney.com*, n.d., Joint Statement section, para. 2, accessed July 21, 2005). In August 2005, the Web site was taken down, effectively ending the historic shareholder revolt known as SaveDisney.

Discussion

In many ways, the case of the SaveDisney campaign follows a very familiar Disney storyline: Good (SaveDisney) conquers evil (Eisner and board). Beyond being simply a good story where good presumably triumphs over evil, this case stands apart due to its innovative route to success. A consideration of the factors that made the SaveDisney campaign successful suggests several insights for scholars and practitioners of public relations.

The case exemplifies the ways in which the use of the Internet has and can alter public relations practice. The most commonly cited impact of Internet use in public relations is the increased ability to share and retrieve information. Part of the success of SaveDisney was its focus on those shareholders who had small numbers of shares. The use of technology allowed R. Disney and Gold to keep all individuals, regardless of how minor, involved.

The Web site's ability to target shareholders both large and small allowed the campaign managers to create what Cozier and Witmer (2001) called an "online social organization" (p. 618). The Internet, with its interactivity features (e.g., discussion posting options and polls) brings previously isolated individuals into contact with one another, strengthening movements. In short, the Internet helps to foster the development of networks by adding to the density and centrality of stakeholder networks (Coombs, 1998). The SaveDisney.com Web site was a virtual place in which various stakeholders were able to come into contact with one another and share their stories. By bringing these voices together and providing avenues for them to be heard, the site allowed them to become part of the SaveDisney campaign. Indeed, we, the authors, felt connected to the campaign as we followed its development. Thus, we analyzed this case both as scholars who stand outside the campaign and as active participants. As such, we believe that the greatest contributor to the success of the campaign was the ability of the campaign's creators to use technology to create dialogue with and generate participation of stakeholders.

Kent and Taylor (1998) identified five characteristics that they argued were essential for a dialogic Web site: (a) the provision for a dialogic loop to allow the free flow of communication in both directions; (b) the inclusion of useful information for all target publics; (c) use of chat rooms, Q&A, and other provisions to keep visitors returning; (d) ease of use; and (e) including only essential links to keep visitors on the site. Our assessment suggests that the SaveDisney.com Web site met all of these criteria at a basic level and exceeded these standards in many ways. In terms of creating a dialogic loop, the Web site featured frequent polls throughout the campaign and encouraged visitors to submit Disney-related news items and personal memories. At its height, those visiting the site could peruse literally hundreds of letters and guest editorials from interested stakeholders. One former cast member letter read as follows (SaveDisney.com, n.d.):

> Roy, I confess after reading that SaveDisney would propose no alternate slate, I figured that meant you had backed down. So, thanks for the Walt (email) quote (which I always read) . . . it made me click on the link to the site and read your message . . . I am a former Cast Member whose lifetime dream was to be part of the Disney vision. . . . . Roy, keep up the fight! Thanks for all you're doing. (Cast Member Letters section, para. 1 and 4, accessed July 21, 2005)

All these features contributed to a dialogue between the campaign managers and its faithful followers.

Other notable Web site features included specific sections targeted at various stakeholder groups. The clean organization of the site allowed visitors to follow tabs to particular sections on the page, thereby fulfilling the ease-of-use criteria. The SaveDisney campaign encouraged return visits both by including weekly postings from R. Disney and through e-mail updates that informed readers of what was happening in the campaign and what they might find on the Web site. The cast member letter previously quoted suggests the success of this strategy, as the writer noted that she visited the site by following the link embedded in the e-mail update. Finally, visitors were encouraged to stay on the site by the inclusion of all pertinent articles directly on the SaveDisney Web site, rather than links to external Web sites. Thus, SaveDisney.com seems to fill a void by providing a model that fulfills the criteria established by Kent and Taylor (1998).

Kent and Taylor's (1998) scheme is an excellent starting point for capitalizing on the dialogic potential of the Internet. However, we argue that there is more to fostering a relationship with stakeholders than attention to the mechanics of a Web site. In particular, the SaveDisney campaign's success seems linked not only to the masterful use of the technology, but also to an effective use of rhetorical strategies for fostering identification with the campaign. Public relations research benefits from adopting a rhetorical perspective (Heath, 2001; Meisenbach & McMillan, 2006).

The postings and weekly updates were written in a personally engaging style that fostered familiarity. Every letter from R. Disney was written in the first person. Many postings began with salutations such as "dear cast members," and "to our faithful supporters." In addition to the familiar language, the postings directly referred to e-mails and letters received from SaveDisney followers. In this way, R. Disney assured supporters that he was hearing them and seeking to respond. The weekly cast member outreach postings serve as a particularly interesting example of this strategy. In a posting from November 2004 (SaveDisney.com, n.d.), the SaveDisney team wrote that "you have told us your concerns so that we may act upon them. You have given us suggestions. . . . and we have catalogued and archived your suggestions" to assist the campaign "in bringing those concerns and suggestions to the forefront of public awareness" (Cast Member Outreach section, para. 2, accessed July 21, 2005). These postings suggest that the managers of the SaveDisney campaign did, in fact, have an interpersonal relationship with every person who visited the site. The many posted responses from SaveDisney visitors suggest that the reverse was also true.

The rhetorical concept of organizational identification also provides some insight into what made the SaveDisney campaign so successful. An individual is identified with an organization when his or her values overlap with the values of the organization (Cheney, 1983). In part, the SaveDisney campaign fostered organizational identification by emphasizing the shared experience or magic that visitors associated with the SaveDisney campaign, contrasting the SaveDisney version of Disney magic with Eisner's corporate greed and creating a unique logo and slogan, Restore the Magic. Taken together, the SaveDisney supporters showed themselves to be highly identified with the SaveDisney campaign. This identification was a key part of the dialogism associated with this public relations campaign and, ultimately, was a large contributor to the success of the campaign.

In using the Internet to create dialogue with shareholders, the SaveDisney campaign also represents a new possibility for incorporating shareholder and public voices in corporate decision-making processes. The unique strategy employed by the SaveDisney campaign and its explosive impacts on the company garnered a great deal of media attention. This case represents a model for other shareholder groups who wish to effect large-scale change (Jones, 2004). Jones pointed out that the success of the campaign was rooted in the strategy of reaching out to small investors and in providing specific explanations on how the voting process worked, even to those who owned shares through larger investment funds. Jones also noted that the SaveDisney campaign was effective in swaying public opinion beyond that of the shareholders. The import of the larger public voice resides in the fact that fund managers are in tune with broader public trends.

The concept of public relations invokes an image in which the public voice occupies a central position in the discussion. Yet, countless examples suggest that the public voice is either absent or limited. With increasing concern about corporate governance and executive accountability, shareholders and publics are seeking the means by which they might impact decisions made by corporate boards. Campaigns, such as SaveDisney, that utilize in savvy ways the capabilities offered by the Internet provide an avenue through which shareholders might be heard in ways never before imagined.

Implications and Conclusion

In terms of developing scholarly understanding of the changing role of public relations in the age of new media, the SaveDisney case demonstrates that many of the theorized impacts of Internet use are being evidenced in large-scale campaigns. Furthermore, this case suggests that it is not simply a matter of focusing on the technical aspects of Web page design (e.g., page load time and accessibility of links) but that the specific message strategies can contribute to the development of a strongly identified and powerful online community.

While much of the discussion amongst public relations scholars is turning toward the dialogic, few examples exist of what this might look like and how it might be effective. This case provides one example of how the dialogic potential of the Internet might be harnessed. The implications for public relations practitioners are twofold. First, practitioners cannot afford to ignore movements that begin on the Internet. This is one of the major flaws in the Walt Disney Company response to the SaveDisney campaign. Few, if any, direct responses to the Web site were offered. In short, many believe that company officials underestimated R. Disney and the impact that his Web-based campaign would have. Second, practitioners should look at this case to glean strategies for engaging their publics in dialogue. SaveDisney offers strategies that can work across organizations. Although some argue that companies do not need to engage in dialogic tactics to the extent that activist organizations do, the fact remains that as more people have access to the Internet and recognize the power and voice it provides, this type of campaign will continue. Particularly in an era of corporate scandals and greater shareholder scrutiny, public relations practitioners must recognize that the Internet changes the dynamics of corporate board and investor relationships. The Internet provides a means for shareholders to send the same message that R. Disney and S. Gold often reminded visitors to the SaveDisney Web site: "We'll be watching."

For consideration

1. In what ways do new media technologies help organizations to adopt a relational or dialogic approach to their publics? In what ways do they hamper those efforts?
2. The Walt Disney Company and its board of directors chose to largely ignore the SaveDisney Web site. Evaluate the merits of this choice.
3. Public relations practitioners are faced with limited resources regardless of the type of organization they work with. Under what circumstances should a Web-based campaign be ignored?
4. Many in the press cite this case as the new model for shareholders to voice their opinions. To what extent is that true?
5. The authors suggest that the rhetorical strategies and tone employed by the campaign contributed to the campaign's success. Which approach is more effective for activist groups: to mimic the tone of the target organization or to communicate counter to it?

For reading

Aftab, P. (2005). The PR professional's role in handling cyberwarfare. *Public Relations Strategist, 11*(3), 28–30.

Cheney, G. (1983). The rhetoric of identification and the study of organizational communication. *Quarterly Journal of Speech, 69*, 143–158.

Jones, D. (2004, March 12). Web-based campaigns a wake-up call for corporations. *IR Web Report*. Retrieved February 2, 2006, from http://www.irwebreport.com/features/features_print/040202.htm

Kent, M.L., & Taylor, M. (1998). Building dialogic relationships through the World Wide Web. *Public Relations Review, 24*(3), 321–334.

Stewart, J.B. (2005). *Disney war*. New York: Simon & Schuster.

Notes

1. There were several versions of the SaveDisney Web site that existed from its December 2003 launch until the page was taken down in August 2005. The content of the pages changed constantly as updates were posted on a regular basis. Most of the content was archived while the main site was active. However, the page is no longer active. Many of the pages can still be accessed via the Wayback Machine (available: http://www.archive.org/web/web.php). The authors visited the SaveDisney Web site frequently throughout the active campaign. In this chapter, the authors have noted the date that they accessed and archived the pages for all references to the Web page.
2. This e-mail was a part of a series of e-mails that was sent to individuals who signed up to be on the SaveDisney electronic mailing list. Both authors received copies of these e-mails.

References

Badaracco, C.H. (1998). The transparent corporation and organized community. *Public Relations Review, 24*(3), 265–272.

The battle for Disney. (2004, August 19). *The Wall Street Journal* (online version). Retrieved August 9, 2006, from http://online.wsj.com/page/2_1070.html

Botan, C. (1997). Ethics in strategic communication campaigns. *Journal of Business Communication, 34*, 188–202.

Cheney, G. (1983). The rhetoric of identification and the study of organizational communication. *Quarterly Journal of Speech, 69*, 143–158.

Coombs, W.T. (1998). The Internet as potential equalizer: New leverage for confronting social irresponsibility. *Public Relations Review, 24*(3), 289–303.

Cozier, Z.R., & Witmer, D.F. (2001). Development of a structuration analysis of new publics in an electronic environment. In R.L. Heath (Ed.), *Handbook of public relations* (pp. 615–624). Thousand Oaks, CA: Sage.

Curtin, P.A., & Gaither, T. K. (2004). International agenda-building in cyberspace: A study of Middle East government English-language websites. *Public Relations Review, 30*(1), 25–36.

Gentile, G. (2005, July 9). Roy Disney, company resolve their disputes. *Washington Post*, p. D1.

Gregory, A. (2004). Scope and structure of public relations: A technology driven view. *Public Relations Review, 30*(3), 245–254.

Grunig, J. (2001). Two-way symmetrical public relations: Past, present, and future. In R.L. Heath (Ed.), *Handbook of public relations* (pp. 11–30). Thousand Oaks, CA: Sage.

Grunig, L. (1992). Activism: How it limits the effectiveness of organizations and how excellent public relations departments respond. In J. E. Grunig (Ed.), *Excellence in public relations and communication management* (pp. 503–530). Hillsdale, NJ: Lawrence Erlbaum.

Heath, R.L. (2001). A rhetorical enactment rationale for public relations: The good organization communicating well. In R.L. Heath (Ed.), *Handbook of public relations* (pp. 31–51). Thousand Oaks, CA: Sage.

Holson, L.M. (2003, December 15). Roy Disney goes online in battle to oust Eisner. *The New York Times*, p. C9.

Howard, P.E.N., Rainie, L., & Jones, S. (2001). Days and nights on the Internet: The impact of a diffusing technology. *American Behavioral Scientist, 45*, 383–404.

Jones, D. (2004, March 12). Web-based campaigns a wake-up call for corporations. *IR Web Report*. Retrieved February 2, 2006, from http://www.irwebreport.com/features/features_print/040202.htm

Kang, S., & Norton, H.E. (2004). Nonprofit organizations' use of the World Wide Web: Are they sufficiently fulfilling organizational goals? *Public Relations Review, 30*(3), 279–284.

Kelly, K. (2001). Stewardship: The fifth step in the public relations process. In R.L. Heath (Ed.), *Handbook of Public Relations* (pp. 279–290). Thousand Oaks, CA: Sage.

Kent, M.L., & Taylor, M. (1998). Building dialogic relationships through the World Wide Web. *Public Relations Review, 24*(3), 321–334.

Kent, M.L., & Taylor, M. (2004). Toward a dialogic theory of public relations. *Public Relations Review, 28*(1), 21–37.

Kent, M.L., Taylor, M., & White, W.J. (2003). The relationship between Web site design and organizational responsiveness to stakeholders. *Public Relations Review, 29*(1), 63–77.

Magill, K. (2004, March 5). Power of Disney's deaniacs grows. *New York Sun*, p. 11.

Meisenbach, R. J., & McMillan, J. J. (2006). Blurring the boundaries: Historical developments and future directions in organizational rhetoric. In C. Beck (Ed.), *Communication Yearbook 30* (pp. 99–104). Mahwah, NJ: Lawrence Erlbaum Associates.

Orwall, B., Steinberg, B., & Lublin, J. S. (2004, March 4). Disney's Eisner steps down from chairman post after protest garners 43% of voted shares. *The Wall Street Journal*, p. A1.

SaveDisney.com. (n.d.). Retrieved February 4, 2004; March 31, 2004; and July 21, 2005, from http://web.archive.org/web/20031210060511/http:/savedisney.com/

SaveDisney rewrites the rules in shareholder activism with Kintera technology. (2004, February 6). *Business Wire*, p. 1.

Springston, J.K. (2001). Public relations and new media technology. In R.L. Heath (Ed.), *Handbook of public relations* (pp. 603–614). Thousand Oaks, CA: Sage.

Stewart, J.B. (2005). *Disney war*. New York: Simon & Schuster.

Taylor, M., Kent, M.L., & White, W.J. (2001). How activist organizations are using the Internet to build relationships. *Public Relations Review, 27*(3), 263–284.

Walt Disney Co.: Withheld support of CEO was stronger than announced. (2004, April 7). *The Wall Street Journal*, p. A7.

The Internet as a Site of Resistance

The Case of the Narmada Bachao Andolan

Mohan J. Dutta and Mahuya Pal

To understand the role of the Internet in activism, this chapter examines Friends of River Narmada (http://www.narmada.org), a Web site dedicated to the cause of Narmada Bachao Andolan (NBA, or Save the Narmada Movement), the struggle against construction of mega-dams on the River Narmada in India. NBA has become symbolic of a global struggle for environmental and social justice. The decade-long protest against the Sardar Sarovar reservoir has become an exemplar of people's movements. A thematic analysis of the Web site based on a grounded theory approach provides an understanding of the discursive space created online by the activist group and the resistance strategies used by activist groups.

The increasing e-presence of networks of activists demanding greater voice in the global world raises critical questions about the power of the Internet in organizing effective political action across national boundaries. Existing research has drawn from case studies to investigate the role of modern computer communication in the context of nonviolent grassroots movements and in the balance of power between citizens, elected officials, and local, national, and international power structures (Danitz & Strobel, 1999; Postmes & Brunsting, 2002). The protests during the November 1999 World Trade Organization (WTO) meeting in Seattle symbolized this trend in global activism spearheaded by the Internet.

Taylor, Kent, and White (2001) regard the Internet as a potential equalizer for activist organizations, because it offers a low cost, controllable communication channel that can create linkages with like-minded stakeholders and bring members of spatially separated activist groups together. Moreover, as the emergence of the global activist public (with diverse identifications, causes, associations, and locations) creates challenges for communication and organization, it becomes necessary to explore the potential of the Internet in garnering a solidarity network for activist groups and in globalizing protest. Of particular interest is the discursive process through which the Internet is mobilized for resisting global structures of power through collective action.

To understand the role of the Internet in activism, this chapter examines http://www.narmada.org (*Friends of River Narmada*, n.d.), a Web site dedicated to the cause of Narmada Bachao Andolan (NBA, or Save the Narmada Movement), the struggle against construction of mega-dams on the River Narmada in India. NBA has become symbolic of a global struggle for environmental and social justice. The decade-long protest against the Sardar Sarovar reservoir has become an exemplar of people's movements. The proposed hydroelectric projects on the Narmada River in western India are hotly contested. NBA has grounded its struggle against the dams in the villages along the Narmada valley, mobilizing indigenous people to resist displacement (Routledge, 2000).

NBA has launched its struggle at several rungs and taken it to nonlocal terrains. For over a decade, the NBA-spearheaded opposition has been successful in mobilizing tribal people in the region, as well as worldwide support (Sen, 1997). It has been immensely successful in developing links with groups outside India, such as Narmada UK, which recently climbed the Millennium Wheel in protest against the Narmada dams, and the International Narmada Campaign, which constitutes interest groups and nongovernmental organizations (NGOs) and lobbies against the World Bank's financial support for the largest Narmada dam, Sardar Sarovar (Udall, 1997). It is important to note that, while the NBA is primarily an empowerment experience encompassing grassroots-level action, the Web site devoted to NBA's cause complements its struggle in generating collective action through the Internet. Though the exact contribution of the Web site is yet to be realized, quantified, or discussed in academic literature or mainstream media circles or by NBA activists, there is no doubting the significant role the Web site has played in contributing to the broader struggle against the construction of the dams on the Narmada.

NBA as an important social movement has attracted substantial academic attention that has primarily focused on its role in empowering indigenous people and its fight for a society based on equity. However, as we have stated, the potential of the Web site Friends of River Narmada (http://www.narmada.org) in contributing to NBA's water wars is yet to be studied adequately. This chapter looks at the possibilities through which the "otherwise socially isolating computer" (Postmes & Brunsting, 2002, p. 290) can contribute to resistance to dominant social structures. A thematic analysis of the Web site based on a grounded theory approach provides an understanding of the discursive space created online by the activist group and the resistance strategies used by the activist group. The chapter begins with a discussion of the growing importance of the Internet in activism and the role of the Internet in fostering resistance, followed by a thematic analysis of the Web site. Specifically, this chapter demonstrates an understanding of the ways in which the Internet is mobilized for resistance.

Activism and the Internet

The Internet as a public sphere has the potential to contribute to the growth of activist networks as well as to mobilize and organize protests on a global scale. The protests during the 1999 WTO meeting in Seattle, where forty thousand people demonstrated for four days, has been recognized as a turning point in activism. It symbolized the upward trend in social actions that had declined in the preceding decade. Routledge describes the phenomenon in terms of the idea of "convergence space" (2000, p. 25), which serves as a common platform for diverse formations. It is a space that provides a common vision for social movements, grassroots initiatives, NGOs, and other formations, "wherein interests, goals, tactics, and strategies converge" (p. 25). Routledge suggests that, along with the material space facilitated by conferences and events, the virtual space, mediated by the Internet, mobilizes protests and actions as well by means of negotiation of different strategies and resistances.

Postmes and Brunsting (2002) claim that the Internet contributes to collective action by opening up new avenues or reinforcing existing forms. Recent literature has documented some of the social struggles stimulated by the Internet. For instance, the movement of the Zapatista freedom fighters in Mexico was a movement that depended heavily on the Internet to sustain it; news providers in Serbia relied on the Internet when faced with censorship on traditional channels of mass communication; the Peoples' Global Action (PGA), a global resistance formation, is forging alliances and action by sharing information on the Internet; and the wave of antiglobalization protests on numerous Web sites provide information about the World Bank, the International Monetary Fund (IMF), the WTO, and a range of other organizations (Postmes & Brunsting, 2002; Routledge, 2000). However, the most compelling case is perhaps the role of the Internet in the pro-democracy case of Burma (Danitz & Strobel, 1999). The Internet became the platform for transmitting information and communicating to members of the Burmese diaspora, scattered around the world, about the human rights abuses in their home country. Organizing and exchanging information on the Internet, activists working on pro-democracy issues in Burma interconnected across nations and strengthened their networks, which successfully culminated in the federal legislative decision to ban new investments by U.S. companies in Burma. The purpose was to drain the military regime in Burma of cash flows that foreign investments would bring in to the regime. Danitz and Strobel note that, without the Internet, it would have been almost impossible for activists to mobilize action and coordinate members dispersed in the

United States and around the world.

Hence, the power of the Internet in facilitating activism is evident. Taylor, Kent, and White (2001) argue that activist organizations need to be explored in public relations because they have unique communication and relationship-building needs. Many activist organizations, operating on minimal budgets, have traditionally relied on public relations as a cost-effective way to reach publics. Consistent with this view, Taylor et al., in particular, emphasize the dialogic potential of the Internet that can be leveraged by activist organizations. In the same vein, Coombs identifies the Internet as a potential equalizer for activist organizations because it offers a "low cost, direct, controllable communication channel" (1998, p. 299). It is important for public relations researchers in the domain of activist organizations to recognize that resistance plays a core function in organizing and coordinating publics across diverse spaces that globalize political and social struggles.

Resistance and the Internet

Castells (1997) emphasizes that the emergence of resistance identities in opposition to economic globalization and the disruption caused by global cultural flows is an inevitable phenomenon of the information age. Activists are generating new networking forms and practices that allow for global resistance. This resistance incorporates diverse models and facilitates an alternative democratic and globally networked society (Juris, 2004). Juris suggests three specific features of what he calls "global justice movements" (p. 345). First, global movements are "global" (p. 345). As activists think of belonging to global struggles, they discursively link their local protests and activities to diverse struggles elsewhere. Second, the movements are "informational" (p. 346). Different tactics circulate through global networks, getting transformed, reproduced, and localized, thereby symbolizing the overarching cultural logic of networking. Third, the movements are organized around flexible, decentralized "networks" (p. 346). With the emergence of the Internet, traditional hierarchical structures (e.g., labor unions and political parties) have been replaced by autonomous units that are much smaller in scale.

Despite the existing digital divide, Mitra observes that the Internet provides a virtual platform in which different social groups can "produce a presence" (2004, p. 492) that might have been otherwise denied. Voice is regarded as a construct that connects one to the larger society and establishes the self in relation to others, thereby emphasizing power relationships (Mitra, 2004; Scrutton, 1983; Smith & Hyde, 1991). Mitra argues that voice on the Internet is inherent in the discourse "constructed and presented" (2004, p. 494) in cyberspace, which has a representational element in the way the message is created. The issues of voice and representation are important, because the Internet has acquired substantial political significance (Bimber, 1998). It is expanding access to politically relevant information and offering citizens new possibilities for political learning as well as action at a rapid speed.

There is growing support among scholars who view the Web as an embodiment of the public sphere (Habermas, 1989). Grossman (1995) and Browning (1996) believe in the power of the Internet to bring a new form of democracy, often regarded as *electronic democracy*. The common thread through these views is that the Internet is leading to a more pluralistic world of decision-making (Fischer, 1998).

Issues related to voice, representation, public sphere, and pluralism in virtual space are linked to what Routledge (2000) regards as the production, exchange, and strategic use of information by networks. In other words, "Globalizing resistance is all about creating networks: solidarity, information sharing, and mutual support" (p. 27). Routledge observes that the Internet allows strategic mobility in activist struggles and facilitates the organization of resistance across diverse spaces. More specifically, place-specific struggles are creating and engaging alliances across borders and boundaries.

These phenomena necessitate an understanding of emerging new practices in cyberspace. One such area is how networking is being used for social and political change, including democratization (Walch, 1999). Juris (2004) argues that networks configure power structures through their interaction. Hence, the collective action of social movements aims to introduce new instructions into the networks' programs.

Juris (2004) observes that resistance to power through and by networks is facilitated by information and communication technologies (ICTs). Using computer-mediated communication, global justice movements operate at both global and local levels; they integrate both online and offline political struggles. Along with e-mail, interactive Web pages are becoming more widespread. Juris illustrates that activist networks, like PGA and World Social Forum (WSF), besides having their own home pages, form temporary Web pages during mobilizations to provide information, resources, and contact lists. These sites are used to post documents and calls to action and to facilitate real-time discussion forums and chat rooms. "Activists are thus using new technologies to physically manifest their political ideals, both within temporary and more sustained spaces" (p. 349).

Hence, an essential challenge of public relations lies in addressing the network structure of social movements in which resistance formation is as decentralized and local/global as the activists themselves. With the emergence of new technologies, network designs are expanding local movements into computer-supported social struggles. Public relations researchers and practitioners need to recognize "the cultural logic of networking" (Juris, 2004, p. 351) as it applies to resistance in activism. This means introducing into public relations discourse the complexities of network-based forms of political organization.

In this chapter, we are going to explore the ways in which resistance is enacted by an activist organization on the Internet. In particular, we will pay attention to the ways in which communication is played out on the Internet in order to resist powerful actors in global systems. As dominant actors within global systems gain more power through the globalization of policies, activist groups at local and global levels seek to challenge this power through various acts of resistance. The interest of this chapter is in understanding how this resistance is carried out on a global scale. That is, what strategies do activist organizations use in the realm of global politics? Of particular relevance here is the role of the Internet as a platform for connecting global activist groups. The NBA case is used to examine the role of the Internet in fostering resistance. In other words, we ask the question, What role does the Internet play in communicating resistance?

Method

This study explored the discursive space of the Web site http://www.narmada.org (*Friends of River Narmada*, n.d.). The Web site belongs to Friends of River Narmada, an international coalition of organizations and individuals that serves as a support and solidarity network for the NBA. A thematic analysis based on a grounded theory approach was used for studying the activist group's Web site. Thematic analysis is a qualitative research method that involves the coding of concepts. The general idea is that, by analyzing an array of *discrete* concepts, a researcher can better understand the theoretical relationships *between* concepts. Grounded theory posits that theory is grounded in themes (Strauss & Corbin, 1998). Hence, the study of Web site themes is appropriate for our interests herein.

Results

The central theme throughout our analysis was the role of the Internet as a site of resistance to the dominant discourses of globalization. The NBA Web site belongs to the Friends of River Narmada, an international coalition of organizations and individuals. An entirely volunteer-based organization, the Friends of River Narmada is a support and solidarity network for the NBA. As a coalition committed to the principles of NBA, it is focused on supporting the struggle of NBA. Maintaining the Narmada Web site is one of the ways by which the Friends of River Narmada expresses solidarity with the movement. Since the Web site is dedicated to the cause of NBA, the phrases *the NBA Web site* and *Friends of River Narmada* have been used synonymously in the chapter. The NBA Web site acts as a gateway to a global movement that challenged the dominant discourse of development embodied in dam construction.

Resistance: Connecting local and global. The NBA Web site connected the local with the global by providing a platform that brought local politics into the global realm and thus built the solidarity of activist groups. Presented in English, the Web site projected an identity to its global publics that connected the local issue (resistance to dam construction) with the broader politics of globalization and the WTO. The use of English demonstrated the desire of the group to communicate with a global public via the Internet. In turn, the Internet served as a conduit to connect local struggles in the Narmada valley with global participants of typically higher socioeconomic segments with access to the resources of mainstream civil society.

Particularly relevant in this geopolitical issue was the presentation of the Indian identity of the organization. Here is how the Web site introduced NBA:

> The Friends of River Narmada is an international coalition of organisations and individuals (mostly of Indian descent). The coalition is a solidarity network for the Narmada Bachao Andolan (Save the Narmada Movement) and other similar grassroots struggles in India. (*Friends of River Narmada*, n.d., Dams on River Narmada section, para. 2)

The geographical roots of the movement were noted as organizations and individuals mostly of Indian descent. Their identity was constructed as a solidarity network committed to the Narmada and other efforts in India. The mention of grassroots struggles grounded the identity of the organization in the realm of local politics within the context of India.

The Indianness of the Web site was connected to the international context of dam construction, and the organization was presented as an international coalition. Furthermore, the site stated that "the struggle against the construction of mega-dams on the River Narmada in India is symbolic of a global struggle for social and environmental justice" (*Friends of River Narmada*, n.d., About Us, Who Are We section, para. 1). Once again, local politics of struggle were connected with a global struggle. Local resistance became a marker of global resistance. The movement communicated how local struggles against globally powerful stakeholders can indeed bring about change.

Resistance: Enacting agency. The Web site communicated the ways in which individual members of the movement exercised their agency: the communication processes through which groups of individuals in marginalized sectors come together seeking to make change. In doing so, they demonstrated the power of marginalized, or subaltern,[1] groups to mobilize themselves and organize to fight global injustices that are often perpetrated by national and international stakeholders at the centers of power. The Web site attested to the possibilities of resistance and offered hope to others through Narmada narratives. One such narrative told of the history of the movement. This sharing of history contributed to the narrative of agency; that is, it explained how displaced people came together for resistance:

> The struggle of the people of the Narmada valley against large dams began when the people to be displaced by SSP [Sardar Sarovar Project] began organizing in 1985–86. Since then the struggle has spread to encompass other major dams in various stages of planning and construction chiefly Maheshwar, Narmada Sagar, Maan, Goi and Jobat. Tawa and Bargi Dams were completed in 1973 and 1989 respectively have seen the affected people organize post-displacement to demand their rights. (*Friends of River Narmada*, n.d., Narmada Dams, Sites of Struggle section, para. 1.)

This narrative documented not only the history of the movement but also how it spread and led to subsequent organizations of resistance. The movement became a model for organizing. The history of the struggle served as a backdrop to the movement's modern struggles amidst Narmada valley politics. Current issues faced by the movement were presented in the form of press releases by NBA. In fact, the site included an archive of press releases that documented the different aspects of the struggle against structural forces. Here is an example:

> As the Narmada Bachao Andolan (NBA) continues to uncover the corruption of hundreds of crores of rupees in the rehabilitation of Sardar Sarovar dam-affected families, the news today is that Ashok Kumar Modi, Assistant Rehabilitation Officer in Manavar, Dhar district, M.P. has been forced to take compulsory retirement, in light of all the charges leveled against him of corruption relating to rehabilitation of affected families. When hundreds of affected families, incensed with the situation of corruption and money-laundering in their rehabilitation, descended on the Manavar tehsil office of the NVDA (Narmada Valley Development Authority) on December 30, 2005, they

> exposed the officials and persons involved in the scam, and also filed FIRs against the guilty persons. The removal of A.K. Modi from office is proof that the Andolan is correct in exposing this sheer squandering of the state's exchequers. (*Friends of River Narmada*, n.d., Press Release section, January 6, 2006, para. 1 and 2).

In this example, the structural impediment was the corruption of the local system that had vested interests in rehabilitating displaced people. The press release discussed the specific steps taken by the families to overcome these impediments, such as organizing a group visit to the local office of the Narmada Valley Development Authority (NVDA) and filing First Investigation Reports (FIRs) against guilty persons. It also discussed the outcome achieved by the organized families in terms of securing the compulsory retirement of a corrupt official. In presenting this incident, the movement once again embodied the agency of subaltern groups and demonstrated the capability of community organizing in bringing about structural changes.

The press release went on to discuss a rally organized in Delhi, the capital of India, where a thousand displaced people organized to protest the injustice and the corruption. The rally was described in terms of the spirits of those who attended: "The freezing cold could not dampen their spirits or their quest for justice. In their meeting with Mrs. Meira Kumar, the Minister of Social Justice and Empowerment, the Minister patiently heard all that the oustees had to say" (*Friends of River Narmada*, n.d., Press Release section, January 6, 2006, para. 3). This depiction celebrated the spirits of the people who attended the rally in spite of freezing temperatures. The rally led to dialogue with the minister of social justice and empowerment, who listened to the concerns of the oustees.

Here is another example of a protest rally held by the members of the movement. Once again, this excerpt illustrates the organizing capacity of subaltern groups who have been marginalized:

> Over 2000 people from the villages to be affected by the Indira Sagar dam demonstrated today at the Khandwa District headquarters. In this rally organized under the aegis of the Narmada Bachao Andolan, the dam-affected people expressed their anger at the injustice being meted out to them. All of them demanded that the gates of the dam be kept open till every single oustee has been rehabilitated. The affected people held a public meeting at the Gandhi Bhavan and held an impressive rally along main roads of the town. District Collector, Khandwa, Superintendent of Police, Khandwa and General Manager, NHDC came to the venue of the public meeting to receive the petition of the people. Representatives of the Maheshwar, Man and Upper Beda participated in the protest program and expressed their solidarity. (*Friends of River Narmada*, n.d., Press Release section, April 15, 2005, para. 1)

The organizing capacity of the movement was communicated not only in terms of the sheer number of people in attendance, but also by the presentation of the demands of the people to the local administration.

The images presented on the Web site further communicated the climate of resistance presented by the NBA. For instance, historic photographs of Satyagraha by NBA activists posted on the Web site depicted subaltern women dressed in saris marching to the police camp in resistance. The Satyagraha march is reminiscent of the famous march of Gandhi during the years of India's freedom struggle. The symbolic analogy between the movement and Gandhi's Satyagraha is an important one: It draws upon the Indian freedom struggle to communicate the agency of people and the possibility of achieving social change through participatory processes.

Another picture depicting subaltern women communicating with the police was titled "Women Questioning the Police at Domkhedi." This picture challenged the dominant subaltern contexts in which subaltern subjects are depicted as bodies to be worked upon—in other words, as targets of top-down communication strategies without really having a say. Instead, it presented the subaltern subject as an active participant, reversing the communicative role of the police and the subaltern by placing the women in the role of the ones asking questions.

In yet another picture, people affected by the dam construction project were shown dancing with activists and supporters of the movement. The participants in the dance were holding hands and moving in circles to the rhythm of the drums. The drum and the folk dance included in the picture have traditionally depicted solidarity and group cohesiveness. As a result, this picture depicted the agency of the subaltern participants.

Resistance: Capacity building. A third element of resistance embodied in the NBA is capacity building. Capacity building refers to the task of developing infrastructures and resources that are required for the operation and survival of communities. Community capacity may be defined in terms of the needs of the community and depends upon how the community is defined. For instance, for the local community of displaced subaltern groups living in the Narmada valley, capacity is defined in terms of the resources that are required at the local level. However, with respect to the global community of stakeholders interested in issues of social justice and social change, community capacity is defined in terms of those resources that are required for the operations of the global community. These multiple local and global needs for capacity building seamlessly played out on the NBA Web site and demonstrated needs shifting from the local necessities to the global ones and getting connected to a broader framework of capacity for global activism.

The NBA Web site epitomized the complex and multilayered nature of community capacity building that interpenetrates the multiple needs and functions of the multiple stakeholders who are engaged in the project. It reflected the complex nature of activism in a global context that must respond to a variety of stakeholder groups, and the Internet served as a platform for connecting these groups.

The Web site also demonstrated the complexity of global activism in a new media context where resources and communication flow in multiple directions rather than in one-way communication frameworks included in the dominant models of public relations. Rather than a centralized actor serving as a diffuser of messages based on power, multiple actors engage in sending and receiving messages simultaneously. In turn, they form a complex web of multidirectional communication. In this sense, the Internet presence of global activist networks offers opportunities for examining the ways in which community dialogues may be created and cocreated through new media platforms, thus connecting the dialectical elements of local and global stakeholders.

For the Friends of Narmada Web site, capacity building focused on the exchange of information regarding resources, policies, strategies, and tactics. This capacity building was achieved in a plethora of platforms and flows in a plethora of ways, serving a multitude of functions simultaneously. It is this simultaneity of capacity building that demonstrates the ways in which resources flow in communicative platforms of global activist organizations. For instance, one aspect of capacity building of the project focused on providing resources about the movement to global stakeholders. In this context, the Web site served as a repository of information, contributing to the capacity of global movements of resistance by documenting the steps, stages, successes, and failures of the movement. This presentation of resources offered opportunities for interested parties to learn about the movement and develop their own information resources regarding the movement. The archives of pictures and press releases provided insights regarding the lessons learned from the movement and became learning tools for other groups interested in similar issues. These resources also become points of solidarity building for other movements, connect these movements with each other, and provide networks of solidarity among movements.

Furthermore, this global-level activism was mobilized for local purposes. For instance, the Web site discussed the ways in which global publics could participate in protesting the building of the dam by signing petitions on the Web, participating in virtual hunger strikes, and participating in rallies and protest marches in their respective locations. In this way, globally dispersed local publics were mobilized for a local cause. The activist capacity of individual local actors was mobilized at the global level to support the capacity of a specific local movement. Ultimately, the Internet served as a platform for community capacity building and for mobilizing resistance.

Discussion

In contrast to the dominant literature in public relations that looks at activism from the organizational standpoint, this chapter suggests the importance of exploring activism from the standpoint of the activist organization(s). The goal here is to understand activism as a communicative phenomenon, with the goal of developing appropriate strategies of resistance to the dominant structures within social systems. The framework for studying public relations therefore shifts from a dominant perspective that seeks to develop strategies for controlling activist publics to a resistive perspective that emphasizes the relevance of communicative spaces that open up alternatives to the dominant models. It is through these alternatives that power and control are resisted, and important points of entry are created for initiatives of social change.

This chapter fundamentally suggests that, in addition to the managerial framework that is typically embodied in the public relations literature, we also need to explore alternative paradigms of public relations theory that offer critical entry points for challenging the dominant discourses of powerful organizations (such as the WTO) in social systems. In other words, there is a need for understanding how communication strategies are played out in order to resist the dominant social structures. We hope that this chapter opens up the literature in public relations to exploration of alternative theoretical and pragmatic frameworks that critique the managerial biases of the dominant framework and open up avenues for participatory social change.

In conclusion, the Web site of the Friends of Narmada embodies the quintessential role of the Internet in connecting movements such as the NBA with the global politics of resistance against transnational hegemony. In this context, the Internet served as resource that connected local and global politics; it brought together multiple stakeholders, who were globally dispersed, to a common platform. It played a key role in facilitating participation and in creating avenues for enacting resistance to the dominant structures of power. The voices of local community members articulated on the site served as points of entry for global advocacy. The active involvement of various global publics was critical to the successes of the campaign, which were achieved through the exchange of information and communication on the Web site.

Our analysis demonstrates support for the subaltern studies framework proposed by Dutta-Bergman in his critique of top-down civil society efforts. Dutta-Bergman (2005) argues that much of public relations scholarship ignores the ways in which various subaltern and marginalized groups participate in communicative processes of social change. He suggests that engaging in dialogue with the cultural participants in underserved segments opens up the discursive space to hitherto unheard voices. Furthermore, members in marginalized settings indeed enact their agency in responding to the constraining contexts within which they live their lives. Listening to the narratives of resistance provides an alternative to the management-driven approach to public relations, thus creating openings for social change.

In this chapter, we have demonstrated that the NBA exemplifies the ways in which local activist groups enact their agency at the global level through new media platforms. The Internet becomes a platform for telling the stories of resistance to a global audience. These stories of resistance not only further catalyze the collective efficacy of the activist group, but also provide critical resources for other activist groups, thus linking activist groups within a network of solidarity. The information and communication posted on the Web site simultaneously served as a repository, a learning resource for global activism, and a catalyst for mobilizing globally dispersed publics to participate locally for a local cause that was situated elsewhere on the globe.

For consideration

1. What are other recent examples that demonstrate how the Internet can be leveraged to facilitate activism?
2. What affects could the increased voice of subaltern groups have on dominant political systems in the next ten years?
3. To what extent can the Internet alone generate political action without there being any offline movement?
4. Despite the issues of digital divide, to what extent does the Internet offer hope for social change by mobilizing global activist publics?

5 To what extent is the Internet actually representing the voice of the local people involved with the NBA?

For reading

Coombs, W. T. (1998). The Internet as potential equalizer: New leverage for confronting social irresponsibility. *Public Relations Review, 24*(3), 289–303.

Dutta-Bergman, M. J. (2005). Civil society and public relations. Not so civil after all. *Journal of Public Relations Research, 17*(3), 267–289.

Routledge, P. (2000). Our resistance will be as transnational as capital: Convergence space and strategy in globalizing resistance. *GeoJournal, 52*, 25–33.

Taylor, M., Kent, L. M., & White, J. W. (2001). How activist organizations are using the Internet to build relationships. *Public Relations Review, 27*(3), 263–284.

Walch, J. (1999). *In the Net: An Internet guide for activists*. New York: St Martin's Press.

Note

1. *Subaltern* refers to those sectors of society that have traditionally been marginalized.

References

Bimber, B. (1998). The Internet and political transformation: Populism, community, and accelerated pluralism. *Polity, 31*, 133–160.

Browning, G. (1996). *Electronic democracy: Using the Internet to influence politics*. Wilton, CT: Online Inc.

Castells, M. (1997). *The power of identity*. Malden, MA: Blackwell.

Coombs, W. T. (1998). The Internet as potential equalizer: New leverage for confronting social irresponsibility. *Public Relations Review, 24*(3), 289–303.

Danitz, T., & Strobel, P. W. (1999). The Internet's impact on activism: The case of Burma. *Studies in Conflict & Terrorism, 22*, 257–269.

Dutta-Bergman, M. J. (2005). Civil society and public relations: Not so civil after all. *Journal of Public Relations Research, 17*(3), 267–289.

Fischer, M. J. M. (1998). Worlding cyberspace: Toward a critical ethnography in time, space and theory. In G. E. Marcus (Ed.), *Critical anthropology now: Unexpected contexts, shifting constituencies, changing agendas* (pp. 245–304). Santa Fe, NM: School of American Research Press.

Friends of River Narmada. (n.d.). Retrieved July 4, 2006, from http://www.narmada.org/

Grossman, L. K. (1995). *The electronic republic: Reshaping democracy in the information age*. New York: Viking.

Habermas, J. (1989). *The structural transformation of the public sphere: An inquiry into a category of bourgeois society*. London: Polity Press.

Juris, J. S. (2004). Networked social movements: Global movements for global justice. In M. Castells (Ed.), *The network society* (pp. 341–362). Northampton, MA: Edward Elgar.

Mitra, A. (2004). Voices of the marginalized on the Internet: Examples from a Website for women of south Asia. *Media, Culture & Society*, *54*, 492–510.

Postmes, T., & Brunsting, S. (2002). Collective action in the age of Internet. *Social Science Computer Review*, *20*, 290–301.

Routledge, P. (2000). Our resistance will be as transnational as capital: Convergence space and strategy in globalizing resistance. *GeoJournal*, *52*, 25–33.

Scrutton, R. (1983). *The aesthetic understanding: Essays in philosophy of art and culture*. Manchester, UK: Carcanet.

Sen, G. (1997, December). *Empowerment as an approach to poverty* (Working paper series, number 97.07; Background paper to the Human Development Report 1997). Retrieved July 4, 2006, from http://www.globalhealth.harvard.edu/hcpds/wpweb/97_07.pdf

Smith, C. R., & Hyde, M. J. (1991). Rethinking the "public": The role of emotion in being with others. *Quarterly Journal of Speech*, *77*, 446–466.

Strauss, A., & Corbin, J. (1998). *Basics of qualitative research*. Thousand Oaks, CA: Sage Publications.

Taylor, M., Kent, L. M., & White, J. W. (2001). How activist organizations are using the Internet to build relationships. *Public Relations Review*, *27*(3), 263–284.

Udall, L. (1997). The international Narmada campaign: A case of sustained advocacy. In Fisher W. F. (Ed.), *Toward sustainable development: Struggling over India's Narmada River* (pp. 201–227). Jaipur, India: Rawat Publications.

Walch, J. (1999). *In the Net: An Internet guide for activists*. New York: St Martin's Press.

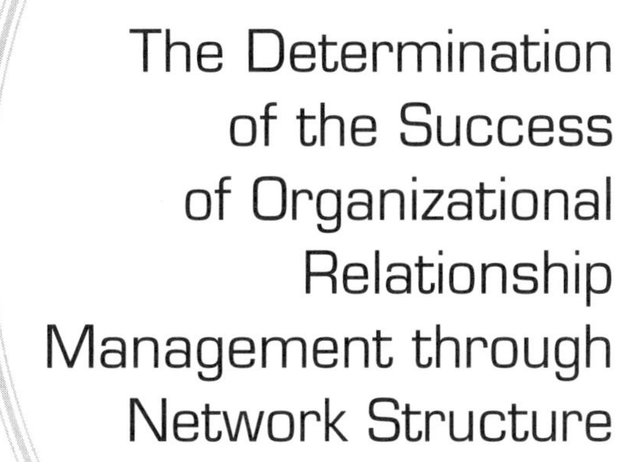

The Determination of the Success of Organizational Relationship Management through Network Structure

The Attac Case

Romy Fröhlich and Jeffrey Wimmer

Upon analyzing previous research in the fields of organizational communication and public relations, one gap can clearly be identified: It is the correlation between networks and the success of organizational relationship management, about which we still know very little. This chapter aims at making a contribution toward answering this question. Firstly, we will define a theoretical framework for the analysis of the requirements of organizational relationship management in the context of sociopolitical and technical change. In connection with that, Castells's concept of the network society plays a key role. Secondly, a concrete case study will be presented: The success in organizational relationship management of Attac—an organization from the field of the so-called counterculture—is analyzed in terms of its very specific network attributes. Our findings show that, in spite of the lack of classic resources, non-established organizations like Attac are capable of achieving successful long-term communication in the political public sphere if they take advantage of sociopolitical and technical change by implementing their own communication strategy and by relying on the advantages of a network organization.

Organizational communication and public relations constitute the connection between an organization and its environment by addressing the relevant groups of an organization both internally and externally, thus establishing a successful network. However, over the last few years, the operating conditions of organizations have radically changed. Sociopolitical and technical changes have played key roles within this process (Beck, Giddens, & Lash, 1994; Beck, Bonss, & Lau, 2003). From the viewpoint of communication, these changes not only have an impact on *external* publics, but also influence *internal* publics.

Against this background, this chapter provides a theoretical framework to serve as the basis for an empirical analysis of the connection between success in organizational communication and organizational structure. In our case, we refer specifically to the structure of an organizational network. In doing so, Castells's (1996, 1997a, 1997b) general sociological analysis of a "network society" is introduced and, by example, transferred to the specific subject of investigation: the organizational communication and public relations of Attac[1] Germany. This chapter explores to what extent Attac Germany has achieved success in communicating with its internal and external publics, particularly as a result of its network organization.

Network Society, Network Structure, and Counterculture

From our perspective, the interrelation between organizational communication and successful communication can only be deduced by referring to theories of social change that take into account the structural attributes inherent in modern society. On the basis of Castells's works (1996, 1997a, 1997b), we argue that societal change and the preconditions contributing to that change—including the role of new communication technologies—point to a new stage of global modernity. Castells's perspective is based on the realization that, by implementing new information and communication technologies (ICTs), society's material basis has fundamentally changed. Instead of labor, land, and capital, nowadays, information and communication are vital factors in the process of creating value.

Castells's main hypothesis states that our modern society is situated around the opposing terms of the "net" and the "self" (1996) to an ever-increasing extent. This means the (Inter)net reflects the fact that the most powerful processes in our economic, political, and media systems are increasingly organized within flexible networks, thereby becoming more and more independent of local and localized realities. To put it in other words, the logic of networks, the "space of flows," is growing in its dominance of the "space of places," which describes the physical reality in which we live (Castells, 1996, p. 412). The space of flows can be thought of as having at least three layers, which Castells uses as his three dimensions of analysis:

1. *Technical:* the circuit of electronic devices that forms the technological infrastructure of the networks.
2. *Geographical:* the topology of the space formed by its nodes and hubs. Hubs are defined by the networks but link them to specific places with specific social and cultural conditions.
3. *Social:* the spatial organization of the people using the network.

This "faceless" functional logic simultaneously undermines the growing number of institutions that have significantly contributed to the development of identity and meaning (Giddens, 1991; Castells, 1997b). In particular, for political organizations, this fact leads to further requirements, since the changed lifestyle of its members makes it necessary to have a different (organizational) policy on identification and non-hierarchical structures (e.g., Bennett, 2003).

What are the repercussions of this change on the political systems? Developments in the field of ICTs, as well as the global and social dominance of capitalism, have led to a kind of new "counterculture" in this area (Castells, 1996). Counterculture is a collective name that includes most nongovernmental organizations (NGOs), new social movements (NSMs)—the best example being the anti-globalization movement—and media (i.e., blog) activists. Counterculture has its roots in the NSMs of the 1960s and 1970s (e.g., student, peace, and environmental movements) from which many organizations have since distanced themselves. At the same time, the traditional differentiation between politics (i.e., state) and non-politics (i.e., all the rest of societal areas) is becoming more blurred (Beck, 1992). Consequently, the political public sphere and its actors are increasingly being decentralized and diversified. For this reason, single collective actors within the counterculture are more capable of participating in the modulation of society outside the official stage of political communication.

Counterculture and Information and Communication Technologies (ICTs)

Castells focuses on the revolution of new technologies, that is, the emergence of ICTs and computer-mediated communication: intranet and Internet, e-mail, mailing lists, community networks, Internet telephony, videoconferences, virtual communities, etc. These technologies have contributed to the development and maintenance of the political public sphere and political actors (Held, McGrew, Goldblatt, & Perraton, 1999). Counterculture organizations (CCOs) rely heavily on Internet communication. One well-known example is the protest against the 1999 World Trade Organization conference in Seattle, which had worldwide media repercussions. New media, in particular, were used to organize the protest (Smith, 2001; Wall, 2002). Besides having the ability to mobilize the masses (Couldry & Curran, 2003), ICTs also provide new venues of articulation for the counterculture (cf. Keck & Sikking, 1998; Siapera, 2004).

ICTs are fundamentally different from classic mass media and internal organizational media (Jenkins & Thorburn, 2003). The new characteristics are based on the fact that communication can be transmitted on a network basis (hypertext). Moreover, it can be combined with a multitude of media (multimedia), and, furthermore, recipients are capable of changing it in any way (interactivity). Digital applications provide a way of reducing distance and time in cyberspace. On a theoretical basis, they enable a nearly limitless amplification of internal and external publics connected to an organization and thereby combine connectivity with interactivity. In this manner, organizational use of the Internet offers possibilities for increasing information.

Though new technical possibilities create new communication frameworks for political organizations, the basis of this thinking is not a theory of technological determinism. That is, the complex communication process cannot be reduced to the exchange of data. Hence, in answering the question about implications and consequences of the *digitalization* of organizational communication and public relations, above all, the specific *context* of it has to be considered. In other words, what are the consequences of communication management in political organizations in light of the changes discussed thus far?

To answer this question, an analysis of CCOs is of special interest because, for various reasons, they have to prove themselves much more vehemently in the global market of opinions than do political actors who are already established, such as executive offices and political parties on national and international levels. Being actors outside of the political mainstream, CCOs have fewer resources at their disposal, including power, employees, and money. Moreover, their access to mass media is disadvantaged if, for example, they do not have a contact person for journalists. This was true for Attac Germany at the time of its foundation. On the other hand, by acting or communicating in a way that too obviously resembles established organizations, CCOs take the risk of losing the generally high moral bonus they enjoy with organizational members and the general public sphere.

All in all, ICTs influence the possibilities for organizational communication and public relations: firstly, by redefining channels and barriers of communication and, secondly, by allowing individuals to have broader reach through media use. Digitalization describes the process in which more and more people can be reached ever faster and easier. Therefore, publics linked to this development are theoretically becoming more open. For organizations, internal, as well as external, publics can be addressed faster. Ultimately, established and non-established political organizations reach more people than was imaginable a decade ago.

Case Study: Attac Germany

Although it was founded in France, Attac has been conceived since its beginnings as an organisation from the counterculture, which is indicated by a number of structures. From this point of view, three aspects are constitutive: Attac enables political processes of learning and experiencing, consolidates various types of emancipatory politics in discussions and common actions, and, in turn, leads to the possibility of acting jointly in commonly defined political fields.

On the basis of the theoretical background presented, the question arises, *What impact do the new ICTs and a network-based organizational structure have on the success of both the internal and external communication of Attac as a prominent representative of the field of counterculture?*

In concrete terms, three organizational areas of Attac Germany have to be taken into account when answering our research question: its organizational structure, internal communication, and external communication.

Referring to the present cognition about the correlation between organizational structure and organizational communication processes, it is to be assumed that a decentralized organizational structure—as is the case with Attac—has a rather negative effect on smooth and effective internal and external communication (cf. Hall, 1997). Our study will show that, surprisingly, this is not the case with Attac. We also demonstrate ways in which modern ICTs are influencing this process.

According to Castells's (1996) three dimensions of analysis, our interpretation of the findings incorporates three contextual levels: (a) the *technical* context providing the real and virtual dimensions of organizational communication, (b) the *geographical* context referring to the spatial nodes of an organizational network, and (c) the *social* context of an organization, referring to the societal framework of an organization. In a network society, these three fundamental dimensions are subjected to powerful changes and therefore also pose sizeable challenges for the field of organizational research (cf. Hiebert, 2005).

Design of the case study. As mentioned, hardly any systematic studies dealing with this research question exist. Studies carried out to date have either focused on specific successes of mobilization in connection with ICTs from the viewpoint of social movement research (e.g., Couldry & Curran, 2003; Donk, Loader, Nixon, & Rucht, 2005) or on the basis of public relations theories dealing with external organizational relationships (e.g., Guiniven, 2002; Taylor, Gabriel, Vasquez, & Doorley, 2003). Thus, we are pursuing our scientific interest in this topic in an exploratory manner.

We chose a multi-step methodological approach: In analyzing Attac's organizational structure and internal communication, we primarily refer to already gathered facts about Attac's contextual dimensions and reinterpret them within the framework of the previously outlined theoretical background and the research question.

In order to measure the success of Attac's external communication, we first conducted a content analysis of German media coverage of the Group of Eight (G8) Summits over the years from 1975 to 2001, followed by an analysis of Attac's public relations materials for the media. In doing so, we took into account each time Attac was mentioned as the author of a statement and/or subject of a media topic.

Our research included two of the most respected national German newspapers: the *Sueddeutsche Zeitung* (*SZ*) (the so-called newspaper of Southern Germany) and the *Frankfurter Allgemeine Zeitung* (*FAZ*) (the so-called general newspaper of Frankfurt), both of which are renowned within the German media landscape. The *SZ* is considered to be a liberal newspaper, and the *FAZ* is viewed as conservative.

We first used content analysis to measure established political actors. These included all organizations (and their representatives) that were named in media coverage as having been *directly* involved in the political decision-making process, such as executive offices and political parties on a national and international level (e.g., the German Government, the Italian Foreign Ministry).

Secondly, we analyzed the frequency of CCOs mentioned in media coverage over the twenty-six-year time frame. These included organizations and individual representatives from the field of counterculture, such as NSMs, protest parties, NGOs, single activists, and, in particular, Attac itself.

As a third step, we carried out a more profound analysis, within the means of *framing analysis*, with open coding, to investigate the question of what influence the special kind of networking between Attac and journalism had on media coverage of the 2001 Genoa Summit. Our analysis included Attac's press releases as well as commentary articles within relevant media coverage of the summit. We selected commentary articles—instead of ordinary news and reports—because commentary is more capable of reflecting medial frames. That is, in this media genre, journalists are generally able to express their own opinions on political issues.

In addition to SZ and FAZ, the alternative daily newspaper *taz* was included in this framing analysis.[2] The *taz* is politically situated left of centre (even more so than SZ). It therefore provides unique (if not critical) angles and opinions on the issue of globalization.

Results

The network structure of Attac Germany and its internal communication. Attac's organizational structures have been established on a national basis in over forty countries; it has been in Germany since 2000. Due to the fact that one of Attac's main characteristics is its conception as a movement, one of Attac's main aims is the development of a broad-based establishment within the population. Attac includes two different models of membership: individuals and organizations. In most European countries, members are not only individuals, but also organizations from other parts of societal politics (e.g., unions).

In Germany, initially, Attac mainly had the character of a network of organizations, which has many advantages. At present, Attac Germany consists of more than 250 local initiatives. In particular, Attac can rely to a large extent on the financial, human, and content resources of its members. In addition, the membership of big and well-established organizations—such as the German workers' union ver.di[3] or the ecological organization Bund für Umwelt und Naturschutz Deutschland (BUND, League for the Environment and Nature Conservation, Germany)—increases the political weight of the network organization.

Although founded in France, Attac has always been conceived as a *transnational* organization that is organized by a number of different structures. Attac has resisted the establishment of a centralized headquarters. Apart from an annual international meeting, a relatively loose form of exchange of experience and planning is assured through bimonthly European meetings. Although Attac is a global operator, its core clearly lies in Europe, where more than 80% of the associated members are established.

Attac describes its organizational philosophy as one that includes guiding principles of a pluralistic view; its conception as a movement; open, decentralized, participatory, and flexible organizational structures; plurality of methods and instruments; and, finally, its search for cooperation and alliances.[4] From an organizational communication and public relations standpoint, Attac's philosophy is normative. That is, Attac endeavors to build a platform that enables political communication while, at the same time, it provides a discussion forum for different political streams to find collaborate ways of acting in areas of shared political interest.

We examined the most vital organizational media for Attac's internal communication. Various mailing networks connect communications of Attac's 250 groups existing on local, national, and international levels. Its Web page, e-mails, and open (to everyone), as well as closed (available only to organizational members), mailing lists have also had an important impact on internal communication by structuring organizational communication.

The links, or connections, of the German Attac Web site to other Web sites reflected two findings: (a) a strong *international network structure* with other Attac organizations in, for example, France, Austria, and Italy and (b) a strong *national network structure* with German membership organizations such as *ver.di*, *World, Economy, Ecology & Development* (WEED),[5] and *Naturschutzbund Deutschland* (NABU).[6]

Le Grignou and Patou (2005) illustrate the prominent role of electronic media for Attac's organizational communication: "Anyone who wants to be committed to ATTAC [*sic*] has no choice but to be connected to the Internet" (p. 170). Bearing this in mind, it is not surprising that Attac's first employee in Germany was a Webmaster. Apparently, from its beginning, the organization was conscious of the importance of digitalized network communication for its existence and development.

However, the question arises of whether this emphasis on new media is also reflected at an organizational level; that is, is it shared by organizational members? In 2001, about 50% of the members and donors were generated via the Internet (Schewe, 2003). An analysis of the daily use of media by Attac's membership showed a very strong affinity toward use of the Internet (Guttenberg et al., 2002, pp. 67–71). Compared to other organizations, Attac's members display a very high rate of Internet usage in the political field.

In contrast to other examples in counterculture, Attac's goal is not to focus on the articulation of a particular political philosophy but rather to supply a technical framework for communication to be filled by the members of Attac themselves. The organization's undogmatic attitude includes openness to new issues, all of which were presented on the Web page. In this way, Attac has transformed from being a single-issue group[7] to a network of concrete political claims flowing from a broad-based approach to the economic aspects of globalization (e.g., financial markets), the world trade system, and activities of transnational enterprises.

According to Le Grignou und Patou (2005), amongst others, one reason for this transformation lies in the way Attac's members communicate digitally. In discussing results gathered from a survey of Attac's members in France, Le Grignou and Patou referred to the Web page, for example, as a "documentary goldmine," being "very convenient, full of good documents" for the activists (p. 170). The Web page's and mailing list's content consisted mainly of local issues and interests, thereby enabling Attac's organizational agenda to diversify rapidly and become more complex. This approach was likewise observed on Attac's German Web page and mailing lists (e.g., criticism of the German labor market policy and discussions of anti-Semitism and German military intervention abroad).

Thus, Bennett ascribes this process to the logic of "click here" (2003, p. 155) inherent in open computer networks. Attac's German Web page allows the most diverse topics to be easily and rapidly shared between numerous local groups. Yet, it is precisely this characteristic that also causes problems of organizational and communicative importance, which are clearly expressed by one of Attac's representatives surveyed by Le Grignou and Patou: "The main problem for Attac today concerns the unification of the movement and the way to give it a more unified content" (2005, p. 172).

The network structure of Attac Germany and its external communication. What, then, is the influence exerted by Attac's network structure on the emergence and quality of its media coverage? We defined Attac's media response as a central indicator of the success of its public relations efforts. For this reason, we measured Attac's German media coverage in regard to the G8 Summits held between 1975 and 2001.

At the beginning, the World Economic Summits of the G8 states consisted of annual meetings coordinating the economic policies of the greatest industrial nations. Later on, topics including the energy crisis, drug trafficking, and current questions about globalization and terrorism were included.

Considering our research question, the following is of particular interest: Before 1992, counterculture actors were very rarely mentioned. However, from 1992 onward, they have been mentioned more frequently in media coverage. Since 1999, they have become a regular and inherent part of the media coverage. In the analysis, we found that Attac was mentioned the most frequently of all CCOs. Furthermore, it was the only CCO that was mentioned several times in media coverage of the World Economic Summits in two consecutive years. Attac was the main subject of reporting in four articles in the year 2000 and in six articles in 2001.

What factors can explain the media coverage? Besides the usual factors influencing media coverage, such as political factors[8] or media events (e.g., scandals), we were particularly interested in *organizational* factors, especially those deriving from Attac's organizational network structure.

In searching for reasons to explain an increase in media coverage of counterculture issues, one must recognize that, over the course of our study's examination period, counterculture communication had been extensively professionalized. That is, organizations were implementing professional tactics of public relations and media relations to a greater degree than was the case in the 1970s and 1980s. In addition, CCOs were holding media events more frequently over this span of time. In doing so, CCOs adapted to production routines within the mass media system. This is especially true for Attac.[9] Its public relations sector is the only organizational area that is regulated centrally; all others are decentralized.

Press coverage during the 2001 Genoa Summit appeared to be the turning point. Attac was, at that time, the first organization from the counterculture to have a rather elaborate and professional Web page about globalization issues. It also established designated press contacts from the very beginning. Journalists were invited to accompany Attac Germany members on the bus en route to the Genoa conference. Once on location, members of the press had a short message service (SMS)[10] providing a continuous stream of up-to-date news about occurrences at the summit. Attac's public relations work rapidly led to new relationships with the press, and the Web page received more publicity when journalists recommended the site as a research tool for interest groups. In turn, a specific mailing list was created, with the needs of journalists in mind, to provide journalists with relevant information (Kolb, 2004).

We analyzed Attac's success in *establishing network structures* with journalists, which it began during the 2001 summit, in more detail using a qualitative framing analysis.[11] Media coverage and Attac's press releases during the 2001 Genoa G8 Summit were analyzed for themes of globalization. According to Entman (1993), three frames need to be differentiated: (a) diagnostic frames (identification of the problem), (b) prognostic frames (proposed solution), and (c) mobilization frames (calls to action). Referring to these three frame types, Table 1 contrasts the frame types analyzed in our research.

In the time preceding the summit (anticipating phase) and during the summit (summit phase), the press drew a fairly identical picture analyzing the problem of globalization (diagnostic framing). Globalization was portrayed as a multilayered process involving positive and negative consequences for different societal spheres (complexity). However, on the contrary, Attac's public relations regarding globalization during the anticipating phase focused on the negative impacts of economic liberalization, or neo-liberalism.[12] During the summit, particularly in light of the impression that violent incidences between police and protestors were making, Attac communicated a negative message of repression: Globalization in a totalitarian system was being controlled by a specific urge for power. This topic was particularly interesting for the press due to police assaults on protestors and the delayed,

Table 1: Dominant Frames within the Media Coverage of the 2001 Genoa Summit and within Attac's Press Releases

	Anticipating Phase	Summit Phase	After-Summit Phase
Diagnostic Framing			
Press	Complexity	Complexity	Repression
Attac's Public Relations	Neo-liberalism	Repression	Neo-liberalism
Prognostic Framing			
Press	Business as usual	—	Business as usual
Attac's Public Relations	Acceleration	Business as usual	Business as usual
Mobilization Framing			
Press	—	Concerted action	—
Attac's Public Relations	Financial reform	—	Jurisdiction

somewhat procrastinated, clarification and investigation of these assaults on the part of Italian authorities hesitant to investigate their policemen.

In the after-summit phase, all three of the analyzed newspapers emphasized the repressive elements connected with the problem of globalization. For the journalists, links to this frame were found in Attac's press releases, as well as in actual events (i.e., the suppression of political and judicial enlightenment about the violent attacks and the policemen's assaults in Genoa). This could explain why, within the diagnostic frame, the media drew heavily on information from Attac's public relations materials.

Despite varying political orientations across the analyzed newspapers, each had a common forecast for the future (prognostic framing): They believed that, similar to other globalization controversies, in the end, the predominant political interests—and not those from counterculture—would be considered (business as usual).

In contrast, Attac warned the public in its public relations materials during this phase that, without precautionary measures, globalization would escalate endlessly (acceleration). However, this frame created by Attac was not adopted by the press. During the summit, the newspapers refrained from giving a forecast; rather, they focused on the demonstrations instead of proposed solutions. Although, in this phase, Attac began by offering a specific solution, it eventually adopted the prior journalistic attitude of the anticipating phase: one of business as usual. During the after-summit phase, the daily newspapers and Attac agreed on this prediction.

In answering the question of how to handle globalization in concrete terms, and how to deal with its consequences (mobilization framing), Attac's dominating mobilization frame in the pre-summit phase was radical financial reform, the central issue of which was the Tobin tax. However, during the summit phase, statements in Attac's press releases relating to mobilization were inconsistent; that is, there was no primary action proposed.

And what about the media? During the anticipation phase, the analyzed newspapers displayed different mobilization frames, depending on the political orientation of the newspaper. During the summit phase, journalists' opinions were more consistent, calling for a common solution to the globalization problem (concerted action).

In the after-summit phase, Attac offered a more consistent, though different, proposed action. It made the more moderate suggestion of pursuing a regulatory and judicial framework to deal with the consequences of globalization (jurisdiction). However, this change in Attac's approach was too late in respect to media coverage. Thus, after the summit, the daily newspapers held to their own established views on judging the globalization phenomenon. A dominating frame could not be identified in the media coverage. Therefore, Attac's proposed solutions to the problem were not reflected by the response of the media.

In conclusion, in the field of mobilization framing, Attac could not register any communicative media success, since the journalists chose to give prognoses, future scenarios, and proposed solutions independent of Attac's point of view. This could be due to the fact that, with respect to mobilization frames, Attac failed to offer consistent statements and a main proposed course of action. This effect is well known from the analysis of media coverage on political election campaigns: When political players choose to communicate unfocused positions, they run a greater risk that related media coverage will not correspond to their own perspectives (Fröhlich & Rüdiger, 2006).

Interpretation and Conclusion

The findings of our content analysis comparing counterculture and established political actors point to a relatively weak media response to the counterculture from a quantitative point of view. Obviously, CCOs have difficulties in breaking through communication and information barriers set up by established political actors. One reason is CCOs' comparative lack of financial resources and insufficient communication know-how. The only exception to the rule is Attac Germany, both in a quantitative as well as in a qualitative way, because it has established a functional—and somehow professional—organizational relationship management practice.

Since Attac in Genoa very seldom organized local press meetings or conferences and instead mainly relied on new decentralized ways of communication with the media, our findings lead us to the assumption that Attac's Web page, mailing lists, and SMS activities formed a vital node between internal and external communication.

Communication creates essential structures and networks within an organization, thus providing a platform for performance and action not previously experienced. By no means do we owe this insight only to the digitalization of communication via ICTs. However, technical and sociopolitical changes have also changed the general public sphere and its boundaries (cf. Calhoun, 1992). This is also true for the communication of political organizations; that is, due to the use of ICTs, the boundaries between internal and external communication are becoming more blurred.[13]

Analysis of the case study of Attac Germany leads to the assumption that CCOs facing this blurring, as a consequence, can now achieve greater communication successes than previously possible. This, in all probability, is because CCOs rely on strong internal communication between their members (networking) fostered through the possibilities of new media like the Internet. The success of Attac's organizational structure likely traces back to its founding as a virtual association utilizing the advantages of digitalized communication.

ICTs have contributed to reducing communication costs, thereby establishing more flexible structures that facilitate interaction between numerous decentralized and localized Attac groups. Additionally, ICTs have also significantly changed communication with journalists and the media. In particular, Attac's success in external communication (media response) with diagnostic frames could be an indicator of the fact that the organization applied new forms of digitalized communication with the media (Web page, SMS, and e-mail). In doing so, it improved relationships with journalists and balanced the weaknesses of a decentralized organizational structure.

Furthermore, perhaps due to its *transnational* network structure, Attac gained a high credibility bonus on the side of the media in comparison to other organizations from the field of counterculture. Journalists attach great importance to authentic, independent, and locally based information in diagnostic framing. In particular, Attac's transnational network structure related to globalization makes Attac one of the few organizations of the counterculture that has been able to establish itself globally.

Attac defines itself as a type of organization "situated between being a network, NGO, and movement" and fights for a pluralistic world (Attac, 2006). Interestingly, Attac's characteristics of being multidimensional and very open regarding new topics and organizational structure do not appear to have had a negative impact on its success to date. This is actually contrary to previous findings on communication problems of decentralized organizations. The data gathered in this study seemingly confirm the advantages of a flexible and decentralized organizational structure—provided that it is combined with the advantages of communicating via modern ICTs.

However, being a mixture of a network *and* an organization has led to conflicts of interest between Attac and its member organizations, especially in those cases where Attac has threatened to undermine the political importance of its member organizations. Furthermore, despite its network structure, Attac is still some distance from reaching its (normative) claim of providing an open communication platform. Within the framework of internal communication, not only access to the organization, but also promotion within Attac's organizational hierarchy requires detailed technical knowledge of the application of digitalized communication. Certain forms of communication, such as open mailing lists and an established discussion forum, display a tendency for information overload. For this reason, Attac's members have complained about the difficulties in reducing this problem (cf. Le Grignou & Patou, 2005).

We have argued that reviewing and developing Castells's recent work on the concept of a network society may help us to better understand the increasingly important role of new ICTs and networks in organizational communication and public relations. To take this study further, our analysis would also have to include other representatives from the field of counterculture as well as be extensively compared to established (political) organizations.

For consideration

1. To what extent do new information and communication technologies add credibility to counterculture organizations and differentiate them from more dominant political actors?
2. What new challenges are arising through social and technical changes that have to be met by (a) internal and (b) external communication management of political organizations?
3. How can nascent as well as established political organizations reach (a) short-term and (b) long-term communication objectives in the political public sphere?
4. How can the relationship between media and activists be described?
5. What are the changing needs of various organizational publics that political organizations should take into account when planning communication activities?

For reading

Holtz, S. (2003). *Public relations on the Net*. New York: AMACOM.

Jones, R. (2002). Challenges to the notion of publics in public relations: Implications of the risk society for the discipline. *Public Relations Review, 28*(1), 49–62.

Kahn, R., & Kellner, D. (2004). New media and Internet activism: From the "battle of Seattle" to blogging. *New Media & Society*, 6(1), 87–95.

Springston, J. K. (2001). Public relations and new media technologies: The impact of the Internet. In R. L. Heath (Ed.), *Handbook of public relations* (pp. 603–614). Thousand Oaks, CA: Sage.

Taylor, M., Kent, M. L., & White, W. J. (2001). How activist organizations are using the Internet to build relationships. *Public Relations Review, 27*(3), 263–284.

Notes

1. The name *Attac* is derived from *Association pour la Taxation des Transactions pour l'Aide aux Citoyens* (Association for the Taxation of Financial Transactions to Aid Citizens).
2. For a more detailed description please refer to Wimmer (2004).
3. The name *ver.di* is an abbreviation of *Vereinigte Dienstleistungsgewerkschaft* (United Union for the German Service Sector).
4. More information available at: http://www.attac.de/interna/selbstverstaendnis011101.pdf (German language document, retrieved June 6, 2006).
5. "Founded in 1990 as an independent non-governmental organisation with offices in Berlin and Bonn. WEED is campaigning for the globalisation of democracy, justice, human rights and environmental sustainability" (WEED, n.d.).
6. German Society for Nature Conservation.
7. As can be seen from the name *Attac*, the first and only political claim for a long period of time was the implementation of the so-called Tobin tax, that is, a tax on exchange transactions applied to all businesses involved in money exchange, making them less lucrative and thereby decreasing an over-liquidity of the money exchange markets while, at the same time, collecting a vast amount of money on an international level to be spent on developmental aid programmes.
8. That is, media system orientation and political decision makers (elite personalities and elite organizations). See Galtung & Ruge, 1965; Östgaard, 1965; Rosengren, 1970; Schulz, 1982; and Staab, 1990.
9. For a detailed overview refer to Kolb (2004).
10. Text messages sent by mobile phone.
11. For a detailed description of the methodological proceedings, please refer to Wimmer (2004).
12. Since the 1990s, activists have used the word *neo-liberalism* for the negative effects of global market liberalism (capitalism) and for free-trade policies.
13. Consequently, the majority of *Attac's* local groups (amounting to a number of more than 250 units) has their own publicly accessible Web pages connected with the main Web page.

References

Attac. (2006). *Zwischen Netzwerk, NGO und Bewegung Das Selbstverständnis von ATTAC 8 Thesen*. Retrieved June 6, 2006, from http://www.attac.de/ueber-attac/was-ist-attac/selbstverstaendnis

Beck, U. (1992). *Risk society: Towards a new modernity*. London: Sage Publications.

Beck, U., Bonss, W., & Lau, C. (2003). The theory of reflexive modernization. Problematic, hypotheses and research programme. *Theory, Culture & Society, 20*(2), 1–33.

Beck, U., Giddens, A., & Lash, S. (1994). *Reflexive modernization: Politics, tradition and aesthetics in the modern social order*. Cambridge: Polity Press.

Bennett, W. L. (2003). Communicating global activism: Strengths and vulnerabilities of networked politics. *Information, Communication & Society*, 6(2), 143–168.

Calhoun, C. (Ed.). (1992). *Habermas and the public sphere*. Cambridge, MA: MIT Press.

Castells, M. (1996). *The rise of the network society. The information age: Economy, society and culture* (Vol. 1). Oxford, UK: Blackwell.

Castells, M. (1997a). *The end of the millennium. The information age: Economy, society and culture* (Vol. 3). Oxford, UK: Blackwell.

Castells, M. (1997b). *The power of identity. The information age: Economy, society and culture* (Vol. 2). Oxford, UK: Blackwell.

Couldry, N., & Curran, J. (Eds.). (2003). *Contesting media power: Alternative media in a networked world*. Boulder, MD: Rowman and Littlefield.

Donk, W. van de, Loader, B. D., Nixon, P. G., & Rucht, D. (Eds.). (2005). *Cyberprotest: New media, citizens and social movements*. London: Routledge.

Entman, R. M. (1993). Framing: Towards clarification of a fractured paradigm. *Journal of Communication*, *43*, 51–58.

Fröhlich, R., & Rüdiger, B. (2006). Framing political public relations: Measuring success of political communication strategies in Germany. *Public Relations Review, 32*(1), 18–25.

Galtung, J., & Ruge, M. H. (1965). The structure of foreign news: The presentation of the Congo, Cuba and Cyprus crises in four Norwegian newspapers. *Journal of Peace Research*, *2*, 64–91.

Giddens, A. (1991). *Modernity and self-identity: Self and society in the late modern age*. Cambridge: Polity Press.

Guiniven, J. L. (2002). Dealing with activism in Canada: An ideal cultural fit for the two-way symmetrical public relations model. *Public Relations Review, 28*(4), 393–402.

Guttenberg, S., Haussmann, C., McFarlane, J., Redler, L., Schlegelmilch, K., Thomas, D., & Wolff, J. (2002). *attac. Die Welt ist keine Ware. Eine Kampagne zum General Agreement on Trade in Services (GATS). Kommunikationsprojekt im Studiengang Gesellschafts- und Wirtschaftkommunikation an der Universität der Künste Berlin* [attac. The world is no good. A campaign about the General Agreement on Trade in Services (GATS). Communication project of the study of societal and economic communication at the Academy of Arts, Berlin]. Retrieved May 29, 2006, from http://www.attac-netzwerk.de/wto/udkprojekt.pdf

Hall, R. D. (1997). *Organizations: Structures, processes, and outcomes*. Upper Saddle River, NJ: Prentice-Hall.

Held, D., McGrew, A., Goldblatt, D., & Perraton, J. (1999). *Global transformations: Politics, economics and culture*. Stanford, CA: Stanford University Press.

Hiebert, R. E. (2005). Commentary: New technologies, public relations and democracy. *Public Relations Review*, *31*(1), 1–9.

Jenkins, H., & Thorburn, D. (Eds.). (2003). *Democracy and new media*. Cambridge, MA: MIT Press.

Keck, M. E., & Sikking, K. (1998). *Activists beyond borders: Advocacy networks in international politics*. Ithaca, NY: Cornell University Press.

Kolb, F. (2004). The impact of transnational protest on social movement organisations: Mass media and the making of ATTAC Germany. In D. della Porta & S. Tarrow (Eds.), *Transnational movements and global activism* (pp. 95–120). Lanham, MD: Rowman and Littlefield.

Le Grignou, B., & Patou, C. (2005). ATTAC(k)ing expertise: Does the Internet really democratise knowledge? In W. van de Donk, B. D. Loader, P. G. Nixon, & D. Rucht (Eds.), *Cyberprotest: New media, citizens and social movements* (pp. 164–179). London: Routledge.

Östgaard, E. (1965). Factors influencing the flow of news. *Journal of Peace Research, 2*, 39–63.

Rosengren, K. E. (1970). International news: Intra and extra media data. *Acta Sociologica, 13*, 96–109.

Schewe, J. (2003). *Netzöffentlichkeit als Alternativöffentlichkeit: Soziale Bewegungen im Internet am Beispiel von "Attac" Deutschland* [Net public as alternative partial public sphere: Social movements on the Internet with the example of Attac Germany]. Unpublished master's thesis, University of Leipzig, Germany.

Schulz, W. (1982). News structure and people's awareness of political events. *Gazette, 30*, 139–153.

Siapera, E. (2004). Asylum politics, the Internet and the public sphere: The case of UK refugee support groups online. *Javnost-The Public, 11*(1), 79–100.

Smith, J. (2001). Globalizing resistance: The battle of Seattle and the future of social movements. *Mobilization*, 6(1), 1–20.

Staab, J. F. (1990). The role of news factors in news selection: A theoretical reconsideration. *European Journal of Communication*, 5, 423–443.

Taylor, M., Gabriel M., Vasquez, G. M., & Doorley, J. (2003). Merck and AIDS activists: Engagement as a framework for extending issues management. *Public Relations Review*, 29(3), 257–270.

Wall, M. A. (2002). The battle in Seattle: How non-governmental organizations used Websites in their challenge to the WTO. In E. Gilboa (Ed.), *Media and conflict: Framing issues, making policy and shaping opinions* (pp. 25–43). Ardsley, NY: Transnational Publishers.

WEED. (n.d.). *Mission statement*. Retrieved July 10, 2006, from http://www.weed-online.org/themen/96267.html

Wimmer, J. (2004). Der Rahmen der Determinierung. Zur Nützlichkeit des Framing-Ansatzes bei der Untersuchung von Beeinflussung zwischen PR und Journalismus am Beispiel des G8-Gipfels in Genua 2001 [The framework of determination. About the usefulness of the framing analysis within the research of the connection between PR and journalism: The case study of Genoa 2001 G8 Summit]. In K.-D. Altmeppen, U. Röttger, & G. Bentele (Eds.), *Schwierige Verhältnisse: Interdependenzen zwischen Journalismus und PR* [Difficult relationships: Interdependence between journalism and PR] (p. 175). Wiesbaden: Verlag für Sozialwissenschaften.

Creating a Dream, Changing Reality

A Brazilian Web Site as a Public Relations Tool for Social Change

Maria de Fátima Oliveira

This chapter discusses how activist groups use new media, specifically Web sites, to promote civic engagement and social change. An example of a successful Brazilian nonprofit organization's Web site is a source of valuable lessons for other activist groups. This chapter also discusses (a) the importance of the Internet to activist groups in general, (b) Web sites as a public relations tool, and (c) the role of framing in the generation of dialogue between organizations and their multiple publics. VivaRio's Web site illustrates how certain frames chosen by the organization, combined with dialogic communication strategies, are able to redefine dialogue about crime, poverty, and unemployment; lessen social difference; and bridge diverse segments of society.

Can a Web site promote social change and fight against pervasive social problems, such as violence, poverty, and unemployment? A reasonable answer would be no; after all, it is just a Web site. Technology by itself is a not a panacea for all the problems of the world. However, some applications of new media suggest new ways to deal with social problems.

Government and activist groups combat social problems through concrete actions; nevertheless, their communication strategies need to go hand in hand with their actions to ensure effective social change.

How activist organizations are using new media, specifically Web sites, to foster social change is the focus of this chapter. As such, a successful example of a Brazilian nonprofit organization's Web site is explored. However, before discussing the Web site and its features, it is necessary to provide background information about the organization.

VivaRio

VivaRio is a nonprofit organization that promotes a nonviolent culture; it fights against crime, poverty, and unemployment through education and job training initiatives. VivaRio was created in 1993 by a group of friends to protest growing violence in Rio de Janeiro. Its first official action mobilized over a thousand people dressed in white to demonstrate for peace. Since then, the organization has achieved an array of accomplishments by supporting inhabitants of Rio de Janeiro's shantytowns.

In 2004, according to the organization's financial reports, VivaRio promoted 1,134 different projects and campaigns, assisting more than 80,000 people in extremely poor neighborhoods (*VivaRio*, n.d., Financial Reports section). Their projects benefit mainly low-income youth, focusing on job and technology training, sports and artistic activities, and counseling on legal and financial issues. Table 1 provides a summary of the projects' activities, audiences, and main goals.

VivaRio relies on several partnerships with businesses to provide job opportunities to individuals attending its training courses. In this way, the organization provides the most basic means to prevent youngsters' involvement with drugs and crime: employment. To put in perspective the problem of criminality in Rio and, consequently, the importance of VivaRio's work, it is worthwhile to compare the city's statistics on violence to some American cities. For example, in 2004, the rate of reported murders in Rio was 42.3 per 100,000 people, whereas, in the same year, per 100,000 inhabitants, New York's rate was only 7.0, Boston's rate was 10.5, Los Angeles' rate was 13.4, and Chicago's rate was 15.5 (Centro de Estudos de Segurança e Cidadania, n.d., Estatísticas de Segurança section).

The organization's capacity to promote social change is strongly associated with its communication strategies. Several activist groups fight against poverty in Brazil; however, VivaRio, with its frequent media presence and effective Web site, is able to attract many investors, partners, and volunteers who support the organization's projects and campaigns (cf. *VivaRio*, n.d., Financial Reports section).

Although the organization's Web site is not a direct way to foster social change, it has a relevant function in raising money and drawing supporters to VivaRio's initiatives. By providing a variety of opportunities for volunteering, the Web site encourages civic engagement and brings together different economic strata of society. To better understand the relevant features of this Web site, the general significance of the Internet and the World Wide Web (WWW) to activist groups will first be discussed.

Table 1: VivaRio Project and Audience Descriptions

Projects	Activities			Audience			Main Goal
	Sports and Artistic Activities	Training Courses	Others	Children	Youth	Adults and Elderly	
Ana and Maria			X		X		To provide support to adolescent mothers during pregnancy.
Children's Hope Space	X	X		X			A place where shantytowns' children can safely play sports.
COAV—Children in organized armed violence		X		X	X		To avoid the involvement of children and youngsters with armed groups.
Citizens' Counseling			X			X	To offer free legal advice to the population.
Community Telecourse		X			X	X	To offer people the means to complete their basic education. Also, to provide computer skills classes.
Disarmament			X	X	X	X	To promote arms destruction campaigns.
Fighting for Peace	X	X			X		To provide boxing training combined with civil rights courses.
Future—computer stations			X	X	X	X	To provide Internet access to shantytowns.
Long-term Job Training		X			X		A seven-month course that prepares young people to work with financial organizations.
Neighborhood Gardeners		X			X		To offer classes in gardening and ecology to adolescents from low-income communities.
Preventing Violence			X	X	X	X	To advise government about crime prevention plans.
Viva Credit			X			X	To facilitate credit access to poor communities' entrepreneurs.
Villa-lobinhos—music project	X			X	X		To provide musical instruction to youngsters from low-income neighborhoods.
Viva Shantytown	X		X	X	X	X	A Web site designed by shantytowns' residents about their communities.
VivaRadio Station			X	X	X	X	A radio station that promotes new talents from poor neighborhoods.

Note. Compiled from information provided on VivaRio's Web site (n.d.) at http://www.vivario.org.br/

Activist Organizations and the World Wide Web

An activist group is described as an organized group of people who share a specific goal and aim to influence other individuals, through actions, in order to promote their objective. This influence may be achieved through educational campaigns and projects, pressure tactics, or force. Certain preconditions facilitate the development of this kind of group. Cultural values and political systems that favor freedom of expression and equal distribution of power are two important factors of activist groups (Smith & Ferguson, 2001; Taylor, Kent, & White, 2001).

Another important aspect of activist groups is reliance on low-cost communication strategies. In general, given their restricted budgets, most activist organizations do not have the resources to sponsor mass media campaigns. The Internet therefore offers a solution to reach a multitude of publics at a low cost (Heath, 1998). Coombs (1998) referred to this as the equalizer factor, affirming that an effective use of the Internet and the WWW makes activist organizations' messages almost as widespread as mass media messages.

Therefore, regardless of economic restrictions, nonprofit organizations, through their Web sites, have similar chances to influence society concerning specific matters, such as social, environmental, and political issues, as large corporations do. Moreover, through the Internet, activist and charitable groups all over the world can be connected, leveraging actions and resources (Orecklin, 2005).

The Internet helps activist organizations build relationships with potential investors, supporters, and volunteers whom they would probably not otherwise meet. As Wallace (2003) pointed out, organizations like What Goes Around[1] (http://www.whatgoesaround.org) are examples of how the Web has the potential to link volunteers and supporters worldwide. Through the What Goes Around Web site, individuals could create a "givelist" of the organizations they supported and could send this list to friends and family, who, in turn, could make donations to organizations in honor of their loved ones. Furthermore, the Internet brings forward issues promoted by organizations, helping groups to survive and excel among a myriad of other activists (Kent & Taylor, 1998; Smith & Ferguson, 2001; Taylor, Kent, & White, 2001). The Internet significantly increases possibilities for fundraising. For example, after Hurricane Katrina, the Humane Society, one of the largest nonprofit organizations dedicated to animal protection, raised about $14 million in relief funds for displaced animals in less than two weeks through a massive Internet campaign (Metz, 2005).

However, merely having a Web site does not provide any of these benefits to activist groups. Not only should organizations' Web sites promote their causes and initiatives, but sites should also foster opportunities to interact and dialogue. The commitment to volunteer and support a group is grounded in trust and belonging. Hence, activist groups' Web sites ought to leverage the characteristics of the Internet by promoting opportunities for dialogue and relationship building between organizations and their potential supporters.

Toward Dialogue

Web sites provide an ample channel of communication for organizations. For nonprofits, Web sites represent the opportunity to provide volumes of information about an organization and its activities to a large number of individuals. Likewise, Web sites provide the chance for community members, investors, partners, volunteers, and curious visitors—the audiences of nonprofit groups—to understand a nonprofit and the scope of its actions (Kent, Taylor, & White, 2003).

Internet users are an active audience who expect more than simple descriptions of an organization and its projects. Users, being attracted to the Internet for its interactive potential, visit an organization's Web site to seek new knowledge and voice opinions. Web sites cannot merely be the reproduction of an informative pamphlet; they need to promote interaction and dialogue.

Kent and Taylor (1998) proposed five principles to promote a dialogic relationship between an organization and its publics through the Web. These principles are a helpful framework to investigate Web sites and their potential as effective tools for public relations.

Principle one: Ease of interface. The first principle indicates that visitors should easily navigate and find information on a site. An organization's Web site should be rich in content, yet succinct and direct in providing information. Site maps, clearly identifiable links, and search engines are examples of this principle.

Web sites should not heavily rely on graphs that take too much time to upload and add little information. Also, if necessary, Web sites should be designed in two versions, a basic one for the average computer and a powerful one for more advanced machines. Such choices emphasize interaction and show that every person can access the organization.

Principle two: Usefulness of information. The second principle states that an organization's Web site will provide useful and trustworthy information to its diverse publics. Rich content is the basis of a useful Web site; however, the hierarchy and structure of information also have to be taken into account.

Not only should diverse publics find interesting information about an organization, but they should also be able to find it easily. Explicative headings and links facilitate people's search for information. A clearly defined mission statement, philosophy, project and campaign descriptions, ways to contribute or volunteer, and links to other relevant organizations are some examples of the types of information that an activist group's Web site should include.

Principle three: Conservation of visitors. Conservation of visitors is grounded in the idea of keeping the audience's interest in a site. Whereas, for commercial sites, the lack of links to other companies is a feature of this principle, for nonprofit organizations, links to other activist groups are necessary to establish credibility.

The initial page of a site plays a major role in attracting and maintaining visitors' attention. The home page should contain relevant, timely, and comprehensible information.

Uploading speed is a paramount aspect of this principle: Users quickly lose interest in Web sites that take too much time to launch.

Principle four: The generation of return visits. In addition to conserving its visitors, an effective Web site should foster a willingness in visitors to return to it often. An attractive Web site is certainly rich in content and constantly updated, but it also needs to include other features. Useful links, a calendar of events, options to have information easily downloaded or regularly sent to an e-mail address, options to bookmark the site, forums, and question-and-answer devices are examples of these features. By generating motives for frequent contact, this principle improves the chances of developing relationships and dialogue between an organization and its Web site visitors.

Principle five: The dialogic loop. If a site follows the other four principles but does not offer opportunities for feedback, its dialogic potential is greatly impaired. Offering ample opportunities for users to get in touch with an organization is the basis of a dialogic relationship. However, the cycle must be completed by timely and professional responses.

It is the organization's responses that show its commitment (or lack of) to deal with its publics' concerns. Opportunities to vote on issues, fill out surveys, and send messages to the institution illustrate this principle. Another way to maintain the dialogic loop is to constantly update information that mirrors public interests and responds to queries.

As Taylor, Kent, and White (2001) pointed out, an organization shows its intention to promote dialogue by following these principles. Nonetheless, the principles are not sufficient to foster a two-way communication process in which the organization and its publics are in a situation of equality. To a large extent, the ways companies position themselves with respect to their publics determine whether a true dialogic relationship will be possible. To determine this mutual positioning, the framing strategies used on organizations' Web sites have to be considered.

Equalizing Positions

Framing means adding salience to selected images and messages about an issue, while downplaying other images, messages, and definitions of the same issue (Entman, 1993). Framing theory emphasizes the ability of any entity—the media, individuals, or organizations—to choose one interpretation of reality and to influence other people's definition of it (Fairhurst & Sarr, 1996; Gitlin, 1980; McCombs & Ghanem, 2001).

The frames picked by activist groups not only define their causes through specific lenses, but also impact their members' perceptions of the matters in question. Indeed, the frames chosen reveal an organization's priorities and goals, as well as influence organization members' understanding of issues. However, framing does not have to be a means of manipulating and persuading people. Instead of being used to manage publics, frames selected by an organization can represent the first step toward a dialogic relationship with publics.

By adopting a dialogic approach to its communication strategies, an organization attempts to elevate its publics to a position of equality. This negotiation of power and hierarchy, however, does not happen by chance (Kent & Taylor, 2002). To a large extent, the ways an organization describes itself through the frames it chooses define whether the public and the organization will stand in an equal position.

Rather than being a way to persuade individuals to be in favor of an organization's objectives, framing can be part of a dialogic model of communication by helping to generate dialogue and nurture relationships between organizations and their constituencies that are based upon respect and trust.

For nonprofit organizations, framing plays a role in determining how the organization's members and supporters will perceive the causes, consequences, and potential solutions regarding an issue. For example, the frames adopted by VivaRio to describe poverty, crimes, and unemployment delineate the organization's main goals, as well as its solutions, presented to potential volunteers, partners, and supporters.

A categorization of frames suggested by Iyengar (1991) is useful in analyzing VivaRio's communication strategies to combat social problems. Iyengar reported that television frames selected for political issues influence audiences' opinions about those issues, primarily as they relate to attribution of responsibility.

Individuals attribute different types of responsibility for problems to other individuals. For example, when one focuses on people as the origin of a difficulty, this person is attributing causal responsibility to the individuals involved in the situation. On the other hand, when someone concentrates on the possible solutions to a situation, this individual is analyzing the difficulty through a treatment responsibility perspective by highlighting ways to overcome it (Iyengar & Simon, 1993).

Different frames will foster the attribution of distinct types of responsibility. Episodic frames adopted by the news media highlight political or social issues in terms of specific events and situations, thereby leading to a causal attribution of responsibility. For example, poverty is often depicted using the image of a homeless individual, which influences the audience to understand the issue as a personal problem. Conversely, thematic frames will describe political or social matters within an abstract context concentrating on feasible solutions. The issues under consideration are not depicted as personal problems; on the contrary, they are defined as something for which the entire society is responsible. Thematic frames lead individuals to examine social issues in accordance with the treatment responsibility approach (Iyengar, 1991; Iyengar & Simon, 1993).

Among the subjects investigated using episodic and thematic frames were crime, poverty, and unemployment. Given VivaRio's focus on these issues, Iyengar's categorization of frames was chosen to study VivaRio's Web site. Grounded in framing and the dialogic approach to public relations, a discussion of VivaRio's Web site and some lessons for activist organizations follows.

Promoting Social Change through Dialogue

VivaRio's Web site is an example of successful use of the WWW to promote dialogue and build relationships between an organization and its multiple publics. This nonprofit takes full advantage of the interaction potential of the Internet by using its Web site as a public relations tool to promote the organization's goals, specifically social change and civic engagement.

The VivaRio site content (*VivaRio*, n.d.) discussed herein was monitored for about one year. Because the main intent was to examine how the organization framed itself in relation to its publics, particular attention was paid to the site's first page, the organization's permanent projects, and the content of sections that have a link on the first page.

Both Portuguese and English versions of the site were used in the study. In some sections of the site, the Portuguese version was richer in content and updated more frequently. For example, the news section of the site was updated every week in the Portuguese version, whereas the English version had a six-month lag. Since VivaRio's publics were mostly composed of Brazilians, the organization's Web site content and communication strategies were evaluated based on the information available—in Portuguese—to the majority of the organization's publics—Brazilians. The five dialogic principles were applied to VivaRio's Web site.

Ease of interface. This feature assures that a Web site is easily navigated and provides useful information in a clear, succinct, and appealing way without excessive reliance on graphs, tables, and pictures.

VivaRio's Web site followed all these guidelines. In a very clear and organized fashion, the site provided a large amount of information to the organization's diverse publics. The initial page had visible links to the main sections of the site, facilitating access to specific information. Furthermore, the site had a search engine with options for basic and advanced searches that provided an additional way to look for facts, events, campaigns, and other information.

By making its Web site rich in content, organized, and easily navigated, VivaRio leveraged the site as an appealing tool to attract potential supporters, investors, and volunteers. A section designed specifically for the media suggested the organization's strategic use of its Web site as a tool to generate future media coverage.

Usefulness of information. This principle implies that relevant, detailed, and updated information will be provided to the multiple publics of an organization. VivaRio tailors its Web site accordingly.

The organization's site presented VivaRio's history and mission statement along with descriptions of projects and campaigns to offer worthwhile information to people with diverse interests in the organization. The site also made VivaRio's financial reports available to target the organization's supporters. Diverse options to donate and volunteer were offered as well.

Conservation of visitors. The main goal of conserving visitors is to generate instances for dialogue and relationship building. However, in order to achieve these aims, a Web site needs to instantly motivate visitors to explore its resources. The initial page of a site is of paramount importance in creating a good impression on visitors. It should provide the most relevant and updated information about the organization and have a reasonable uploading time.

The initial page of VivaRio's Web site presented information about the main aspects of the organization, which were identified through clear headings and links. For example, a visitor could find links to the organization's mission statement, history, policies, advisory board and staff, project descriptions, campaigns, donation and volunteer options, and contact information. In addition, individuals saw the most recent news about the organization and its activities, financial reports, and specified media contacts.

The entire site relied mostly on text and had a reasonable uploading time (six to seven seconds), thus avoiding the bother of long waits.

Generation of return visits. VivaRio's Web site also followed Kent and Taylor's (1998) fourth dialogic principle. Aiming to build relationships with the public, VivaRio structured its site to make visitors interested enough about the organization to want to return continually to the Web site.

Examples of how VivaRio made its Web site appealing and useful for its publics included having links to other activist groups, an updated calendar of events, and many opportunities to interact with the organization. In addition to the Contact Us section, other parts of VivaRio's Web site, such as Donations and Volunteer, presented links to contact the organization. Within VivaRio's project descriptions, users could find links to obtain more information about the projects, as well as to contact the organization. Visitors were encouraged to sign up to receive information about the organization and its activities via e-mail. A gallery of photos and awards also played a role in motivating visitors to become part of the organization.

Additionally, VivaRio promoted several online surveys and petitions where visitors could voice their opinions about relevant social issues. Some of these initiatives became proposals for laws that were sent to and discussed at the Brazilian Congress. For example, in response to a long campaign led by VivaRio against firearms, the Brazilian Congress proposed a referendum about the subject. In October 2005, Brazilians voted on the matter. Although the specific form of the bill proposed by VivaRio and its legislative consultants was not approved, the event raised public awareness of the necessity for stricter regulation of firearms (Toledo, 2005).

Dialogic loop. The last dialogic principle encompasses all the others. Even when a Web site offers useful, updated, and well-tailored information for an organization's multiple publics, if it does not promote opportunities for questions and prompt answers to public concerns, its dialogic potential is diminished. Not only did VivaRio's Web site provide plenty of opportunities for publics to communicate with the organization, but it also responded to these questions quickly. On two separate occasions, requests submitted by the researcher (i.e., to be added to the listserv and to receive information about volunteering) were responded to by the organization in less than forty-eight hours.

Another aspect of VivaRio's Web site that needs to be considered is its main audiences: potential investors, partners, and volunteers. VivaRio has provided computers and Internet access to thousands of disadvantaged individuals (*VivaRio*, n.d., Financial Reports section), although the organization is aware that the majority of its beneficiaries still does not have access to this medium. The organization's site is carefully tailored to attract investors, partners, and volunteers who can help meet the needs of disadvantaged communities.

Nevertheless, VivaRio did not use its Web site only as a magnet for financial support. The organization voiced the needs of low-income neighborhoods—actors normally excluded from any kind of public dialogue. VivaRio's Web site mirrored the organization's belief that investor, partner, and volunteer concerns were as important as those expressed by disadvantaged communities. By highlighting the relevance of partnerships with community associations and focusing on feasible solutions that originated from the neighborhoods' inhabitants, VivaRio's Web site indicated that both disadvantaged communities and wealthy ones—represented by supporters—are brought together as equals in the orbit of the organization. In other words, the programs offered by VivaRio are not merely the ideas of rich people trying to help a poor neighborhood but rather create a space shared by the people assisted and the ones committed to help.

Besides meeting the five dialogic principles previously discussed, VivaRio's Web site also provided an example of how framing can be incorporated in a dialogic approach to public relations. Instead of being used as a means of manipulation, certain frames can describe social problems, such as poverty, crime, and unemployment, in terms of their potential solutions, thereby fostering social change and civic engagement. Next, the frames adopted by the organization are discussed.

Social Problems: Each Individual Has His or Her Share

VivaRio's Web site was analyzed using Iyengar's (1991) categorization of frames. Throughout the site, the organization adopted thematic frames that described social problems in terms of possible solutions. Rather than describing poverty, crime, and unemployment through the exploration of personal stories (which are appealing but may connect these difficulties to personal weaknesses), VivaRio's Web site concentrated on combating the causes of and presenting solutions for such problems that should be pursued at a societal level.

The organization's mission statement not only epitomized VivaRio's main goals, but also was a clear example of the organization's focus (and thematic frame) on solutions. It affirmed the following:

> Mission—to integrate a divided society and develop a culture of peace, interacting with civil society and public policies, working at grassroots and internationally, through: designing and testing solutions to social problems, consultancies, advocacy, training (projects), campaigns, and communication (initiatives). (*VivaRio*, n.d., Financial Reports section)

Social change is achieved through VivaRio's concrete actions in the assisted communities. According to the organization's online financial reports, VivaRio helped more than 140,000 individuals between 2001 and 2004. VivaRio's campaigns and projects promoting job and technology training, sports and arts activities, and legal and financial counseling offer the means for low-income people to change their lives.

However, none of this would be achieved without resources and support from volunteers and investors. VivaRio's Web site communication strategies are well tailored to call the attention of Brazilian society to its initiatives and to motivate donations and volunteering. The amount of media coverage received by the organization is an example of the importance placed on communication. In 2004, VivaRio received a daily average of more than three minutes of TV coverage and nearly seventeen column inches of print media coverage (*VivaRio*, n.d., Financial Reports section). Moreover, its Web site was accessed by more than 450,000 people.

The way VivaRio framed community needs in terms of feasible solutions set the ground for a constructive way to discuss crime, poverty, and unemployment. Through education, job training, and strong partnerships with business, VivaRio provides the means for social change. Through communication strategies that promote solutions to social problems and attribute the responsibility for these matters to society, the organization fosters civic engagement.

Lessons to Be Learned

VivaRio's Web site is a useful example of leveraging characteristics of the WWW as a public relations tool to generate dialogue and build relationships with an organization's publics. However, it is not flawless. For example, the organization's staff and Web site developers did not provide information about evaluation and measurement of the site's efficiency, and it is not possible to say whether the organization has data to validate its communication strategies.

Whereas the site seemed to be the result of carefully planned public relations strategies, attempts to verify whether VivaRio investigates the site's actual outcomes failed, calling attention to one of the most important lessons in public relations: the necessity of evaluation. VivaRio's Web site offers valuable lessons about applying a dialogic approach to an organization's Web site; nevertheless, the efficiency of the strategies adopted could not be verified. Thus, the Web site's apparent potential as a public relations tool could not be affirmed.

Regardless of its significant limitations, VivaRio's Web site offers—especially for nonprofit organizations—a valuable example of how to take advantage of the characteristics of the Internet and foster dialogue between an organization and its publics. The site provided a vast amount of information about the organization and could serve as a valuable resource in attracting visitors with different interests. In this fashion, the site had the low-cost potential to widely promulgate the organization's goals and activities and enabled it to overcome budget constraints on publicity in mainstream media.

As previously stated, VivaRio promotes social change via concrete actions; however, as claimed by the organization, the results accomplished by its projects and campaigns would not be possible if more privileged segments of the Brazilian society did not support such activities. The organization's Web site is a tool not only to promote VivaRio's initiatives, but also to promote civic engagement and, indirectly, social change.

Following the guidelines of a dialogic approach to communication, VivaRio's Web site was rich in content yet easily navigated. The site provided in-depth information about several aspects of the organization and its actions for different publics. Moreover, the site presented necessary features to keep visitors interested and motivated to return. Finally, VivaRio's Web site offered several opportunities for visitors to voice their opinions and contact the organization. Completing the dialogic loop, VivaRio's staff frequently updated site information to reflect the issues voiced by visitors.

The organization's site is also an example of how framing can be a tool for creating dialogue and neutralizing social differences. This function sheds a new light on our understanding of framing theory, which is commonly associated with manipulative goals and persuasion. Framing poverty, crime, and unemployment through the lens of education and technology training, VivaRio's Web site promoted a novel dialogue about social matters.

The frames adopted by the organization set the ground for a position of equality between assisted communities and wealthy supporters. VivaRio's activities showed that the relationships between supporters and assisted communities should be beneficial for both. Low-income individuals have their voices heard and their necessities met, for example, by receiving education and technical training sponsored by several companies. By the same token, such firms have a source of well-trained personnel and, by preventing the involvement of shantytowns' inhabitants with crime and drugs, also have safer environments for their stores, offices, and industries.

VivaRio changes the lives of thousands of people through concrete actions, yet its Web site is a relevant tool to create a forum for a new dialogue about social problems and attainable solutions. The organization's Web site offers a good example of WWW use that other activist groups can emulate. Furthermore, exploring framing from a dialogic perspective on communication, the site illustrates the potential of the WWW to neutralize social differences and bridge different strata of society. This combination of framing with the dialogic approach to public relations offers a novel perspective to scholars, practitioners, and students for the analysis of Web sites and other organizational communication strategies.

For consideration

1. What are the potential ethical implications of framing?
2. What values must an organization embrace in order to fully adopt the five dialogic principles described in this chapter?
3. Is equality in dialogue between organizations and publics more likely through mediated means of communication? Why or why not?
4. What are examples of for-profit organizations that use framing to equalize themselves with their publics and to promote an open dialogue?

5 To what extent could framing enhance and/or diminish an organization's level of transparency with publics?

For reading

Iyengar, S. (1991). *Is anyone responsible? How television frames political issues*. Chicago: University of Chicago Press.

Kent, M., & Taylor, M. (1998). Building dialogic relationships through the World Wide Web. *Public Relations Review, 24*(3), 321–334.

Metz, C. (2005, November). Online donations to the rescue. *PC Magazine, 24*(21), p. 83.

Orecklin, M. (2005, July). A net for volunteers. *Time, 165*(10), p. 68.

Wallace, N. (2003). Web site offers charity gift options. *Chronicle of Philanthropy, 16*(5), 29.

Note

1. At the time of the publication of this chapter, the organization What Goes Around and its respective URL (http://whatgoesaround.org) were no longer active. Nevertheless, the service offered by the organization still is a valid example of the potential of the Internet for activist organizations.

References

Centro de Estudos de Segurança e Cidadania. (n.d.). Retrieved May 13, 2006, from http://www.ucamcesec.com.br/arquivos/estatisticas/ev012005_p04.xls

Coombs, W. T. (1998). The Internet as potential equalizer: New leverage for confronting social responsibility. *Public Relations Review, 24*(3), 289–303.

Entman, R. M. (1993). Framing: Toward clarification of a fractured paradigm. *Journal of Communication, 43*(4), 51–58.

Fairhurst, G. T., & Sarr, R. A. (1996). *The art of framing: Managing the language of leadership*. San Francisco: Jossey-Bass Publishers.

Gitlin, T. (1980). *The whole world is watching: Mass media in the making and unmaking of the New Left*. Berkeley, CA: University of California Press.

Heath, R. (1998). New communication technologies: An issues management point of view. *Public Relations Review, 24*(3), 273–288.

Iyengar, S. (1991). *Is anyone responsible? How television frames political issues*. Chicago: University of Chicago Press.

Iyengar, S., & Simon, A. (1993). News coverage of the Gulf crisis and public opinion: A study of agenda-setting, priming, and framing. *Communication Research, 20*(3), 365–383.

Kent, M., & Taylor, M. (1998). Building dialogic relationships through the Word Wide Web. *Public Relations Review, 24*(3), 321–334.

Kent, M., & Taylor, M. (2002). Toward a dialogic theory of public relations. *Public Relations Review, 28*(1), 21–37.

Kent, M., Taylor, M., & White, W. (2003). The relationship between Web site design and organizational responsiveness to stakeholders. *Public Relations Review, 29*(1), 63–77.

McCombs, M., & Ghanem, S. (2001). The convergence of agenda setting and framing. In S. D. Reese, O. H. Gandy, Jr., & A. E. Grant (Eds.), *Framing public life: Perspectives on the media and our understandings of the social world* (pp. 67–82). Mahwah, NJ: Lawrence Erlbaum Associates.

Metz, C. (2005, November). Online donations to the rescue. *PC Magazine, 24*(21), p. 83.

Orecklin, M. (2005, July). A net for volunteers. *Time, 165*(10), p. 68.

Smith, M. F., & Ferguson, D. P. (2001). Activism. In R. L. Heath & G. Vasquez (Eds.), *Handbook of public relations* (pp. 291–300). Thousand Oaks, CA: Sage Publications.

Taylor, M., Kent, M., & White, W. (2001). How activist organizations are using the Internet to build relationships. *Public Relations Review, 27*(3), 263–284.

Toledo, D. (2005, October 17). Referendo opõe visões distintas de combate à violência [Referendum opposes visions different from combat to the violence]. *BBC Brasil*. Retrieved July 14, 2006, from http://www.bbc.co.uk/portuguese/reporterbbc/story/2005/10/051017_referendodieg02ro.shtml

VivaRio. (n.d.). Retrieved April 16, 2006, from http://www.vivario.org.br/

Wallace, N. (2003). Web site offers charity gift options. *Chronicle of Philanthropy, 16*(5), 29.

Engendering the Arabic Internet

Modern Challenges in the Information Society

Mahmoud Eid

The Arab world faces a wide range of modern challenges to develop the Internet in the global information society. This chapter investigates the political environments prevailing in the Arab countries and sheds light on their establishment of telecommunications industries and new media. It demonstrates the many differences among the Arab countries that influence Internet access and development, as well as the huge dual digital divide—between the Arab world and the rest of the whole world, and within the Arab countries themselves. Leading countries in the Arab world have played a major role in shaping and enhancing the Arabic Internet. However, the Arabic Internet is still too far from achieving any significant accomplishments on the global and even the Arab levels. The chapter discusses the various experiences and examples of Arabizing and Islamizing the Internet and provides backgrounds and explanations of the major challenges and obstacles that face Arabic Web sites. Finally, it highlights a group of recommendations that may help the Arabic Internet take better shape and content, enabling it to better perform regionally and globally in the information society. The types of democracy prevailing throughout the Arab countries constitute a major factor affecting the use and production of new media and have a fundamental, significant impact on how public relations can be practised by both governments and businesses.

Democracy Surviving

Democracy cannot survive without the achievements of a group of presuppositions. For any political system to be democratic, there has to be a considerable understanding of the process of democratization. In many areas in the world, democracy does not seem to be able to find suitable environments to flourish. In our era of globalization, with increasing reliance on the information society and new media, healthier environments for democracy become a necessity. The Arab world is an area where the appropriateness of democracy is hugely debated, which, in turn, appears to have an affect on its usage of the Internet and other forms of new media. In order to assess the functioning of such new media in this area, it is important to highlight how democracy survives.

Democracy presupposes self-discipline and political tolerance, without which it cannot survive. Popular participation is an indispensable part of a democratic framework. Public deliberation is essential to democracy in order to ensure that the public's policy preferences—upon which democratic decisions are based—are informed, enlightened, and authentic. Democracy requires the construction of a vibrant, vigorous, and pluralistic civil society. Without such a civil society, democracy cannot be developed and secured. Democratic theory not only has to recover the notions of civil society and social spaces free from state interference, but it also has to understand the cultural and expressive resources of a free and independent citizenry. The democratic revolution is the cumulative achievement of citizens who become actively involved in civic movements and independent media. Democracy involves the widening of access to government information and the freedom to debate political issues at both a personal level and through independent media. In sum, the democratization process aims at bringing decision-making closer to the people affected by those decisions and diffusing power in society. Furthermore, it includes notions of access and accessibility, equal opportunity, fairness, and equity in social relations. Simply, democracy is not just a form of government. It is a pattern of thought and a way of life (Bruck, 1990; Diamond, 1992; Langdon, 1999; Ojo, 1999; Page, 1996; Raboy, 1990; Raffer & Salih, 1992).

With regard to the usage of telecommunications and new media, according to Winseck and Cuthbert (1997), there are three faces of democracy: *limited/technical* democracy, *pragmatic* democracy, and *communicative* democracy. Although democracy is often thought to be a very expansive concept, in practice, it mainly refers to the periodic right to vote, the ability of citizens to change their political leadership, participation through/as representation, rule by expert systems and policy agencies, and the institutional separation of the economy, government, legal system, and civil society on the basis of formalized rules/laws and functionally based competencies. This theory of democracy also separates private interests from public interests, attempting to expand the boundaries of the former while tightly circumscribing the scope of the latter. The technical view of democracy privileges *stable* forms of *representative government*, which are derived from clear distinctions between a concept of participation rooted in periodical voting and representation, on the one hand, and the actual day-to-day governance of societies, on the other. On the first side of the distinction is the domain of the people, while on the other side of the divide are the rulers and experts who carry out governance and policy making on a day-to-day basis. From this view, democracy, by nature, must be *limited democracy*.

From the limited democracy perspective, communication systems are privately financed so as to secure autonomy from state interference, based on a set of negative freedoms[1] clearly delineated in a constitution so that people do not confuse the realm of the possible with what is actually on offer, and representative social relations. Representative media contain clear distinctions between the majority of people who receive and consume information and a smaller class of media professionals who represent the public interest and mediate the relations between the state, economy, and civil society.

In contrast to the previous view, others have offered more expansive views of democracy that are not so much rule based and rule maintaining as they are rule altering and transformative. *Pragmatic* democracy means "the integration of capital, state, community, and labour into formal structures of problem resolution and 'class harmonization'" (Winseck, 2001, p. 145). The *communicative* dimension of democracy is crucial not only to the legitimation of power and authority, but also to the cultivation of democratic minds.

It is argued here that most Arab countries follow the *limited* democracy type in their communication policies, while only few follow that of *pragmatic* democracy. Based on Winseck and Cuthbert's (1997) explanation of the three types of democracy, when applied to Arab countries, one can see that *limited* democracy in most Arab communication systems is reflected in the high number of rules, low importance of social change, scarcity of citizen participation, censorship, and understanding of communication as a political knowledge. At the same time, *pragmatic* democracy can be seen in a few Arab communication systems, reflected in a moderate number of rules, moderate importance of social change, regular but elite-guided participation of citizens, mixed negative and positive freedoms, and an understanding of communication as an integration of mediated participation.

Although many Arab countries have experienced a degree of political liberalism at some point in their contemporary history—most notably Egypt, Iraq, Jordan, Kuwait, Lebanon, Morocco, and Syria—none of these experiences has given rise to fully fledged democratic systems (Aliboni & Guazzone, 2004). The failure of political regimes in the Arab world to deliver on democracy is coupled with some other failures, such as not delivering on development nor responding to rapid social change. In fact, "the system could not be democratic if a democratic principle could be used to usurp the individual rights of expression and association on which the whole system was dependant" (Adam, 1992, p. 13). Arab journalists still face many problems and challenges, including the political, cultural, and economic environments where the Arab media function. Arab journalists, publishers, and other media practitioners continue to be victims of harassment and political pressures, including dismissal, censorship, restrictions on travel and passport withdrawals, physical assault and torture, arrests and detentions, abduction, and exile. Along with such problems as low salaries, a lack of adequate legal protection, excessive bureaucracy, and administrative constraints that affect journalists' performance and make them vulnerable to possible conflicts of interest and outright corruption, more general economic, political, and cultural concerns also limit their freedom of expression.

The Internet as a new medium in the Arab world is facing obstacles similar to those from which the traditional Arab media are still suffering. Major obstacles are related to freedom of expression, control and ownership, legal regulations, privatization, censorship, and so on. In Arab countries, most notably, censorship is targeting new media the same way it has done with traditional media. "Whether newspapers are government owned or private, they are subject to considerable influence, and even censorship, by authorities. This is also true of broadcasting, though more stations are government-owned or -controlled than private throughout the region" (Ogan, 1995, p. 202). In spite of the fact that residents of the Middle East are going online in increasing numbers, many governments in that region are hoping to control access to sensitive political and religious discussion, as well as sex-related material (Sorensen, 1996).

Governments have adopted various means to restrict the flow of information online ("Internet in the Mideast," 1999). Kuwait ensures that no pornography or politically subversive commentary is available. Abu Dhabi's Internet clubs ban sexual, religious, and political materials on the Internet to respect local laws. Bahrain, which went online in December 1995 through the government-run phone company, Batelco, installed an expensive system to block access to certain Internet-related sites. Jordan authorities asked GlobeNet, a U.S. firm that won a contract to provide Internet service in 1995, to install a special screening facility to control sexually explicit material. Saudi Arabia confines Internet access to universities and hospitals and inspects all local accounts through the Ministry of Interior, claiming to be protecting people from pornographic and other harmful effects of the Internet. Morocco procures Internet service and governs all aspects of the Internet's operations. Internet service is targeted at the banking and insurance sectors, universities, and multinational corporations. Most Arab governments justify their measures to restrict Internet access with the necessity to protect cultural identity.

Into the Information Society

The globalization of media has great influence on the development, policy, and regulation of communication systems all over the world. Information technology (IT) and the telecommunications industry are unique examples of areas that have been greatly influenced by globalization. Until the early 1980s, as Martinez (1996) explains, telecommunications equipment was manufactured primarily by companies located in developed countries, producing mostly for captive local markets and for developing countries as well. In recent years, developing countries have contributed to and then taken over a surprising portion of this arena. In fact, "the development of new technologies and services have [*sic*] given developing countries the possibility to 'leap-frog' their infrastructure problems" (*Telecommunication Policies*, 1996, para. 1).

Digitally based communication media technologies have catapulted modern Arab societies into globalization, placing further pressures on them to cope with the imperatives of the new information age. The digital communications revolution sweeping through the Arab world has stimulated intellectual and political debates, spawning numerous views on the social, economic, and cultural implications of new media. The new media are opening public access to international communication. The Web gives instant access to international content, and e-mails facilitate interpersonal communication across borders. These tools are more individual centred and less government controlled than old pathways to international communication. Digitization and new media have led to an unprecedented democratization of international communication and empowerment of the growing number of people who have access to them (Ayish, 2001; Chalaby, 2005).

The convergence of computers, telecommunications, and IT is expected to bring new telecommunication products and services progressively within reach of larger proportions of the global population. This trend has led to new customer needs, suppliers, and operators, and it is exerting tremendous pressure for changes in the traditional policy and regulatory framework of the telecommunication sector in most countries of the world. Indeed, the Arab world is not an exemption. As a result of an increasing vision of telecommunications as a key factor for Arab economic development, the first International Telecommunications Union (ITU) Arab Regional Telecommunication Development Conference adopted a resolution on the restructuring of the telecommunications sector in the Arab states. This resolution urged countries to study and propose appropriate national information and telecommunications policies that will cover the regulation and operation of the sector (*Telecommunication Policies*, 1996).

In Saudi Arabia, for example, "following 40 years of slow growth, the telecommunications industry expanded with remarkable speed, from a rudimentary network of 188,000 working lines in 1978 to about 858,000 lines only five years later—an increase of 455%" (Kayal, 1997, p. 163). Jordan has taken several steps toward improving its telecommunications sector; however, "these initial steps will have to be accelerated if the country is to achieve its vision as a center of international trade and finance" (Vivekanand & Kollar, 1997, p. 157). In Yemen,

> Because the two parts of Yemen were under two fundamentally different political and economic systems until May 1990, telecommunications in the two parts developed differently. In general, the North has an advanced and more extensive national and international telephone system, whereas the South, despite a better and vaster inheritance from Britain, has a technologically backward, inadequate, and poorly performing system. (Chowdary, 1997, p. 172)

Comparatively, Bahrain

> has shown an extraordinary capability for acquiring and using advanced telecommunications. Consequently, the island nation has become a vital regional telecommunications center in the Middle East. The field of telecommunications may be similar to computers, in which relatively wealthy developing countries like Bahrain can successfully leapfrog evolutionary stages of tech-

> nology diffusion by capitalizing on technology transfer and effective assimilation of advanced technology. (Winterford & Looney, 1997, p. 223)

The ongoing developments in telecommunications have paved the road for Arab countries to go online and join the global information superhighway. There are three categories of Internet providers in the Arab world: countries with a single provider, most often a government post, telegraph, and telephone (PTT) provider, such as Jordan, the United Arab Emirates, Bahrain, Qatar, Kuwait, and Oman; countries with multiple providers, such as Lebanon, Egypt, and Morocco; and countries where Internet services are confined to research centers and universities or are not available at all, such as Iraq, Syria, Sudan, and Libya (Ayish, 2001, p. 121). The level of development cannot be assumed to be homogeneous, as individual countries differ greatly in educational standards, financial strength, and willingness to innovate. The level of political acceptance of the new medium also varies; consequently, some relatively wealthy countries with a lot of high-tech potential only have a few Internet ports, whereas the number of users is growing much faster in other structurally weaker countries (Kirchner, 2001).

The development of telecommunications has differed greatly from one country to another in the Arab region. Although the twenty-two Arab countries are similar in religion, customs and values, history, and language, they differ[2] in many other aspects, including wealth, population, area, geographical location, political directions, and foreign relations. Differences among these countries have been clearly reflected in such things as fields of telecommunication infrastructures, information and media productions, communication policies, and cultural industries. For example, the wealth of the various countries is very different, populations vary from less than four hundred thousand to over fifty million, and country areas vary from 688 to over 2.5 million square kilometers. Such differences have an impact on the financial resources required for providing teleservices throughout a country and the rapidity with which demand can be met (*Telecommunication Policies*, 1996).

Arabs on the Internet

In the last decade, the Arab world has witnessed developments in the fields of IT, telecommunications systems, the Internet, and satellite broadcasting that have each caused some dramatic changes in Arab economics, culture, and politics. However, the picture is not as bright and optimistic as some may imagine. In the Arab region, people do not have sufficient access to media and information technologies compared to global rates, other countries in the region, or in proportion to the population of the Arab world. Statistics show that access to the Internet in the Arab world is restricted to the elite,[3] who have the skills and financial power necessary to take advantage of this medium. The current picture of the Arab telecommunications sector, media and information production, and cultural industries demonstrates that, while the Arab world may appear to have the technological qualifications, there are still many deficiencies that stop it from playing any significant role in the era of globalization.

Despite the fact that the adaptation of new media, technology, and telecommunications in the Arab world is faster and increasing at a higher rate than that of traditional media—daily newspapers, radio, and television—the public usage is much lower than in many other areas of the world. Moreover, there are even significant differences in such usages among the Arab states. There are clear cultural constraints on the use of the Internet in a group of Arab countries, given that the twenty-two Arab countries are not equal in accessing or using the Internet. This group suffers from restrictions that place them in an unequal situation not only with other Arab countries, but also, more significantly, with the rest of the world. Some of these constraints come from religious conservation, and others come from social norms and traditions. Aladwani finds the following:

> It is no exaggeration to assert that we, as an information technology community of researchers and practitioners, know very little about the state of the Internet in the Arab world and more so in some parts of it. (2003, p. 9)

Throughout the Arab world, the Internet is more commonly a child of telecommunications concerns and the media arms of commercial conglomerates. New technologies, the convergence of satellite broadcasting and the Internet, a multiplication of actors in a widening public realm, and generational change are bringing forward the region's first cohort raised on television, with increasing numbers adept in the Internet (Anderson & Eickelman, 1999). In 1999, all Arab countries, except Libya, Iraq, and Syria, allowed their publics to access the Internet through local service providers ("Internet in the Mideast," 1999). Even in poor countries, such as Algeria, Morocco, and Tunisia, leased lines and dial-up access to the Internet and Internet gateway host sites registered with the top level domain name system (DNS) were offered. However, the level of government support for developing network services varies considerably among the three states (Kavanaugh, 1998). Berenger notes the following:

> While an estimated billion global citizens were hooked up to the Internet by 2006, 5 billion were not. At any given time in 2005, two thirds of all North Americans were connected to the Internet, and while usage doubled from 2000 to 2005, growth around the world was more dramatic, possibly because places like Africa, the Middle East, and Latin America, have more ground to make up. (2006, p. 56)

This digital divide can be noticed even more clearly when we look at statistics in the Arab world. Table 1 was created based on 2006 data drawn from the Internet World Stats: Usage and Population Statistics Web site ("Internet Usage Statistics," 2006). It lists the twenty-two Arab countries[4] (originally classified under regions of Africa and the Middle East) to demonstrate the huge digital divide between the Arab world and the whole world in terms of Internet usage in proportion to population. As shown in Table 1, the entire Arab world percentage of world Internet usage is only 1.9%. That is, among the more than 316 million Arab peoples (4.9% of the world population) in twenty-two countries, only 19 million have access to the Internet, with a penetration rate (6.1%) much less than the world

Table 1: Internet Usage and Population Statistics in Arab Countries

Regions	Population (2006 Est.)	Population % of World	Internet Usage (Latest Data)	% Population (Penetration)	Usage % of World
Arab Countries					
Algeria	33,033,546	0.51%	845,000	2.6%	0.083%
Bahrain	723,039	0.01%	152,700	21.1%	0.015%
Comoros	666,044	0.01%	8,000	1.2%	0.001%
Djibouti	779,684	0.01%	9,000	1.2%	0.001%
Egypt	71,236,631	1.10%	5,000,000	7.0%	0.489%
Iraq	26,628,187	0.41%	36,000	0.1%	0.004%
Jordan	5,282,558	0.08%	600,000	11.4%	0.059%
Kuwait	2,630,775	0.04%	600,000	22.8%	0.059%
Lebanon	4,509,678	0.07%	600,000	13.3%	0.059%
Libya	6,135,578	0.09%	205,000	3.3%	0.020%
Mauritania	2,897,787	0.04%	14,000	0.5%	0.001%
Morocco	30,182,038	0.46%	3,500,000	11.6%	0.342%
Oman	2,424,422	0.04%	245,000	10.1%	0.024%
Palestine (West Bank)	3,259,363	0.05%	160,000	4.9%	0.016%
Qatar	795,585	0.01%	165,000	20.7%	0.016%
Saudi Arabia	23,595,634	0.36%	2,540,000	10.8%	0.248%
Somalia	12,206,142	0.19%	89,000	0.7%	0.009%
Sudan	35,847,407	0.55%	1,140,000	3.2%	0.111%
Syria	19,046,520	0.29%	800,000	4.2%	0.078%
Tunisia	10,228,604	0.16%	835,000	8.2%	0.082%
United Arab Emirates	3,870,936	0.06%	1,384,800	35.8%	0.135%
Yemen	20,764,630	0.32%	220,000	1.1%	0.022%
Arab Total	**316,744,788**	**4.9%**	**19,148,500**	**6.1%**	**1.9%**
Africa	915,210,928	14.1%	23,649,000	2.6%	2.3%
Asia	3,667,774,066	56.4%	364,270,713	9.9%	35.6%
Europe	807,289,020	12.4%	291,600,898	36.1%	28.5%
Middle East	190,084,161	2.9%	18,203,500	9.6%	1.8%
North America	331,473,276	5.1%	227,303,680	68.6%	22.2%
Latin America / Caribbean	553,908,632	8.5%	79,962,809	14.4%	7.8%
Oceania/Australia	33,956,977	0.5%	17,872,707	52.6%	1.7%
World Total	**6,499,697,060**	**100.0%**	**1,022,863,307**	**15.7%**	**100.0%**

Note. Table was produced based on data retrieved July 15, 2006, from Internet World Stats: Usage and Population Statistics at http://www.internetworldstats.com/stats.htm

average penetration rate (15.7%). This picture was not clearly obvious when the different Arab countries were classified by Internet World Stats under continents and regions, such as Africa and the Middle East.

Arab countries with the highest rates of Internet access are Egypt (0.489%), Morocco (0.342%), and Saudi Arabia (0.248%), while those with the lowest rates are Comoros (0.001%), Djibouti (0.001%), and Mauritania (0.001%). However, the Arab countries with the highest percentages of Internet usage do not have the highest penetration rates (i.e., the percentage of those who are using the Internet in proportion to the total population in these countries). Arab countries that have the highest penetration rates are United Arab Emirates (35.8%), Kuwait (22.8%), Bahrain (21.1%), and Qatar (20.7%). In sum, there is no one country in the Arab world that has a high percentage of using the Internet and, at the same time, is on top with regard to penetration rates, albeit Morocco (0.342%; 11.6%) and Saudi Arabia (0.248%; 10.8%) are doing well on both levels of global usage and penetration, respectively.

It is true that the digital divide is quite obvious among the Arab countries themselves, but, most importantly, it is very evident between the whole Arab world and all other regions in the world. The lowest rates of accessing the Internet in relation to population size are in Africa (2.6%) and the Middle East (9.6%)—where all Arab countries were originally classified by Internet World Stats—while all other regions of the world have much higher rates, chiefly North America, Australia, and Europe. However, some individual countries in the Arab world are expanding the Internet at higher rates than others. For example, "Egypt's Internet growth rate is estimated to be an annual 500% . . . Egypt has maintained this 500% growth rate for the past four years, mainly due to major efforts made by the government to spread the new technology" (Abdulla, 2006, p. 94). Overall, the collective Arabic access to the Internet is dismal: It is the worst in the world.

Leaders

One example of a highly qualified Arab country in the telecommunications sector is the United Arab Emirates. It subscribes to the Arab Satellite Communications Organization (Arabsat) and has upgraded its entire telecommunications service with digital technology. Emirates Telecommunications (Etisalat), the United Arab Emirates' federal telecommunications authority, launched its Emirate Internet and Multimedia unit after a major organizational overhaul. Other e-services into which Etisalat is moving include Warraq, an online encyclopedia of Arabic literature, and a bilingual search engine enhancing Arabic-language Internet browsing ("United Arab Emirates," 2001, p. 34). Egypt is another example:

> Telecommunications in Egypt is now controlled by Telecom Egypt, a joint-stock company in which the Egyptian government owns the majority shares. With over 9 million subscribers, Telecom Egypt is the biggest telecommunications operation in the Arab world. . . . Egypt is also witnessing a boom in computer ownership, despite the relative high cost of computers. (Abdulla, 2006, pp. 93–94)

Furthermore, Egypt and Saudi Arabia are the best examples of major players in the Arab region with regard to the adoption and production of new media technologies. Egypt is an Arab-world leader in the development of broadcasting and has influenced radio and television development in the region; it is also the primary television program producer in the area. Egypt has been known as a leader in media content and production in the Arab world. Cairo—known as the Hollywood of the Middle East—is the Arab-world film capital. Arabic programs, movies, music, and dramas are mainly produced in Egypt, and Egyptians play most of the roles in this process. These cultural products are exported in high volumes to all Arab countries and make up the biggest portion of Arabic media content. As a result, the Egyptian dialect of colloquial Arabic is most widely understood in the Arab world. Saudi Arabia is a wealthy, large, and politically important country that has introduced Western technology and cultural forms with great caution because of the conservative nature of its Islamic culture.

A look into the Internet experiences of these two countries uncovers two major prevailing tendencies in the Arab world: *Arabizing* and *Islamizing* the Internet. Egypt has been the best example of provoking the awareness and usage of the Arabic Internet as well as Arabizing its content, while Saudi Arabia is the best example of promoting the Islamist Internet.

The Internet was introduced to Egypt in October 1993 by the Cabinet Information and Decision Support Center. In January 2002, the government started a plan to increase Internet connectivity. Access to the network, through any of the sixty-eight Internet service providers (ISPs) then serving the Egyptian market, became free for all. Users only had to pay the small price of a telephone call while connected to the Internet. Telecom Egypt paid 70% of the telephone revenues from Internet use to the respective ISPs. Since this initiative was introduced, ISPs have competed in offering more access lines and better services to prospective customers to entice them to dial up through their access numbers. Others introduced high-speed broadband Internet service so that they could have a competitive edge over the free service. There are now more than 180 ISPs in Egypt, besides a multitude of public libraries and cybercafés (Abdulla, 2006).

Through the efforts of the United Nations Development Program, Technology Access Community Centers (TACCs) were launched in March 1999 in two cities in the Sharkiya Governorate, about eighty kilometers north of Cairo. The "cybercafé for the poor" project aimed at providing affordable public access to the Internet and other communication technologies to residents of rural and remote communities. The project also provided free training on the use of such technologies. More than 1,400 Web pages were developed in Arabic, including a much needed database of local medical practitioners that was later adopted and further developed by the Ministry of Health and Population. The project introduced the concept of e-commerce to the Sharkiya Governorate and encouraged the establishment of IT-related businesses, cybercafés, and Web development projects. The project's success encouraged its replication in two other Egyptian areas: Siwa and Luxor. It is estimated that each TACC is visited by an average of six hundred users per month. On the broader Arab level, replication is under way in Jordan, Syria, and Yemen (Abdulla, 2006).

Among the more than nine hundred IT clubs registered in Egypt, 21st Century Kids Clubs were established as a joint effort between the government, nongovernmental organization, and private sectors. First established in 1997, the clubs provide Internet access, software, and IT trainers to help children navigate their way through technology. In 2001, there were forty kids' clubs available in seventeen governorates. There are now over four hundred clubs throughout Egypt's twenty-six governorates (Abdulla, 2006).

Egypt's online government portal is a well-developed site that has recently started to offer several e-government services. These include automobile license renewals, traffic ticket payments, phone bill payments, electric bill payments, filing for taxes, and university applications. E-commerce has started to emerge in Egyptian society, albeit at a slow pace. Efforts in this regard include establishing Web sites for major Egyptian financial institutions, including the Central Bank of Egypt and the National Bank of Egypt. Banks have started aggressive marketing campaigns to spread the adoption and use of credit cards in the Egyptian cash-oriented society. A project to further spread the use of asynchronous transfer mode (ATM) networks, and therefore encourage the use of banking cards, has been under way with a budget of $600 million. At the end of 2005, ten banks were offering e-banking services online (Abdulla, 2006).

Other Egyptian sites have also tried to find creative ways to answer customer demands and overcome e-commerce obstacles. For example, Otlob.com allowed users to place online orders for a variety of foods, pharmaceuticals, flowers, and video rentals, which they would receive in thirty minutes to an hour. The site offered access to over three hundred restaurants and had over four thousand registered users, averaging 25,000 hits per day. Egypt's biggest portal, Masrawy (http://www.masrawy.com), offered a database of contact information to more than 12,000 businesses in Egypt. The site also offered real-time information on the Egyptian and international stock markets. Other portals, such as Yallabina (http://www.yallabina.com), offered information from an Egyptian cultural perspective (Abdulla, 2006).

The Internet has also allowed for media convergence, the medium itself being used as a news provider (online newspapers and magazines) as well as an online radio and television broadcaster. Throughout the years, Egypt has led the Arab world in introducing such media Web sites and Arabizing their content. These efforts started in 1996 when Egypt's Information Highway Project was launched, putting the first Arabic Web pages on the Internet. The Egyptian newspaper *Al Gomhuria* was the first Arab publication to launch an electronic version in 1996. The *Al Ahram* newspaper, which has been Egypt's most widely distributed daily since 1876, followed with its own electronic version in 1998. Today, most Arabic- and English-language Egyptian newspapers and magazines have electronic versions on the Web. One Web site, Sahafa Online (http://www.sahafa.com), listed at least 250 online publications within the Arab world (Abdulla, 2006).

The Islamist[5] Internet embodies many traditional characteristics of online media as well as a wide range of views on the place of politics in Islam. It displays a geographic fluidity that bypasses national borders but remains bound by the imbalances of the global information infrastructure. Its fairly low costs of entry and production allow for a diversity of voices.[6]

However, authoritative institutions and governments offer innovative attempts to restrict content and monopolize discourse as well as limit the Islamist Internet as a source of information. As a result, the Islamist Internet may not offer a comprehensive library of Muslim thought. It often distorts Islamic discourse through the prisms of sectarian Muslim technophiles (Maguire, 2005).

In general, there are four main clusters (religious modernism, Saudi opposition, pro-Saudi traditionalists, and jihad participants and enthusiasts) that identify major themes and examples of the Islamist Web sites. These clusters are united by a common political focus, even though the specific content, location, and format may differ. The *religious modernism* Web sites (e.g., http://www.islamicity.com, http://www.islamonline.net) reflect a willingness to participate in Western political culture and encourage a more active civil society within the Muslim world. As well, they maintain a strong religious component (e.g., online access to core Islamic texts, fatwa and counseling services, and question-and-answer sessions with Muslim scholars) and attempt to strike a balance that often lands them in controversy. The *Saudi opposition* Web sites (e.g., http://islah.tv and the now-inactive http://www.cdlr.net,) represent the exiled, Islamist opposition to the Saudi monarchy. Organizations, such as the Committee for Defense of Legitimate Rights (CDLR) and the Movement for Islamic Reform in Arabia (MIRA), pioneered the use of "small" media (i.e., Web sites) to expose corruption in the Kingdom of Saudi Arabia. The *pro-Saudi traditionalists* Web sites (e.g., http://www.salafipublications.com, http://www.salafitalk.net, http://www.troid.org) focus on reform of the Muslim world based on careful adherence to core Islamic texts and practices. They promote the Salafi Manhaj (the methodology of the early Muslims) and rigorously oppose Bid'ah (innovated religious beliefs and rites). Although politics is not a vital component of their content, these sites actively discourage rebellion against rulers, particularly the royal family of Saudi Arabia. The *jihad participants and enthusiasts* Web sites (e.g., http://www.as-sahwah.com, http://www.jusonenews.com, http://kavkaz.org.uk), probably the most notorious and contested cluster of the Islamist Internet, offer news, analysis, and discussion forums that focus on jihad.[7] The recent conflict points in the Muslim world (Palestine, Chechnya, Afghanistan, Iraq, Kashmir, etc.) establish, to a great extent, valid conditions for jihad. Yet, the ideological orientation, sectarian allegiance, and combat tactics of various groups lead to a great deal of controversy (Maguire, 2005).

Modern Challenges

According to Rugh,

> Arab states feel bound together by strong cultural and psychological ties. The vast majority of them regard Arabic as their mother tongue; most of them share a single culture, language, and religion, and their sense of a common destiny is very strong. Nationalism, both in the pan-Arab sense and as felt toward the newer individual nation-states, separate and distinct from Western or any other identity, is a powerful force. And despite the differences in wealth, the Arabs are all living in a developing world environment of rapid economic and political change in which high priority is given to modernization. (2004, pp. 1–2)

Arabic is the official language of the twenty-two Arab countries. Each of these nations has a pressing need to adapt Arabic to the demands of modern science and technology. Although language planning in the Arab world purportedly enjoys support from government, education, and business, this professed support is often no more substantial than the ephemeral political unions of various Arab countries. Language planning is an issue of contemporary concern around the globe. Every sovereign nation wishes to preserve its national tongue and maintain its status as a preferred medium of communication. However, the phenomenon of globalization, coupled with the increasing hegemony of English, has motivated many nations to revisit their language planning policies with a view of ensuring and strengthening the preeminence of their own languages. The Arabic-speaking countries, while recalling with pride their historical dominance in the medieval scientific arena, are now struggling to prevent the language from an inundation of foreign, modern terminology. The main goals of the official agencies of language planning in the Arab world have always been the regeneration of Arabic as an effective communication medium for modern science and technology as well as the preservation of the purity of the language. All proposals for change are carefully scrutinized to ensure compatibility with the phonological, syntactic, and morphological structure of Arabic. The majority of Arabic planners show considerable reluctance to tamper with the fundamental linguistic and grammatical principles of the language. Although the Arab countries have strong practical and economic reasons to collaborate on scientific and technological issues, including terminology and standardization, the lack of inter-Arab cooperation has stunted this potential route for development (Elkhafaifi, 2002).

Figure 1 demonstrates the fact that Arabic does not even place among the top ten languages used on the Internet. English (30.6%) is the dominant language on the Internet, followed by the growing languages of Chinese (13%), Japanese (8.5%), and Spanish (7.9%) ("Internet Users by Language," 2006). Thus, in general, the claim that Arab media are unique in terms of content and style may garner more negatives than positives. In the era of globalization, it is fairly incorrect to describe the Arab media as *global*, given Arabic's uncompetitive situation with other languages in the media of globalization. Some (e.g., Sabry, 2005) even argue that it is sometimes inaccurate to describe the media in the Arab world as *Arab*, given their dependence on Western content, style, and sources.

The Arab media cannot easily be described as global, because they cannot transcend nation-state boundaries and language communities, use English (the language of globalization), or attract a cross section of international audiences that is not limited to the rich and influential. They do not even have sufficient access to the resources and means of production necessary to compete at a global level. Arab media are hampered by restrictive state policies and also a lack of acceptance of the new media by state authorities. They are mostly nation bound in terms of regulation, programming, and power structures. In addition, the dependence of Arab media is extraordinary and multidimensional. That is, for technology, more than 90% of all media and communication requirements are imported from the developed world. For production style, most Arab media follow Western style. For content, the majority of television programs and other media contents is imported from Western sources and is mostly entertainment, leaving little space for cultural and current affairs for

Figure 1: Top Ten Languages Used on the Web

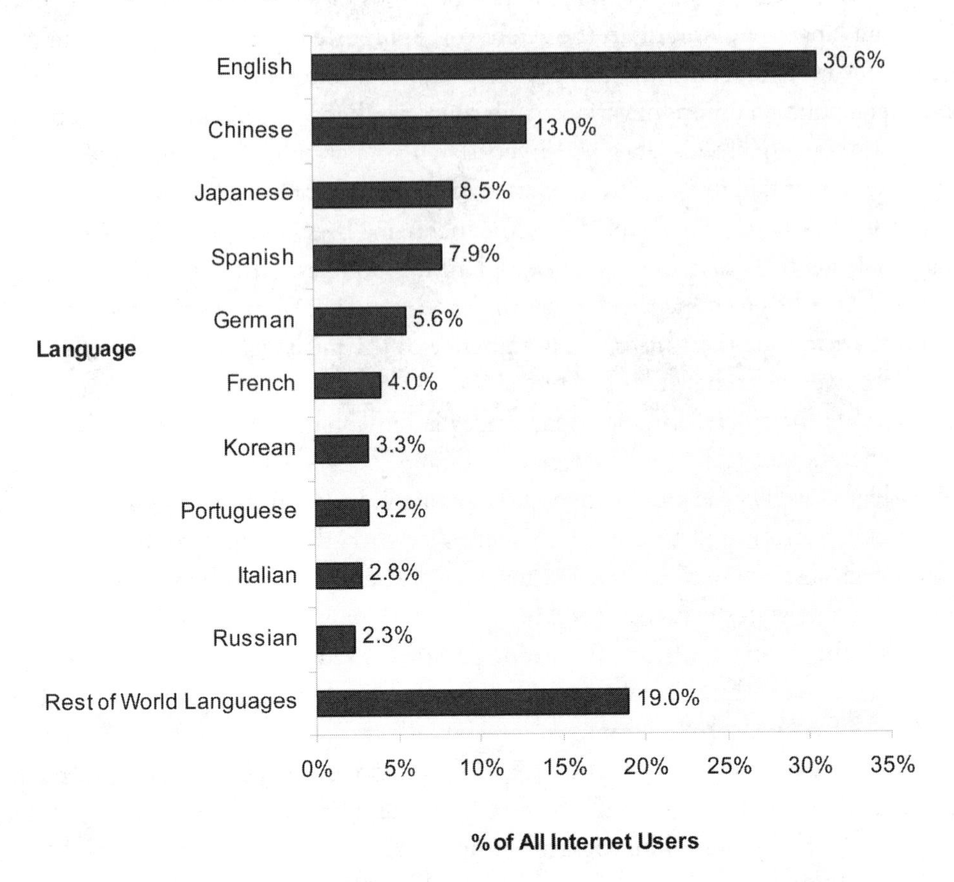

Note. Chart was produced based on data retrieved July 15, 2006, from Internet World Stats: Usage and Population Statistics at http://www.internetworldstats.com/stats7.htm

Arab audiences. The stylization and standardization of Arab media into that of its Western counterparts mean incorporating and mimicking cultural and economic structures that are deeply embedded in structures of Western modernity. In addition, censorship continues to be a major problem that renders the Arab media ideologically dependent (Sabry, 2005).

Kitchin explains:

> New technologies[8] alter the structure of our interests: the things we think about. They alter the character of our symbols: the things we think with. And they alter the nature of community: the arena in which thoughts develop. (1998, p. 74)

The features of the Arabic language are reflected in the design, layout, format, structure, and commercial activities in the major Arabic e-mail, chat, and news Web sites. Language requirements; various Arabic language accents; preferences of pictures, fonts, and colors; an Arabic style of goods consumption; and so on are examples of factors that formulate or determine the nature of such Arabic Web sites. Given the fact that the Arabic language, which does not use the Latin alphabet, needs appropriate software to be reproduced on users' computers, some Arabic language speakers use English format in their Arabic chat Web sites due to the pervasiveness of the English language or global Internet speak. In their text communication through e-mails, chat rooms, or cellular messages, Arab chatters are influenced by the nature of the Internet. For example, in chat rooms, when Arab participants use speech-like patterns in their online English textual communication, they follow an informal, socially agreed upon system, or style, of characters. These characters are written in English but either have Arabic-meaning readability (e.g., *salam* is the Arabic word for "peace"), have similarities with some Arabic letters (e.g., number 7 looks like the Arabic letter *ha'a*), or are newly innovated abbreviations to speed up communication (e.g., ASAWRAWB, which is an abbreviation of the main greeting among Arab Muslims: *Al-Salamo Alykom Wa Rahmato Allah Wa Barakatoh*).

The complexity of the Arabic language's different styles provides both challenges and opportunities for Egyptian designers of hypermedia. The styles of language used in various Web pages indicate whether a site is meant to be official or personal, broadly Arab or more specifically Egyptian, or cultured or popular. Most corporate, government, and education Web sites are written strictly in modern standard Arabic. Most of these Web sites do not use colloquial language as a general practice; rather, they sprinkle popular idioms throughout predominantly modern, standard Arabic Web sites. In particular, they use colloquialisms to determine site names. The name of the Web site Masrawy (http://www.masrawy.com), for example, derives from the colloquial Egyptian word *masrawi*, which signifies "Cairene, an inhabitant of Cairo" (of people, with the implication of sophistication). It is a variant of the modern standard Arabic word for "Cairene," *masri*. Spoken Arabic in Egypt, for example, falls into a continuum of dialects, but these differences are expressed through a language ideology that asserts Arabic as a diglossic[9] system divided between high (*fusha* or "elegant") and low (*il-'ammiyya*, "common," or *masri*, "Egyptian"). The highest form of Arabic, that of the Qur'an, is not the spoken language of any of the world's community of Muslims, yet every believing Muslim must learn it at least to a sufficient degree to carry out his or her religious obligations. Classical Arabic is the language of almost all that is written, and efforts to modernize the language are ongoing among linguists, poets, religious leaders, and others. Modernized *fusha*—the language of news media, among others—is often seen as lower than classical *fusha*, and some versions of the colloquial are seen as more appropriate than others. There is an elaborated model of diglossia that splits the vernacular into three levels (illiterate, enlightened, and educated) and *fusha* into two levels (modern and classical). Using one or another of these types of Arabic differs based on context. Modern standard Arabic is the language of nearly all written texts and is the spoken language of news and sermons, while Cairene colloquial is the language of movies, radio, popular music, and most other genres of media. Yet, code switching is common and can be used to great effect by gifted communicators. New media open entirely new realms for language use. E-mail users in Egypt, for example, typically write in English but with frequent code switching into *'ammiyya*, sometimes using numbers to represent Arabic characters that do not appear on standard keyboards (Peterson & Panovic, 2004).

Although the Internet undoubtedly offers the Arab world a chance for a democratic, less restricted flow of information and a better civic society in which citizens can be publishers—not merely receivers—of information, it has not yet made the change because of challenges—mainly economic and language in origin—to penetration rates in the Arab world (Abdulla, 2006). For example, cybercafés have been introduced to the Egyptian market to make the Internet accessible to a new base of customers. However, there are some challenges that face cybercafé development in Egypt, including many legal, technical, and operational difficulties (e.g., the type of data connection, licensing, and a lack of properly trained technicians, engineers, trainers, and consultants) (El-Gody, 2000).

Assembling the Loose Window

The Arabic window onto the Internet seems to provide a distorted and incoherent view. The Internet in the Arab world is in great need of a wide range of tasks that are required to enhance its usage (quantitatively and qualitatively), development, and production in order to appear, continue, and compete effectively in the global arena of our information society.

The types of democracy prevailing throughout the Arab countries constitute a major factor affecting the use and production of new media, telecommunications, and new technology. Mostly *limited*, and rarely *pragmatic*, democracies weaken Arab communication policies and inhibit most trials for innovation; consequently, Arab countries will tend to imitate rather than innovate. Although the usage of new media technologies in the Arab world has been accepted as necessary in the era of globalization, it remains under a high level of censorship. This, in effect, will exhaust a large portion of development programs' budgets and, instead of helping to promote creation and progress, will continue to irritate the problems of freedom of expression and the conditioned right-to-know policy. Therefore, there has been a growing need for a potential *communicative* democracy that can give rise to powerful communication policies, enabling Arab countries to create new media technologies and improve access to information for their citizens.

The democratic environment in the Arab world requires the construction of a pluralistic civil society, the widening of access to government information, the freedom of expression and debate over political issues, the spread of social equity, and the power of a multivoice independent media. The Arab political environment requires removal of measures restricting the media's abilities to do their jobs, such as governmental pressures, censorship, abuse of media personnel, regulations against privatization, and limits of freedom of expression. The smooth and free flow of information should be encouraged with the right directions toward development and progress. Government fears regarding the spread of sensitive political and religious discussions through new media, along with its desire to protect cultural identities, should not exist as justification to control or pan access to information. Instead, these fears and desires should be properly discussed and openly communicated between governments and their publics. Internet usage should be open to all fields in Arab societies and should never be limited to specific purposes.

Arab countries have many tasks ahead of them in order to achieve the development and expansion of telecommunications services. They need to develop specific strategies for extending services to rural and less privileged areas, establish the necessary legislation and regulatory framework to facilitate the investments required for service development and network expansion, remove constraints on telecommunication operations to make service providers responsive to needs, and define appropriate terms and conditions to allow domestic, regional, and international private sector participation in investment and service provision (*Telecommunication Policies*, 1996).

The Arab world also needs to take many steps toward contributing to the global development and production of new media technologies and services. Intellectual and political discussions and debates about the digital communications revolution should increase. Arabs should have open access to international communication and new media. Investments in telecommunication technologies and infrastructures should receive a much higher level of attention from Arab governments and institutions. Rates in the Arab world's ability to access the Internet should grow, so that access is not restricted to elites but, rather, open for all. This should be coupled with the removal of cultural constraints on Internet use. Arab governments and institutions should meet the challenge of bridging the digital divide in two equal ways: between the Arab world and the rest of the world and within the Arab countries themselves. For it is important for the Arab world to be equal to the rest of the world in its ability to access new media and the Internet, just as it is important that *all* Arab countries should contribute to formalizing Arabic contributions.

Religion should be separated from language on the Arabic Internet. That is, plans for enhancing the Arabic language should not be mixed with Islamic directions in various activities to create and introduce the Arabic Internet globally. The Islamist Internet should represent Muslims all over the world, not only those in the Arab world. Thus, it could be presented in all languages all over the world, including Arabic. By the same token, Arabic can represent various religions, not just Islam. This separation between Arabic and Islam can result in more focus on language and can pave the road for Arabic to become one of the main languages used globally on the Internet. This can only happen if efforts are made to fix the technical and cultural problems of using Arabic on the Internet, instead of getting into religious debates and inter-Arab political conflicts. The focus on the Arabic language will strengthen the cultural and psychological ties in the Arab world and will positively respond to the demands of modern science and technology. The lack of inter-Arab cooperation in this regard and coordination in the production of new media technologies are major problems that require serious efforts from Arab governments and institutions to solve. Rugh argues the following:

> Indeed, the Arabic language is an especially crucial element linking the Arabs with each other and with their culture; it is inseparable from Arab culture, history, [and] tradition. . . . The best definition of who is an Arab is not in terms of religion or geography but of language and consciousness, that is, one who speaks Arabic and considers himself an Arab. Arabic is extremely important to Arabs; they pay considerable attention to the language, and it shapes their thinking in many ways. (2004, p. 19)

A major part of improving the Arabic Internet is related to production and design. Similar to conventional media, the Arabic Internet should be independent in content and format. The tendency of many Internet users in the Arab world to use non-Arab Web sites and databases not only reflects these sites' higher technological capabilities, but also, more importantly, reflects the fact that the content and format of the Arabic Internet do not gratify users. Moreover, there is, in many cases, a strong disconnect or miscommunication between Arab policy decision makers or Web designers and their targeted audiences.

For example, in their examination of how Egyptian Web producers at the turn of the millennium (1999–2001) sought to design Web portals that would allow the "typical" Egyptian to easily access the World Wide Web, Peterson and Panovic (2004) argued that Egyptian Web producers are deeply influenced by national and international discourses that frame IT as a national mission for socioeconomic development. Using ethnography,[10] Peterson and Panovic found that, in the absence of clear definitions of the Web audience, Web producers imagined a "typical" Egyptian that contradicted their own experiences as users of the Web. Furthermore, producers largely borrowed preexisting models, using design elements to "inflect" their sites with an Egyptian motif. Building on nationalist discourses of development, Web producers were able to offer investors a compelling vision of a culturally "Egyptian" Web site that would bring Egyptian consumers to the Internet. Web producers attempted to mobilize this audience by designing sites that mixed English and Arabic, employed Egyptian colloquial Arabic, and used traditional and ancient symbols to generate an Egyptian "feel." However, the conceptual models of access and related design strategies created by Egyptian Web producers were out of touch with Egyptian social realities, as there was a significant disconnect between the potential audience imagined by Web producers and those Egyptians who actually had access to the Internet, contributing to a collapse of most Web portal projects in 2001.

Another dimension of the failure of Arab policy decision makers to enhance the Arabic Internet was their unsuccessful educational programs for users to understand the real benefits of the Internet and to direct them toward constructive ways of usage. For example, customers of cybercafés in Egypt, according to El-Gody (2000), use computers for discussion groups, chatting, and socializing with others via the Internet (62%); playing games (i.e., they consider cybercafés a substitute for video games and arcade stores [25%]); e-mailing purposes (9%); and searching the net for news and information or to download data (4%). This last small percentage is a major indicator that cybercafés play a limited role in developing computer and Internet literacy among Egyptians. Citizens of other Arab countries are not at all different. Therefore, a wide range of strategies to educate Arabs in better ways of using the Internet and directing their enthusiasm toward useful developmental projects is required.

In addition, instead of imitating and relying on the Western Internet in content and format, the Arabic Internet should be distinct and reflexive to the Arabic culture. It is very logical that users use non-Arabic Web sites rather than Arabic Web sites if the latter are very similar to the former. But the users will be more interested in using the Arabic Internet if it has a distinct nature that reflects real Arabic culture, deals with their actual realities, uses their own sources, and solves their own problems. This helps to enhance not only the Arab aspect of the Arabic Internet, but also its global aspect. A distinct Arabic Internet that relies heavily on distinct Arab culture, content, and format will attract more Arabic users in diasporas as well as non-Arabic users who seek various sources of information in the context of global interactions.

For consideration

1 To what extent do political, cultural, and economic factors affect the practise of public relations?
2 To what extent can public relations practitioners enable the free flow of information, as promoted by the Public Relations Society of America (PRSA) Code of Ethics, in a country that practises censorship?
3 How can new media and public relations aid in the push for a more open, communicative Arab world?
4 How can new media contribute to increased understanding between Arab and Western cultures?
5 What characteristics of new media could be contributing to government fears of its use in the Arab world?

For reading

Gher, L. A., & Amin, H. Y. (Eds.). (2000). *Civic discourse and digital age communications in the Middle East*. Stamford, CT: Ablex Publishing.

Hafez, K. (Ed.). (2001). *Mass media, politics, and society in the Middle East*. Cresskill, NJ: Hampton Press.

Internet world stats: Usage and population statistics. (n.d.). Available at http://www.internetworldstats.com

Noam, E. M. (Ed.). (1997). *Telecommunications in Western Asia and the Middle East*. New York: Oxford University Press.

Telecommunication policies for the Arab region (The Arab book). (1996, November 11–15). Paper presented at the regional telecommunication development conference for the Arab states, Beirut, Lebanon. Geneva, Switzerland: ITU, Telecommunication Development Bureau. Retrieved July 27, 2006, from http://www.itu.int/itudoc/itu-d/rtdc96/010v2e_ww2.doc

Notes

1. Positive freedom provides people with possibilities of exercising political power, while negative freedom takes these possibilities away from some people and gives them to others. Positive freedom gives possibilities for people to act and to take control, while negative freedom makes actions available to them in this negative sense that there are no difficulties. Institutions and key members or leaders in collective societies can have positive freedom, while normal individuals can have negative freedoms.

2. To understand reasons behind differences in media usage among Arab countries, it is useful to understand the nature of the economic structure of the Arab world:

The Arab world is divided into four categories: (i) low-income countries that are poor in natural resources, manpower skills and financial capacities; (ii) non-oil-exporting middle-income countries that, in general, have large populations and skills but limited natural resources and small financial capacities; (iii) oil-exporting countries without large financial surpluses, using all their income plus additional borrowing to finance development; (iv) oil-rich exporting countries with relatively small populations, which are the main sources of aid and finance for many development projects in other Arab countries. (Raffer & Salih, 1992, p. 1)

3. See, for example, Peterson & Panovic (2004, pp. 215–216) and Sabry (2005, p. 42).
4. The twenty-two Arab countries are divided into four geographical categories: (a) Arab Gulf countries (Bahrain, Kuwait, Oman, Qatar, Saudi Arabia, and United Arab Emirates), (b) Arab North African countries (Algeria, Egypt, Libya, Mauritania, Morocco, and Tunisia), (c) Arab heartland countries (Iraq, Jordan, Lebanon, Palestine, and Syria), and (d) Arab East African countries and Yemen (Comoro Islands, Djibouti, Somalia, Sudan, and Yemen).
5. The term *Islamist* refers to modern political movements in which Islam plays a central ideological role. However, the term often holds a pejorative connotation. It implies that Islamist movements are, in fact, distinct from Islam itself, that they superficially invoke Islam for political ends, without any real understanding of, or commitment to, the religion (Maguire, 2005, p. 121).
6. Islamist groups "have begun to use modern print and electronic media as tools for their aims, as was the case with Ayatollah Khomeini's spread of his revolutionary message through videotapes before the Iranian revolution in 1978–79" (Hafez, 2001, p. 11).
7. A spiritual or physical striving, against evil in the way of God, that can take various forms, which can range from a daily inner struggle to be a better person to an armed struggle fought in defense of Islam.
8. The Internet, for example, offers users a range of interactions allowing them to explore the world beyond their home. Users can browse information stored on other computers, exchange electronic mail, participate in discussion groups on a variety of topics, transfer files, search databases, take part in real-time conferences and games, and run software on distant computers (Kitchin, 1998, p. 3).
9. "In linguistics, diglossia is a situation where, in a given society, there are two (often) closely-related languages, one of high prestige, which is generally used by the government and in formal texts, and one of low prestige, which is usually the spoken vernacular tongue. The high-prestige language tends to be the more formalised, and its forms and vocabulary often 'filter down' into the vernacular, though often in a changed form" ("Diglossia," n.d.).
10. A qualitative research method that combines multiple data-gathering techniques with a commitment to detailed description of the structures of everyday life generated by a participant-observer physically present at the research site and actively engaged with the subjects of the research.

References

Abdulla, R. A. (2006). An overview of media developments in Egypt: Does the Internet make a difference? *Global Media Journal: Mediterranean Edition, 1*(1), 88–97.

Adam, G. Stuart. (1992). Truth, the state, and democracy: The scope of the legal right of free expression. *Canadian Journal of Communication, 17*(3), 1–17. Retrieved July 27, 2006, from http://www.cjc-online.ca/viewarticle.php?id=102&layout=html

Aladwani, A. M. (2003). Key Internet characteristics and e-commerce issues in Arab countries. *Information Technology & People, 16*(1), 9–20.

Aliboni, R., & Guazzone, L. (2004). Democracy in the Arab countries and the West. *Mediterranean Politics, 9*(1), 82–93.

Anderson, J. W., & Eickelman, D. F. (1999). Media convergence and its consequences. *Middle East Insight, 14*(2), 59–61. Retrieved July 27, 2006, from http://www.georgetown.edu/research/arabtech/converges.htm

Ayish, M. I. (2001). The changing face of Arab communications: Media survival in the Information Age. In K. Hafez (Ed.), *Mass media, politics, and society in the Middle East* (pp. 111–136). Cresskill, NJ: Hampton Press.

Berenger, R. D. (2006). Political will and closing the digital divide. *Global Media Journal: Mediterranean Edition, 1*(1), 56–70.

Bruck, P. A. (1990). Communication and the democratization of culture: Strategies for social theory, strategies for dialogue. In S. Splichal, J. Hochheimer, & K. Jakubowicz (Eds.), *Democratization and the media: An East-West dialogue* (pp. 56–72). Yugoslavia: Communication and Culture Colloquia.

Chalaby, J. K. (2005). From internationalization to transnationalization. *Global Media and Communication, 1*(1), 28–33.

Chowdary, T. H. (1997). Telecommunications in Yemen republic. In E. M. Noam (Ed.), *Telecommunications in Western Asia and the Middle East* (pp. 172–184). New York: Oxford University Press.

Diamond, L. (1992). Civil society and the struggle for democracy. In L. Diamond (Ed.), *The democratic revolution: Struggles for freedom and pluralism in the developing world* (pp. 1–27). New York: Freedom House.

Diglossia. (n.d.). *Wikipedia*. Retrieved August 24, 2006, from http://en.wikipedia.org/wiki/Diglossia

El-Gody, A. M. (2000). The role of cyber cafés in developing Internet literacy in Egypt. In L. A. Gher & H. Y. Amin (Eds.), *Civic discourse and digital age communications in the Middle East* (pp. 271–273). Stamford, CT: Ablex Publishing.

Elkhafaifi, H. M. (2002). Arabic language planning in the age of globalization. *Language Problems & Language Planning, 26*(3), 253–269.

Hafez, K. (2001). Mass media in the Middle East: Patterns of political and societal change. In K. Hafez (Ed.), *Mass media, politics, and society in the Middle East* (pp. 1–20). Cresskill, NJ: Hampton Press, Inc.

The Internet in the Mideast and North Africa: Free expression and censorship. (1999). *Human Rights Watch*. Retrieved July 27, 2006, from http://www.hrw.org/advocacy/internet/mena/index.htm

Internet usage statistics—The big picture. (2006). *Internet world stats: Usage and population statistics*. Retrieved July 15, 2006, from http://www.Internetworldstats.com/stats.htm

Internet users by language. (2006). *Internet world stats: Usage and population statistics*. Retrieved July 15, 2006, from http://www.Internetworldstats.com/stats7.htm

Kavanaugh, A. L. (1998). *The social control of technology in North Africa: Information in the global economy*. Westport, London: Praeger.

Kayal, A. D. (1997). Telecommunications in Saudi Arabia: A paradigm of rapid progress. In E. M. Noam (Ed.), *Telecommunications in Western Asia and the Middle East* (pp. 163–171). New York: Oxford University Press.

Kirchner, H. (2001). Internet in the Arab World: A step towards "Information Society." In K. Hafez (Ed.), *Mass media, politics, and society in the Middle East* (pp. 137–158). Cresskill, NJ: Hampton Press.

Kitchin, R. (1998). *Cyberspace: The world in the wires*. West Sussex, England: John Wiley & Sons Ltd.

Langdon, S. (1999). *Global poverty, democracy and North-South Change*. Toronto: Garamond Press.

Maguire, T. E. R. (2005). Islamist Web sites. *Global Media and Communication, 1*(1), 121–123.

Martinez, Andrea. (1996). The telecommunications equipment market: Globalization or selective regionalization. *Canadian Journal of Communication, 21*(2), 1–17. Retrieved July 27, 2006, from http://www.-cjc-online.ca/viewarticle.php?id=363&layout=html

Ogan, C. (1995). The Middle East and North Africa. In J. C. Merrill (Ed.), *Global journalism: Survey of international communication* (pp. 189–207). New York: Longman.

Ojo, B. A. (1999). The military and the democratization process in Africa. In B. A. Ojo (Ed.), *Contemporary African politics: A comparative study of political transition to democratic legitimacy* (pp. 51–63). Lanham, MD: University Press of America.

Page, B. I. (1996). *Who deliberates? Mass media in modern democracy.* London: The University of Chicago Press.

Peterson, M. A., & Panovic, I. (2004). Accessing Egypt: Making myths and producing Web sites in cyber-Cairo. *New Review of Hypermedia and Multimedia, 10*(2), 199–219.

Raboy, M. (1990). Policy-making and democratization: The case of Canadian broadcasting. In S. Splichal, J. Hochheimer, & K. Jakubowicz (Eds.), *Democratization and the media: An East-West dialogue* (pp. 108–120). Yugoslavia: Communication and Culture Colloquia.

Raffer, K., & Salih, M. A. M. (1992). Rich Arabs and poor Arabs: An introduction to intra-Arab issues. In K. Raffer & M. A. Mohamed Salih (Eds.), *The least developed and the oil-rich Arab countries: Dependence, interdependence or patronage?* (pp. 1–12). New York: St. Martin's Press.

Rugh, W. A. (2004). *Arab mass media: Newspapers, radio, and television in Arab politics.* Westport, CT: Praeger.

Sabry, T. (2005). What is "global" about Arab media? *Global Media and Communication, 1*(1), 41–46.

Sorensen, K. (1996). Silencing the Net: The threat to freedom of expression online. *Human Rights Watch,* 8(2), 1–24. Retrieved July 27, 2006, from http://www.epic.org/free_speech/intl/hrw_report_5_96.html

Telecommunication policies for the Arab region (The Arab book). (1996, November 11–15). Paper presented at the regional telecommunication development conference for the Arab states, Beirut, Lebanon. Geneva, Switzerland: ITU, Telecommunication Development Bureau. Retrieved July 27, 2006, from http://www.itu.int/itudoc/itu-d/rtdc96/010v2e_ww2.doc

United Arab Emirates. (2001). *World of Information Business Intelligence Reports, 1*(1), 1–58.

Vivekanand, P. V., & Kollar, John E. (1997). Telecommunications in Jordan. In E. M. Noam (Ed.), *Telecommunications in Western Asia and the Middle East* (pp. 157–162). New York: Oxford University Press.

Winseck, D. (2001). *Reconvergence: A political economy of telecommunications in Canada.* Cresskill, NJ: Hampton Press.

Winseck, D., & Cuthbert, M. (1997). From communication to democratic norms: Reflections on the normative dimensions of international communication policy. *Gazette,* 59(1), 1–20.

Winterford, D., & Looney, R. E. (1997). Advanced telecommunications and the economic diversification of Bahrain. In E. M. Noam (Ed.), *Telecommunications in Western Asia and the Middle East* (pp. 223–235). New York: Oxford University Press.

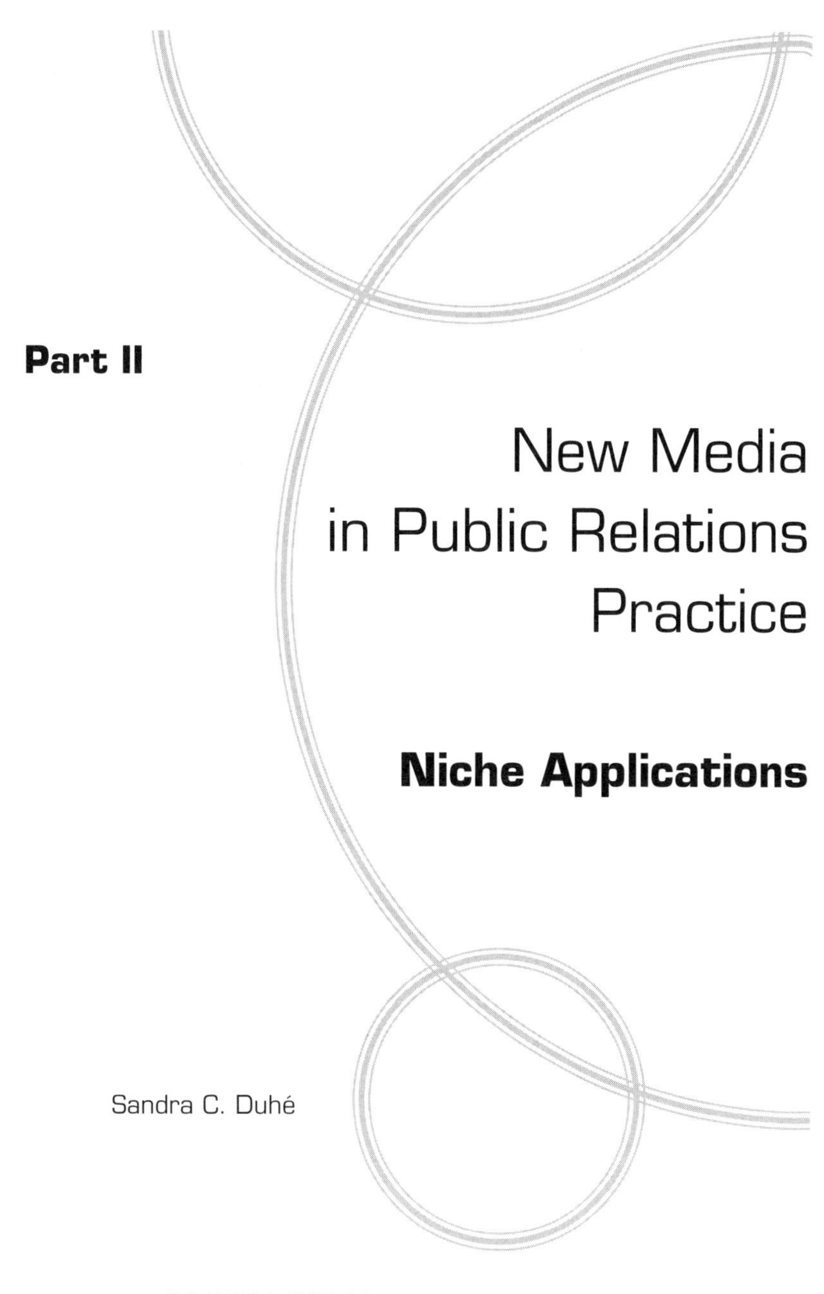

Part II

New Media in Public Relations Practice

Niche Applications

Sandra C. Duhé

OVERVIEW

This section includes a broad array of new media applications in business, nonprofit, and educational sectors. Although the settings differ, each of the eight chapters highlights an organizational desire to engage publics interactively through mediated means.

From an insider's point of view, Cassandra Imfeld, Glenn Scott, and Glen Feighery uncover the best practices employed by one of the largest banks in the United States to fight the insidious crime of phishing, whereby scammers masquerade as bank officials to deceive customers online for their personal gain. Calin Gurau reports on an extensive study of how biotech companies are using both marketing and public relations messages in their online communications and examines the overlapping relationship between them. The emerging practice of e-philanthropy is discussed in Richard Waters's chapter on fundraising, in which he outlines how nonprofits can use the Internet not only to efficiently reach more donors, but also to demonstrate accountability in how funds are utilized. Jordi Xifra discusses how several museums creatively apply the concept of interactivity to attract viewing publics and use information and communication technologies to engage visitors in the cocreation of online exhibits. Ric Jensen focuses on how a range of organizations, including corporations, special interest groups, and government agencies, can experience successes and challenges when using computer-mediated public relations to advocate positions on environmental issues. Nete Nørgaard Kristensen draws on media theory and her own empirical research to provide a look into how the Internet is becoming an increasingly important tool in Danish media relations. Brigitta Brunner and Mary Helen Brown then explore how Web "sights" can create (sometimes inaccurate) impressions of diversity and inclusivity on university campuses in their investigation into how colleges can better attract Generation M students. This section concludes with Denise Bortree's chapter on the youngest of publics, in which she examines relationship management theory in light of how organizations build relationships with children through nutrition Web sites.

In each of these niche applications, there resides an opportunity for deception, thereby making the public relations practitioner's role as "the ethical conscience of an organization" (per Ivy Ledbetter Lee's description) particularly important in online communications.

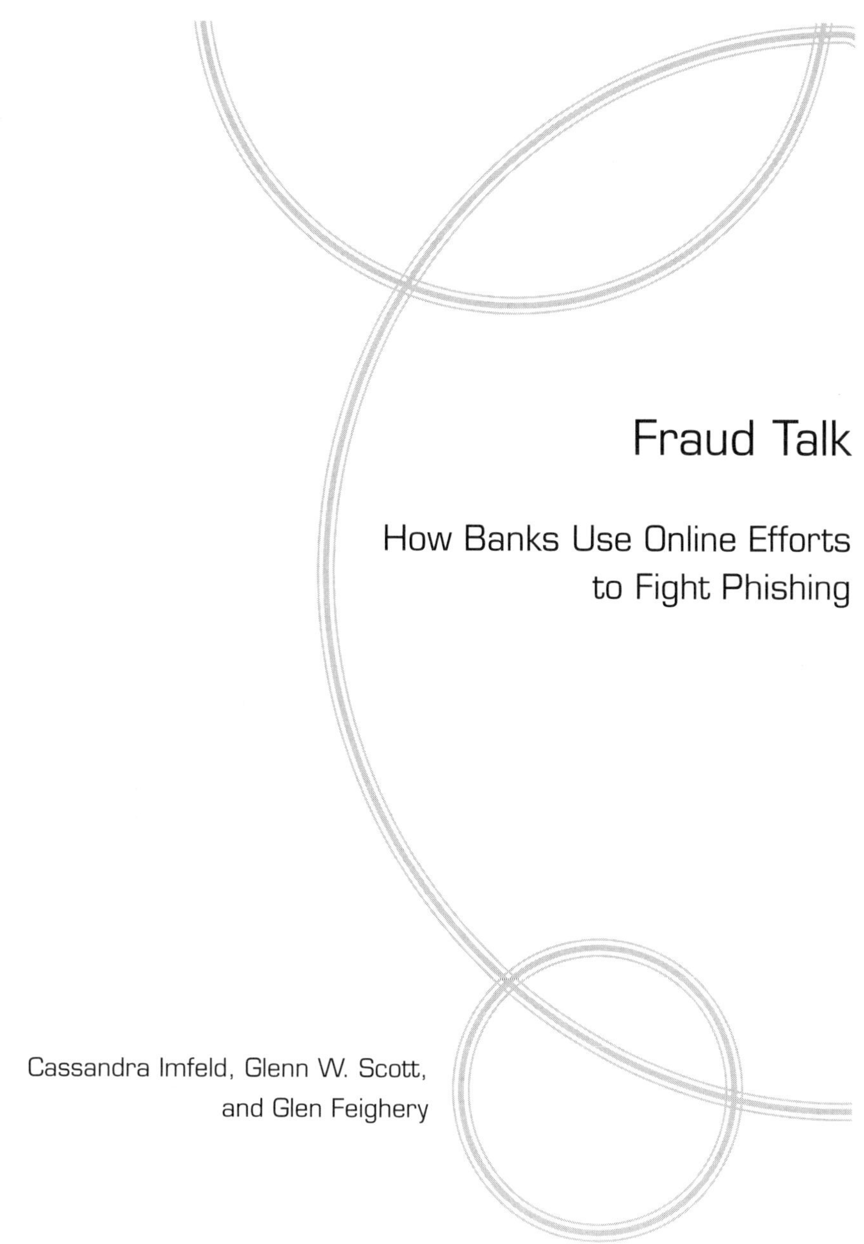

Fraud Talk

How Banks Use Online Efforts to Fight Phishing

Cassandra Imfeld, Glenn W. Scott,
and Glen Feighery

Waves of digital fraud attacks, known as phishing, have buffeted financial institutions, costing them billions and threatening to undermine relationships with a crucial public—their customers. Driven by these pressures, banks have realized that, although the Internet presents the problem, its capacity for open, two-way communication also provides part of the solution. Scholars who have promoted dialogue as the basis of an evolving relational theory for public relations did not anticipate the details of phishing. These scams have tested not only banks but also the capacity of dialogic theory itself to provide direction as technology and fraud evolve. This chapter examines nineteen major banks' online strategies and provides a case study exemplifying best practices at one bank. Generally, the more effort to involve users in defeating fraud, the better. Consistent with dialogic theory, some of the most prevalent features in banks' online discourse are also some of the most salient for relationship building.

A bank customer gets an e-mail alerting her of an unauthorized transaction in her account and asking her to click on a link to verify her information. She does, but, a few weeks later, she discovers her account is overdrawn. She—and her bank—have become victims of online fraud. Such scams pose a fast-growing threat to organizations involved in financial transactions (Garretson, 2006; Hallahan, 2004), costing them money and credibility (Kawamoto, 2004; McGuire, 2004).

This insidious innovation in digital fraud comes with its own suitably misleading name: *phishing*. In phishing attacks, scammers send mass e-mails masquerading as official communications from well-known institutions. They typically solicit users to volunteer confidential information, such as passwords and account numbers, which can be harvested for criminal gain (Bielski, 2004). Banks have been the primary target. An industry consortium, the Anti-Phishing Working Group (2006), found that 89% of reported phishing incidents in April 2006 targeted financial services companies.

These attacks have had double-barreled effects. First, they have led to massive theft: Banks have absorbed losses in the billions. Second, they have undermined consumer confidence by introducing new layers of doubt about online interaction with financial organizations (Garretson, 2006; Hines, 2004). As public relations practitioners pursue more open, two-way relationships with key online publics, phishing attacks are clearly adding new stresses. However, the Internet provides tools for crisis management, even as it becomes a target for new crises (DiNardo, 2002; Hallahan, 2004).

Dialogue as a Public Relations Ethic

As scholars, such as Botan and Taylor (2004), point out, theoretical considerations in the public relations field have evolved over the past two decades from functional perspectives to aspects of relationship building. Functional approaches helped connect theory to how practitioners worked toward effective programs and campaigns. They implied tactical applications to achieve organizational goals. Today, researchers are more focused on involving groups and organizations to share in processes of meaning making. Botan and Taylor call this cocreational. The emphasis is less on how practitioners accomplish their immediate missions and more on broader questions of how organizations interact with key publics to negotiate or achieve productive and ethical paths. Relationships are not only a means but conceivably an end, in the sense that they produce the kinds of trust and credibility that advance mutually accepted activities. Cocreational perspectives can be traced through Grunig (1992); Broom, Casey, and Ritchey (1997); Ledingham and Bruning (2000); and Marsh (2001), among others.

Relationship building comes with requirements for open flows of communication and the primacy of dialogue. Thus, "dialogue elevates publics to the status of communication equal with the organization" (Botan, 1997, p. 196). Dialogic theory promises to raise an organization's interaction with its publics from instrumental—that is, a means to an end—to enlightened. From such relationships, positive-sum outcomes are possible. A dialogic perspective also acknowledges the fluidity of relationships and participants—organizations, publics, and the contexts in which they interact (Botan & Taylor, 2004). Dialogic theory provides a useful lens to view how banks have sought to strengthen relationships with their customers, given the damage caused by what might be termed the *trust shock* of phishing attacks.

Kent and Taylor (2002) have applied versions of dialogic theory to develop ethical and feasible public relations practices. Others have applied the concept to online activities. Curtin and Gaither looked at the "dialogic index" (2004, p. 30) of Web sites of Middle Eastern governments, rating the capacity of sites to supply information and to promote transcultural information flows. Ryan (2003) considered how public relations professionals approach online communication, finding that training poses more of a barrier to dialogic activities than technology. Kent, Taylor, and White studied the "general dialogic capacity" (2003, p. 63) of online sites. They concluded that organizations most dependent upon their publics should be most concerned with adopting dialogic features, calling them "imperative for the survival of highly stakeholder-dependent organizations" (p. 75).

Pressured by phishing, banks risk losing their good names and also what Hallahan (2004) deemed a crucial tool for crisis management. Therefore, it is imperative for organizations to use Web sites not only instrumentally but also for their capacities to connect with key groups. Bank Web sites can be both functional and relational. They can process economic transactions and encourage communication with publics alienated by phishing attacks. This chapter considers what kinds of content major banks have provided on their Web sites to prompt dialogue and to build relationships with online publics. Through a case study, it shows specifically how such online content—if guided by a broader dialogic orientation—can effectively manage relationships attenuated by fraud.

How Institutions Are Fighting Back

Descriptive Web site analysis. This section briefly evaluates Web sites of the top nineteen financial institutions targeted by fraud. The list, from the Internet security company FraudWatch International (2005), included twenty banks. Since the record was compiled, one of the top twenty banks merged with another (unlisted) bank. Thus, nineteen banks were examined. Communications on the financial institutions' Web sites were evaluated for twenty-seven variables (see Appendix, p. 281). These variables exhibit best practices and offer the most educational value to clients. Two reports by the Federal Deposit Insurance Corporation (FDIC) further outlined communication strategies to protect against online fraud:

- Examples of spoofed e-mails, including graphics.
- Examples of spoofed e-mail subject lines.
- Toll-free numbers for reporting details about identity theft.

- E-mail addresses for sharing information.
- Links to the Federal Trade Commission and other agencies.
- Consumer alerts about new developments. (Federal Deposit Insurance Corporation, 2004, p. 20; Federal Deposit Insurance Corporation, 2005, p. 27)

These communication elements are echoed by the private firm Forrester Research, which recommends that Web sites contain information on how to identify phishing, a listing of valid URLs, a means to report e-mail fraud, pop-up prevention tips, and a fraud prevention hotline (Graeber, 2004, p. 5). In addition, the Anti-Phishing Working Group recommends "an alert on the front page of your company's public-facing Web site and setting up an e-mail address for customers to report phishing attempts" (as cited in Delio, 2005, para. 30).

Complementing government and industry recommendations are news articles identifying vulnerabilities stemming from a lack of education and the need for certain communications. For example, because the latest fraudulent e-mails sometimes involve scripts that silently rewrite code to send users to spoofed sites (Ramsaran, 2004, p. 13), financial institutions have been urged to provide up-to-date information and remedies, such as anti-spyware. Users, meanwhile, are encouraged to vigilantly update their browsers. These recommendations aim to limit fraudulent e-mails and pop-up windows.

Content analysis revealed the most common features provided on the nineteen bank Web sites.[1] These features took different forms, reflecting organizational presentation styles and branding strategies. Table 1 summarizes the six most common features. Topping the list were two features found on sixteen sites. One listed tips on how to protect against phishing. These included maintaining passwords, signing off computers, and keeping software up-to-date. The other feature was a policy statement assuring users that banks would never ask for personal information in e-mails. Overall, the most common features provided some of the most fundamental measures of communication. Adding links on the banks' home pages revealed that most of these organizations gave high priority to communicating with customers about online fraud. However, four of the nineteen banks did not do this. Other

Table 1: Most Commonly Used Antifraud Features

Rank	Feature/Content	N
1.	Tips advising users how to protect against online fraud	16
1.	Statement that bank will never ask for personal account information via e-mail	16
3.	Telephone number for users to report online fraud	15
4.	Guidance for users who provided information in response to a fraudulent appeal	14
4.	Link on home page for content about online fraud	14
6.	Suggestion that responsibility lies with institution-customer partnership	13

Note. Total features evaluated = 27. Repeated numbers in left column indicate ties.

Table 2: Least Commonly Used Antifraud Features

Rank	Feature/content	N
1.	How to identify legitimate e-mail messages	0
1.	Screen shots of legitimate e-mail messages	0
1.	List of legitimate URLs for bank	1
4.	Textual description of spoofed Web site	1
4.	How to stop pop-up windows	1
4.	How to identify a legitimate Web site	1

Note. Total features evaluated = 27. Repeated numbers in left column indicate ties.

common features were advising users on steps to take if they had inadvertently volunteered information and offering e-mail addresses and telephone numbers that users could use to ask questions or report fraud.

Table 2 lists the least common features. These were more technical options whose dialogic function was less clear than their didactic functions. Lists of legitimate URLs and screen shots of legitimate e-mail messages aimed to illustrate how proper online activities should look. Although banks might have argued that more electronically knowledgeable users would be more confident online, few institutions chose to provide such knowledge in a detailed manner.

In terms of scores assigned to each bank, a wide distribution was evident, suggesting no great consensus as to what information banks should communicate. However, there were similarities in overt efforts to keep in touch with users and customers. Table 3 ranks financial institutions, with the top bank using twenty-two out of twenty-seven variables. The last bank had only one feature on its Web site.

Because the costs of phishing are high, more communication is better in efforts to share information that helps customers avoid such scams. Protective features also provide the potential for relationship building to the extent that they solicit feedback, offer personal contact, and empower customers to interact online. Interaction is key. Banks can employ many of the specific communications listed, but these are insufficient by themselves. Although banks are required by law to protect clients' personal and account information, they cannot prevent clients from disclosing account and credit card numbers or other personal data and becoming victims of online fraud. Banks have assumed losses from online fraud, even when their clients have voluntarily provided information. Given these stark realities, many banks have reached out to make their customers active partners in foiling phishing.

An examination of one bank. How can these features be most effectively used? The following case study of one of the country's largest banks, one of the nineteen banks studied, demonstrates how it used specific objectives, tactics, and evaluation criteria during a two-year period. The authors enjoyed access to this bank's planning and decision-making processes, and the goal in this section is to provide a discreet level of insight that honors proprietary considerations and customer privacy.

Table 3: Ranking of Financial Institutions by Number of Antifraud Features on Web Site

Rank	Financial Institution	Number of Features
1.	Citibank	22
2.	SunTrust	18
3.	Wells Fargo	16
4.	Commerce	15
5.	Halifax	14
6.	South Trust	13
7.	Lloyds	12
7.	U.S. Bank	12
7.	Washington Mutual	12
10.	Charter One	11
10.	Citizens	11
10.	Key Bank	11
10.	TCF	11
14.	Visa	10
15.	Regions	9
16.	BB&T	8
16.	HSBC	8
18.	Westpac	7
19.	Barclays	1

Note. Total features evaluated = 27. Repeated numbers in left column indicate ties.

Starting in August 2003, this bank, based in the Southeast, became the target of fraudulent e-mails and spoofed Web sites. Initially, the e-mails and Web sites were amateurish and included typographical errors and misspellings. The bank's clients paid little heed to these obviously fraudulent e-mails. However, a year later, scammers became more sophisticated and began to include the bank's logo, images from its Web site, flawless grammar, masked hyperlinks, and the names of actual bank employees in their e-mails. It became increasingly difficult to determine whether an e-mail that appeared to be from the bank was legitimate. The impact was massive. During a phishing attack in the fall of 2004, the bank's call center fielded more than one thousand phone calls in one day. As the months passed, a direct correlation between the number of phishing attacks and financial losses emerged. With financial pressures escalating and complaints soaring, the bank made consumer education a high priority.

In response, the bank formed a corporate Online Fraud Task Force. The purpose of this group was to identify and classify all aspects of online fraud to mitigate losses and damage to the bank's brand. The task force included teams focusing on communications (people), technology, and process. These teams were told to implement strategies to stem the escalating losses. This bank, like all others, assumed the liability for online fraud, even though it was the clients who provided their personal information to scammers.

One of the communications team's first actions was to create a Web page specifically dedicated to online fraud. It was here that the problem became part of the solution: The same channel scammers used to deceive clients could become the first line of defense. Web pages could quickly and inexpensively reach thousands of people, and information could be frequently updated in response to the latest scams and information. The Web pages aimed to demonstrate that the bank was aware of the fraudulent activities, educate clients and non-clients about online fraud, provide users with a sense of control over the security of their accounts, and build dialogue with online users.

The online fraud page was launched in mid-2004 with basic information. This included the fact that online fraud was an industry-wide concern, as well as tips on identification, reporting, and protection against such fraud. Text was supplemented with images of actual phishing attacks, and these examples were often submitted by clients. The communications team reviewed incoming reports of fraud and posted screenshots of the most effective and damaging phishing attacks.

Early on, the communications team realized that protecting clients against online fraud required a partnership. The team needed customers to submit reports so that it could warn other clients and the community. Online users needed information from the bank on how to protect themselves. Accordingly, the team's Web site included an e-mail address and online form to report fraud. By providing users with information about online fraud and communication channels to talk to the bank, the team hoped clients would perceive the institution as trying to protect them. This would fulfill the bank's branding mission: "Doing what's right for the client."

To drive traffic to this site, the communications team placed a fraud alert link in the top half of the bank's home page. By placing the link in prime online real estate, the team hoped to convey to clients that this was a serious issue that merited immediate attention. To visually grab visitors' notice, the alert was red—a significant deviation from the company's style guidelines. However, this was considered necessary to educate and protect clients from becoming victims of online fraud. In the latter half of 2004, traffic to the Web site mirrored the increasing number of phishing attacks against the bank, with incidents of online fraud and traffic to the Web pages peaking in November 2004.

As phishing attacks continued, the communications team updated the examples of scams on the bank's Web pages. The fraud pages later included external, third-party links to additional resources, such as the Federal Trade Commission. At the same time, the team monitored the bank's call centers, asking phone representatives to provide a list of words callers used to describe online fraud and a list of the most frequently asked questions. Based on the language callers used, the communications team updated the Web site's search functionality. For example, if a user went to the Web site's search feature and typed in "fake e-mail"—a phrase callers frequently used—the search would provide the page about online fraud. The communications team took the same approach of using the callers' language to create a list of frequently asked questions.

At the peak of phishing activities targeting the bank, the communications team added its e-mail policy to the fraud Web page. This policy stated that the bank "will never send unsolicited e-mails asking clients to provide, update or verify personal or account information, such as passwords, Social Security Numbers, PINs, credit or Check Card numbers, or other confidential information." This Client Commitment was carefully crafted to reflect the types of information scammers sought and included words scammers used, such as "provide," "update," or "verify." The communications team's goal was to actively educate clients—and non-clients—about the types of e-mails they could expect from the bank, while also alerting those who had responded to phishing attacks and provided personal information. In early 2007, the Client Commitment continued to be included in almost all offline marketing and communications.

Although the bank's information about online fraud was comparable to that of other financial institutions, the communications team recognized that not much in-depth information was available. Because online fraud was evolving quickly, the team wanted to provide dynamic, comprehensive education. The team researched best industry practices and evaluated customer calls, e-mails, and search engine queries to the bank's Web site to help the team revamp the Web site's entire online fraud and security section. The communications team believed that detailed information could not only help protect clients and other online publics, but also show online users that the bank understood concerns about fraud and was committed to protecting clients.

One of the initial results of this undertaking was an Online Fraud and Identity Theft Guide. To create the guide, the communications team again worked with members from the bank's call centers and fraud resolution and technology teams. Members from the different areas of the bank shared their experiences in handling clients and non-clients who had experienced online fraud. For example, individuals not only wanted specific information about online fraud, but they also wanted to know what to do if they determined that someone had stolen their identities. Thus, the guide included a step-by-step plan of what to do if a user became a victim of identity theft. Consistent with the bank's other steps, this tool was designed to be frequently updated.

Another product of research was relaunching the online fraud and security Web pages in early 2005. Although the peak of online fraud appeared to be past, the bank was about to complete a merger, introduce a new logo, and unveil a new Web site. The communications teams anticipated that scammers would exploit the transition period because clients and employees would be uncertain and have questions about the changes. The team believed it was critical to expand the online fraud and security Web pages. The additional content largely reflected client feedback as well as competitive analyses. This feedback incorporated e-mails, phone calls, search engine queries, letters from online users, and information from branch employees who had handled clients who believed they had been phished.

One example of new content was a section called How We Help Protect You. This addressed one of the biggest complaints from clients during the phishing attacks: They didn't know how the bank was helping them. Clients expected the bank to protect them from online fraud and to stop the e-mails. The new section addressed these concerns and reiterated that preventing online fraud was a partnership. The site stated, "We're committed to keeping your accounts and personal information secure. Our goal is to help you protect yourself against fraud and identity theft." To address requests from clients and non-clients that the bank stop the fake e-mails—a feat no target of phishing can technically do—the communications team created a new frequently asked questions section. Online users wanted to know how the bank (actually, the scammers) obtained e-mail addresses. To answer this and to assure online users that the e-mails were fake, the communications team explained the bank's e-mail policy and provided a link to its privacy policy.

Meanwhile, the team also heeded client requests for content to help people in a hurry. The team included a link titled Quick Tips to Help Protect Yourself. This offered a number of bulleted safety tips for online banking and shopping, Internet browsers, e-mail, credit cards, and automated teller machines (ATMs). Finally, the communications team added a section to the online fraud page about fraud in general. During discussions with call center representatives and branch employees, the communications team recognized that the issue of fraud was much broader—even though online scams were the bank's focus at the moment. For example, call center representatives related how several clients had fallen victim to phone schemes that told clients that the bank was the official bank of the lottery and that they had won. To collect the winnings, clients were told, they needed to provide the bank with their check routing number, Social Security number, and date of birth, and the money would be deposited into their accounts. The communications team recognized an opportunity not only to educate external clients but also to provide internal clients with a resource. The Telephone and Print Fraud section included examples of telephone fraud received by call center representatives and fraudulent newspaper ads that clients had brought to the branches.

Whereas initial efforts of the Online Fraud Task Force were to educate employees and clients about online fraud, it shifted focus in 2006 to improving communications relating to processes. After an internal survey of bank employees and feedback from clients who had been scammed, the task force realized that there were more than a dozen different communication channels to report fraud, leaving clients and employees confused. To streamline reporting processes and ensure that employees knew how best to help victims, the communications team prepared to redesign the bank's intranet site dedicated to fraud. The changes would allow branch and call center representatives to visit the intranet site and find links with answers to questions such as, What do I do if my client has responded to a phishing e-mail? or, What do I do if my client received a letter stating he won the lottery? As with earlier steps, these questions incorporated employee and client feedback.

In the bank task force's first two years, it engaged with various publics to learn how to most effectively combine people, process, and technology initiatives and update online communications to reflect ever-changing online assaults. Although later communication efforts about fraud were not as urgent as at the beginning, the task force continued to identify opportunities for actively reaching out to internal and external clients about online fraud.

Conclusions: The Need to Negotiate

This chapter has explained the financial and credibility pressures associated with online fraud, has described the most common features on Web sites of major banks fighting fraud, and has examined how one bank used a dialogic approach to inform, direct, and assess its efforts to manage relationships. Consistent with dialogic theory, we find that some of the most prevalent features in online discourse are among the most salient for relationship building. For example, the case study shows how strong invitations can initiate communication with insecure customers. It is useful to revisit a related point advanced by Kent and Taylor in which they observe that, to build dialogue, "organizations must create website locations, telephone access, and public forums where the public can actually engage other human beings in discussions about organizational issues" (2002, p. 31).

How can dialogic practice improve? The case study of one bank showed how it relentlessly pursued a free flow of communication through multiple channels—including the very channel used to attack it. However, there is room for even more dynamism and transparency in these relationships. Invitations to talk via phone or e-mail are helpful, but banks can promote dialogue in less restrictive forms—for example, via hosted discussion forums, chats, and Web logs where bank representatives can carry on conversations about fraud and other important topics.

The overall orientation remains the same. One evident strength of dialogic theory is its robustness in the face of conflict. Theorists who have promoted dialogue as the basis of an evolving relational theory for public relations did not anticipate the exact details of the current phishing scams. These attacks have tested not only banks but also the capacity of the theory itself to provide meaningful direction as conditions change. This study highlights one of dialogic theory's key traits, its elasticity, which flows from the primacy of human negotiation to overcome conflicts. Kent and Taylor point out that, although public relations is often useful in avoiding risks, the occasional need for "dialogic risk" is also sensible (2002, p. 29). This is especially true in a situation like phishing, which involves profound risks to financial institutions.

In summary, public relations practitioners, especially those working with banks, should be prepared to apply dialogic and other principles to new situations as they arise and evolve. Theories that call for the strong maintenance of relationships are useful when applied to cases as unexpected and fluid as phishing. Banks have benefited greatly from the ability to reach clients online, but the phishing threat—and related scams sure to follow—also shows that the Web remains an insecure space. Public relations professionals can benefit from this study and learn to emphasize quick action and constant vigilance against threats to mobilize ways to maintain dynamic ties with key groups. To do less is to risk losing the enormous benefits of online communication.

For consideration

1. Financial scams are nothing new. What makes phishing so threatening to the reputation and brand of banks and financial institutions? What are the broader societal impacts of phishing?
2. What are the consequences if banks lose their customers' trust in using the banks' Web sites? What about noncustomers?

3 Given that dialogic theory is rooted in open flows of communication, how well does the theory apply to online interaction?
4 Banks have a vested interest in protecting clients from identity theft. If banks did not have to assume losses from online fraud, how far should banks go in protecting their clients from scams?
5 To what extent can trust between an institution and its clients be established through only mediated means?

For reading

Anti-Phishing Working Group. (n.d.). Retrieved May 26, 2006, from http://www.antiphishing.org

Federal Trade Commission. (n.d.). Retrieved July 11, 2006, from http://www.ftc.gov

Garretson, C. (2006, March 31). MIT spam conference focuses on phishing. Computerworld. Retrieved May 23, 2006, from http://www.computerworld.com/action/article.do?command=viewArticleCoverage&articleId=110039&continuingCoverageId=1010&intsrc=article_cc_feat_bot

Gartner, Inc. (n.d.). Retrieved July 11, 2006, from http://www.gartner.com

Kent, M. L., & Taylor, M. (2002). Toward a dialogic theory of public relations. *Public Relations Review, 28*(1), 21–37.

Note

1. The nineteen banks in this analysis were evaluated twice, and an intercoder reliability test of five randomly chosen sites yielded an acceptable Scott's pi coefficient of 0.94.

References

Anti-Phishing Working Group. (2006, April). Phishing activity trends report. Retrieved May 26, 2006, from http://www.antiphishing.org

Bielski, L. (2004, September). Phishing phace-off: Online fraudsters and vendors do battle to gain control of electronic information. *ABA Banking Journal*, 96(9), 46–54.

Botan, C. (1997). Ethics in strategic communication campaigns: The case for a new approach to public relations. *Journal of Business Communication*, 34(2), 188–202.

Botan, C., & Taylor, M. (2004). Public relations: State of the field. *Journal of Communication*, 54(4), 645–661.

Broom, G. M., Casey, S., & Ritchey, J. (1997). Toward a concept and theory of organization-public relationships. *Journal of Public Relations Research*, 9(2), 83–98.

Curtin, P. A., & Gaither, T. K. (2004). International agenda-setting in cyberspace: A study of Middle East government English-language Websites. *Public Relations Review*, 30(1), 25–36.

Delio, M. (2005, February 11). IT tackles phishing. Information Age. Retrieved July 7, 2005, from http://www.infoage.idg.com.au/index.php/id;1765797087;fp;512;fpid;144956388

DiNardo, A. M. (2002). The Internet as a crisis management tool: A critique of banking sites during Y2K. *Public Relations Review*, 28(4), 367–378.

Federal Deposit Insurance Corporation. (2004, December 14). Putting an end to account-hijacking identity theft. Retrieved July 7, 2005, from http://www.fdic.gov/consumers/consumer/idtheftstudy/index.html

Federal Deposit Insurance Corporation. (2005, June 17). Putting an end to account-hijacking identity theft study supplement. Retrieved July 7, 2005, from http://www.fdic.gov/consumers/consumer/idtheftstudysupp/index.html

FraudWatch International. (2005). Company phishing index. Retrieved May 30, 2005, from http://www.fraudwatchinternational.com/phishing/company_index.php

Garretson, C. (2006, March 31). MIT spam conference focuses on phishing. Retrieved May 23, 2006, from http://www.computerworld.com/action/article.do?command=viewArticleCoverage&articleId=110039&continuingCoverageId=1010&intsrc=article_cc_feat_bot

Graeber, C. (2004). Phishing concerns impact consumer online financial behavior. Retrieved January 27, 2005, from http://www.forrester.com/Research/Document/Excerpt/0,7211,35677,00.html

Grunig, J. E. (1992). Excellence in public relations and communication management. Hillsdale, NJ: Erlbaum.

Hallahan, K. (2004). Protecting an organization's digital public relations assets. *Public Relations Review, 30*(3), 255–268.

Hines, M. (2004, June 15). Gartner: Phishing on the rise in U.S. CNET News. Retrieved March 23, 2005, from http://news.com.com/Gartner+Phishing+on+the+rise+in+U.S./2100–7349_3–5234155.html

Kawamoto, D. (2004, May 6). U.S. hit by rise in "phishing" attacks. CNET News. Retrieved March 23, 2005, from http://news.com.com/U.S.+hit+by+rise+in+phishing+attacks/2100–7355_3–5207297.html

Kent, M. L., & Taylor, M. (2002). Toward a dialogic theory of public relations. *Public Relations Review, 28*(1), 21–37.

Kent, M. L., Taylor, M., & White, W. J. (2003). The relationship between web site design and organizational responsiveness to stakeholders. *Public Relations Review*, 29(1), 63–77.

Ledingham, J. A., & Bruning, S. D. (2000). *Public relations as relationship management: A relational approach to the study and practice of public relations*. Hillsdale, NJ: Erlbaum.

Marsh, C. W. (2001). Public relations ethics: Contrasting models from the rhetorics of Plato, Aristotle, and Isocrates. *Journal of Mass Media Ethics*, 16(2 & 3), 78–98.

McGuire, D. (2004, July 12). Senate bill targets "phishers." *The Washington Post*. Retrieved March 27, 2005, from http://www.washingtonpost.com/wp-dyn/articles/A44826–2004Jul12.html

Ramsaran, C. (2004). Catch of the day: Banks face new phishing scams. Bank Systems and Technology. Retrieved January 27, 2005, from http://www.banktech.com/showArticle.jhtml?articleID=54200898

Ryan, M. (2003). Public relations and the Web: Organizational problems, gender, and institution type. *Public Relations Review*, 29(3), 335–349.

Appendix: Bank Antifraud Features

Coding variables

1. Link on organization's home page to content about online fraud.
2. Location of link on top half of home page.
3. Graphic examples of fraudulent e-mails.
4. Textual descriptions of fraudulent e-mails.
5. Graphic example of a spoofed Web site.
6. Textual description of a spoofed Web site.
7. Information on how to identify fraudulent e-mails.
8. Information on how to identity a spoofed Web site.
9. Information on how to identify legitimate e-mail.
10. Information on how to identify a legitimate Web site.
11. Information on how to identify a legitimate Web site / graphics.
12. Information on how to identify a legitimate Web site / text.
13. Information on how users can protect against online fraud.
14. Tips on protection against viruses and/or spyware.
15. E-mail addresses for users to report online fraud.
16. Telephone numbers to report online fraud.
17. A form for users to report online fraud.
18. Information on what to do if users have provided personal or account data in response to a fraudulent appeal.
19. References to partnerships to fight online fraud.
20. Links to third parties for more information about online fraud.
21. Links to software providers.
22. Suggestion for responsibility for protection against online fraud.
23. List of fraudulent e-mail subject lines or addresses.
24. List of legitimate URLs for the bank.
25. Information on blocking pop-up windows.
26. Information on updating browsers.
27. Discussion of bank's policies for sending e-mail seeking personal information.

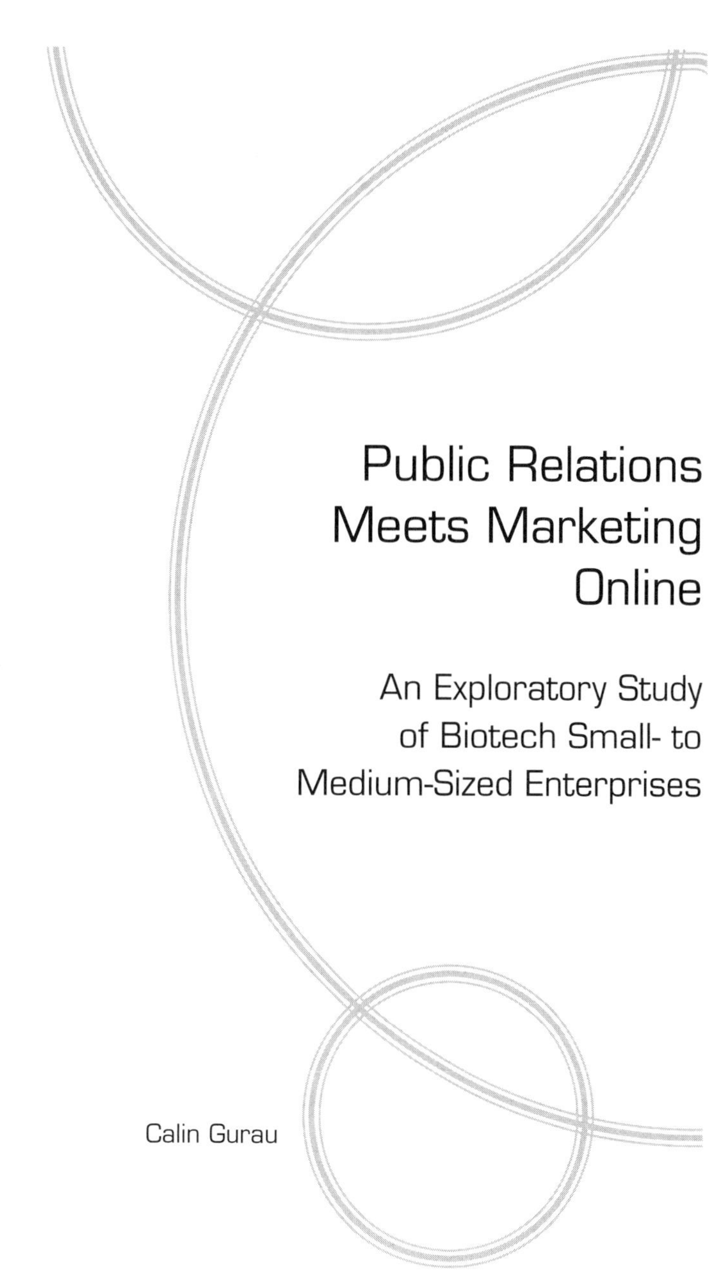

Public Relations Meets Marketing Online

An Exploratory Study of Biotech Small- to Medium-Sized Enterprises

Calin Gurau

This study presents, based on extensive data collection and processing, the public relations communication strategy applied by biopharmaceutical small- and medium-sized enterprises (SMEs) on their Web sites. The operating environment of these firms: negative public perception about biotech activities, long product development cycles, lack of direct revenues, intense competition, and their prolonged dependence on investment funds, requires a proactive online communication policy. The study identifies and presents the main communication themes that are developed by these firms, their main target audiences, and the relationship between marketing and public relations communication in the online environment.

The Internet is rapidly changing the infrastructure of information, creating a turbulent and sometimes confusing communication environment. In these new circumstances, companies are forced to alter the way they communicate, both internally and externally, in order to take full advantage of the new communication channels. Internet communication provides a series of major advantages for public relations professionals: increased ability to access and transmit information, interactive capabilities, and higher flexibility in targeting audiences. On the other hand, online communication is transparent to multiple audiences and cannot be fully controlled, because all Internet users can create and publish information on the Web.

The New Online Communication Model

The rapid development of the Internet in the last ten years has had a profound impact on traditional business paradigms and practices. But, most importantly, the Internet has changed classical communication procedures (Blattberg & Deighton 1991; Holtz, 1999) because of three specific and coexistent characteristics that differentiate it from any other communication channel:

1 *Interactivity:* The Internet offers multiple possibilities of interactive communication, acting not only as an interface, but also as a communication agent, allowing a direct interaction between individuals and software applications.
2 *Transparency:* The information published online can be accessed and viewed by any Internet user, unless this information is password protected.
3 *Memory:* The Web is a channel not only for transmitting information, but also for storing information. In other words, the information published on the Web remains in the memory of the network until it is erased.

These options are significantly transforming the profile and behaviour of online audiences. Marketing communication practitioners should therefore adapt to the new realities of how audiences get and use information.

The audience is connected to organisations. The traditional communication channel was often unidirectional: The institutions communicated, and the audiences consumed the information. Even when communication was considered a two-way process, organisations had the resources to send information to audiences through a very wide channel, while the audiences had only a minuscule pipeline for communicating back to the organisations (Ihator, 2001).

Now, the communication channel can be described as a network. All the actors of the communication process: the company, its chief executive officer (CEO), its communications manager, and the external communication agency, are only one click away from the audience. In the new model, communicators must engage members of the audience on a one-to-one basis (Holtz, 1999).

The audience is connected to each other. Considering the nature of the network, if the audience is one click away from the institution, it is also one click away from other members of the audience. Today, a company's activity can be discussed and debated over the Internet, and, in many cases, the organisation is not even aware of this parallel debate. In the new environment, everybody is a communicator, and the institution is just a part of the network.

The audience has access to other information. Today, it takes a matter of minutes to access multiple sources of information over the Internet. Any statement made can be dissected, analysed, discussed, and challenged by interested individuals. In the connected world, information does not exist in a vacuum.

Audiences pull information. Until recently, television offered only a few channels. People communicated with one another by mail and by phone. In these conditions, it was easier for a marketing communication practitioner to make a message stand out.

The networked world has increased exponentially the number of available channels of communication (Holtz, 1999). However, the networked environment provides audiences with a new advantage, in that they no longer accept every message a communicator wants to push to them but rather are able to select and pull information that suits their interests and needs (Rowley, 2001, 2004).

Taking advantage of the various online resources requires strategic thinking that recognizes that these aspects of the networked world coexist. In turn, these aspects must be coordinated to achieve specific, measurable objectives that are consistent with the goals of any marketing communications effort.

Public Relations in the New Communication Model

The main role of public relations is to present a positive image of an organisation to all its publics (Greener, 1990). The word *publics* should not be limited to the general public or external organisations; in fact, the target audiences of an organisation can be extremely varied, including employees, business suppliers, associates and partners, the media, trade networks, industry regulators, and financial organisations, as well as a whole host of influential groups of people known as *stakeholders*.

Managing public relations is an activity that affects and impinges on virtually every aspect of an organisation's operations. Within commercial organisations, the major thrust of public relations activity is likely to be directed toward products or services marketing, either directly or indirectly. This is the area in which public relations activities have the most clear and direct contribution to the organisation's profits, and, therefore, they are more easily justified in terms of budget spend. Because of this, it is not surprising that the so-called marketing public relations is often the first form of planned public relations to which an organisation commits itself. However, it is misleading to say that public relations activity is confined to marketing support (Black, 1993).

Kotler and Mindak (1978) developed a model that describes the relationship between public relations and marketing. The model comprises five alternative arrangements: (a) separate but equal functions; (b) separate but overlapping functions; (c) marketing as the dominant function; (d) public relations as the dominant function; and (e) marketing and public relations as converging functions. In the last arrangement, marketing and public relations are fully integrated with equal status and no subservience.

Much debate has taken place (Grunig & Grunig, 1998; Lauzen, 1991, 1992; Lewton, 1991) about the relationship between public relations and marketing communication within the framework of traditional communication channels. However, little empirical research has been undertaken to explore the impact of the Internet on a company's communication mix.

This study investigates the use of public relations and the relationship between public relations and marketing within the Internet communication strategy of biotechnology firms. To provide a general overview of these issues, three research objectives are addressed:

1. To identify the categories of online messages used by biotechnology firms for public relations and marketing communication and the main target audience(s) for each type of message.
2. To investigate the relationship between online public relations and marketing communication messages.
3. To analyse the use of online public relations and marketing communication messages by biotech firms to enhance their corporate image.

After a presentation of the main characteristics of biotech firms and activities, the chapter briefly addresses the role of corporate image as a core concept of an organisation's communication strategy and the various theories explaining the mechanism of its formation. Then, the research methodology applied to investigate the previously stated objectives is described, as well as the categories of online data collected and analysed. Research findings are discussed in relation to existing theory. The study concludes with a summary and propositions for further research.

The Biotechnology Sector: Definition and Characteristics

Biotechnology is the systematic industrial use of biological processes and organisms to manufacture medical, agricultural, and consumer products (Oakey, Faulkner, Cooper, & Walsh, 1990). Depending on their applicability, biotechnology techniques have been implemented and integrated in a series of traditional industries in order to produce novel products or to improve the quality of existing ones (Daly, 1985).

The full development of a new biotechnology product can take as long as six to eight years and typically costs about £250 million (Gracie, 1998). The complete value-added chain of activities between the generation of a new idea and the commercialisation of the final product can be divided into three main stages: product innovation, product development, and product commercialisation (Walsh, 1993). Each of these stages comprises a series of essential activities and processes (Chandrasekar, Helliar, Lonie, Nixon, & Power, 1999). Meeting all these requirements is difficult for small- or even medium-sized companies, which are usually characterised by scarce resources and a strong focus on their distinctive competitive advantage.

For these companies, a clear presentation of success factors in corporate communication messages is paramount for business development. Although the main target audience and, therefore, the focus of the corporate message will differ depending on their profile of activity (community of investors for the dedicated research and development companies, organisational clients for suppliers), an effective communication strategy will differentiate and favourably position these organisations in relation to their competitors and the general market environment.

Biotechnology is perceived to be a high-risk sector. This statement can be interpreted in two different ways. Firstly, biotechnology is usually characterised by a high level of business risk, because it requires large investments with uncertain outcomes (Nelson & Mukherji, 1998). Secondly, many biotechnology activities are perceived as having a high level of safety risk for the human body and environment, especially in the long term.

Because of its perceived risk, biotechnology is a highly scrutinized sector with extensive media coverage. A variety of ethical and moral problems (Rothstein, 1996) related to many biotechnology activities (e.g., human cloning and genetic improvement) is a permanent subject of debate at national and international levels (Nelkin, 1995).

Many analysts (McKelvey, 1996; Russel, 1988) consider that this situation results from a communication crisis between companies and stakeholders, a crisis that can be solved through the implementation of a better communication strategy. Given its present efficiency and popularity for business communication, the Internet represents an important opportunity for biotech firms to present and project a positive corporate image toward all audiences and, more specifically, toward investors and customers.

The Role of Corporate Image in Biotechnology

Previous research in the process of corporate image formation has concentrated both on the process of corporate image creation and projection by the organisation and on the process of corporate image reception by the members of the audience (Kazoleas, Kim, & Moffitt, 2001).

The research oriented toward marketing, advertising, and consumer behaviour has suggested that commercial organisations create images in order to foster increased sales. One of the main findings of the consumer behaviour orientation is that multiple images are used by various segments of customers and that these images are variable and subject to constant change (Ackerman, 1990; Cottle, 1988; Dowling, 1986; Garbett, 1988; Gray & Smeltzer, 1987; Knoll & Tankersley, 1991).

On the other hand, business management research privileges the term *corporate identity* and argues that this identity is primarily a form of social identification and association between employees and organisations (Ashforth & Mael, 1989; Carlivati, 1990; Kovach, 1985; Pratt & Foreman, 2000).

Finally, the public relations research has argued that image is created by the interaction between the organisation and its audiences during a complex communication process (Alvesson, 1990; Fombrun & Shanley, 1990). One particular conceptualisation of image formation is built on the cultural model of meaning, which acknowledges that meanings (i.e., images) are generated not only by multiple kinds of factors, but also through the intersection or struggle among these factors. This vision emphasises the dynamic, flexible, and conflictual nature of corporate image formation (Moffitt, 1994a, 1994b; Williams & Moffitt, 1997).

An example of how good public relations strategy can improve the corporate image of biotech companies, as well as public perception of the biotechnology sector, is the public relations transfer process modelled and explained by Jefkins (1995). Jefkins argues that when a negative situation is converted into a positive achievement, through knowledge, the result is the primary objective of public relations: a better understanding. The potential existence of four negative states: hostility, prejudice, apathy, and ignorance, means that before planning, budgeting, and recommending a public relations programme, communication professionals must uncover the extent and nature of these problems in an organisational context. Thereafter, the negative situation of hostility can be transferred to a more positive achievement of sympathy, prejudice can be transferred to acceptance, apathy can be transferred to interest, and ignorance can be transferred to knowledge.

Research Methodology

Five hundred biotechnology companies were investigated between March 2005 and May 2005 to analyse public relations strategies used on the Internet, as well as the types of messages published online. The five hundred companies studied were randomly selected from the countries with the most developed biotechnology sectors at a global level (Ernst and Young, 2000; Persidis, 1998): 186 U.S. companies, 137 UK companies, and 177 German companies. Regarding their profile of activity, 341 of these firms can be defined as dedicated R&D SMEs; 159 are classified as biotech supplier companies.

The study focused on small-sized (firms with up to fifty employees) and medium-sized (firms that have between fifty-one and 250 employees) biotech SMEs for the simple reason that large companies have a level of resources that permits them to develop extremely complex and complete Web sites in addition to highly developed communications for traditional media channels.

The Internet presence of these five hundred biotech SMEs was identified and studied using search engines and databases that specialised in providing information about biotech companies (*BIO.COM*, n.d.; *BioIndustry Association*, n.d.; *Institut für Informatik*, n.d.). To increase understanding of Internet communication strategies used by biotechnology companies, and the relationship between marketing and public relations messages, the author collected and analysed the following categories of primary data: (a) the main categories of messages used on the Internet; (b) the balance between marketing communication and public relations messages, as well as the relationship between them in Internet communication strategies; (c) the main categories of target audiences for Internet messages; (d) the extent to which companies' sites took advantage of Web interactivity in their communication strategies; and (e) the relationship between Internet communication strategies and the level of interaction allowed by company sites. The data were analysed using SPSS (Statistical Package for the Social Sciences) software to conduct basic data processing (frequencies and cross-tabulations) and simple statistical tests (chi-square).

Research Findings

To differentiate between marketing communication and public relations messages, the author applied the criterion of the most probable effect of information on the audience. That is, messages that provided information to directly facilitate the commercial activities of the company were categorized as marketing communication messages, while messages designed to increase audience understanding of the organisation and its activities, as well as about biotechnology activities and products in general, were considered public relations messages. Each is discussed, beginning with public relations messages.

The main categories of public relations messages. The main categories of public relations messages identified on biotechnology company Web sites (presented as percentages of the five hundred investigated Web sites) were the following: general profile (97.7%), newsletter (43%), educational information (37.8%), external links (22.4%), investor relations (21%), special events (12.6%), and discussion forum (1.4%).

General profile. This category of public relations message was most frequently used on the Internet. The general profile usually comprised a short, concise, general description of the company and its activities. In a few instances, only the name, location, and main activity of the company, supplemented with a few elements of organisational history (e.g., year of incorporation, main stages of development), were provided. Other times, the information provided was extremely detailed, including an extensive presentation of the company's history and achievements, as well as a list of the company's board of directors and scientific advisors. The general profile message was directed to all categories of stakeholders by providing a clear and simple source of information about the company and its activities.

Newsletter. This category provided audiences the opportunity to access a series of press releases about the company and its activities. In some cases, this category appeared under the label "News" when it also comprised elements of marketing communication (e.g., promotional offers, information about updated catalogues or new products) or other categories of public relations messages (e.g., special events, investor relations). This category can have a positive effect on both the general public (who can have a better understanding of the company's achievements and evolution) and on potential clients (who can evaluate better the strengths of the company and the quality of its products/services).

Educational information. This category usually explained biotechnology products and activities at a level of understanding for the general public. To increase the attractiveness of the presentation, the organisation sometimes displayed simple diagrams, tables, and schemes to complement the textual information.

External links. This category represented hypertext connections to other related Internet sites, which were either sources of general information or sites of similar organisations with which the firm had forged business partnerships.

Investor relations. This category is on the border between marketing and public relations messages. It usually incorporated general and financial reports about company activities, financing sources, and, in the case of firms listed on the stock exchange market, links to specialised sites that offered quotes of the company's shares.

Special events. This category gave information about an organisation's professional celebrations or its participation in various exhibitions, trade fairs, and conferences. Messages were particularly useful for potential clients or partners of the firm, who could plan to contact the firm's representatives during the special events.

Discussion forum. The implementation of a discussion forum on a company Internet site allows direct interaction not only between the company and its audiences, but also between different members of publics. Different persons can access the site and participate in the forum by sending e-mail messages. These messages are then processed by the company and displayed on the Internet site, fuelling the thread of virtual interaction. This is one of the best opportunities for a company to identify and understand existing public opinion about a certain topic of interest and to efficiently increase and influence, through direct messages, various participants' understanding of topics. Unfortunately, very few of the surveyed companies (seven, or 1.4%, of five hundred firms) used this opportunity, perhaps because discussion forums require management and moderation.

The main categories of marketing messages. The main categories of marketing messages identified on biotechnology company Web sites (presented as percentages of the five hundred investigated Web sites) were the following: general activity (91.2%), job opportunities (49.4%), distribution network (19.6%), and online orders (13%).

General activity. This category of marketing messages had a function similar to the public relations category of general profile in that it familiarised Internet users with the main activity/expertise of the company. In this case, the information had a commercial nature by providing the first step toward a possible future transaction and was the most frequently used marketing message. Many times, however, this was the only Internet marketing message used by biotechnology companies, being complemented by the company's contact information (physical address, telephone and fax number, and e-mail address).

Job opportunities. Although the job opportunities category had no direct relationship with the commercial transactions of a company, it was considered a more appropriate fit for marketing messages. This classification was justified by the fact that these messages reflected some of the company's ethos and encouraged potential applicants to apply for posts.

Distribution network. This category usually comprised a description of the distribution strategy used by the company and a list of its agents/distributors and/or subsidiary offices. The main purpose of this information category was to provide potential customers with quick and easy contact to the closest distributor. The sometimes extensive distribution network of many biotechnology organisations represented an organisational response to the global reach of the Internet and, subsequently, to the global scattering of clients.

Online orders. This category offered potential clients the opportunity to make direct transactions on the Internet. This option usually provided an online catalogue of the company's products, with quality, quantity, and price specifications, and a detailed presentation of the sales conditions (including information about the method of payment and the physical distribution method).

Communication strategies. The online presence of these biotechnology firms was classified as marketing oriented when the site contained only marketing messages, public relations oriented when the site included only public relations messages, or mixed when marketing and public relations categories of messages coexisted on the same site.

Only seventeen, or 3.4%, of the five hundred biotechnology sites were purely marketing oriented. On the other hand, nearly one-fourth (23.4%) of biotech companies used the Internet purely for public relations communication.

These results should be interpreted from the specific context of the biotechnology sector. Since biotech commercial operations are almost exclusively business-to-business transactions, the information required by clients is highly technical and objective. From this point of view, the best option is a combination of marketing and public relations messages that can be accessed by different audiences depending on their objectives and interests. Most (73.2%) of the biotechnology companies included in the study used a mix of marketing and public relations messages on their Web sites.

Regarding level of interaction, biotech sites were classified as having a *low* level of interaction if only the address and telephone number of the firm were provided, a *medium* level of interaction if e-mail capability was also provided, or a *high* level of interaction if the site included both e-mail capability and a discussion forum. Most of the companies used the

Internet as a communication pipeline, either passively with a low level of interaction (9.2%) or interactively with a medium level of interaction (89.4%). Only seven, or 1.4%, of the five hundred company sites utilized a discussion forum to enable networked communication at company-to-public and public-to-public levels.

Table 1 shows that there was a statistically significant relationship between low, medium, or high levels of interaction allowed by a company Internet site and the communication orientation (marketing, public relations, or mixed) of that site. Most companies had a medium level of interaction and pursued a mixed communication strategy, or orientation, on their Web sites. In terms of Internet usage, these results indicate that most biotech companies have grasped the idea that a mixed public relations/marketing orientation offers the best avenue for communicating with their various stakeholders on a global basis.

The relationship between the marketing and public relations functions of online messages. The characteristics of the Internet (ubiquity, flexibility, memory, networked interaction, speed, global reach, and time independence) impose a dynamic format for presenting information. In these conditions, all information has a dual role, providing both a marketing and a public relations function, with one of them being predominant and the other being secondary (see Figure 1). The predominance will depend upon the communication objectives and the targeted audiences of a particular firm.

In the physical world, marketing communication can be more easily separated from public relations communication, because the functions often use different channels and formats. On the Internet, marketing and public relations functions and objectives support and reinforce each other and are thereby forged into an integrated online communication strategy. This premise is supported by the large number of sites with mixed communication characteristics (366, or 73.2%, of the 500 companies studied).

Concluding Remarks

The biotechnology sector presents specific characteristics that maximise the importance of effective communication, especially for small- and medium-sized firms. Although highly dynamic and innovative, these firms are engaged in a long-term, highly risky product development process, often without any source of funding except external investments. In these conditions, the development and very survival of these firms depend on their capacity to create, project, and manage a positive corporate image. The Internet represents for these firms both an inexpensive and effective communication channel that permits direct and timely communication with various publics. A large proportion of these biotech firms, those defined as dedicated R&D firms, do not initially have any product or service for sale; their online communication has a predominant public relations orientation.

The study of five hundred randomly selected biotech SME Web sites offered insights into the main categories of online messages published by these firms. Analysis of the collected data identified the main categories of public relations and marketing messages, the

Table 1: Cross-tabulation between Level of Interaction and the Main Orientation of Biotechnology Internet Sites

Level of Interaction	Marketing Oriented		Public Relations Oriented		Mixed		Total	
	N	%	N	%	N	%	N	%
Low	5	29.4	31	26.5	9	2.5	46	9.8
Medium	12	70.6	83	60.9	353	96.4	447	89.0
High	0	0.0	3	2.6	4	1.1	7	1.2
Total	17	100.0	117	100.0	366	100.0	500	100.0

Note. Chi-square = 73.82; $p < 0.0001$.

predominance of a specific online communication orientation (marketing, public relations, or mixed), and the relationship between the level of interactivity offered by the site and the communication orientation of the site.

The results show that the level of site interaction tends to increase in more complex sites in which the functions of marketing and public relations communication coexist in an integrated structure. However, the interactive dimension of the Internet as a multi-user communication channel is not yet fully exploited by biotechnology firms. This may be an effect of rigidly translating communications strategies from the real world to the online environment, without taking into consideration the specific advantages offered by the Internet.

On the other hand, the Internet, as a common communication channel for public relations and marketing messages that are accessible to various publics, forces firms to adopt an integrated model of online communication. From this perspective, messages published online cannot be considered as purely public relations or marketing communication, but rather as information having a dual function and effect, both public relations and market-

Figure 1: Model of an Integrated Online Communication Strategy

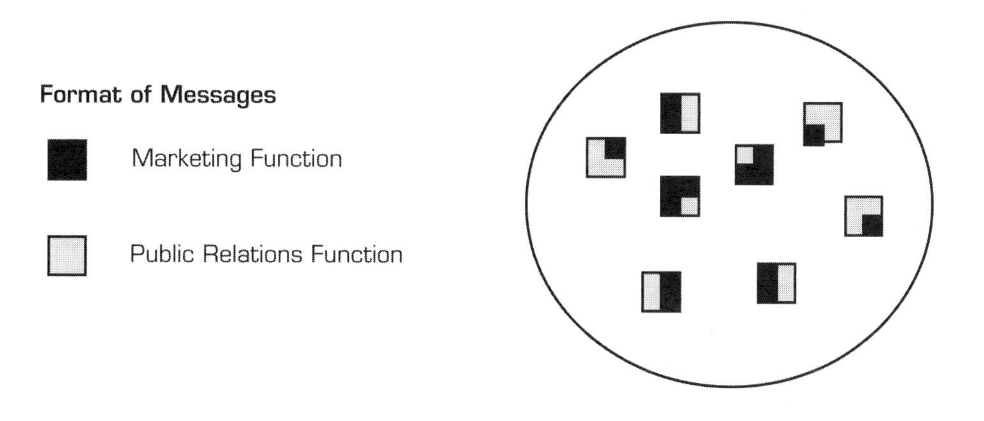

ing, with one of these functions often being considered as dominant.

This study has a number of limitations. The sample size is relatively small, and there is scope for further expansion of this research study to biotechnology firms from other major national markets. This would allow an interesting comparison between the communication styles of different biotech companies, integrating a cultural dimension, which was not considered in this study. On the other hand, the online communication strategy of biotech firms cannot be fully assessed without measuring its effect on target audiences. Building on the basis of this study, further research could address complementary issues regarding corporate image and identity, such as an analysis of various sources of information about the activities of a specific firm (newspapers, financial reports, and press releases) or a measurement of the impact of online corporate messages on target audiences.

For consideration

1. What is the best relationship between marketing and public relations communications in a business organisation?
2. To what extent do "new economy" or technology-oriented firms have an advantage over more mature industries in using the Internet to communicate with their audiences?
3. Which is a more critical investment for start-up firms: marketing or public relations?
4. How specifically could biotech SMEs develop a positive corporate image through new media?
5. Which is more important for effective online communication: content or interactivity?

For reading

Grunig, J. E., & Grunig, L. A. (1998). The relationship between public relations and marketing in excellent organizations: Evidence from the IABC study. *Journal of Marketing Communications*, *4*, 141–162.

Gurau, C., & McLaren, Y. (2003). Corporate reputations in UK biotechnology: An analysis of online "company profile" texts. *Journal of Marketing Communication*, 9(4), 241–256.

Holtz, S. (1999). *Public relations on the Net*. New York: AMACOM.

Oakey, R., Faulkner, W., Cooper, S., & Walsh, V. (1990). *New firms in the biotechnology industry*. London: Pinter Publishers.

Williams, S. L., & Moffitt, M. A. (1997). Corporate image as an impression formation process: Prioritizing personal, organizational, and environmental audience factors. *Journal of Public Relations Research*, 9(4), 237–258.

References

Ackerman, L. D. (1990). Identity in action. *IABC Communication World*, 7(9), 33–50.

Alvesson, M. (1990). Organization: From substance to image? *Organization Studies*, *11*(3), 373–394.

Ashforth, B. E., & Mael, F. (1989). Social identity theory and the organization. *Academy of Management Review, 14*, 20–39.

BIO.COM. (n.d.). Retrieved July 2006, from http://www.bio.com

The BioIndustry Association. (n.d.). Retrieved July 2006, from http://www.bioindustry.org

Black, S. (1993). *The essentials of public relations*. London: Kogan Page.

Blattberg, R. C., & Deighton, J. (1991). Interactive marketing: Exploiting the age of addressability. *Sloan Management Review, 33*(1), 5–14.

Carlivati, P. A. (1990). Measuring your image. *Association Management, 42*, 49–52.

Chandrasekar, C. R., Helliar, C. V., Lonie, A. A., Nixon, W. A., & Power, D. M. (1999). Strategic management of innovation risk in the biopharmaceutical industry: A UK perspective. *International Journal of Healthcare Technology and Management, 1*(1/2), 62–76.

Cottle, D. W. (1988). How firms can develop and project a winning image. *The Practical Accountant, 21*, 46–50.

Daly, P. (1985). *The biotechnology business: A strategic analysis*. London: Pinter Publishers.

Dowling, G. R. (1986). Managing your corporate images. *Industrial Marketing Management, 15*, 109–115.

Ernst and Young. (2000). *Evolution: Seventh annual European life sciences report 2000*. London: Ernst & Young International.

Fombrun, C., & Shanley, M. (1990). What's in a name? Reputation building and corporate strategy. *Academy of Management Journal, 33*, 233–258.

Garbett, T. F. (1988). *How to build a corporation's identity and project its image*. Lanham, MD: Lexington Books.

Gracie, S. (1998, July 5). Boffin's brave new world: The enterprise network. *The Sunday Times*, Business section, p. 13.

Gray, E. R., & Smeltzer, L. R. (1987). Planning a face-lift: Implementing a corporate image. *Journal of Business Strategy, 8*, 4–10.

Greener, T. (1990). *The secrets of successful public relations and image-making*. Oxford: Heinemann Professional Publishing.

Grunig, J. E., & Grunig, L. A. (1998). The relationship between public relations and marketing in excellent organizations: Evidence from the IABC study. *Journal of Marketing Communications*, 4(3), 141–162.

Holtz, S. (1999). *Public relations on the Net*. New York: AMACOM.

Ihator, A. S. (2001). Communication style in the information age. *Corporate Communications: An International Journal*, 6(4), 199–204.

Institut für Informatik. (n.d.). Retrieved July 2006, from http://www.informatik.uni-rostock.de

Jefkins, F. (1995). *Public relations techniques* (2nd ed.). Oxford: Butterworth-Heinemann.

Kazoleas, D., Kim, Y., & Moffitt, M. A. (2001). Institutional image: A case study. *Corporate Communications: An International Journal*, 6(4), 205–216.

Knoll, H. E., & Tankersley, C. B. (1991). Building a better image. *Sales and Marketing Management, 143*, 70–78.

Kotler, P., & Mindak, W. (1978). Marketing and public relations: Should they be partners or rivals? *Journal of Marketing, 42*(10), 13–20.

Kovach, J. L. (1985, August 19). Corporate identity. *Industry Week, 226*, pp. 21–22.

Lauzen, M. M. (1991). Imperialism and encroachment in public relations. *Public Relations Review, 17*(3), 245–256.

Lauzen, M. M. (1992). Public relations roles, intraorganizational power, and encroachment. *Journal of Public Relations Research*, *4*(2), 61–80.

Lewton, K. L. (1991). *Public relations in health care: A guide for professionals*. Chicago: American Hospital Publishing.

McKelvey, M. D. (1996). *Evolutionary innovations*. Oxford: Oxford University Press.

Moffitt, M. A. (1994a). Collapsing and integrating concepts of "public" and "image" into a new theory. *Public Relations Review*, *20*(2), 159–170.

Moffitt, M. A. (1994b). A cultural studies perspective toward understanding corporate image: A case study of State Farm Insurance. *Journal of Public Relations Research*, 6(1), 41–66.

Nelkin, D. (1995). Forms of intrusion: Comparing resistance to information technology and biotechnology in the USA. In M. Bauer (Ed.), *Resistance to new technology* (pp. 379–390). Cambridge: Cambridge University Press.

Nelson, R. T., & Mukherji, A. (1998, June). Valuing biotechnology assets. *Nature Biotechnology*, *16*(6), 525–529.

Oakey, R., Faulkner, W., Cooper, S., & Walsh, V. (1990). *New firms in the biotechnology industry*. London: Pinter Publishers.

Persidis, A. (1998, May). Bioentrepreneurship around the world. *Nature Biotechnology Supplement*, *16*(6), 3–4.

Pratt, M. G., & Foreman, P. O. (2000). Classifying managerial responses to multiple organizational identities. *Academy of Management Review*, *25*(1), 18–42.

Rothstein, M. A. (1996). Ethical issues surrounding the new technology as applied to health care. In F. B. Rudolph & L. V. McIntire (Eds.), *Biotechnology: Science, engineering, and ethical challenges for the twenty-first century* (pp. 199–207). Washington, DC: Joseph Henry Press.

Rowley, J. (2001). Remodelling marketing communications in an Internet environment. *Internet Research: Electronic Networking Applications and Policy*, *11*(3), 203–212.

Rowley, J. (2004). Just another channel? Marketing communications in e-business. *Marketing Intelligence and Planning*, *22*(1), 24–41.

Russel, A. M. (1988). *The biotechnology revolution: An international perspective*. Brighton, Sussex: Wheatsheaf Books.

Walsh, V. (1993). Demand, public markets, and innovation in biotechnology. *Science and Public Policy*, *20*(3), 138–156.

Williams, S. L., & Moffitt, M. A. (1997). Corporate image as an impression formation process: Prioritizing personal, organizational, and environmental audience factors. *Journal of Public Relations Research*, 9(4), 237–258.

Building the Nonprofit Organization–Donor Relationship Online

The Increasing Importance of E-philanthropy

Richard D. Waters

The rapid growth of e-philanthropy in the past ten years has impacted the nonprofit sector and the manner in which it conducts fundraising operations. Though often viewed solely as an opportunity to raise money, the Internet offers nonprofits a unique chance to build relationships with new, current, and past donors while providing valuable information to both foundations and venture capitalists. The Internet provides a relatively inexpensive means for nonprofits to become more transparent and accountable to their key stakeholders. Additionally, the incorporation of e-philanthropy strategies can allow the organization to shift its financial resources to programs and services because of the financial savings.

Professional fundraisers often are expected to be familiar with a wide variety of fundraising strategies, ranging from face-to-face major gift negotiations to preparing direct mail for an annual campaign to the entire donor database. The growth of e-philanthropy requires that fundraisers must also now develop skills to manage relationships with donors in a virtual setting as well. Most fundraisers agree that e-philanthropy will never replace traditional fundraising channels (Hart, 2002); however, there is no doubt that it is changing nonprofit organizations' fundraising strategies.

Though some may view e-philanthropy as a specialty area that needs to be administered separately from other fundraising techniques, doing so would diminish its overall effectiveness. This chapter will provide an overview of e-philanthropy techniques that can be incorporated into the fundraising program to build and maintain relationships with key publics and to solicit contributions online.

Defining E-philanthropy

E-philanthropy has been defined differently by individuals with different social scientific perspectives. Network for Good (2006), a nonprofit organization dedicated to connecting people to charities all over the world, views e-philanthropy as the use of the Internet and technology to manage fundraising, volunteer, and advocacy efforts. Similarly, fundraising practitioner James Greenfield, former Association for Fundraising Professionals' Outstanding Fundraising Executive award winner, advocated that the multi-tiered definition used by the ePhilanthropy Foundation best describes the phenomenon:

- ePhilanthropy is the building and enhancing of relationships with supporters of nonprofit organizations using an Internet-based platform.
- ePhilanthropy is the online contribution of cash or real property or the purchase of products or services to benefit a nonprofit organization.
- ePhilanthropy is the storage of and usage of electronic data or use of electronic methods to support fund raising activities. (2001, p. 320)

Hart defines e-philanthropy as "a set of efficiency-building techniques that can be employed to build and enhance relationships with stakeholders interested in the success of a nonprofit organization" (2002, p. 360).

Though these definitions—proposed by practitioners and education scholars—reference the ability to build and cultivate relationships, they fail to capture the true essence of relationship management by highlighting tactical aspects, such as the storage of data or the techniques used to raise funds. Viewing e-philanthropy from a public relations perspective, a more suitable definition is proposed: *E-philanthropy is the cultivation and management of relationships with key stakeholders of nonprofit organizations using the Internet.*

This definition allows for the examination of e-philanthropy practices from a public relations perspective, because success ultimately does not come from an emphasis on technology. Instead, the incorporation of interactive communication strategies will benefit nonprofits by ensuring that the organization is meeting the needs of its online community.

The Rapid Growth of E-philanthropy

In 1996, Allen, Warwick, and Stein introduced fundraisers to the concept of raising money over the Internet when they published *Fundraising on the Internet: Recruiting and Renewing Donors Online*. However, organizations did not rush to the Internet during the technology boom of the 1990s like their for-profit counterparts. Instead, they continued to raise funds through traditional channels.

But change is inevitable, and organizations began seeing trends with their own research. In 1996, the American Red Cross was not collecting credit card donations online despite internal research that showed nearly 30% of donors who called the toll-free donation number said they learned of the phone number and need for donations on the Internet. Other nationally known organizations, such as the Salvation Army and Habitat for Humanity, were noticing similar donor patterns (Demko & Moore, 1998).

The Red Cross was the first nonprofit that conducted an official campaign directing donors to its Web site to donate money. During the Kosovo Conflict in 1999, the Red Cross used media interviews and advertisements to direct donors to its Web site (http://www.redcross.org) if they were interested in making a donation to help the wounded and displaced Kosovo population. The campaign raised more than $2.6 million online from an estimated 22,000 donors. This amount was less than 2% of the $172 million the Red Cross raised for disasters that year, but it showed the organization that donors would use technology to help in times of need (Wallace, 1999).

The *Chronicle of Philanthropy* (Wallace, 1999) noted the success of the Kosovo Internet campaign, and other organizations began to explore the feasibility of incorporating Web-based fundraising strategies into their own campaigns. For many organizations, these systems would face their most demanding test two years later on September 11, 2001.

The devastating results of the September 11 terrorist attacks caused many to question why the events unfolded in that manner and left many uncertain as to what would happen next. But they also caused a tremendous outpouring of concern for the victims and their families. Shortly after the attacks, people turned to the Internet and made donations to relief agencies. More than $215 million was donated to the relief efforts by approximately two million donors (Wallace, 2001). Though most of the donations went to the Red Cross, many other organizations experienced surges in online giving.

The Red Cross noted that, for the first time in its history, online donations outnumbered those made via their toll-free number. There were so many Web-based donations that they actually outnumbered telephone donations by a three-to-one margin. A Red Cross spokesperson noted, "Clearly, the power of the Internet is huge" (Christensen, 2001, p. H1).

In the days and weeks following the attacks, the level of philanthropic activity became so extensive that it became a defining moment in U.S. philanthropy. Before the terrorist attacks, the largest effort to raise money online by a single organization was $3 million. Though early fundraising efforts were driven by war and terrorism relief efforts, online giving grew for mainstream nonprofits as well (DiPerna, 2003).

Online giving has continued to grow since 2000. In a recent *Chronicle of Philanthropy* survey of the top nonprofit organizations in the United States, organizations responding to the survey reported that they received more than $911 million from online donors (Wallace, 2006). A few of the organizations that participated in the seventh annual survey of online giving are highlighted in Table 1, which underscores the growth rate in the amount of donations made via the Internet.

Indeed, after evaluating the growth of e-philanthropy after its first five years of practice, one scholar made the following observation:

> The e-Philanthropy revolution is here to stay, and it will transform charitable giving in as profound a way as technology is changing the commercial world. Charities that have dismissed e-philanthropy as a fad, or run from it in confusion, will sooner or later need to become reconciled to it. If they don't, they risk losing touch with donors and imperiling the vitality of their work. (Austin, 2001, p. 72)

Table 1: Growth in Internet Donations Received by Select Charities, 2001—2005

Charity	Number of Donations					Percentage Change: 2001–2005
	2001	2002	2003	2004	2005	
American Heart Association	1,100,000	1,060,863	5,807,700	9,850,133	14,552,245	1,222.9
American Red Cross (National)	3,860,000	65,940,736	1,889,058	3,319,668	157,700,000	3,985.5
Catholic Relief Services	593,356	635,609	762,534	879,466	15,261,321	2,472.0
Marine Toys for Tots Foundation	251,400	313,538	392,431	423,924	706,736	181.1
Multiple Sclerosis Society (National)	233,400	2,500,000	10,000,000	16,500,000	26,200,000	11,125.4
University of Georgia	16,632	219,973	169,219	114,044	168,602	913.7
World Vision	1,758,000	2,879,327	5,700,000	8,500,000	37,100,000	2,010.4

Note. Data were obtained from the Chronicle of Philanthropy's seventh annual survey of online fundraising.

Relationships and Fundraising

Though there are numerous books written about fundraising techniques, very few have taken a scholarly approach to understand either the theoretical underpinnings of the process or the dynamics of the donor-nonprofit relationship. Kelly (1991) formally proposed that fundraising be considered a specialization of public relations, and she further demonstrated how crucial public relations practices, including strategic communication, play an important role in developing relationships with donors in her second book, *Effective Fund-Raising Management* (1998). In discussing the public relations process, Kelly (2001) noted the parallels between fundraising and public relations when she discussed the importance of stewardship.

Kelly (2001) advocates that fundraising practitioners must incorporate four elements of stewardship into the organization's official fundraising plan: reciprocity, which allows the organization to demonstrate its gratitude for the gift; responsibility, which means that the organization uses the gift in a socially responsible manner; reporting, which includes the basic principles of demonstrating accountability; and relationship nurturing, which includes regular communication and cultivation activities. These principles help the organization and fundraisers maintain ethical standards as well as ensure continued fundraising success. Worth (2002) says that these principles can result in increased donor loyalty and trust.

Recently, public relations scholars have begun focusing on refining methods to measure the impact of relationships. A study that looked at the donor-nonprofit relationship (Waters, 2006) revealed that people who made multiple charitable gifts to nonprofit organizations had statistically significant higher evaluations of the relationship in regards to their commitment to, satisfaction with, and trust of the organization than individuals who only donated once to the organization. When trying to create a model that would predict who would be a repeat donor to an organization, the variables concerning a donor's trust of an organization were the ones with the most predictive power.

Hart and Johnston (2002) insist that trust can be developed in a virtual setting just as easily as it can be developed in the physical world. They point out that many donors develop a sense of trust from a distance. While major gift donors often are involved in face-to-face meetings with organizational representatives, donors at smaller levels often develop a sense of trust in an organization as a result of direct mail solicitations, which are frequently followed by thank-you letters, newsletters, and annual reports. Following Kelly's (2001) principle of stewardship, these printed pieces allow organizations to demonstrate their appreciation for the gift as well as be accountable for proper use of the donation.

Building trust online. Nonprofit organizations have multiple ways of developing trust via the Internet. Perhaps the simplest method is to design the site with the needs and expectations of those publics in mind. Highlighting new and ongoing programs and services provides the opportunity to see how the nonprofit is serving the community. Other donors may be more interested in how the organization is managing its finances. The inclusion of annual reports and updates on fundraising campaigns can serve these needs. Still other publics may want to learn how to volunteer for the nonprofit or get directions to the location's office.

By using either formal (e.g., surveys and focus groups) or informal (e.g., casual conversations) research, the organization can find out what its key stakeholders want to see on its Web site. The research does not have to be costly. It can be conducted in person or even via the Internet by e-mailing key stakeholders and asking for their opinions of a new Web site design or if they would like to see any other pieces of information about the organization on the site.

The nonprofit sector is often accused of *mission creep*, a term that describes organizations that change their programs and services depending on the funding that is available. Similar trends frequently occur on nonprofit organizations' Internet sites. But instead of morphing for funding, their sites often keep expanding to include information that may not be relevant to the stated mission of the organization. By keeping the Web site's content in line with the organization's mission, communication can be used to enhance the individual's previous experience with the organization. It also helps to grow the trust between the organization and that individual.

Hart and Johnston (2002) suggest that, because Web site content is often very complex and changing on a regular basis, organizations should include a site navigator or site plan on the home page. These maps help novice browsers find the material for which they are searching in a timely manner. Additionally, the inclusion of a search feature can also be beneficial for browsers who are looking to find very specific material that may not be easily found by using a site plan. For example, the Muscular Dystrophy Association's (n.d.) Web site (http://www.mda.org) has a site map that graphically represents the contents of the site in a simplified manner. It shows what main pages in the site are available, but it does not show where someone could find the organization's most recent annual report. However, because they have included a search tool on the home page, the annual report and most recent IRS 990 tax form are easily available for people seeking that information. The provision of the site map and search feature allows the public to develop a sense of trust that the Muscular Dystrophy Association and other nonprofits are open and transparent in their virtual and physical operations.

In defining e-philanthropy, it was necessary to emphasize that the subject included more than simply raising money online. The public must have trust in an organization before it will see money being donated via its Web site. Looking specifically at the fundraising function, there are three main strategies that an organization can use to demonstrate that it is trustworthy. First, nonprofits are always raising money for their programs and services. Their stakeholders—employees, volunteers, donors, clients, and community residents—know that a nonprofit organization cannot exist without raising money. Therefore, nonprofits with the most successful e-philanthropy strategies always include a hyperlink on the organization's home page that sends visitors directly to a Donate Now page. By directing site visitors to the Donate Now page, nonprofits are able to openly demonstrate their need for charitable gifts.

Once someone has clicked through to the donation page, the organization can continue to build trust online by demonstrating its safety precautions. By using online security services, donors are reassured that their privacy will not be violated when they make a donation with their credit cards. Nonprofit organizations frequently outsource Internet security to trusted companies, such as VeriSign or TRUSTe. These companies manage site security features to make certain that information, such as names and credit card numbers, is carefully handled, so that identification is not stolen online. Frequently, these companies allow their partners to place a seal of approval graphic on their Web sites. These seals indicate that the company has verified that its safety principles are carried out at the organization.

For nonprofit organizations, the inclusion of a seal on the Donate Now page may not be enough. As discussed earlier, donating online is still relatively new. Many still prefer to donate through traditional channels. To build virtual trust with this group, it may be necessary to state a donor privacy and safety policy on the page as well. Hart and Johnston encourage nonprofit organizations to use the following language to help build a donor's trust and confidence in the organization:

> [Name of organization] takes the confidentiality of your information very seriously. For this reason, we use the highest level of security that technology provides when dealing with your credit card information. To guard against fraud, your information is securely encrypted and automatically passed to the financial network for processing. At no time is this information made available to anyone. (2002, pp. 5–6)

Stories from fundraising practitioner literature and studies from public relations scholars both support the idea that trust is one of the fundamental components a nonprofit needs to secure lasting relationships with its donors. However, trust alone will not suffice. In the physical world, fundraisers often hold special events to bring donors to the organization and get them involved by volunteering, touring an open house, or even meeting clients who benefited from their charitable gifts. By engaging the donor, the fundraiser is able to cement a solid relationship. For e-philanthropy to be successful, efforts must be made to ensure that donors and Web site visitors are encouraged to participate.

Engaging the Donor

Communication and education. To expand organizations' online donor databases or volunteer pools, nonprofits must first decide how they are going to use their Web sites. A communication strategy that solely involves the Internet will not produce results that will lead to a communication revolution. Despite the encouragement of the initial growth of e-philanthropy, most practitioners stress that online techniques will never replace traditional approaches to soliciting support. Instead, a combined approach of virtual and physical communication strategies is the best route.

Nonprofits must incorporate an integrated approach that takes advantage of the opportunities available in offline and online fundraising and communication strategies. Stakeholders should be encouraged to visit the organization's Web site. The URL should be noted in brochures, newsletters, and even news releases. By driving visitors to the Web site, organizations can demonstrate their openness by regularly updating site content and keeping it relevant to their stated mission. If a Web site does not contain up-to-date information, then it gives the impression that the organization has nothing new to share.

Whether it be the need for volunteers with specific sets of skills, the need to announce new program hours, or the need to raise money for a specific program, nonprofit organizations frequently have news to share. The Internet can engage the organization's supporters by offering them the chance to register for e-newsletters or regular updates about the organization and its programs and services. E-newsletters may not completely replace the traditional newsletter, though many organizations have now begun making Adobe portable document format (PDF) files of their newsletters available for downloading. Providing both a printed and digital format of the newsletter helps educate the maximum number of individuals about the organization.

Organizations can also engage stakeholders to participate by providing Web site visitors with the ability to contact an organization with an issue or concern or by offering them the chance to request additional information. The provision of Contact Us links and feedback forms also provides an outlet for the organization to keep track of who is visiting the site. With a timely response, the nonprofit has the ability to develop a relationship with the donor by practicing proper stewardship.

Stewardship. Returning to Kelly's (2001) four principles of stewardship—reciprocity, responsibility, reporting, and relationship building—the Internet provides organizations with an opportunity to do all of these things efficiently and in a cost-effective manner. Ultimately, donors want to have their gifts acknowledged by the organization (Prince & File, 2001). Recent research has shown that donors and volunteers who conduct business with nonprofits over the Internet want to be acknowledged in less than one business day (Waters, 2006). To demonstrate the organization's gratitude, fundraisers must make sure to send thank you e-mails to the donors as quickly as possible. Similarly, volunteers, special event attendees, and individuals with influence in the community (e.g., local and regional politicians) should also be thanked appropriately, whether through the Internet or traditional means, in a timely manner.

The second principle of responsibility refers to an organization's behavior and how it uses the charitable gifts it receives. The Internet provides an opportunity to demonstrate its transparency to all individuals interested in the organization's operations. Providing appropriate information about the nonprofit's programs, number of clients served, and where donations were used helps demonstrate this trustworthiness. Additionally, nonprofit organizations that utilize Internet software to register clients for conferences or services or to manage a schedule of volunteer hours can show that they respect the individual's preferences by following up with appropriate acknowledgment.

When it comes to the stewardship principle of reporting, an Internet presence can result in a significant amount of financial savings for nonprofits. As required by law, nonprofits must provide anyone with a copy of their most recent annual report when requested.

Many organizations print costly full-color annual reports that highlight success stories, the year's financial information, and updates on organizational management and leadership. This same information can be converted to a digital format and made available on the Internet for downloading. Many organizations will still produce hard copies of the report for foundation leaders, major gift donors, and politicians, but offering a digital format to everyone else helps cultivate relationships with donors on every level, volunteers, and other important stakeholders.

Finally, relationship nurturing involves communication and cultivation activities that are carried out separately from those described previously. For philanthropic volunteers, this may include an online orientation before being allowed to work with the organization, and it may be as simple as participating in a virtual activity, such as those described in the next section.

Participation. Nonprofit organizations can build stronger relationships with stakeholders by directly involving them in the virtual environment. There are several different methods of virtual participation, including blogging, personal networking, and raising money from friends and family. Nonprofit organizations must learn to take advantage of the interactive nature of new media. By becoming more engaged with their stakeholders, organizations have the potential to develop stronger relationships with them.

Blogging. Nonprofit organizations can also involve their publics in a virtual setting by entering into the blogosphere. A recent study found that nearly 30% of all Internet users regularly read blogs (Pew Center, 2005), and current estimates place the number of active blogs at more than eight million (Project for Excellence in Journalism, 2006). Blogs have been growing in popularity as a means of delivering messages. News media often turn to blogs for legitimate news stories, and individuals frequently use them to communicate with like-minded individuals. However, the nonprofit community has largely ignored blogs.

One organization that has benefited greatly from its blog is Oceana (2005), an environmental nonprofit in Washington state. This 501(c)3 organization uses its blog to communicate with its stakeholders on a variety of issues pertaining to environmental protection and the organization itself. The Oceana blog is updated several times a week and provides frequent updates about topics of concern to its readership. Shortly after the launch of its blog, Oceana reported a donation of several thousand dollars from a donor who was impressed with the organization's use of technology to educate the public. The organization also has stated that its blog has been helpful in recruiting volunteers and activists, as it was able to communicate with other participants before the organization's events, thereby increasing excitement about the activities (*Oceana Network*, 2006).

The interactive nature of blogs and chat rooms allows an organization to communicate with interested individuals. Dialogic communication is the essence of two-way symmetrical communication, which has been advocated as the most ethical communication style. Blogs provide organizations with an opportunity to engage publics in conversation. Though Grunig (2001) says that organizations have made considerable efforts in improving the nature of their communication, they have not yet truly become symmetrical. Taylor, Kent, and

White (2001) say that the blog offers organizational communication efforts the opportunity to become symmetrical. Perhaps nonprofit organizations can use blogs to develop conversations with their stakeholders on how to best address problems.

Personal networking. Many nonprofits recognize that their stakeholders like to share items of interest with their own peer groups. One e-philanthropy strategy that helps publicize the organization involves tapping into the personal networks of involved publics. For example, the Red Cross Web site has an extensive list of news articles and press releases about its national and international relief efforts. If a volunteer reads a press release that says the organization is seeking additional help in a certain region, the individual has the opportunity to send an e-mail to his or her personal network of friends, family, coworkers, and acquaintances from the Red Cross site.

These e-mails frequently have links back to the organization's own Web site so the reader can obtain more information about the topic. By allowing the site's readers to share information with others, additional conversations are created. Individuals with little or no experience with the organization may become more interested and explore other sections of the Web site. They may not donate that day or offer to be a volunteer, but the initial contact has been made. Now, it is up to the organization to develop a relationship with that individual and work to develop the level of trust to a point where the individual wants to pursue a relationship with the organization.

Raising money. A recent study found that almost 95% of the top four hundred nonprofit organizations in the United States were raising money online (Waters, 2005). These organizations often have the financial backing to pursue risky endeavors that smaller nonprofits lack. However, smaller nonprofits can gain valuable insight into how to develop these fundraising strategies for their own publics based on the successes and failures of their larger competitors.

Over the past five years, many nonprofits have begun asking individuals to raise money online for the organization. The most common example of this type of fundraising strategy involves many of the different run/walk events, such as the National AIDS Marathon, the American Cancer Society's Relay for Life, or the Breast Cancer 3-Day Walk. Each of these organizations requires that its participants raise a set amount of money to participate in the event. However, the organization provides an infrastructure to support the individual's fundraising efforts.

For example, the Breast Cancer 3-Day Walk required each individual to raise $2,200 to participate in the 2006 walks. After registering for the walk, participants were directed to the organization's Web site, which allowed the walkers to send announcements indicating their participation to friends and family. In the e-mail, the walkers asked for support in the form of a tax-deductible donation, which could be made over the Internet. The e-mail also included a hyperlink back to the Web site, where the donation could be made in a safe, secure virtual environment (Breast Cancer 3-Day, n.d.).

In 2005, the Salvation Army began using a similar technique during the busy end-of-the-year holiday season. After receiving notices that many retailers were no longer going to allow any soliciting on their properties, the Salvation Army established virtual kettles. In a similar process to that used by the run/walk participants, anyone who was interested in establishing a virtual kettle could register at the Salvation Army Web site. Then, the individual was encouraged to send e-mails to everyone he or she knew to invite others to help the organization by making a donation on the Salvation Army's secure site (Salvation Army, n.d.).

Implications for Relationship Management

E-philanthropy has only recently emerged in the American nonprofit sector with any significant presence in the last five years. Though initial philanthropic efforts on the Internet stirred from the American response to help others during times of crises, e-philanthropy has had an impact on traditional charitable giving and volunteering as well.

Fundraisers have always sought to develop close relationships with major gift donors because of the greater likelihood of receiving a significant gift for the organization. The Internet has made it possible to reach more individuals. Because of the relatively low costs of sending e-mails and e-newsletters, organizations are able to reallocate funds traditionally spent on paper copies to improve their programs and services. More time can also be spent with donors and volunteers from a wide variety of backgrounds. Through blogging and the interactive features of Web sites, nonprofit organizations can practice excellent public relations with their stakeholders by incorporating two-way symmetrical communication into the overall communication plan.

Additionally, the Internet has created more avenues for the organization to show that it uses donations in a cost-efficient and effective manner. In light of the recent scandals involving donation misuse with the American Red Cross and the September 11th Liberty Fund or the United Way / William Aramony scandal of the 1990s, donors are demanding that organizations be more accountable for the way they raise and spend donations. The Internet provides an inexpensive way to relay this information in the form of downloadable annual reports and IRS 990 tax forms.

As e-philanthropy continues to grow and become more common in the nonprofit setting, organizations are going to insist that future fundraisers and public relations practitioners have a solid understanding of the Internet and how it can be used to best reach the organizations' key stakeholders, whether they are in the same community, the same country, or across the world. With the Internet actively bringing together diverse people from across the globe, nonprofit organizations never know who will be actively browsing and seeking information on their Web sites. However, one thing is for certain: The organization has the potential to turn that casual Internet surfer into a dedicated stakeholder by incorporating the e-philanthropy strategies outlined in this chapter.

For consideration

1. What potential harms exist for nonprofit organizations that decide to incorporate e-philanthropy into their fundraising toolkits?
2. One of the most crucial elements of e-philanthropy success involves knowing who visits an organization's Web site and how to reach them. What are appropriate ethical boundaries for this type of data gathering?
3. What information might a donor want to see on a nonprofit organization's Web site? What information might a volunteer, a client, or a local government official want to see? How could an organization proceed in putting together a Web site to serve such diverse publics while trying to maintain a consistent image and feel to the site?
4. As easy as it is to delete e-mails, will e-mailed solicitations ever completely replace direct mails for annual giving? Why or why not?
5. What types of safeguards or precautions could a nonprofit employ to avoid mission creep?

For reading

Hart, T. R. (2002). ePhilanthropy: Using the Internet to build support. *International Journal of Nonprofit and Voluntary Sector Marketing, 7*(4), 353–360.

Hart, T. R., Greenfield, J. M., & Johnston, M. (2005). *Nonprofit internet strategies: Best practices for marketing, communications, and fundraising success*. Hoboken, NJ: John Wiley & Sons.

Johnston, M. (2001). Fund raising on the Net. In J. M. Greenfield (Ed.), *The nonprofit handbook: Fundraising* (pp. 518–536). New York: John Wiley and Sons.

Sargeant, A. (2001). Fundraising on the Web: Opportunity or hype? *New Directions for Philanthropic Fundraising, 33*, 39–58.

Warwick, M., Hart, T., & Allen, N. (2001). *Fundraising on the Internet: The ePhilanthropyFoundation.org's guide to success online*. San Francisco: Jossey Bass.

References

Allen, N., Warwick, M., & Stein, M. (1996). *Fundraising on the Internet: Recruiting and renewing donors online*. Berkeley, CA: Strathmoor Press.

Austin, J. (2001, March 8). The e-philanthropy revolution is here to stay. *Chronicle of Philanthropy, 13*(10), pp. 72–73.

Breast Cancer 3-Day. (n.d.). *How we support you*. Retrieved July 3, 2006, from http://www.the3day.org/site/pp.asp?c=pmL6Jn08KzE&b=2182527

Christensen, J. (2001, September 26). Tools for the aftermath: Relief agencies retool to handle online flood. *The New York Times*, p. H1.

Demko, P., & Moore, J. (1998, October 8). Charities put the Web to work: Non-profit groups broaden their view of what the Internet can do for them. *Chronicle of Philanthropy, 10*(24), pp. 11–15.

DiPerna, P. (2003). Media, charity, and philanthropy in the aftermath of September 11. In *September 11: Perspectives from the field of philanthropy* (Volume 2, pp. 159–173). New York: Foundation Center.

Greenfield, J. M. (2001). *The nonprofit handbook: Fundraising*. New York: John Wiley and Sons.

Grunig, J. E. (2001). Two-way symmetrical public relations: Past, present, and future. In R. L. Heath (Ed.), *Handbook of public relations* (pp. 11–30). Thousand Oaks, CA: Sage Publications.

Hart, T. (2002). Ephilanthropy: Using the Internet to build support. *International Journal of Nonprofit and Voluntary Sector Marketing, 7*(4), 353–360.

Hart, T., & Johnston, M. (2002). Building trust online. In M. Warwick, T. Hart, & N. Allen (Eds.), *Fundraising on the internet* (pp. 3–10). San Francisco: Jossey Bass.

Kelly, K. S. (1991). *Fund raising and public relations: A critical analysis*. Mahwah, NJ: Lawrence Erlbaum Associates.

Kelly, K. S. (1998). *Effective fund-raising management*. Mahwah, NJ: Lawrence Erlbaum Associates.

Kelly, K. S. (2001). ROPES: A model of the fund-raising process. In J. M. Greenfield (Ed.), *The nonprofit handbook: Fundraising* (pp. 96–116). New York: John Wiley and Sons.

Muscular Dystrophy Association. (n.d.). *Site map*. Retrieved July 3, 2006, from http://www.mda.org/sitemap.html

Network for Good. (2006). *About us*. Retrieved July 3, 2006, from http://www.networkforgood.org/about

Oceana. (2005). Retrieved July 3, 2006, from http://www.oceana.org/

Oceana Network. (2006). Retrieved July 3, 2006, from http://community.oceana.org/

Pew Center. (2005, January). *The Pew Internet and American life project: The state of blogging*. Retrieved March 27, 2006, from http://www.pewInternet.org/pdfs/PIP_blogging_data.pdf

Prince, R. A., & File, K. M. (2001). *The seven faces of philanthropy: A new approach to cultivating major donors*. San Francisco: Jossey-Bass.

Project for Excellence in Journalism. (2006, March). *The state of the news media 2006: An annual report on American journalism*. Retrieved March 28, 2006, from http://www.stateofthenewsmedia.org/2006/narrative_online_intro.asp?media=4

The Salvation Army. (n.d.). *Tell a friend*. Retrieved July 3, 2006, from http://www.salvationist.org/contactus.nsf/fm_recommend?openform&link=https://secure.salvationarmy.org/donations.nsf/donate?openform

Taylor, M., Kent, M. L., & White, W. J. (2001). How activist organizations are using the Internet to build relationships. *Public Relations Review, 27*(3), 263–284.

Wallace, N. (1999, September 9). Red Cross sees jump in Internet donations. *Chronicle of Philanthropy*, p. 40.

Wallace, N. (2001, October 4). Online giving soars as donors turn to the Internet following attacks. *Chronicle of Philanthropy, 13*(24), p. 22.

Wallace, N. (2006, June 15). Charities make faster connections: Fund raising for disaster relief helped lift online totals in 2006 [Electronic version]. *Chronicle of Philanthropy*. Retrieved August 16, 2006, from http://philanthropy.com/premium/articles/v18/i17/17001901.htm

Waters, R. D. (2005, August). *Fundraising on the Internet: A content analysis of ePhilanthropy trends on the Internet sites of the organizations on the Philanthropy 400*. Paper presented to the Public Relations Division, AEJMC Annual Conference, San Antonio, TX.

Waters, R. D. (2006, August). *Measuring the donor-nonprofit organization relationship: The impact of relationship cultivation on donor renewal.* Paper presented to the Public Relations Division, AEJMC Annual Conference, San Francisco, CA.

Worth, M. J. (2002). *New strategies for educational fund raising.* Westport, CT: Praeger Publishers.

Building Relationships through Interactivity

A Cocreational Model for Museum Public Relations

Jordi Xifra

One of the most innovative aspects that information and communication technologies (ICTs) provide to how online museums can now project themselves is the possibility for the user to interact with the content on display. There are new rules of interaction between museums and their publics. Although the concept of interactivity may seem obvious, detailed analysis shows that the concept has a wide range of meanings, covering numerous intentions that affect relationship building and symmetrical communication. Analysis of what interactivity involves should consider various factors: the type of public it addresses, the definition of its aims (education, participation, evaluation, etc.), and the technological accessibility of content from the equipment of potential users. The aim of this chapter is to propose a typology for interactive resources of this kind, to take a look at different examples of interactivity that can be found on museum sites, and to evaluate their potential and intentions for relationship management from a cocreational perspective.

The recent and dramatic rise of information and communication technologies (ICTs) in Western society has brought about a new paradigm in constructing and transferring knowledge. This pervasive transformation has also had an impact on the world of culture, and, in the last few years, technology has been seized as a new opportunity with which museums can accomplish their missions. Unarguably, ICTs offer museums a new way of relating to their publics—both actual and potential—forcing them to rethink their exhibits and broaden the range of products and services they offer at their physical locations. That is, ICTs present a new way of building relationships between museums and their publics. If we regard the Internet as a huge shop window, museums must compete within this marketplace to attract a very wide, but selective, public. Merely publicizing a collection at a museum's physical headquarters is not enough; a complementary online service is needed that will capture visitors' attention and interest.

Within this new paradigm, one of the most innovative aspects of ICTs in relation to new online museum discourse is the chance for visitors to interact with the exhibits. ICTs allow visitors to have continual presence and participation, blurring the traditional roles between the museum—the exhibitor of collections—and the visitor—the passive receiver of these collections. The Internet is therefore an opportunity for museums to build and maintain relationships based upon real dialogue with their visiting publics.

Nevertheless, it is important to clarify what is meant by this theoretical capacity to interact with a Web site. Not everyone has the same understanding of interactivity, and we often find very different activities and resources with highly divergent purposes being lumped together.

The next section of this chapter explores the various types of interactive resources recently available on some museum Web sites and analyzes their potential and purpose in terms of relationship building.

Interactivity, World Wide Web, and Relationship Building: A Museum Approach

As Jo and Kim point out, interactivity can play a crucial role on the Internet, setting it apart from traditional media. In consequence, "the intrinsic interactivity of the Web can enhance the mutual relationship and collaboration between the message sender (the organization) and the receiver (the public)" (2003, p. 202).

Public relations is a complex phenomenon that goes beyond the linear sender/receiver relationship. Grunig (1992) showed that excellent public relations is that in which communication is two-way symmetrical. Feedback and proactivity from publics, which are an increasing part of public relations practice, are also of great significance. Botan and Taylor argued that symmetrical communication and excellence theory remain the most researched of the cocreational approaches:

> The co-creational perspective sees publics as co-creators of meaning and communication as what makes it possible to agree to shared meanings, interpretations, and goals. This perspective is long term in its orientation and focuses on relationships among publics and organizations. (2004, p. 652)

Kent and Taylor (1998) showed that public power is related to the interactivity of media for better two-way communication and relationship building. Thus, the advent of the Internet, as an interactive medium, can have important effects on organization-public relationship building. From this standpoint, Kent and Taylor suggested that it was possible to build dialogic relationships through the Internet and proposed five principles to be followed in building dialogic Web sites: (a) create a dialogic loop that allows publics to query organizations and, more essentially, "offers organizations the opportunity to respond to questions, concerns and problems" (p. 326); (b) be sure that information on the Web is useful for all publics; (c) generate return visits by members of the public; (d) ensure ease of interface with the site; and (e) conserve visitors. These principles are closely linked with interactivity.

According to Vidal (1998), there are three forms of interactivity underpinned by three different criteria. Firstly, it is necessary to examine if the relationship is with a programmed mechanical device or with another human being. This criterion allows us to distinguish between the concepts of *interactivity* and *interaction*. The former concept refers to a *dialogue activity* between a person and a machine by means of a technical device. In the latter, the concept signifies the *reciprocal action* that can take place between a sender and a receiver and that may lead to role reversal. Applying this criterion allows us to draw a qualitative line between resources, according to whether the person is following a preprogrammed pattern in order to use a machine or is really interacting with other people, either the owners/senders of the Web site or other service users.

A second, quantitative criterion relates to the extent of the impact involved, with users able to interact with the Web site on various levels. Level 1 would offer contact mechanisms between the user and the Webmaster. At level 2, the user could take part in some kind of activity proposed by the senders following a set of predefined patterns. Level 3 would enable the user to create contents under the control of the Webmaster, whilst, at level 4,

Table 1: Levels of Potential Interactivity

Level 1	The user is able to contact the Webmasters, although there is no guarantee that the message receivers will respond. Interaction is confined to a private domain between the sender and receiver, but the Webmasters have no prior control over what the user produces (for instance, e-mails).
Level 2	The user can take part in an activity that has been previously defined and programmed by the Webmasters. Therefore, the Webmaster has prior control over the activity and determines what the user can and cannot do (for instance, online games).
Level 3	The user is able to create contents under the control of the Webmasters. This is the case with opinion forums, for example, whereby users can express their opinions, but the forum moderator has the power to delete or not publish a message if it is deemed inappropriate.
Level 4	The user can become coauthor of the Web site and modify or extend the information initially provided by the Webmasters. The Webmaster has no prior control, and the user's audience is the general public.

From "L'interactivité des sites Web de musées" [Museum Web sites' interactivity], by G. Vidal, 1998, *Publics et Musées*, *13*, pp. 89–107. Adapted with permission.

the user could become a coauthor of the Web site and could freely modify all information contained within (see Table 1).

Thirdly, it is also necessary to consider if the outcome of the interaction is confined to a private circle or, on the contrary, will reach the public sphere. This third criterion draws a distinction between interactive resources with a public domain and those that are aimed at a private sphere alone. The former includes user activities in which the results can be viewed by anyone who visits the Web site, whilst, in the latter case, the results may only be accessed by the user and the Webmasters. Table 2 shows a summary of interactive resources currently available on the Web.

Table 2: Types of Interactive Resources

Level	Interactivity	Interaction	Sphere
Level 1		E-mail	Private sphere
Level 2	Online games; personalized contents	Surveys	Private sphere
Level 3		Discussion forums; guest books	Public sphere
Level 4		Content creation	Public sphere

From "L'interactivité des sites Web de musées" [Museum Web sites' interactivity], by G. Vidal, 1998, *Publics et Musées*, *13*, pp. 89–107. Adapted with permission.

Interactivity in Online Exhibitions

Exhibitions are at the heart of dissemination for any museum. According to Solanilla (2002), ICTs have given rise to new narrative and communicative aspects in museum exhibitions and, as a result, a new means of generating exhibition discourse that is brimming with potential and open to experimentation.

One of the most significant upshots of this new discourse is that exhibition creation has now become a two-way, interactive process. Traditionally, exhibitions entailed closed, static discourse, conceptualized and generated by the museum itself, with the visitor playing the role of passive contemplator. However, in the mid-1980s, this concept of the museum and its one-way discourse began to be questioned (Kotler & Kotler, 1998). The result of this critical assessment, accompanied by a desire to prioritize communications and relationships in museums, was that multimedia were brought into the rooms of the museum as a complement to the linear, one-way discourse of the exhibition. However, the real revolution in conceptual approaches is more recent, pushed forward by the spread of ICTs into everyday life.

In order to consider the museum in terms of ICTs, the traditional roles played out by each party must be rethought. Several online experiences already allow the visiting public to create knowledge, thereby making a significant contribution to the exhibition discourse generated by the institution. Such resources are aimed at developing individual knowledge in the private sphere of the user, whereas others have a public dimension and may result in generating collective knowledge.

The private sphere. The most basic interactive resource that almost all museum Web sites now have is an e-mail address to contact the Webmaster or institution. This tool, known to all and very widely used, allows the entity involved to set up an affordable hotline for all users through which they can send their complaints, demands, and suggestions. If the institution so desires, it can also enter into interpersonal communication with the user. However, electronic mail enables very simple (sender/receiver) interactivity that relies on the will of both parties to ensure a flow of communication.

At the second level of interactivity are online games and activities in which users follow a series of rules that have been established by the organizers. This type of activity is reasonably popular and generally relates to the educational and learning aims of the exhibition. The activities themselves vary widely and are aimed at a broad spectrum of people, to which we shall return.

The first resource highlighted is the Museum of American History's Star-Spangled Banner Web site (n.d.). This online exhibition told the tale behind the first American flag, which inspired the national anthem. There were two interactive tools of interest here. One was a Test Your Knowledge section on the contents of the Web site, requiring the visitor to read through the information provided in order to get the right answers. The other was a game (You Solve the Mystery) in which documentary sources must be consulted (articles, letters, and photographs) to answer a set of questions. These games were not aimed exclusively at schoolchildren but rather at a broad range of ages; indeed, their level of difficulty suggested that the target audience was actually adults or university students.

The panoply of games included in the British Museum's (Trustees of the British Museum, 1999) exclusively virtual Ancient Egypt exhibition is certainly worth a special mention. This online exhibition taught visitors what life was like for an Egyptian pharaoh. Each subtopic was split into four parts: a general introduction, a more in-depth explanation of that area (entitled Story), a second part consisting of questions to which the visitor must find the answers (entitled Explore), and, lastly, a section entitled Challenge with interactive online games related to each topic area. Some of these included learning to play senet (an ancient Egyptian board game), deciphering hieroglyphics, and passing tests and obstacles with the help of the gods.

These games were a structural component of the Web site itself, rather than mere extras, since the approach taken to learning was the result of active involvement on the part of the visiting public. The progression through the levels, the architecture of the site—allowing visitors to situate themselves within the discourse at all times—and the use of games to boost personal involvement in the search meant that information was transferred efficiently and imperceptibly to the user. Learning through play is a good way of transmitting knowledge, and linking games to education is a new way for museums to build relationships with their audiences. This innovation certainly breaks with the image of learning and museums as being academic, dull, and lifeless.

Lastly, a new tool beginning to emerge on museum Web sites allows visitors to create a personalized site by choosing from the digital resources available on the online database. Service users can build their own online museum by adding their favorite pieces and can even write their own comments, as could be seen under the My Personal Museum section at the Virtual Museum of Canada's (n.d.) portal. This option allowed visitors to personalize the portal according to their own preferences: Up to thirty images could be selected from those offered in the image gallery, and personal comments could be included. Every time visitors went to the site, they could look at their own selected objects and continue adding comments and information whenever they wanted. This is museum building à la carte, enabling visitors to create unique Web sites based on their own interests and likings.

The public sphere. The examples previously discussed refer to individual users and results that remain in the private sphere. However, there are other initiatives on the Web that take this interactivity one step further by opening the dialogue to a public forum and reaching new levels of interactivity and relations. These levels allow the user to create contents under varying degrees of control by the sender and to display them publicly, thereby potentially motivating others to do the same.

The most common resource of this type is an opinion forum on a specific topic in which the public can give its opinion on an issue raised by the organization and debate it with other users. Alternatively, visitors can simply say what they think about the Web site or exhibition in an electronic visitor's book. In both cases, unless the content is deemed to be offensive or insulting (and the organization sees fit to delete it), users have complete freedom to express their opinions. Some of these forums are a real boost to Web site content, in that they breathe a collective life into the site that perhaps even the organizers had not envisaged.

Let us first look at the Voyage of the *St. Louis* exhibition to analyze this in more detail. This online exhibition, set up by the United Status Holocaust Memorial Museum (n.d.) in Washington, DC, explored the forced exile of 937 German Jewish refugees who set sail aboard the *St. Louis* in search of asylum in Cuba and the United States. The site was built upon an empathetic discourse and told the tales of some of the passengers. A type of interactive game was used in which the visitor had to analyze clues contained in original period documents to reconstruct these true life stories. What needs to be underscored about this Web site, however, was its appeal for public collaboration, be it from survivors, descendents, or acquaintances of the *St. Louis* passengers. Anyone who had any testimonials or graphic documentation related to the journey and its passengers was asked to submit it, in order that new personal stories could be included on the site. Hence, the content of this Web site was developed by the users as they collectively built knowledge and memories.

A similar, but not parallel, example can be found in the Without Sanctuary: Lynching Photography in America exhibition. The New York Historical Society (n.d.-b) exhibited James Allen's private photo collection of the lynching of black people until the 1930s in America. Both the physical and online exhibitions stoked such strong feelings that testimonials with new images started emerging and, in some cases, descendents of the dead recognized their relatives in the photographs. A heated public debate on this all-but-forgotten shameful collective past got under way. Even after the physical exhibition had closed, the Web site continued to grow through individual contributions. This phenomenon differs from the previous example in that the organizers had neither anticipated nor requested this public response, although they were cognizant enough to realize that the online exhibition could act as a stimulus to collective feeling. So, in this case, the online exhibition brought a new dimension of communications and relationships to the physical exhibition, which enabled a community to build its own collective memory. Although the physical exhibition was taken down years ago, the debate forum was still very much active at the time of this book's publication and was used by teachers from various disciplines to teach students about a part of their history that is so often buried. In doing so, teachers and students can explore their own collective memory.

Nevertheless, not all forums are met with the same amount of success as the experiences already mentioned. Despite the best efforts of organizers to make them work, forums very often become the dead pages of a Web site. The two examples discussed herein have been popular because there was enormous interest in the subject matter of the exhibitions. When it comes to nonspecialist, or more generalized, publics, this often occurs when the topic of the exhibition touches upon some aspect of their private emotional lives and is only possible when users are given the opportunity to share an experience, be it their own or that of someone close to them. Therefore, interaction with an online exhibition is likely to be higher if it deals with a contemporary issue for which personal stories can be the framework for the exhibition discourse. Specialist publics, on the other hand, may also participate if the topic is related to their profession or field of study, regardless of any topic trends.

There is one other level of interactivity in which visitors are asked to collaborate directly in generating Web site content. This tool is not very widespread but is starting to emerge on certain sites that have various features in common. All deal with contemporary history and have a strong emotional or empathetic appeal to visitors, either because they refer to situations that visitors themselves have experienced or because they belong to a given community's personal or collective memory. This is the most intriguing aspect because of its innovative potential for building knowledge. Indeed, it takes the possibilities of the World Wide Web to its very limits. That said, no Web site as yet is made up of contents generated exclusively by Internet users, the norm being to set up an online exhibition on a chosen topic and ask visitors to make personal contributions to it.

A good example to illustrate this is the Choosing to Participate: Facing History and Ourselves exhibition, run by the New York Historical Society (n.d.-a). This exhibition, devoted to the civil rights struggle in the United States, established a narrative discourse motivated by the premise that individuals can change the world. Visitors were invited to scrutinize modern democracy and the importance of individual citizens in preserving it. Three examples of people who made a difference in the fight for civil rights were provided, and visitors were asked to share their own reflections on socially unjust situations.

These testimonials (including images, texts, documents, drawings, and so on) were displayed on the site so that anyone could access them, making this exhibition an open project that was shaped and modified by visitors' input. Users felt a sense of belonging to the community that created the site and could manage and generate experiences in the same way as Webmasters.

The last example is an exhibition based entirely on the concept of interactivity. Collected Visions: An Interactive Archive of Stories and Snapshots invited visitors to write true or invented stories about the images contained in the Center for Creative Photography's (n.d.) searchable archive. These family snapshots, sent in by over 350 people to construct a discourse based on identity, allowed real or imaginary collective perspectives to be taken on everyday life. It was an experiment in microhistory that transcended the anecdote to give amazing insight into human sentiments and feelings at the end of the century. The institutional intervention was minimal, amounting merely to provision of the tools needed to create the contents. In this sense, Collected Visions was a foray into new ways of communicating and establishing identity by means of ICTs.

Interactivity, then, can be seen as the opportunity to strike relationships between ICTs and dissemination of our heritage. In this context of relationships, it is only by creating new languages and forms of user intervention in exhibition discourse that we will be able to achieve the same degree of communication and relationships as online exhibitions. The Positive Visions section of the Collected Visions Web site is worth singling out for attention here as it related to people with AIDS—an issue capable of creating strong bonds between both patients and their families.

Implications for Relationship Management

That the Internet has the potential to revolutionize communications between museums and their publics and to create relationships based on active public participation with knowledge transfer in mind is patent from the cases that have been discussed. The interactivity afforded by ICTs is changing the voice of museums in dialogue with an infinitely more genuine world.

As we have seen in these examples, the concept of interactivity is not clear-cut. It entails multiple meanings and conceals multiple purposes. An interactive resource on a Web site is efficient when it meets the expectations of those who have created it. Theoretically, the ultimate aim of any interactive resource, whatever it involves, is that it can be used by an Internet user. However, because the ways in which these resources can be used are manifold, if, in practice, an interactive resource is not utilized, it is probably badly designed or at odds with users' needs.

In the case of museum Web sites, the most popular interactive resources, aside from e-mails sent to the organization, are games (Durbin, 2003; Hooper-Greenhill, 2004). This is no coincidence but part of a deliberate drive to present the museum as an informal space for relationships and learning. Providing entertainment resources makes knowledge transfer light hearted and enjoyable for the user whilst also fostering and reinforcing the relationships between the museum and its audience.

Nevertheless, although games are unarguably a very powerful tool for learning in the field of heritage, there are other types of interactive resources that should not be overlooked. Experiences noted here are proof that an interactive approach to the Web, even in terms of content generation, is a practical reality. In order for these resources to reach their potential, there has to be some kind of empathy between the user and the subject of the exhibition. In other words, a partnership must be created between the museum and its visitors (Durbin, 2003; Ockuly, 2003; Tzanavari, Vogiatzis, Zembylas, Retalis, & Lalos, 2005).

For museum public relations specialists, these experiences do not have to be limited to a specific type of audience—any sector of the public can become comanagers and, as a result, cocreators of online content. Take, for example, forums aimed at members of a professional sector (specialist public) in which the museum acts as the catalyst for experience exchanges and as a meeting place for researchers and academics. This situation follows Botan and Taylor's thoughts on the cocreational perspective of public relations, which

> places an implicit value on relationships going beyond the achievement of an organizational goal. That is, in the cocreational perspective, publics are not just a means to an end. Publics are not instrumentalized but instead are partners in the meaning-making process (2004, p. 652).

Numerous factors must be taken into account when analyzing this type of resource: (a) the target public and its expectations and needs, (b) the aims (education, participation, feedback, and so forth) that the Web site managers hope to achieve by providing a given resource, and (c) the technological accessibility of the resource with regard to potential users' equipment (most users will be unable to utilize a resource requiring state-of-the-art software).

The possibilities afforded by interactive resources in virtual environments are only just beginning to emerge. As the medium itself and the new language it generates come of age, light will be shed on new options and new types of relationships between various stakeholders. What we are seeing is an unstoppable process; it is our duty to think critically about the meaning and potential of this new language in the knowledge society and the role that public relations plays in managing, transmitting, and creating knowledge.

Indeed, the Internet will change the relationship between museums and their publics. The Web allows and arguably requires museums to enter increasingly into dialogue with their users in new ways. The one-way presentation of content, accompanied by the unassailable voice of museum commentary, will need to evolve into a more interactive and responsive relationship through new electronic media, problematic though that may be (Jackson, Bazley, Patten, & King, 1998).

A number of museums have deliberately sought and incorporated the perspectives of visitors from within their physical walls. Examples in this chapter include allowing the selection and presentation of art exhibitions to be engaged in by visitors and providing space for individuals to display their own collections of objects.

As Jackson, Bazley, Patten, and King (1998) noted, while it is possible for the public to occupy and affect the physical exhibition space of a museum, it is much easier and cheaper to do so online. In addition, the capacity of the Internet for communication makes it possible to make contact and develop long-term relationships with the public.

The Internet gives museums access to global audiences in their own homes. It can reach people who may visit museums as well as those who do not. Interactive media are clearly appealing to younger audiences, and, as MacDonald and Alsford (1997) have remarked, museums must keep pace with technological progress if they want to attract twenty-first century audiences. In effect, tomorrow's museum visitors will be people who have grown up with computers and multimedia as a major factor in their lives.

However, beyond enlarging the geographical scope and audience age range, can the Internet actually change the socioeconomic, educational, and/or cultural profile of museum visitors? Although online museum visitor numbers are burgeoning, demographic data on users of the World Wide Web still indicate that their profile is similar, in terms of income and education, to that of traditional museum visitors (MacDonald and Alsford, 1997; Keene, 1998).

ICTs also play a central role in creating audience-focused museums, which is the current trend. A raft of specific audience-focused strategies outlined in this chapter illustrates how interactivity afforded by ICTs helped to build and maintain museums' relationships with their visitors. These include (a) adding visitors' stories and perspectives to the interpretational process; (b) linking activity contents with visitors' lives; (c) connecting objects to people, places, and purposes; (d) connecting people with other people and resources; (e) facilitating and promoting entertainment; (f) making messages more personal through stories and narratives; (g) asking visitors to make decisions and choices and to give opinions; (h) providing multiple perspectives and points of view; (i) creating receptive environments; and (j) supplying information that is relevant to the user.

In the museum domain, Anderson finds that

> We live now in a time of changing paradigms: from communicating to (that is, at) the public, to inviting their contribution; from a concept of excellence that is focussed on product (that is, the object) to excellence of process and experience as well as product; from giving prominence to "cool," intellectual spaces to mixing "hot" lived as well as "cool" spaces; from directing to enabling; from linearity to multiplicity; from concern with the profession to concern with the public; from what's wanted to what's needed; from site to network; from the conceived to the experienced; from public passivity to public creativity; from data and information to learning. (2000, para. 2)

Applying and using ICTs in this way means bringing people to the forefront, thereby championing software that focuses on individuals, creates social relationships, and promotes participation by including the audience's points of view and experiences. For Jackson, such "knowledge creation in collaboration" (1998, p. 3) is useful because it allows the public to have a say on specific sections of museum collections. This is a new cocreational model for managing the relationships between organizations and their visitors based on knowledge management.

Implementing this model means that knowledge building is no longer in the hands of the institution but rather is the upshot of collective action. Thus, relationship building and knowledge building become two parallel activities in a context in which public relations takes on a new social function—building, maintaining, and transferring knowledge.

In the year 2000, Anderson raised the following questions:

> For this is the learning age. Over 90% of adults say they believe they learn every day of their lives. They say they mainly learn through the cultural sector—the media broadcasting, libraries and museums—rather than formal educational institutions such as colleges. Most say they prefer to learn independently and self-directly, pursuing areas of their own interest, often socially, with like minded people, rather than formally in a classroom where they are taught. And the great majority say they enjoy learning. So why does the museums sector have such difficulty in saying that its primary purpose is public learning? Where is the democratic accountability in this? It is what our users want and expect. (para. 5)

Today it is possible to argue that public relations, in its role as knowledge cocreator, fulfills the function of satisfying the expectations of citizens in the learning age.

For consideration

1. How has the emergence of new media changed the way public relations practitioners must think about two-way symmetrical communication? How has the definition of *interactivity* changed over the past ten years?
2. This chapter has explored museum Web sites that are mainly concerned with history. Does this kind of museum lend itself more to interactivity than other nonhistoric museums? Why or why not?
3. What role do graphic and audiovisual resources play in boosting online relationships for museums?
4. To what extent can the cocreational model discussed in this chapter be applied to other organizations in different spheres?
5. What is the role of public relations in the knowledge society, beyond the domain of the museum? What does it mean to be *cocreators* of knowledge?

For reading

Botan, G. M., & Taylor, M. (2004). Public relations: State of the field. *Journal of Communication, 54*(4), 645–661.

Falk, J. H. (2000). *Learning from museums: Visitor experiences and the making of meaning.* Oxford: Altamira Press.

Kotler, N., & Kotler, P. (1998). *Museum strategy and marketing.* San Francisco, CA: Jossey-Bass.

McLean, K. (2001). *Planning for people in museum exhibitions.* Washington, DC: Association of Science-Technology Centers.

Thomas, S., & Mintz, A. (Eds.). (1998). *The virtual and the real: Media in the museum.* Washington, DC: American Association of Museums.

References

Anderson, D. (2000, October). Networked museums in the learning age. *Cultivate Interactive, 2.* Retrieved April 15, 2006, from http://www.cultivate-int.org

Botan, G. M., & Taylor, M. (2004). Public relations: State of the field. *Journal of Communication, 54*(4), 645–661.

Center for Creative Photography. (n.d.). *Collected visions. An interactive archive of stories and snapshots.* Retrieved February 19, 2006, from http://www.cvisions.cat.nyu.edu/

Durbin, G. (2003, March). *Using the Web for participation and interactivity.* Paper presented to the annual conference of Archives & Museums Informatics (Museums and the Web), Charlotte, NC.

Grunig, J. E. (Ed.). (1992). *Excellence in public relations and communication management.* Hillsdale, NJ: Lawrence Erlbaum.

Hooper-Greenhill, E. (2004). Changing values in art museum: Rethinking communication and learning. In B. M. Carbonell (Ed.), *Museum studies: An anthology of contexts* (pp. 556–574). Malden, MA: Blackwell Publishing.

Jackson, R. (1998). Whatever happened to the "C" in ICT? In *The cultural grid: Content & connections: Museum Documentation Association conference proceedings* (pp. 3–6). Cambridge, UK: Museum Documentation Association.

Jackson, R., Bazley, M., Patten, D., & King, M. (1998, March). Using the Web to change the relation between a museum and its users. *Museums and the Web.* Retrieved April 4, 2006, from http://www.archimuse.com/mw98/papers/jackson/jackson_paper.html

Jo, S., & Kim, Y. (2003). The effect of Web characteristics on relationship building. *Journal of Public Relations Research, 15*(3), 199–223.

Keene, S. (1998). *Digital collections: Museums and the information age.* Oxford: Butterworth-Heinemann.

Kent, M. L., & Taylor, M. (1998). Building dialogic relationships through the World Wide Web. *Public Relations Review, 24*(3), 321–334.

Kotler, N., & Kotler, P. (1998). *Museum strategy and marketing.* San Francisco, CA: Jossey-Bass.

MacDonald, G. F., & Alsford, S. (1997). Towards the meta museum. In K. Jones-Garmil (Ed.), *The wired museum: Emerging technology and changing paradigms* (pp. 35–43). Washington, DC: American Association of Museums.

Museum of American History. (n.d.). *The star-spangled banner.* Retrieved February 12, 2006, from http://www.americanhistory.si.edu/ssb

New York Historical Society. (n.d.-a). *Choosing to participate: Facing history and ourselves.* Retrieved January 15, 2006, from http://www.facinghistory.org/campus/reslib.nsf/sub/aboutus/whatwedo/ctp

New York Historical Society. (n.d.-b). *Without sanctuary: Lynching photography in America.* Retrieved March 2, 2006, from http://withoutsanctuary.org

Ockuly, J. (2003, March). *What clicks? An interim report on audience research.* Paper presented to the annual conference of Archives & Museums Informatics (Museums and the Web), Charlotte, NC.

Solanilla, L. (2002, April). Què volem dir quan parlem d'interactivitat? [What do we mean by interactive?] *Digithum, 4.* Retrieved January 23, 2006, from http://www.uoc.edu/humfil/digithum/digithum4/catala/tecnica/index.html

The Trustees of the British Museum. (1999). *Ancient Egypt.* Retrieved February 12, 2006, from http://www.ancientegypt.co.uk/menu.html

Tzanavari, A., Vogiatzis, D., Zembylas, M., Retalis, S., & Lalos, P. (2005). Affective aspects of Web museums. *Studies in Communication Sciences, 5*(1), 39–56.

United Status Holocaust Memorial Museum. (n.d.). *Voyage of the St. Louis.* Retrieved February 15, 2006, from http://www.ushmm.org/museum/exhibit/online/stlouis/

Vidal, G. (1998). L'interactivité des sites Web de musées [Museum Web sites' interactivity]. *Publics et Musées, 13,* 89–107.

Virtual Museum of Canada. (n.d.). My *personal museum*. Retrieved February 15, 2006, from http://www.virtualmuseum.ca/English/Personal/index.html

Going Green

Using Computer-Mediated Public Relations to Communicate Environmental Causes

Ric Jensen

A large number of diverse organizations—including government agencies, special interest groups, nonprofits, and corporations—want to have a say about important environmental issues. They often turn to formal and informal public relations measures, ranging from grassroots education campaigns to complex media relations programs. This chapter provides insights into the extent to which practitioners engaged in environmental public relations have used computer-mediated public relations methods, such as blogs, RSS feeds, podcasts, digital video, and other new technologies. It also presents examples of some of the success stories and challenges associated with the use of computer-mediated public relations to communicate environmental issues.

Many corporations, agencies, and organizations that deal with the environment are now urging the need for business leaders to adopt sustainable environmental practices and interact with stakeholders in a transparent and open manner if they expect to build public trust.

In a recent article published in *WorldWatch*, the author contends that "snazzy new public relations initiatives" (Assadourian, 2006, p. 16) will not be sufficient to help corporations flourish in today's marketplace. Companies that incorporate the best environmental practices into their operations and then honestly communicate their efforts to the public through blogs and other means will likely improve their reputations.

In 2001, a white paper published by the Environmental Section of the Public Relations Society of America (Lee, 2001) demonstrated how corporations benefit financially when the public feels they are acting in an environmentally responsible manner. PR Watch, an organization that critically monitors the intersection between public relations and potential conflicts of interest with business and industry, observes that "corporations can improve their own images by 'partnering' with nongovernmental organizations and governments on beneficial social projects" and by making these partnerships visible through public relations activities (Rampton, 2002, para. 11).

Dutta-Bergman (2004) suggests that environmental activists expect the organizations they support to provide extensive information about issues of concern and follow environmentally friendly practices. The challenge for public relations professionals is to communicate the progress that organizations are making through the use of traditional and computer-mediated methods. Even though environmental activists desire that comprehensive and factual information be presented to them in a timely manner, Taylor, Kent, and White (2001) suggest this ideal is not being fully met. Their survey of the computer-mediated practices of one hundred environmental organizations showed that many of these groups did not provide the communications tools on their Web sites that are needed to establish long-term relationships or to encourage people to regularly visit these sites.

According to a recent report by the Council of Public Relations Firms,

> Consumer-to-consumer conversations are changing traditional marketing in fundamental ways and public relations firms can either join the revolution or risk being overcome by it. . . . Consumers are increasingly participating in and creating their own media. . . . They are using vehicles such as blogs, wikis and podcasts to reach small groups of people who share interests about which they are passionate. (as cited in Cripps, 2005, p. 1)

The article suggests that the future of public relations will include delivering content on multiple platforms. To increase public trust, organizations should focus on building their brand reputations through the use of advanced Internet-based technologies. Kent and Taylor (2002) suggested that organizations should reinforce their commitment to dialog and should foster more interaction with stakeholders by using such mass-mediated communications channels as e-mail, the Web, and emerging Internet technologies. They suggest that the Web comes closest to emulating the traits found in personal face-to-face communication, thus creating opportunities for dialog.

In 2001, Hurme suggested that computer-mediated technologies are shifting the way in which public relations is practiced. In the past, public relations efforts focused on sending messages through the mass media; now, computer-mediated technologies allow communicators to provide information directly to stakeholders. Hurme advocated that computer-mediated public relations should supplement both face-to-face communications and the traditional use of one-way communications, such as press releases and brochures. Hurme said, "If companies doing business on the internet can't figure out what the web is, they may soon self-destruct" (p. 74).

Suh, Couchman, and Park (2003) developed a theoretical model to explain how individuals act in a computer-mediated communications environment. They suggest that people who utilize computer-mediated Web sites become part of a virtual community where corporations and consumers interact. Therefore, Web site design must be focused on building effective communications and dialog that reflect the principles of face-to-face interaction, rather than simply being driven by the latest technology.

In some cases, public relations firms are using computer-mediated technologies and practices to communicate environmental news and public relations programming to the mass media and targeted stakeholders. For example, E-Wire (2006) uses targeted e-mails and Web sites to deliver environment-related news, research, services, and product information from its clients to major news organizations, publications, government agencies, databases, and environmental professionals around the world. The E-Wire strategy emphasizes the use of daily distribution of news releases, the syndication of its news releases to major wire services, and the posting and archiving of news releases to environmental Web sites to reach nearly ten thousand journalists and more than seventy thousand industry subscribers.

Enhancing the Capabilities of Web Sites

In many cases, corporations and environmental organizations are creating Web sites specifically to communicate issues of concern and are utilizing high-tech tools to prompt interaction with stakeholders.

The best of today's Web sites allows users to subscribe to and receive timely updates through Really Simple System (RSS) feeds or to send e-mail alerts or e-newsletters. RSS allows content providers to build awareness and create interactive relationships with users. RSS subscribers quickly and automatically obtain updated Web site content and spend less time Web surfing. Barnes suggests that

> RSS presents a huge opportunity to create communication efficiencies inside the [business] enterprise by usurping the politics that bog email down and expediting the process associated with team communications. . . . Users determine their favorite websites and a properly configured RSS aggregator will syndicate selected lists of hyperlinks and headlines along with other information about the websites. (2004, p. 1)

In addition, many organizations that present environmental information (e.g., the Sierra Club, Global Exchange, and Moving Forward) allow users to download RSS feeds in extensible markup language (XML) format so that their information can be easily read through an RSS aggregator.

Corporate efforts. In 2006, the General Electric Company (GE) began promoting its environmental efforts in the Ecomagination public relations campaign: an effort that relied heavily on computer-mediated practices. For example, users could go to the Ecomagination Web site (General Electric Company, 2006) and download wallpaper for their computer desktops, view animated ads that use flash technology, and read an online annual report linked to digital video interviews. According to the company's Ecomagination report,

> True transparency cannot exist solely as one-way communication or issuing of documents: its clarity comes from the give and take between interested parties. . . . GE continuously engages customers and stakeholders on its ecomagination commitments. (General Electric Company, 2005, p. 26)

To reach out to stakeholders, GE used its Web site, special engagements, conferences, and "dreaming sessions" where heads of utilities meet with GE and tell the company how they believe environmental concerns can best be resolved.

The Walt Disney Company (n.d.) provides another example of how specialized Web sites can be created to address specific environmental topics. For example, the company created Web sites describing the Disney Wildlife Conservation Fund and offered daily "eco news" e-mail updates through Disney's Environmentality Web site. These sites let users determine their "carbon footprint" and learn about environmental efforts that Disney supports.

Environmental organizations. Environmental organizations and agencies throughout the globe are increasing the amount of information presented through innovative media on their Web sites. Two prominent international examples of the use of computer-mediated public relations involve the Food and Agriculture Organization (FAO) of the United Nations and the World Business Council for Sustainable Development.

The FAO Web site (Food and Agricultural Organization, 2006) enabled users to ask questions in nontechnical terms about the organization's programs. Once the question was entered, the FAO Web site search engine queried whether any branch of this international agency had produced information for the press or technical data about this topic. FAO has begun efforts to initiate live Internet-based chats between reporters and agency administrators and scientists. FAO developed and released through the Internet a series of detailed maps that displayed the locations most damaged by tsunamis in the Indian Ocean.

The World Business Council for Sustainable Development (2006) enabled people to subscribe free of charge to any of ten e-newsletters covering specific environmental topics. Users could also view digital video interviews of environmental leaders with feeds to accommodate low- or high-speed Internet connections. The Council worked to develop the Partnerships Central Web site that provided comprehensive links related to sustainable development initiatives in areas including water and sanitation, biodiversity, agriculture, and oceans and fisheries. Partnership Central allowed users to subscribe to free e-newsletters.

The Sierra Club (n.d.) enabled users to receive e-news alerts with specialized Web site content. For example, the Sierra Club's *Raw* newsletter provided first-person accounts and opinions about timely environmental issues, such as oil exploration in sensitive areas, natural habitat preservation, and climate change. Similarly, the organization's *Insider* and *Currents* newsletters provided regular updates of environmental news, current events, and publications. The main Web site also provided links to state news and organizations, as well as clicks to specific environmental topics. Reber and Berger (2005) caution that most of the information presented by the Sierra Club simply frames the viewpoint of the organization, rather than presenting a balanced view that also incorporates the views of other organizations. In other words, if one is looking for balanced information, it may likely not be found at the Sierra Club Web site.

In 2006, a Web site was created by a diverse group of partners including the Natural Resources Defense Council, the National Council of Churches, the Sierra Club, and the Union of Concerned Scientists, among others, to develop a "virtual march" to stop global warming (StopGlobalWarming.org, 2006). The site was promoted by viral marketing, by which environmentally minded organizations and individuals were encouraged to help promote the event by sending e-mails urging other people to take part or by hosting a banner, video, sign-up form, or one of many other assets available on their Web site or blog. More than 375,000 people took part. The Web site allowed users to determine if their local elected officials had passed resolutions to fight climate change, to view and link with others in their region who also took part in the march, and receive news updates on the progress of the march via live messaging. The Web site described the effort this way:

> Through the power of a grassroots "virtual march"—an online movement that grows as it's empowered by all of us—we will demand that our government pay attention. With the support of leading scientists, political, religious, cultural, and business leaders, as well as everyday citizens, the March is moving across the country via the internet showing evidence of global warming's alarming effects and offering solutions along the way. (Home Page section)

The Nature Conservancy (2006) Web site offered video showing habitat they were working to preserve, as well as the *Great Places* e-newsletter that provided updated information about issues related to preserving diverse habitats. The Web site also alerted subscribers when opportunities arose to take part in Internet chats with experts and allowed subscribers to send e-posters depicting ecosystem and habitat issues.

The Los Angeles Department of Water and Power (LADWP) (n.d.) encouraged 275,000 of its 1.4 million customers to register at its Web site. As a result, the utility had the capability to send announcements and communications to them via e-mail. LADWP also offered online workshops and PowerPoint presentations on its Web site.

The American Chemistry Council (2005) provided links to digital video interviews of its leaders addressing environmental concerns, e-newsletters, magazines, and an education campaign that explained how chemistry is vital to the environment and quality of life. The Web site also provided an easy-to-use, automated way to develop letters that can be sent to members of the U.S. Congress.

The Texas Water Resources Institute (n.d.) established a policy of creating Web sites for each of its externally funded projects. As a result, the institute developed Web sites about topics including water needs along the Texas-Mexico border, the beneficial use of compost to reduce sources of water pollution, and training materials for small water systems, among other things. Users could subscribe to e-newsletters, and scientists could ask to be alerted about opportunities for research funding.

Blogs

The use of personal web logs or *blogs* can provide a number of public relations benefits for corporations and organizations. Blogging consultant John Cass (Backbone Media, n.d.) suggests that blogs that are well done excel in allowing organizations to connect and build trust with key publics on a personal level, especially if opportunities for feedback are provided. The word of mouth generated by conversations about blogs can also enhance viral marketing. As a result, several environmental organizations and corporations have embraced blogging as a public relations strategy.

According to Internet consultant Steve Rubel (2006), blogging has replaced the paradigm of the traditional mass media (i.e., television and newspapers) with a blogosphere that has created new distribution channels that can be accessed by mobile devices, such as personal digital assistants (PDAs), smart cell phones, and iPods. Rubel contends that blogs provide fresh content that is uniquely suited to building relationships with target audiences. He estimates that 28% of all Internet users regularly read blogs or download content from them.

Several blogs have been developed by corporations, government agencies, organizations, and individuals that provide a forum to discuss environmental issues. In some cases, it is difficult to determine if these blogs simply communicate environmental points of view or are tools used in formal public relations campaigns.

McDonald's Corporation (n.d.) created a blog focusing specifically on corporate responsibility. The blog, Open for Discussion, was written by Bob Langert, the company's senior director for corporate social responsibility. Some of Langert's postings discussed sustainable agricultural production and the development and use of alternative refrigerants that do not include hydrofluorocarbons (which are thought to contribute to global warming). Langert wrote, "We want to open our doors to corporate social responsibility—to share what we're doing and learn what you think. That's the purpose of this blog."

Stonyfield Farm (n.d.), a company that produces organic yogurt, used its Web site to explain the environmental practices the firm embraces. The Web site described how Stonyfield utilizes the industry's best environmental practices, supports organic agriculture, and prompts customers on how they can take sustainable "earth actions."

Global Exchange (2005) is a Web-based international organization dedicated to promoting social and environmental justice throughout the world. At the 2006 World Social Forum in Venezuela, Global Exchange helped 180 conference goers blog their experiences. The organization also featured podcasts of featured speakers and let users sign up for any of eighteen customized e-mail lists that discussed topics like clean cars and global environmental justice.

Several blogs comment about current public relations and marketing campaigns. The Inspired Protagonist (2006) provided a forum to learn about sustainable strategies, such as environmental economics, clean energy, and green methods to clean up pollution. Similarly, Triple Pundit (n.d.) focused on the nexus of environmental issues, business economics, and society; the blog commented on the efforts of the Competitive Enterprise Institute and

ExxonMobil to develop the We Call it Life public relations campaign to offset concerns about global climate change.

Podcasts and Digital Video

The Woodrow Wilson International Center for Scholars (2006), a foundation in Washington, D.C., incorporates digital video and audio as part its public relations strategies to communicate the importance of protecting water quality. The Center's Web site featured weekly online video and audio news reports, live video coverage for special events like World Water Day, and regular e-mail updates.

TreeHugger (2006) is a Web magazine dedicated to providing information about environmentally responsible products and services. Its Web site featured a weekly newscast and digital videos that could be downloaded as podcasts. Some topics presented on TreeHugger included the development and use of biofuels, hybrid cars, and organic wines, as well as efforts to preserve the Everglades. TreeHugger also offered e-newsletters and blogs.

Wikis

A wiki is a type of Web site that allows users to easily add or edit comments. Although not yet widely used by public relations firms, wikis have been used to monitor public relations efforts related to the environment. For example, SourceWatch (Center for Media and Democracy, n.d.), a project of the Center for Media and Democracy, used wikis to encourage its readers to post critical comments about the use of front groups in public relations, potential conflicts of interests between environmental groups and corporations, and the use of public relations for greenwashing. For example, SourceWatch used wiki technology to create a critical analysis of efforts between DuPont and Environmental Defense to reduce environmental risks.

Flash Mobs and Meet-Ups

Flash mobs and meet-ups are related concepts that utilize Internet tools to bring people together for public events. In a flash mob, Internet technologies, such as text messaging, e-mail, and Web sites, are used to suddenly and spontaneously assemble people in a public place (Shmueli, 2003). Similarly, meet-ups bring people together by the use of Web sites where individuals can search for people in their area with similar interests. The meet-up strategy was effectively used by the Howard Dean presidential campaign in 2004 as a way to organize and energize the candidate's supporters (Heiferman, 2006).

Moblogs

Moblogging is the practice of taking photos, digital video, and text and transmitting this information from cell phones and other digital devices so that it can be posted on Web sites (Wikipedia, 2006). *Business Week* ("Blogging on the Go," 2005) reported that moblogging was used to instantly post breaking news and photos about the 2005 tsunami. Environmental agencies in the United States and Europe use a form of moblogging when field personnel report pollution incidents. The environmental group Friends of the Earth has urged its members to use moblogging methods to report instances where habitats and ecosystems are being degraded. In essence, moblog technology has the potential to present timely ecological information from remote sites, even if mass media are not there to cover news events.

Using Computer-Mediated Methods to Address Public Relations Challenges

In a few instances, public relations professionals have turned to the use of Web sites and other computer-mediated technologies to resolve environmental crises.

In 2005, the Lockheed Martin Corporation (2006) developed a Web site to inform and regularly update residents of the Florida town of Tallevast about the company's efforts to treat perchlorate and other groundwater pollutants. The pollution had taken place when Loral Metals operated a precision metalworking plant in Tallevast from 1961 to 1996. Later, Lockheed Martin purchased the site and inherited this pollution concern. To communicate with local residents, Lockheed Martin developed a Web site that provided updated information about the progress of the cleanup efforts. The site also let individuals sign up for e-mail alerts.

In 2002, environmental activists launched a public relations campaign that criticized Dell Inc. for not sufficiently encouraging customers to recycle old computers (Computer Take Back Campaign, 2002). Environmental advocacy groups, including EcoPledge and the Grassroots Recycling Network, created a public relations campaign called Dude, Why Won't They Take Back My Old Dell? that encouraged stakeholders to write letters to local newspaper editors, e-mail Dell using automatically generated form letters, and flood the company's customer service telephone lines. In addition, the Calvert Group (an investment firm that advocates environmentally responsible stocks and mutual funds) pressured Dell to develop a more comprehensive recycling program.

To resolve this public relations crisis, Dell began partnering with the Computer Take Back campaign and the National Recycling Coalition, Inc., to sponsor and host free workshops for recycling professionals. The Computer Take Back campaign featured a clickable map where consumers could locate recyclers in their area. A Dell Web site provided information about what the company was doing to recycle old computers and how consumers could participate. As a result of Dell's publicity efforts, more than one thousand tons of unwanted computers were collected. In 2004, *Business Ethics* magazine awarded Dell its Environmental Progress Award because of the company's recycling initiatives (*Business Ethics Awards*, 2005).

Similarly, Apple Computer bore the brunt of criticism for not encouraging recycling and was the target of an activist Web site (*Computer Take Back Campaign*, n.d.). In response, Apple created specialized Web sites that incorporated downloadable digital video, e-mail updates, and other computer-mediated tools to provide information to stakeholders.

In 1998, Touchstone Pictures released the movie *A Civil Action* that described widespread groundwater pollution at sites near Woburn, Massachusetts (Wikipedia, 2006). The movie was based on a best-selling book by Jonathan Harr. The book alleged that W. R. Grace & Company, a chemical manufacturer, was not only responsible for much of the contamination, but was also still acting in ways that would cause future pollution events. To respond, W. R. Grace developed a Web site as part of a comprehensive public relations strategy to send the message that the company was environmentally responsible and worked with the community to resolve groundwater quality issues (McGuinness, 1999). Although some critics scoffed that the Web site just tried to greenwash the issue, several journalists began providing more balanced coverage and now often note that W. R. Grace has improved its environmental record.

In 2003, the United States Forest Service (Armstrong, 2003) decided to temporarily ban e-mails that appeared to be automatically generated by Web sites. Agency leaders said that the use of these technologies was similar to preprinted postcards and form letters that do not provide any meaningful input into the comment process. Activists argued that restricting these e-mails prevented the ability of average citizens to influence public policy and protect the environment.

Might Computer-Mediated Public Relations Be Misused?

In some cases, it seems to be unclear whether computer-mediated public relations will facilitate greater transparency or if it will provide another vehicle for front groups to confuse the public.

According to many environmental groups, several corporations in the United States can rightly be accused of greenwashing. SourceWatch describes greenwashing as "disinformation disseminated by an organization so as to present an environmentally responsible public image" (Center for Media and Democracy, n.d., Greenwashing section, para. 2). Johnson (2005) compiled information about corporations that were thought to misuse public relations and image advertising to greenwash their images and make their entities appear to be more environmentally friendly. For more than a decade, both Corp Watch (an environmental watchdog organization) and SourceWatch (Rampton, 2002) have been tracking the extent to which corporations are misusing greenwashing in their promotions, public relations, and marketing programs (*Greenwash + 10*, 2002).

It is difficult to determine the extent to which environmental claims made by various Web sites are truthful or whether they are merely attempts by corporations and organizations to spin an issue. Ironically, the same computer-mediated public relations tools that can be used to increase transparency (e.g., Web sites and blogs) can also be used to confuse, mislead, or even deceive the public.

At an international level, LobbyWatch.org (n.d.) monitors the extent to which it believes corporations may be misrepresenting their environmental records. LobbyWatch.org has been especially critical of the European Science and Environment Forum, which they say advertises itself as an independent voice for the environment but is really funded by corporate interests. As a result, LobbyWatch.org charges that the forum advocates the interests of corporations, not the welfare of the public or the environment.

The Green Life (n.d.) blog advertises itself as an interactive watchdog campaign that publishes e-newsletters and annual reports to explain, expose, and erase corporate greenwashing in public relations, advertising, and labeling. Its focus is to identify public relations efforts that promote corporations as being environmentally responsible when their actual practices degrade the environment.

At first glance, The Council on Water Quality (2006) Web site would have seemed to be a resource for information about how to reduce pollution. In reality, a close look showed that the Web site was funded by chemical manufacturing companies. Its primary aim was to convince readers that perchlorate contamination is not a serious environmental threat that needs to be regulated.

Similarly, a quick look at the Save Our Species Alliance (2006) Web site may have led readers to believe its goal was to preserve habitat and/or bring back threatened plant and animal species. In fact, the Web site stated that its goal was to

> Work across the country to promote common sense, balanced and scientifically-supported changes to the ESA [Endangered Species Act] which will . . . strengthen the act to make it more . . . efficient in recovering and saving species at risk. (Home page section)

However, after looking at the Web site in more detail, it became apparent that the aim of the Web site was to argue that the Endangered Species Act needs to be repealed.

Conclusion

This chapter describes some of the issues associated with the use of computer-mediated public relations for environmental causes and presents examples about the extent to which corporations and environmental organizations are using these methods. Computer-mediated public relations is becoming a more common practice for environmental organizations, especially with the use of e-mails and e-newsletters and the distribution of Web content through RSS feeds.

Although the use of computer-mediated public relations is increasing in the environmental arena, readers need to be reminded that these high-tech methods will only supplement, and not replace, traditional media relations efforts. Because of the user-friendly nature of Internet-based tools, computer-mediated public relations can be used by any individual or organization, not just by those whose intent is to present all sides of an issue or to tell the truth. Consequently, it may be challenging for the public to differentiate between factual information and half-truths, and it may become difficult for public relations professionals to counter falsehoods that are spread about the environmental performance of their organizations.

The use of computer-mediated public relations offers an opportunity for communicators to automatically provide regularly updated information to stakeholders. It also provides round-the-clock access to information. Corporations and organizations can choose how transparent they wish to be. Although this provides a better way to reach and interact with stakeholders, it is not clear if these practices actually improve media relations, as reporters might not use these services (see Appendix, pp. 338–339).

Computer-mediated technologies empower public relations professionals with the ability to meet the information needs of individuals and organizations. Socially, computer-mediated public relations can give more people information automatically and remotely, as portable computing technologies are more widely used. Economically, computer-mediated public relations can help corporations prevent a loss of reputation by allowing them to be transparent and responsive to the needs of reporters and other publics. Politically, government agencies and those who receive federal and state funding are being asked to provide meaningful information that has resulted from funds provided by taxpayers.

The bottom-line message is that the general public and stakeholders want to access and learn about the environmental practices of organizations in the corporate and nonprofit sectors. At the same time, stakeholders are interested in scanning the work of organizations they support and oppose. Due to the increased popularity of Internet-based search tools, publics want to follow environmental news as it occurs. They also expect to gain access to transparent information about the actions of the environmental groups involved.

For consideration

1. How can individuals and organizations determine if information provided by computer-mediated efforts is truthful, accurate, reliable, and credible?
2. How effective are new media in setting public agendas, compared to conventional media sources?
3. To what extent are agencies and corporations increasingly obligated to be transparent in providing information about their environmental actions to stakeholders? Should there be a limit to environmental transparency?
4. What are the most effective computer-mediated public relations strategies for environmental organizations? To what extent do they differ for radical versus more moderate groups?

5 What can public relations practitioners do if their organization's environmental activities (whether positive or negative) are falsely portrayed in new media venues? Does every inaccuracy warrant a response?

For reading

Lee, S. (2001). *From Main Street to Wall Street: Environmental performance is good for the environment and shareholders alike as companies seek to become more sustainable enterprises*. Report published by the Environmental Section of the Public Relations Society of America. Retrieved March 17, 2007, from http://www.prsa.org/networking/sections/environment/documents/PDF/lee.pdf

Munshi, D., & Kurian, P. (2005). Imperializing spin cycles: A postcolonial look at public relations, greenwashing, and the separation of publics. *Public Relations Review, 31*(4), 513–520.

Murphy, P., & Dee, J. (1996). Reconciling the preferences of environmental activists and corporate policymakers. *Journal of Public Relations Research*, 8(1), 1–33.

Reber, B., & Berger, B. (2005). Framing analysis of activist rhetoric: How the Sierra Club succeeds or fails at creating salient messages. *Public Relations Review, 31*(2), 185–195.

Taylor, M., Kent, M. L., & White, W. J. (2001). How activist organizations are using the Internet to build relationships." *Public Relations Review, 27*(3), 263–284.

References

American Chemistry Council. (2005). Americanchemistry.com. Retrieved July 6, 2006, from http://www.americanchemistry.com

Armstrong, L. (2003). U.S. Forest Service bars mass political e-mails. *On the Horizon* (PRSA Environmental Section newsletter), *11*(2), 3.

Assadourian, E. (2006, March/April). The evolving corporation: Next steps for the business community. *WorldWatch*, pp. 16–20.

Backbone Media. (n.d.). Retrieved July 7, 2006, from http://www.backbonemedia.com/

Barnes, T. (2004, July). *RSS: The next big thing on line*. White paper presented to Mediathink, Atlanta, GA. Retrieved July 7, 2006, from http://www.mediathink.com/rss/mediathink_rss_white_paper.pdf

Blogging on the go. (2005, June 22). *Business Week*. Retrieved July 7, 2006, at http://www.businessweek.com/technology/content/jun2005/tc20050622_2628_tc_212.htm

Business Ethics Awards. (2005). Retrieved August 17, 2006, from http://www.business-ethics.com/annual.htm#Dell,%20Inc

Center for Media and Democracy. (n.d.). *SourceWatch*. Retrieved July 7, 2006, from http://www.sourcewatch.org/index.php?title=SourceWatch

Computer Take Back Campaign. (n.d.). Retrieved August 17, 2006, from http://www.computertakeback.com/bad_apple/bad_apple_biz.cfm

Computer Take Back Campaign. (2002, March). *Dude, why won't they take back my old Dell?* Retrieved July 7, 2006, from http://www.grrn.org/e-scrap/Dell_TakeBack_Report.pdf

Council on Water Quality. (2006). *Facts about perchlorate*. Retrieved July 7, 2006, from http://www.councilonwaterquality.org/facts/key_facts.html

Cripps, K. (2005, October). *Insights from the critical issues forum*. Report presented to the Council on Public Relations Firms, New York, NY. Retrieved July 7, 2006, from http://www.prfirms.org/docs/council_publications/2005/Critical_Issues_Forum_Highlights.pdf

Dutta-Bergman, M. (2004). Describing volunteerism: Theory of unified responsibility. *Journal of Public Relations Research, 16*(4), 353–369.

E-Wire. (2006). *Environmental media distribution*. Retrieved July 7, 2006, from http://www.ewire.com/Benefits.cfm/Page/Distributionlist

Food and Agriculture Organization of the United Nations. (2006). *Ask FAO*. Retrieved July 7, 2006, from http://www.fao.org/askfao/home.do?lang=en

General Electric Company. (2005). *GE 2005 Ecomagination report: Taking on big challenges*. Retrieved July 7, 2006, from http://www.ge.com/ecoreport/files/ge_2005_ecomagination_report.pdf

General Electric Company. (2006). *Ecomagination*. Retrieved July 7, 2006, from http://ge.ecomagination.com

Global Exchange. (2005). *Reality tours: Past participants share their experience*. Retrieved July 7, 2006, from http://www.globalexchange.org/tours/saying.html

The green life. (n.d.). Retrieved March 17, 2007, from http://sierraclub.typepad.com/greenlife/

Greenwash + 10: The UN's global compact, corporate accountability and the Johannesburg Earth summit report. (2002). Retrieved July 7, 2006, from http://www.corpwatch.org/downloads/gw10.pdf

Heiferman, S. (2006, May 29). See you offline. *Newsweek*. Retrieved July 7, 2006, from http://www.msnbc.msn.com/id/12875814/site/newsweek/

Hurme, P. (2001). Online PR: Emerging organisational practice. *Journal of Corporate Communications, 6*, 71–75.

The inspired protagonist. (2006). Retrieved July 7, 2006, from http://www.inspiredprotagonist.com/

Johnson, G. (2005, June 8). *Don't be fooled: America's ten worst greenwashers*. Retrieved March 17, 2007, from http://www.sustainablemarketing.com/content/view/121/80/

Kent, M., & Taylor, M. (2002). Toward a dialogic theory of public relations. *Public Relations Review, 28*(1), 21–37.

Lee, S. (2001). *From Main Street to Wall Street: Environmental performance is good for the environment and shareholders alike as companies seek to become more sustainable enterprises*. Report published by the Environmental Section of the Public Relations Society of America. Retrieved March 17, 2007, from http://www.prsa.org/networking/sections/environment/documents/PDF/lee.pdf

LobbyWatch.org. (n.d.). Retrieved July 7, 2006, from http://www.lobbywatch.org/p1temp.asp?pid=29&page=1

Lockheed Martin. (2006). *Tallevast.info*. Retrieved August 17, 2006, from http://www.tallevast.info/

Los Angeles Department of Water and Power. (n.d.). Retrieved July 7, 2006, from http://www.ladwp.com/ladwp/homepage.jsp

McDonald's Corporation. (n.d.). *Open for discussion*. Retrieved July 7, 2006, from http://www.csr.blogs.mcdonalds.com

McGuinness, J. (1999). Update on A Civil Action. *On the Horizon* (newsletter published by Environmental Section of PRSA). Retrieved July 7, 2006, from http://www.prsa.org/_Networking/environment/civil.asp?ident=en4

The Nature Conservancy. (2006). Retrieved July 7, 2006, from http://www.nature.org/

Rampton, S. (2002). *Rio + 10, environment zero*. *PR Watch*. Retrieved July 7, 2006, from http://www.prwatch.org/ prwissues/2002Q3/rio.html

Reber, B., & Berger, B. (2005). Framing analysis of activist rhetoric: How the Sierra Club succeeds or fails at creating salient messages. *Public Relations Review, 31*(2), 185–195.

Rubel, S. (2006). Blogging's impact on public relations. *Micro Persuasion*. Retrieved July 7, 2006, from http://www.micropersuasion.com/

Save Our Species Alliance. (2006). Retrieved March 17, 2007, from http://www.sourcewatch.org/index.php?title=Save_Our_Species_Alliance

Shmueli, S. (2003, August 8). *Flash mob craze spreads*. Retrieved March 17, 2007, from http://www.cnn.com/2003/TECH/internet/08/04/flash.mob/

Sierra Club. (n.d.). Retrieved July 7, 2006, from http://www.sierraclub.org/

Stonyfield Farm. (n.d.). *Earth actions*. Retrieved July 7, 2006, from http://www.stonyfield.com/EarthActions/

StopGlobalWarming.org. (2006). Retrieved July 7, 2006, from http://www.stopglobalwarming.org/default.asp

Suh, K., Couchman, P., & Park, J. (2003, July). A Web-mediated model based on activity theory. Paper presented at the 7th World Conference on Systemics, Cybernetic and Informatics, Orlando, FL.

Taylor, M., Kent, M. L., & White, W. J. (2001). How activist organizations are using the Internet to build relationships. *Public Relations Review, 27*(3), 263–284.

Texas Water Resources Institute. (n.d.). Retrieved July 6, 2006, from http://twri.tamu.edu/

TreeHugger. (2006). Retrieved July 7, 2006, from http://www.treehugger.com/

Triple pundit. (n.d.). Retrieved July 7, 2006, from http://www.triplepundit.com/

The Walt Disney Company. (n.d.). *Disney's environmentality*. Retrieved July 7, 2006, from http://corporate.disney.go.com/environmentality/index.html

Wikipedia. (2006). Retrieved August 17, 2006, from http://www.wikipedia.org/

The Woodrow Wilson International Center for Scholars. (2006). *Environmental Change and Security Program: Water*. Retrieved July 7, 2006, from http://www.wilsoncenter.org/index.cfm?topic_id=1413&fuseaction=topics.categoryview&categoryid=A84F71E7–65BF-E7DC-4E3D65C4974A971A

World Business Council for Sustainable Development. (2006). Retrieved August 17, 2006, from http://www.wbcsd.ch/templates/TemplateWBCSD5/layout.asp?MenuID=1

Appendix: Twenty-Five Practitioners Discuss Computer-Mediated Public Relations

To develop information about this topic, chapter author Ric Jensen surveyed twenty-five environmental public relations professionals, including members of the Environmental Section of the Public Relations Society of America, via e-mail in May and June, 2006. A majority of the respondents worked for government agencies or corporations. A summary of the twenty-five responses is presented here:

Q: How long have you used Web sites and the Internet for computer-mediated public relations?

A: The average response was four years.

Q: How have public relations efforts changed since using computer-mediated public relations?

A: The most common answer was that computer-mediated public relations provides a mechanism to instantly transmit updated information to stakeholders who are highly interested in a company or an organization.

Q: Has computer-mediated public relations improved communications with stakeholders?

A: Respondents said that computer-mediated public relations lets their organizations reach out to and connect with more people and builds better relationships as stakeholders discover all the information that is available on Web sites.

Q: What is the biggest advantage of computer-mediated public relations?

A: Major advantages include the ability to make information available throughout the 24/7 news cycle and to direct reporters to specific Web site content that can provide background information.

Q: What are the main shortcomings of computer-mediated public relations?

A: Overwhelmingly, the most common answer was that even the best computer-mediated efforts are not a substitute for face-to-face communication. Others responded that reporters do not pursue news stories based on information they see on Web sites. Also, it is difficult to know which people are visiting a Web site and how they react when they read Web site content.

Q: Has the use of computer-mediated public relations increased the dialog between your organization and stakeholders?

A: Most respondents said that more people are contacting their organizations, wanting to learn more, and asking questions as a result of computer-mediated public relations efforts.

Q: Is computer-mediated public relations increasing transparency among corporations and organizations?

A: Only a few people responded to this question. Those that did said that computer-mediated public relations can allow organizations and corporations to disclose comprehensive information to the public. On the other hand, some people feared that the use of blogs and specialized Web sites merely gives corporations another tool to greenwash environmental shortcomings.

Q: What computer-mediated technologies do you use in public relations?

A: Nearly all respondents use RSS feeds. The next most commonly used methods include (in order) the automatic distribution of e-newsletters, Web-based conferencing, podcasts, and blogs. Automatically sending subscribers e-newsletters and updates was thought to be the most effective computer-mediated technology.

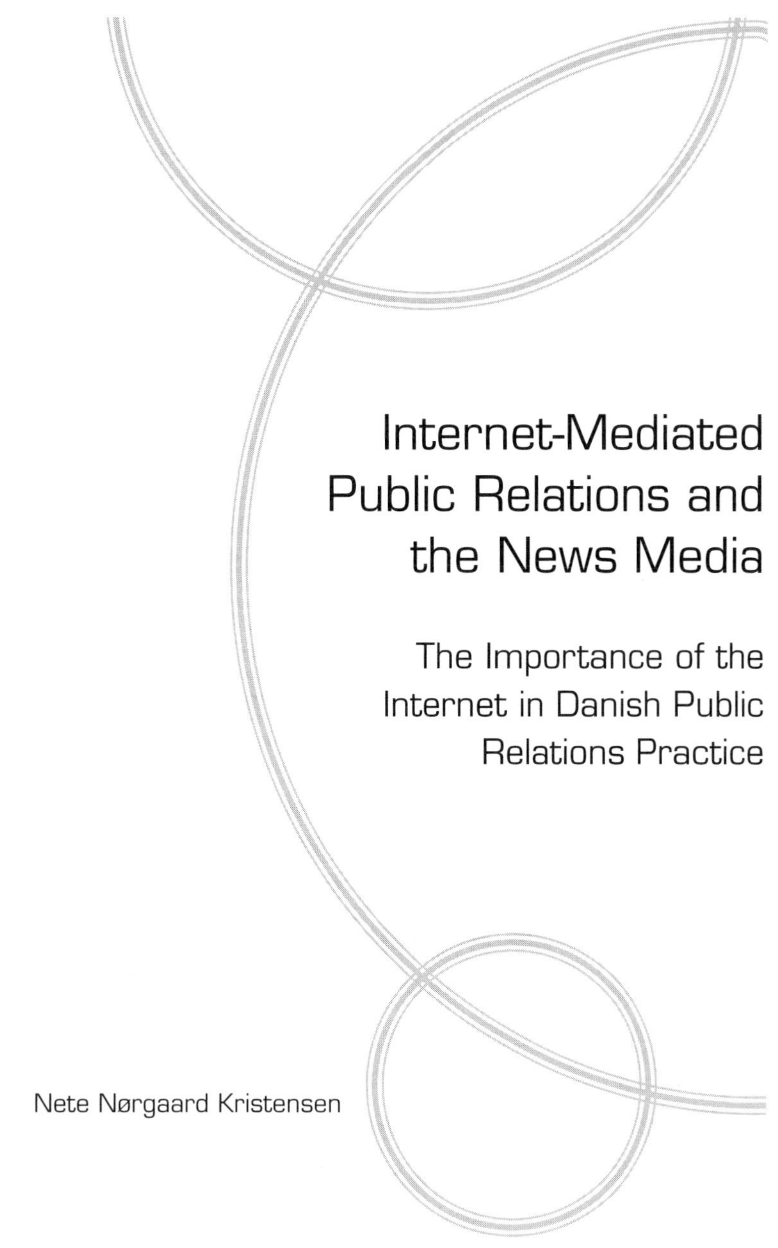

Internet-Mediated Public Relations and the News Media

The Importance of the Internet in Danish Public Relations Practice

Nete Nørgaard Kristensen

This chapter discusses how the Internet is increasingly becoming a central and valuable tool in Danish public relations practice, especially in media communications. The chapter will empirically support how, theoretically examine why, and analytically discuss the consequences for the conventional practice of public relations by drawing on theories of mediatisation and news management as well as empirical results from the first study of the relationship between the Danish public relations industry and the media.

This chapter analyses how the Internet is increasingly becoming a central and valuable tool in Danish public relations practitioners' communication with journalists. Obviously, the media are not the only target of public relations messages communicated by a Web site. These messages are aimed at customers, employees, business partners, competitors, and political institutions, among others.[1] It can, however, be argued that journalists are *especially* significant for online communication—as both a target group and as a communication channel to other target groups. This chapter will examine how and why journalists are significant for online communication in this way.

More specifically, the chapter will empirically support the increasingly significant role of Internet communication in Danish public relations practice in relation to journalists as a target group; it will theoretically examine why this is the case in light of social, economic, and cultural changes; and it will analytically discuss how this affects, expands, and transforms conventional public relations practice across space and time.

The chapter draws on theories of mediatisation and news management and is empirically based on the first Danish study of the relationship between the Danish public relations industry and the media (Kristensen, 2005). Aspects investigated in this study were Danish public relations practitioners' use of the Internet when communicating with journalists, the changing position of this new medium in the practitioners' daily work, and its impact on the media messages.

The chapter thus integrates theoretical, empirical, and practical perspectives based on new data. Consequently, this work points to how the Internet is increasingly becoming a central instrument in public relations and thereby indicates a *mediatisation* of parts of the profession and a change in the daily routines—and power—of public relations practice.

Theoretical Context: News Management and Mediatisation

Although public relations practitioners might prefer the term *media relations*, the concept *news management* has, for a long time, been central to strategic communication (Falkheimer, 2004). The term covers the many strategies and tactics of the public relations industry on behalf of organisations, employers, and customers to control and influence the flow of (news) information and, thus, the agenda and messages communicated by the news media (Tulloch, 1993).

Generally speaking, news management consists of two main communication strategies: (a) a proactive strategy and (b) a reactive strategy, which will often be at play simultaneously. Whereas the first is characterised by initiative from the public relations industry toward the media (e.g., a press release or media event), the latter is characterised by a lack of action on the part of the sender, giving the media the lead in the communication process (e.g., by letting the journalist make contact or even by diverting the attention of the media). Though both strategies aim at influencing the media agenda, the main difference is who plays the proactive part—the public relations practitioner or the journalist? Put in marketing terms, it is a question of *pull* or *push strategies;* that is, whether the public relations industry pushes information and agendas onto the media or whether journalists pull information from their sources (Tuchman, 2002).

As it will be argued and developed in this chapter, Internet-mediated information can be seen as an *implicit* proactive public relations strategy, placing itself between the apparent proactive/reactive distinctions of news management and the traditional pull/push distinctions of marketing. On the one hand, the communication is planned with the media, among others, in mind. On the other hand, the information is not imposed on media representatives (as, for example, a press release). Journalists have to search for information and messages on the sender's Web site. Thus, it can be said that Internet communication is a means for the public relations industry to implicitly push information to the media by letting journalists pull information from a Web site.

To continue this line of thinking, news management practiced by way of the Internet can be interpreted as part of a more general *mediatisation* of society. Though mediatisation is often related to political communication (Asp, 1990; Amnå, 1999; Jackson & Lilleker, 2004), it is not restricted to political phenomena. It characterises the interplay between the media and many different social institutions and agents, such as business life, public institutions, and nongovernmental organisations (NGOs). These establishments are increasingly forced to adapt to the logic of the media's form, content, and deadlines in order to attain visibility, influence, and power.

However, as emphasised by Schultz (2004), mediatisation is not just a question of *adaptation* to the media's logic. It also involves processes of *extension* and *substitution*, both of which are especially relevant to Internet-mediated public relations. That is, the Internet bridges the spatial and temporal distances characterising human communication and creates new forums of communication that enable dialogue between consumers and producers, citizens and politicians, and media representatives and public relations practitioners via home pages, e-mail, chat rooms, and so on. Likewise, the Internet substitutes and alters social actions that have not been traditionally mediated. For example, online communication can replace or change the personal dialogue and interaction between public relations practitioners and journalists that are so important in relations management. Theoretically, Internet-mediated public relations can thus be interpreted as a development, expansion, or transformation—a mediatisation—of traditional proactive, pushy news management strategies.

Methodology: Quantitative and Qualitative Approaches

The importance, expansion, and effect of Internet-mediated public relations in relation to the news media were analysed quantitatively and qualitatively from a Danish perspective in the spring of 2005 (Kristensen, 2005).

The quantitative part of the study—a survey—consisted of a thirty-six-item questionnaire distributed by post in March and April 2005. The instrument was sent to the 839 members of The Danish Union of Journalists (DJ) working in strategic communication[2] and organised in the Communication Group of DJ.[3] Of 774 members who were relevant for the study, 407 completed the questionnaire, resulting in a satisfactory "adjusted response rate" (Weaver & Wilhoit, 1996, p. 250) of 52.6%.[4] The data were processed with SPSS (Statistical Package for the Social Sciences) software, and the sample proved representative of public relations practitioners in DJ as to sex, age, and geographical distribution.[5]

In May and June 2005, the quantitative results were elaborated by thirteen in-depth interviews with representatives of different parts of the Danish public relations industry (e.g., private/public/political institutions, NGOs, consultants, freelancers, academics and craftsmen, leaders, and employees). The interviews were conducted with a relatively fixed set of questions that focused on the professional experiences and viewpoints of those interviewed. Though a discourse analysis was not performed on these statements, excerpts will, in the following discussion, be quoted word for word, since expressions and choices of words reflect the professionals' self-understanding and interpretation. Hereby, emphasis can be placed on the double hermeneutics (i.e., the relationship between the articulations and interpretations of the interviewees and the presentation and readings of the researcher) of the research process (Giddens, 1984).

Thus, in combination, the studies provided both an overall knowledge of professional experiences, self-understandings, and attitudes across a heterogeneous group[6] and in-depth information on some of the tendencies pointed out by the quantitative data regarding Danish public relations practitioners' use of the Internet when communicating with journalists.

The Increasing Position of the Internet

As to the position of the Internet in the daily work of Danish public relations practitioners, the quantitative survey showed that Web-based communication, on average, takes up 16% of the respondents' weekly work hours[7] (see Figure 1).

In public institutions, Internet-related tasks consume even more of the weekly work time—20%—which is double that of public relations advisers in agencies that spend only 10% of their weekly work hours on Web-related assignments. This indicates that Internet-based communication is relatively important and prioritised in public contexts. One expla-

Figure 1: Percentage of Weekly Work Hours Dedicated to Various Public Relations Tasks (*N* = 339 to 362)[8]

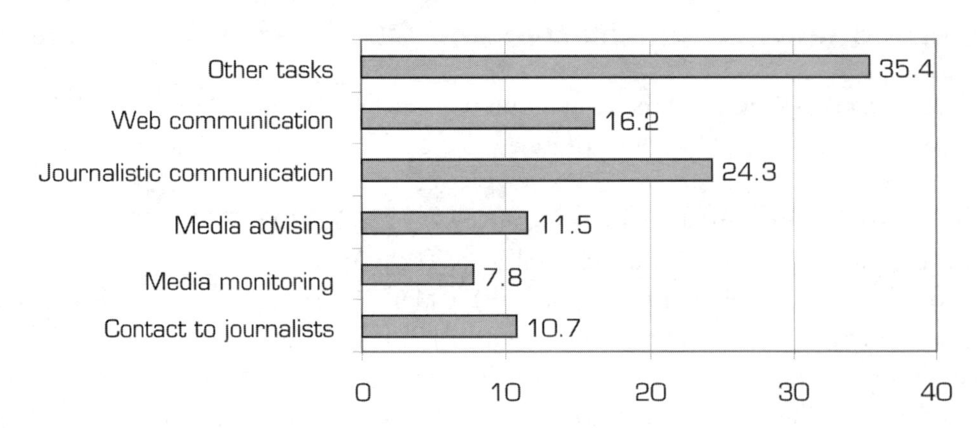

Figure 2: Change in Weekly Work Hours Dedicated to Various Tasks Compared to When Respondents Began Working in Strategic Communication (*N* = 407)

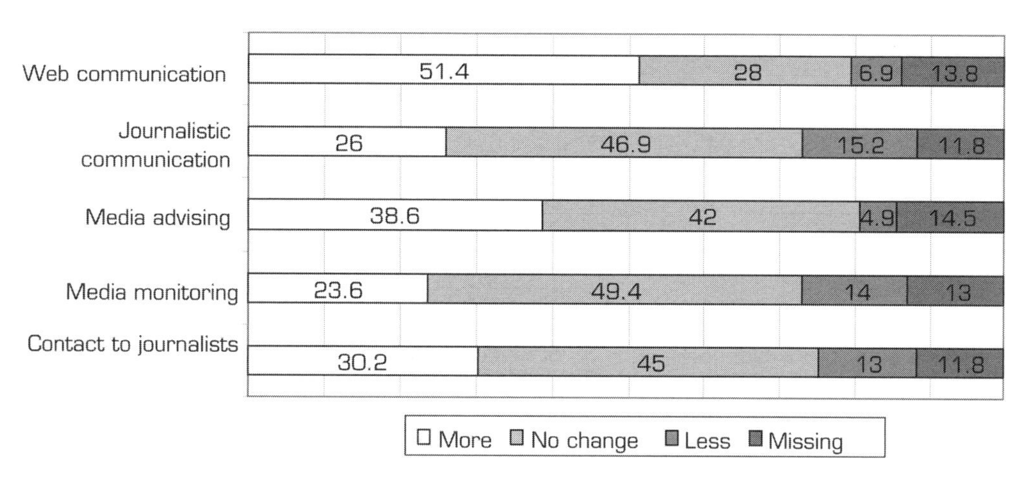

nation is that public institutions have an obligation to inform the public, including the media, because of their association with the democratic process in which the public plays an important part, compared to commercial contexts or public relations bureaus. Pedersen points out that

> Public communication differs from all other sorts of communication because public communication—contrary to, for example, commercials and instructions/directions—is closely related to democracy. (2004, p. 7, translated from Danish).

Daily communication of information via the Internet requires a nuanced, inside knowledge of the communicating institution—a knowledge that an external adviser from a public relations agency does not necessarily possess. Thus, another explanation of the variations between the different sectors of the public relations industry might be that organisational Web communication—or, for that matter, corporate (Web) communication—is best handled by communication officers working inside the organisation.

The survey further showed that tasks related to the Internet consume more of public relations practitioners' daily work today compared to ten years ago,[9] at least for a considerable number of respondents (see Figure 2). This indicates that the medium has become an essential and more significant channel of communication for many companies and organisations.

Moreover, the survey pointed to public institutions and NGOs, in particular, as having experienced increased use of the Internet over time (see Figure 3). This confirms the previously mentioned point regarding the obligation to inform in public and noncommercial institutions—an obligation with which Web-based communication can comply.

There was a considerable difference found between the expanded use of Web communication in a public context compared to consultant agencies and small public relations firms

Figure 3: Percentage of Respondents Experiencing a Change in the Portion of the Workday Dedicated to Web Communication since Beginning Work in Strategic Communication, by Organisation Type (*N* = 352)

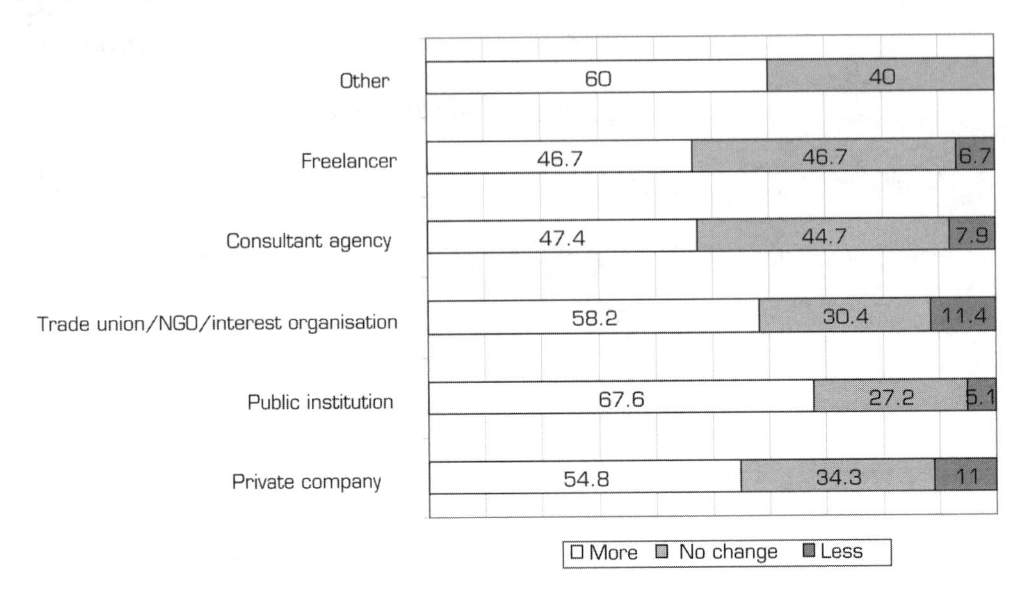

with only one or a few employees. These organisations have also experienced an increase in using the Internet as a communication tool but not to the same extent as public institutions.

In addition, the survey showed that 43.5% of the respondents offer Web-based information to the media on a daily or weekly basis. Only press releases are offered almost as frequently (41%). Furthermore, 36.6% answered that they offer the media "tips/information by e-mail or electronic newsletters" more frequently today compared to when they started working in strategic communication, while 47.7% answered that they offer the media "other kinds of Web-based information" more frequently. This again confirms the increased use and status of the Internet as not only a direct channel of communication for a variety of groups, but also a medium for mediated communication through the news media.

Again, public relations practitioners in public institutions and NGOs were shown to use different kinds of Web-based communication and distribution of information more frequently compared to other respondents. Furthermore, public relations practitioners in these organisations have experienced the largest increase in the distribution of information to media and journalists by the Internet. More specifically, 58% of the respondents working in trade unions or NGOs make Web-based information available to the media on a daily or weekly basis, and 58.5% of this group answered that they have increased their distribution of this kind of information. Similarly, 58.4% of the respondents working in public institutions make Web-based information available to the media on a daily or weekly basis, and 64% answered that they have increased the distribution of this kind of information.

Explanations: Multitargeting and Journalistic Research Tool

The Internet provides senders with the opportunity to communicate with many different target groups through one channel instead of using many different media with different messages customised for specific target groups. During the interviews, an information officer for a public, political organisation noted, "In the past, our communication was adjusted at guiding the consumer—with magazines, folders, etc., but since the end of the '90s, the Internet has become our primary information channel. In addition, there is the press."

Journalists use the Internet, especially the Web sites of institutions and organisations, in their daily research of news stories, which helps to explain why "press rooms" are commonly found on organisations' home pages. Lipinski and Neddenriep's (2004) study of U.S. Congress members' use of home pages showed that all members of Congress had a home page, and three out of four of these home pages had sections targeting the press. They further emphasized that instruction in how to use the Internet for research is a central part of American journalists' education. This might be explained by increasing competition in the global media industry affecting editorial research resources and production of journalistic content, thus creating added pressure on the editorial staff.

Some of these same tendencies characterise the Danish media industry (Kristensen, 2004). Sparre's (2005) study of Danish journalists' use of the Internet as research tool, specifically in relation to local politics, showed that journalists use the Internet to a considerable extent in their daily work to (a) monitor and identify potential stories, (b) get access to facts, and (c) find sources and information. Thus, the Internet represents more than a new medium of communication and information in itself. By providing journalists with input to their stories, Internet-mediated information can also give rise to traditional, mediated communication—in the form of editorial copy (Lipinski & Neddenriep, 2004).

From a public relations perspective, this is imperative for several reasons. Journalists, as opinion makers, can influence the attitudes and actions of politicians, publics, and other interest groups. Although the results of agenda-setting research are inconsistent, at least some of them point to a correlation between the agendas of the media, the public, and politicians regarding both *what to think about* (e.g., subjects, persons) and *what to think* (i.e., opinions) (Dearing & Rogers, 1996).

Furthermore, media texts—or editorial copy—are attributed more credibility than the advertisement and information campaigns of companies and organisations because of the journalists' professional selection of a story or event as newsworthy in their role as gatekeepers to the public; that is, due to their intervention as neutral third parties, who, from a professional perspective, estimate specific information to be interesting, relevant, and significant. This, in fact, is paradoxical, because the media are constantly criticised by the public and academia for their lack of reflection or construction of a fair and credible picture of reality. At the same time, politicians, business entities, and public institutions increasingly compete for media coverage (cf. Aubenas & Benasayag, 1999).

Anyhow, the trustworthiness of public relations information and messages, communicated by the Internet, can be increased by the fact that journalists find and process it.

Consequently, communication of information on Web sites can be a means to (try to) secure the control of specific (self-)representations in relation to both the media and other target groups. As one of the interviewed public relations practitioners pointed out: "The Internet is the place where the [public, political organisation] can present the cases in its own words."

Summing up, communication with an essential target group, such as the media, via the Internet has several strategic advantages from a public relations perspective. The Internet represents an important source of information and serves as a research tool for journalists, who can, in turn, ascribe its contents substantial credibility. This, however, points to a paradox regarding journalists' use of Internet-mediated information.

Journalists' Paradoxical Use of Internet-Based Public Relations Information

Some of the public relations practitioners interviewed for this study pointed to a perceived contradiction in Danish journalists' understanding and use of the information that they find on organisations' Web sites. Based on their experience, they believe that some journalists seem to think that information communicated by an organisation on the Internet is more legitimate or less intended than information communicated in a press release. An in-house editor for a professional association, interviewed for the study of the relationship between the Danish public relations industry and the media, noted the following:

> Our homepage has one part targeted at the members and one part targeted at the press on the front page, and we can see that the journalists find stories there, that they use it a lot. . . . It seems that they differentiate between, for instance, a press release and information that they think they themselves have discovered, for example, on the home page—although Web communication is controlled, intended, and thought through. It is a funny mechanism, but that's how it is, a least for the moment.

Similarly, another interviewee, a journalist working in a trade union, pointed out that

> Some journalists constantly visit our homepage searching for news. . . . Normally, we prefer making an Internet article or something like it instead of a press release because you can package a press release in an article and increase the distribution of the story. Some of the national media might use the story, and many other media as well—free papers, local papers, and now millions of home pages. They can very quickly copy a story and publish it once more.

It appears that some journalists seem to overlook or even ignore the fact that, even though they regard Web-based information as unbiased, the information was originally communicated with a specific intention.

This paradox was confirmed, a least partly, by the quantitative survey, which showed that 38% of the respondents find journalists less critical of information communicated on a Web site than in a press release, although these different kinds of information, first of all, are likely to be equally intentional and, secondly, often overlap.

Press releases will often be an integrated part of press communications on a Web site. According to Lipinski and Neddenriep's (2004) aforementioned study, 96.4% of the American congressional members' home pages with a press room section also included press releases. This indicates that the press release—a fundamental, proactive, and intended means of communication in public relations—is a very significant part of Internet communication targeted at the media. A former spin doctor interviewed in the Danish study stated implicitly that the communication value of the press release can be increased by the Internet:

> We also produced press releases but they were not very useful. They can be put on the home page so that people visiting the page can orientate themselves about the opinion of the party in relation to a specific subject.

The differences in journalistic source critique depending on means of communication used by public relations practitioners are likely due to the fact that the press release is openly targeted at the media, while Internet communication, as already mentioned, is aimed at many target groups and, thus, to a lesser degree, signals an explicit attempt to influence media content. The respondents in the survey do not, however, completely agree, since 46.4% do not experience a source critique from journalists depending on the form of communication.

The Internet—A Supplement, Not a Substitution of Personal Relations

Despite variations between the different sectors of the Danish public relations industry, it is important to accentuate that, across the industry, the increased use of the Internet reflected in the study should not be interpreted as a substitution of other forms of communication with journalists, but rather as a supplement. Personal dialogue and networking with journalists are still important, more so now compared to previous years. On average, 10.7% of the respondents' weekly work hours is dedicated to contact with journalists, and 30.2% of the respondents spend more time on this task compared to when they began working in strategic communication (see also Figures 1 and 2).

From a media perspective, as Sparre (2005) emphasises in her study, the human factor, including personal dialogue with sources, is still fundamental to journalists' work and cannot be replaced by Internet-mediated research of information. This might be explained by the fact that the media are increasingly flooded with uninvited information from news sources, not the least of which are public relations practitioners working on behalf of different interests. According to a survey of Danish journalists' relations to their news sources (Kristensen, 2004), nine of ten journalists were found to experience this trend. In order to penetrate this enlarged stream of information and achieve media exposure, the news sources must use personal channels in the media either before or after distributing their information or story—a task for which the public relations practitioner is typically responsible. Relationships with journalists and editors are crucial to public relations practice.

According to Jackson and Lilleker (2004), these relationships can contribute to realising the goal of symmetric two-way communication, originally proposed by Grunig and Hunt (1984), in modern public relations. At the very least, informal relations can generate goodwill and so-called favour banks (Schneider, as cited in Jackson & Lilleker, 2004) with the media. This goodwill can be used in a situation of crisis to restore legitimacy within and throughout the media by drawing on both formal and informal media contacts.

A former public relations officer in a pharmaceutical company interviewed in the Danish study confirms the importance of relationship management—using "an account of goodwill" as a metaphor:

> It is my opinion—at least in my job—that "you get what you deserve." . . . We figuratively operate with an "account" to which you can earn points. If you experience head wind or a problematic situation which is difficult to explain, but have "a positive account," you will be treated fair. But if you have "a negative account," you must count on being in "deep [expletive removed]."

Internet-mediated communication will not replace personal communication and dialogue with journalists; rather, it complements it.

These findings confirm, to some extent, the theory of mediatisation in regard to communication media, such as the Internet, increasingly changing and expanding other kinds of communication, such as interpersonal dialogue. At the same time, it contradicts the prediction that mediated communication will substitute interpersonal communication, at least when it comes to communication between journalists and public relations practitioners.

The Impact of Internet-Mediated Public Relations Information and Messages

It is one thing for Danish public relations practitioners to attempt to influence media messages implicitly or explicitly via the Internet. Another matter is the actual impact achieved by these proactive/reactive or pull/push strategies.

The Danish survey showed that, from the perspective of the public relations industry, information distributed by the Internet seems to have a considerable impact on the media. More specifically, 24.6% of the respondents experienced that tips distributed by e-mail, newsletters, or the like are used by journalists on a daily or weekly basis. Likewise, 24.8% experienced that other kinds of Web-based information are used by journalists on a daily or weekly basis. Only press releases—again—are used more frequently, pointing to the continuous, though altered, importance of this tool in public relations.

Furthermore, this impact seems to have increased in recent years. Thirty-one percent of the respondents experienced that journalists increasingly use tips offered to them by e-mail or electronic newsletters, and 34.9% experienced that journalists increasingly use other kinds of Web-communicated information.

The Internet is therefore not only a useful channel of communication from the sender's perspective for self-representation to various target groups, but also a useful and valuable source of information from the receiver's perspective when journalists are the target group. This assertion is confirmed by earlier studies (cf. Sparre, 2005), and this Danish study indicates that journalists are not necessarily as suspect of Internet-mediated messages as they are of a press release or an event planned for the media. Overall, this work supports the argument that Internet-mediated communication can help to advance traditional media coverage.

Conclusion

As demonstrated by the empirical data presented in this chapter, the increasing use of Web-based communication with journalists in the Danish public relations industry is part of larger societal changes in which new media are playing an increasingly central role. That is, there exists an ongoing mediatisation of society, including public relations practice, in which interpersonal dialogue and exchange of information are perhaps not substituted, but at least expanded upon—and thus changed by—Internet-mediated communication.

The increasing use of Web-based communication with journalists in public relations can be interpreted as a relatively new strategy within, and thus a development of, traditional news management. Because the Internet is an essential research tool for journalists, this new medium can be a means to obtain, or at least inspire, traditional media coverage, leaving considerable room for the public relations practitioner to control or manage the presentation of messages, institutions, and individuals. In addition, the journalistic process can provide credibility to Internet-mediated public relations information.

The proactive push strategy, inherent in Internet-communicated public relations information, also seems to be effective. Journalists apparently have the impression that they take the lead in the communication process when they search the Web pages of organisations for information and stories—despite the communicative intentions behind this information.

Therefore, regardless of the considerable power of journalists and the news media as to the construction and communication of agendas, images, and credibility, the Internet is an important platform in public relations at both (a) a practical level (as to managing information to obtain specific media coverage and to communicate with target groups through the media across space and time) and (b) a professional level (as to the development of a profession with considerable influence or power). A question for further research is the importance of the Internet in relation to public relations tasks and target groups beyond the media, including, for example, how the Internet can be used to evade or sidestep the media agenda when messages are to be communicated to specific target groups.

For consideration

1. What are the strengths, weaknesses, opportunities, and threats of Internet-mediated public relations—from a public relations perspective?
2. What are the strengths, weaknesses, opportunities, and threats of Internet-medi-

ated public relations—from a news media perspective?

3 To what extent does Internet-mediated public relations represent a professionalisation of public relations practice? To what extent does it detract from it?

4 How might results differ if this study were replicated in other countries? How could culture affect the results?

5 What ethical principles should guide Internet-mediated public relations? Are there any limits to how it should be practised?

For reading

Davis, A. (2002). *Public relations democracy: Public relations, politics and the mass media in Britain*. Manchester: Manchester University Press.

Falkheimer, J. (2005). Formation of a region: Source strategies and media images of the Sweden–Danish Öresund Region. *Public Relations Review, 31*(2), 293–295.

Jackson, N. A., & Lilleker, D. G. (2004). Just public relations or an attempt at interaction? *European Journal of Communication, 19*(4), 507–533.

Lipinski, D., & Neddenriep, G. (2004). Using "new" media to get "old" media coverage. *Press/Politics*, 9(1), 7–21.

Schultz, W. (2004). Reconstructing mediatization as an analytical concept. *European Journal of Communication, 19*(1), 87–101.

Notes

1. The many professional public relations tasks discussed in the international literature (cf. Davis, 2002; Pieczka, 2002; Theaker, 2001), such as product promotion, branding, improvement of (media) profile, communication of corporate identity/culture, information campaigns, and lobbying, will not be discussed in this chapter due to the extent of such a discussion.

2. In a Danish context, public relations has many names—PR, professional communication, strategic communication, spin, and information work, among others. This pluralism is, among other things, related to the fact that it is a profession still in development. The Danish public relations industry is relatively new compared to the United States, since public relations was not integrated in the private and public sectors until the 1960s and 1970s. In this chapter, the term *strategic communication* is used interchangeably with *public relations* because it does not demarcate itself from specific professional, organisational, or communicative contexts or specific dimensions of communication work. Instead, it conveys that this line of work theoretically draws on many different traditions (marketing, management, communication, media science, journalism, etc.) and, in praxis, consists of many varied tasks applied in many different organisational contexts with different target groups. Furthermore, the term *strategic* points to the fact that the communication is intended, planned, and has a purpose—namely, to attend to specific interests and/or achieve an effect, for example, on the actions of other people.

3. Contrary to the Norwegian Union of Journalists, from which public relations professionals were excluded in 1997 (Allern, 2001), DJ handles the professional interests of both journalists in the media and professionals working in strategic communication (Danish Union of Journalists, n.d.).

4. The sixty-five members not relevant for the survey were unemployed, on leave, or no longer work-

ing in public relations. Compared to similar studies, the response rate was satisfactory as well. The response rate of a 2000 study of the members of the Danish Association of Communication Professionals (DACP) was 38.8%.

5. More specifically, about 50% were men, indicating that public relations is neither a male- nor a female-dominated profession in Denmark. Geographically, more than one in two lives in or around the Danish capital, Copenhagen. Finally, Danish public relations practitioners are, on average, 42.8 years old.

6. According to Ørberg (2002), the Danish public relations industry counted approximately three thousand professionals in 2002—a figure that has probably increased by three hundred to four hundred people since. These professionals represent various backgrounds, power, and authority, including in-house employees and external consultants in bureaus, large communication agencies and small firms with only one or few employees, practitioners in private corporations and the public sector, journalists and professionals with communication degrees, as well as academics and craftsmen or self-made professionals.

7. The different public relations tasks were initially constructed on the basis of the professional public relations literature, including a Danish study (Danish Association of Communication Professionals, 2000). These theoretical categories were then discussed with the board of the Communications Group of DJ and, finally, were tested in a pilot study with the aim of ensuring understandable, suitable, exhaustive, and differentiable categories.

8. The average work time used on the different tasks is calculated on the basis of the number of respondents who filled out the questionnaire in relation to the specific tasks and not on the basis of the total number of respondents in the survey. Therefore, the totals do not equal one hundred.

9. The respondents had, on average, worked in strategic communication for ten years.

References

Allern, S. (2001). Kildene og Mediemakten. In M. Eide (Ed.), *Til dagsorden! Journalistikk, makt og demokrati* [To the Agenda! Journalism, power and democracy] (pp. 273–303). Oslo: Gyldendal Norsk Forlag.

Amnå, E. (1999). *Politikens medialisering* [The mediatisation of politics]. Stockholm: SOU/Demokratiutredningen Forskarvolym 3.

Asp, K. (1990). Medialisering, medielogik, mediekrati [Mediatisation, media logic and mediacracy]. *Nordicom-Information, 4*, 7–12.

Aubenas, F., & Benasayag, M. (1999). *La fabrication de l'information. Les journalistes et l'idéologie de la communication* [The fabrication of information. Journalists and the ideology of communication]. Paris: La Decouverte.

Danish Association of Communication Professionals & PLS Consult A/S. (2000). *Dansk Kommunikationsforenings medlemsundersøgelse: Arbejdsvilkår og–ønsker for professionelle kommunikatører* [Member survey of Danish Association of Communication Professionals: Work conditions and wishes of communication professionals]. Retrieved May 5, 2005, from http://www.kommunikationsforening.dk/db/files/file9f61ef05246_1.pdf

Danish Union of Journalists. (n.d.). Retrieved April 12, 2005, from www.journalistforbundet.dk/sw2175.asp

Davis, A. (2002). *Public relations democracy: Public relations, politics and the mass media in Britain*. Manchester: Manchester University Press.

Dearing, J. W., & Rogers, E. M. (1996). *Agenda-setting*. Thousand Oaks, CA: Sage.

Falkheimer, J. (2004). *Att gestalta en region* [To form a region]. Göteborg: Makdam Förlag.

Giddens, A. (1984). *The Constitution of society*. Berkeley, CA: University of California Press.

Grunig, J. E., & Hunt, T. (1984). *Managing public relations*. New York: Holt, Rinehart & Winston.

Jackson, N. A., & Lilleker, D. G. (2004). Just public relations or an attempt at interaction? *European Journal of Communication, 19*(4), 507–533.

Kristensen, N. N. (2004). *Journalister og kilder—slinger i valsen?* [Journalists and news sources—is there a hitch?]. Århus: CFJE/Ajour.

Kristensen, N. N. (2005). *Kommunikationsbranchens medierelationer—professionelle netværk, kommunikationsfaglige perspektiver og mediedemokratiske konsekvenser* [The media relations of the communication industry—professional networks, professional perspectives and media democratic consequences]. Working paper no. 19. København: MODINET.

Lipinski, D., & Neddenriep, G. (2004). Using "new" media to get "old" media coverage. *Press/Politics*, 9(1), 7–21.

Ørberg, E. (2002). Uddannelserne får vink med vognstang. *Kommunikatøren*, 5. Retrieved August 19, 2006, from http://www.kommunikationsforening.dk/2000029

Pedersen, K. (2004). *Offentlig kommunikation—i teori og praksis* [Public communication—in theory and practice]. Copenhagen: Djøf Forlag A/S.

Pieczka, M. (2002). Public relations expertise deconstructed. *Media, Culture & Society, 24*, 301–323.

Schultz, W. (2004). Reconstructing mediatization as an analytical concept. *European Journal of Communication, 19*(1), 87–101.

Sparre, K. (2005). *Journalisters brug af internettet til research på lokalpolitik* [Journalists' use of the Internet on local politics]. Working paper no. 17. København: MODINET.

Theaker, A. (Ed.). (2001). *The public relations handbook*. London: Routledge.

Tuchman, G. (2002). The production of news. In K. B. Jensen (Ed.), *A handbook of media and communication research: Qualitative and quantitative methodologies* (pp. 78–90). London: Routledge.

Tulloch, J. (1993). Policing the public sphere—the British machinery of news management. *Media, Culture and Society, 15*, 363–384.

Weaver, D. H., & Wilhoit, G. C. (1996). *The American journalist in the 1990s: U.S. news people at the end of an era*. Mahwah, NJ: Lawrence Erlbaum Associates.

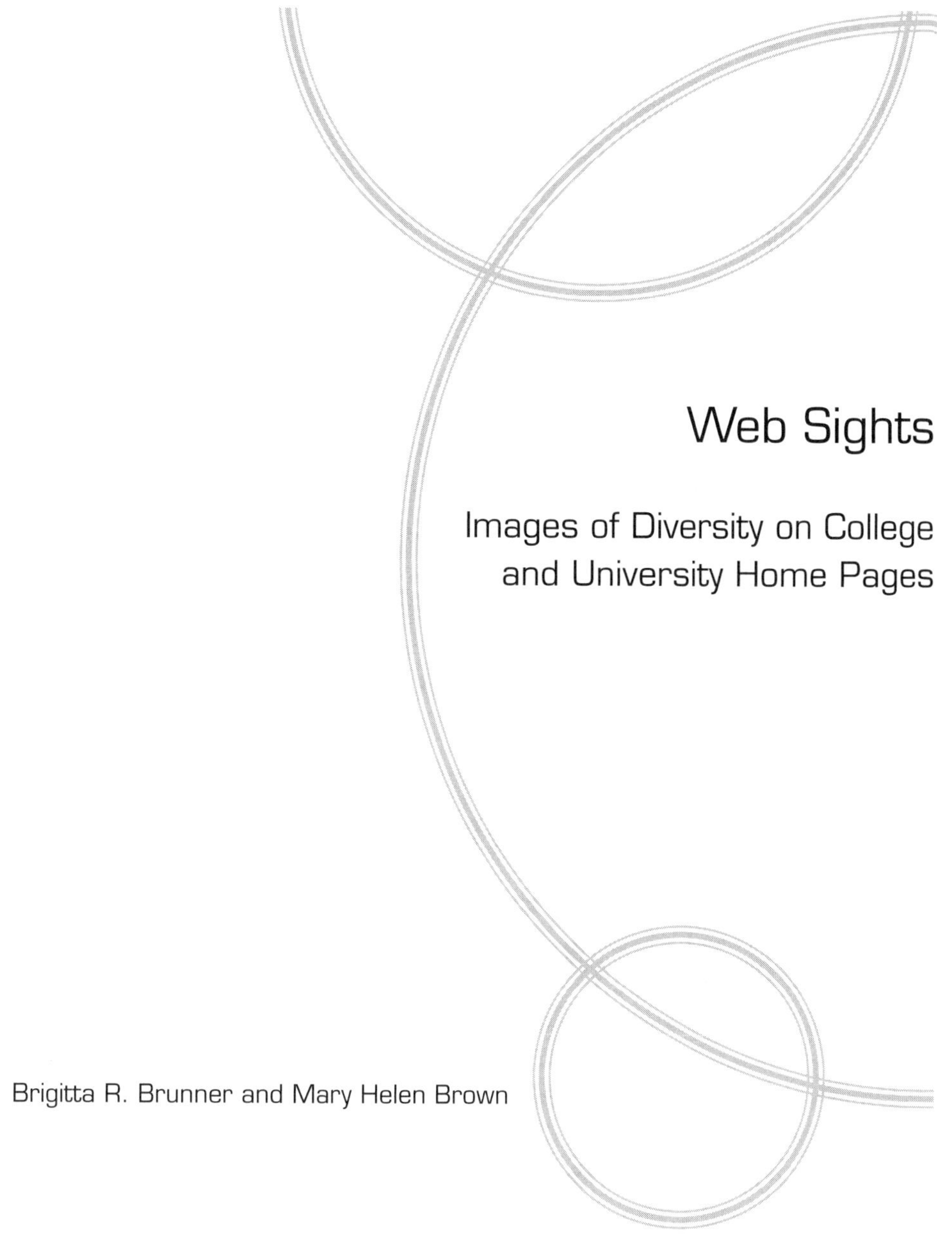

Web Sights

Images of Diversity on College and University Home Pages

Brigitta R. Brunner and Mary Helen Brown

University Web sites are often the first means of relationship building with an important constituency: students. Increasingly, students use the Internet to find out about universities rather than using campus visits and viewbooks. In addition, potential students have the luxury of revisiting the Web site, thereby solidifying a certain image of the university in their minds. However, are universities realizing the potential and power of this tool to form impressions and build relationships? This chapter will examine the home pages of the National Collegiate Athletic Association's (NCAA's) Division I schools, allowing for a broad range of school types for a sample. Using content analysis, the researchers examined each university's home page for visual elements that showcased the university's attention to diversity and inclusivity.

Relationships are key to organizational success in the twenty-first century (Wilson, 2001; DeSanto & Garner, 2001). Managers should use communication, especially public relations, as a tool to build and maintain relationships (Wilson, 2001; Ledingham, 2003).

The term *relationship* is defined in public relations literature as "a complex phenomenon for which few practitioners and scholars share a common definition and set of measurements" (Broom, Casey, & Ritchey, 1997, p. 86). Some basic principles of relational communication, regardless of setting, are that relationships follow cycles and require (a) interest, (b) interaction, (c) trust and risk, and (d) maintenance (Taylor, Kent, & White, 2001).

In this chapter, we examine an organization-public relationship, which exists whenever a public's behaviors affect the organization or vice versa (Hon & Grunig, 1999). Organization-public relationships not only influence publics' attitudes, perceptions, and behaviors, but also enhance the organization's bottom line (Bruning, 2002). Therefore, organizations should build and maintain positive relationships with their publics. In addition, public relations practitioners must remind publics that they are in a relationship with an organization (Bruning, 2002), so that publics will pay attention to the actions of an organization and be involved in its outcomes. Publics in a relationship with organizations are more likely to remain loyal (Bruning, 2000). If an organization has loyal publics, it can better serve them, while also enhancing its own image and bottom line.

Image

Image is a controversial term in public relations because of its many negative connotations. Grunig stated that he never used the term "image" in textbooks or classes because he "did not know what the term really means" (1994, p. 124). Image is a catchall, umbrella phrase related to all corporate communications that may imply that public relations deals with unreality. The idea of image suggests that public relations practitioners manipulate, polish, project, tarnish, dent, bolster, and/or boost something that is not real (Grunig). However, because image combines the symbolic (perceptions, attitudes, and schemas formed about an organization by publics) and the organization's behavioral relationships (interaction between organizations and their publics), it remains a subject of inquiry (Grunig).

Image depends on organizational activities: how management and publics interact, how the organization is doing financially, how leaders behave, and day-to-day operations. Image results from the messages sent by an organization, both intentional and unintentional, as well as social, historical, and lived experiences with the organization (Kazoleas, Kim, & Moffitt, 2001). Image is the total impression that the entity makes on a public's perceptions (Dichter, 1985; Druckenmiller, 1993; Theus, 1993). Image is a mosaic made of all the conclusions different publics make after being exposed to an organization's public relations efforts (Sauerhaft & Atkins, 1989). Any public relations efforts to enhance image should reflect the organization's mission, culture, values, and personality (Jump, 2004). In addition, public relations efforts of promoting a positive image or reputation have a positive effect on relationships with members of an uninvolved public (Grunig & Hung, 2002).

Image and Universities

Few researchers have focused their attention on the importance of image to educational institutions, but the concept is clearly relevant (Arapan, Raney, & Zivnuska, 2003). Universities depend on their image to survive (Kazoleas, Kim, & Moffitt, 2001). Theus argues that one of a university's most important assets is its image, defined by

> entering-student test scores, quality of faculty, expenditures per student, size of endowment, number of volumes in the library, admissions selectivity, volume of gifts and grants, accomplishments of alumni, quality of facilities, size of operating budget, peer rankings, reputation for innovation, and quality of leadership. (1993, p. 281)

Because universities compete for resources, the best students, and excellent faculty, image enters the equation. Publics, such as legislators, students, and faculty, consider these factors when making choices about resource allocation, education, and employment. Universities and colleges are similar to businesses, then, because they must use public relations, marketing, lobbying, and image to acquire the resources necessary for survival (Dill, 1982; Pelletier & McNamara, 1985; Theus, 1993).

Image plays an important role for universities (Immerwahr & Harvey, 1995). Universities must communicate the benefits and values of attending a particular institution to potential students and show how they differ from others; they must demonstrate their competitive strengths while paying attention to students' wants and needs (Johnston, 2003; Pulley, 2003; Washburn & Petroshius, 2004). Once a university has communicated these values, it builds its image and relationships.

Strong image for a university makes for a strong relationship with publics (Pulley, 2003). Therefore, institutions of higher learning should promote positive images with their constituents, so that strong relationships are built. Once a positive image and a strong, committed relationship exist, the university is more apt to attract the best students and faculty along with the resources it needs to operate.

If academic institutions want to build strong relationships with prospective and current students and other constituents, they must send clear, consistent messages about their values and image (Moore, 2004). A university's image becomes a part of students' individual identities (Moore). They want to be identified as belonging to the institution, its values, culture, and image. They become relationship builders.

A university's public artifacts, such as publications and Web sites, promote an organization's image and values (Hatch, 1993; Higgins & Mcallaster, 2004). The first impression that universities make with these artifacts is critical: Potential students want to know if they will fit in and if students there are similar to them (Glass, 2004). If the answer to their questions is no, they seek other options. The university may, in turn, lose valuable resources. Therefore, universities should carefully monitor the content contained on Web sites, especially home pages, because they are so visible and widely accessed and may provide the first image of the university that a student experiences.

Universities, Image, Relationships, and the Web

The American Council on Education and the University of California at Los Angeles' freshman survey, the most comprehensive assessment of student attitudes, first examined Internet use in 1998 (Sax, Austin, Korn, & Mahoney, 1998). Sax et al. stated, "Our findings show that the Internet has become a way of life for the majority of students" (para. 3). At that time, 82.9% of freshmen reported using the Internet for research and homework purposes, and 65.9% stated that they communicated via e-mail. College students use the Internet for many academic and social reasons (Jones, 2002). Approximately 57% of prospective students use university Web sites to learn more about academic institutions (Tucker, 2006).

Institutions of higher education use the Web as a way to reach prospective students (Green, 1996; Klein & Niles, 2005; McKay, 1997; Stoner, 2004). The Web allows institutions to showcase their resources in an efficient and cost-effective manner (Foster, 2003; McKay). It establishes communication and builds relationships between a university and its publics (Padgett, 2006; Taylor, Kent, & White, 2001). Because the Web is a top source used by prospective students to learn more about an institution (Stoner, 2004), universities must pay attention to the visual messages they send and the image their Web sites project. The Web often gives prospective students their first impression of a university; as such, the Web is the first step in building a positive image and a good relationship.

Diversity Issues in Higher Education

Successful relationship management with diverse populations is of particular concern because of the changing demographics of the United States and the world. If a university ignores diversity issues, it risks losing an opportunity. "Diversity is one of the largest, most urgent challenges facing higher education today. It is also one of the most difficult challenges colleges have ever faced" (Levine, 1991, p. 4). With changes in demographics, different cultural norms and values may emerge and potentially cause tension (B. J. Allen, 1995; Henderson, 1992). Therefore, as higher education moves into the twenty-first century, learning about and interacting with diverse people will be more important (Rowan, 1997).

Universities must make campuses places where students from different backgrounds take part in conversations and share experiences. In this way, they learn to understand the perspectives of others and become better world citizens (Adelma, 1997; Gurin, 1999). Improving diversity and, subsequently, diverse relationships on campus is not just an altruistic, politically correct goal; it is necessary for both social and economic reasons (Carnevale, 1999). Attention to diversity is a component of organizational image (Theus, 1993; Fombrun & Shanley, 1990). Universities can demonstrate their academic quality (Sevier, 2003) and enhance organizational image. Diverse relationships give some institutions an advantage over others (Hon & Brunner, 2000).

For example, the University of Nebraska made use of tractor trailers in a marketing campaign (Klein, 2004). Slogans and images covered the trucks, and the university's message was delivered across the nation. The university hoped to reach publics with which it traditionally did not communicate. This good start could be improved. One criticism of the campaign was that the diversity of the ads seemed artificial, because no Asian Americans and few African Americans were depicted. Although the approach was innovative, it did not incorporate diversity into the message as well as it might. The university's image did not convey inclusiveness. As such, the campaign may not have built relationships with prospective minority students.

A university should create a welcoming image in people's minds (Hon & Brunner, 2000). When people are visible members of an organization, their identification and support for the organization increase (Toma, Dubrow, & Hartley, 2005). The publicity materials used by a university, such as home pages, influence future students (Ralph & Boxall, 2005). Therefore, if a home page includes diversity, it might appeal to a greater number of people, project a stronger image, and build better relationships. In addition, publics lend more support to an organization that has values similar to their own, and, if those values disagree, nothing can be done to resolve the matter (Toma, Dubrow, & Hartley). If no or little representation of certain groups appears, then these marginalized groups may believe their values are not important to the organization, and no relationships will be built.

Traditionally, marginalized groups do not share in elements of the dominant culture, such as symbols, artifacts, norms, and values (Toma, Dubrow, & Hartley, 2005). Therefore, universities must select images that will be inviting and will build a sense of community. Unless the perspectives of subgroups are considered, institutional culture can become exclusive rather than inclusive (Toma, Dubrow, & Hartley). Individuals, such as the students featured on a university's home page, serve as a symbol or icon of that university. These symbols become the university and encourage publics to identify with them. When these images are exclusionary, identification may not result.

Therefore, showing a diverse student body is a way for a university to bolster its image and to build a relationship with diverse publics. However, universities may not use the Web to promote an image of diversity and inclusion. The visuals presented may represent a homogeneous student body with a few token photos of minority students in isolation, thus alienating them from the rest of the community. On the other hand, universities may be doing their best to show that they are inclusive and that minorities are a welcome and active part of the community.

In this chapter, we look at many universities' home pages to discover how they present diversity in their initial images. A university's home page is a type of cultural artifact that represents the university's values and images. Logically, a university would consider a home page to be a place to put its "best foot forward" and to make a favorable impression about priorities and values. Thus, if a university considers diversity to be an important value and wants to develop solid relationships with diverse publics, then diversity should be featured on the home page. This chapter, then, focuses on university home pages with special attention paid to the extent to which universities promote images of women and people of color. To do this, we performed a content analysis of university home pages.

Content analysis can be used to study many areas of the behavioral and social sciences and is used frequently in journalism, communication, public relations, and business. It is a systematic and objective method of analysis (Stacks, 2002); replicable and valid inferences can be made from data analyzed using content analysis (Krippendorf, 1980). In addition, content analysis is an ideal research method for the analysis of documents, because it provides a logical basis for understanding how messages are constructed (Stacks, 2002).

The Study

The authors examined the visual images presented on the home pages of NCAA Division I schools. The researchers went to each Web site's home page and counted instances of relevant visuals: the number of people, women, and minorities shown. In addition, the researchers coded whether minorities were included in a photo with nonminorities or if minorities were isolated in photos. Only the initial photos appearing when the Web site was first launched were coded. Therefore, if a Web site had images that changed while someone looked at the home page, or if different photos were displayed with each click on the Web site, not all the images were included in this research. Thus, the unit of analysis was the first page shown immediately upon accessing a university's Web site. We chose this procedure to reflect the likely behaviors of visitors to home pages. We entered the Web addresses and accessed the home pages. We did not repeatedly enter the home pages to access all possible images. In this way, we received realistic first impressions of the Web sites.

Results and Discussion

Three-hundred and twenty-six home pages were examined from the list of member schools of the NCAA (*National Collegiate Athletic Association*, 2005). Approximately 42% of the institutions were public, 32% were private, 17% were land grant, 3% were historically black colleges and universities (HBCUs), 3% were HBCU land grant, 2% were Ivy League, and 1% were military. About 34% of the institutions were located in the Southeast, 23% were located in the Northeast, 19% were located in the Midwest, 8% were located in the Southwest, 8% were located in the Northwest, 7% were located in California, and 1% were located in Hawaii.

The vast majority of home pages (86%; n = 281) contained visual images of people. Approximately 80% (n = 226) of the home pages with visual images of people contained images of women. Furthermore, 63% (n = 178) of the home pages with visual images of people presented images of people of color. Judging from these results alone, it appears that many university administrators and, presumably, public relations officers make a conscious decision to promote images of diversity in race and gender on their home pages.

These images represent the notion that elements of diversity in gender and race are an important consideration for these universities and colleges and that they are making an effort to present an image of inclusiveness in their relationships with their publics. This effort should be considered as an attempt at attaining the benefits associated with increased diversity. As noted, some of these benefits include a better-quality student body, enhanced fundraising, more competitive alumni, and so on. Looking at these images alone would lead one to believe that universities do a good job of promoting positive, diverse images and building relationships with diverse publics. The images show a level of cultural sensitivity that is inclusive of all constituents.

However, a closer examination of these images reveals something else in terms of inclusiveness. The most telling result was that only 35% (n = 97) of the images with people depicted people of different races interacting together. In other words, for the most part, white people only appeared with white people, and people of color only appeared with people of color. In short, the people who appeared on these pages were visually segregated in this virtual world.

This sort of visual segregation may be telling about the university's true valuing of diversity in its population. Clearly, universities and colleges have a choice about the images to present on their Web pages. Home page images should be selected purposively, and two important purposes are to represent the university or college in a positive manner and to create a good first impression. Public relations officers know that image building is important in developing and maintaining good relationships. As such, these images give the impression that the universities are more interested in promoting images of diversity than diversity itself. Thus, the universities may select photographs that show women and minorities on campus but do not value the inclusiveness that marks a truly diverse campus.

In other words, the universities and colleges may be following an informal formula for promoting an image of diversity. Following this formula implies that, by including images of people of color or women on a home page, a university will then be seen as promoting diverse, inclusive relationships. In fact, some universities may carry out this formula to extremes. For instance, some universities have doctored photographs in official publications to show diversity ("University of Idaho," 2000). By separating individuals of different races on the home pages, universities ignore the value of inclusiveness in favor of a formulaic approach to promoting a particular image without regard to the reality present on the campus. If the university does not locate images of inclusive relationships to place on its home page, a location over which it has great control, then that attitude and reality may be present in all other aspects of university life with regard to inclusiveness and diversity. That is, the visual segregation on the home page may represent a true value of the university or college: that expedience in creating and maintaining an image of inclusiveness is more important than actual inclusiveness.

On the other hand, the observation that universities and colleges visually segregate minorities is not necessarily completely negative. These institutions make an effort to include women and people of color in their public images. Indeed, more than a third of the images depicted people of differing races. Furthermore, these results do not mean that colleges and universities should immediately go out and find images of diverse people interacting on their campuses. The visual segregation present may be nothing more than coincidence, or it may, in fact, show reality. Some researchers state that younger Americans, especially those of college age, want to be among people with similar backgrounds and of the same ethnicity (Goodheart, 2004). The younger generation may prefer to retain distinctive cultures due to their exposure to multiculturalism at a young age. In addition, researchers contend that this situation may last a few generations and is not necessarily a bad circumstance (Goodheart).

However, given that the artifacts of an organization often represent its values, colleges and universities must examine the values underlying their Web site design. The organization should consider whether the images selected for the Web site have been chosen to create an *impression* of inclusiveness in relationships or rather to reflect the *reality* of inclusiveness in relationships. In either case, the Web site performs a valuable service. If the university performs a self-analysis and determines that it represents an expedient image, then it should consider ways to promote true diversity and relationships among its members. If it forgoes a self-analysis, the contradictions between the image and the reality may result in more damage than benefit to the university. Conversely, if the university truly promotes diversity and inclusiveness, then the images it portrays should serve as an accurate reflection of this value. In either case, universities and colleges should carefully examine their Web sites to identify the ways in which these tools create, maintain, and reflect diverse, inclusive relationships among members.

Conclusion

In today's society, technology and media are inescapable. Publics are bombarded with messages and information. Therefore, public relations practitioners must work at a faster pace and for longer hours (D'Aprix, 2005). However, new technology can also be seen as an evolution of communication tools (Stoff, 2005).

New technology allows practitioners to tune messages for very specific publics (Weiner, 2005). This application of technology is very important for reaching Generation M, the millennial generation, now entering college ("Admission Podcasts," 2005; Tucker, 2006; Zeller, 2006). This generation relies on technology, not face-to-face interaction, for communication (Tucker). For example, Mansfield University successfully uses podcasts to reach prospective students and their parents ("Admission Podcasts"). By embracing this new technology, Mansfield attracts thousands of new visitors to its Web site without cost, and delivers a targeted message to an important public. Other universities would be wise to follow this example and use new technologies to reach Generation M.

The advent of new technology forced many public relations practitioners to jettison traditional media (Fernando, 2004). This action, however, may be exactly what is needed to attract Generation M, because they do not trust and often do not use traditional media (Zeller, 2006).

Generation M is the most connected, most protected, and most marketed-to generation (Abate, 2006: Tucker, 2006). Some say that this generation *is* the Web, because it is so entrenched in technology (Tucker). They want information and access on their own terms and on their own time. They want instant gratification, and they want information customized through instant messaging, the Web, e-mail, text messaging, and so on (R. Allen, 2006; Howard, 2006).

These demands, preferences, and expectations can only be met with new technologies; traditional media cannot compete. Therefore, public relations practitioners must incorporate new technology and build relationships through technology.

Relationships require mutual interest, interaction, trust, risk, and maintenance, and new technology has built buzz for the importance of word-of-mouth relationships. Although this process sounds like interpersonal communication, it marks a digital dialogue. Universities must be mindful of the images and messages presented on Web sites, because they may be reaching exponentially increasing publics. Prospective students may talk about universities on blogs, text messages, instant messages, Web sites, and e-mails. Word of mouth then reaches a greater audience with the capability of connecting to even greater audiences. If a university presents a positive image that is helpful in creating relationships, word of mouth will help its cause. However, if it does not, word of mouth could have a devastating effect on its recruitment efforts.

Finally, universities do a good job of showing diversity on their Web sites by incorporating women and people of color. By showcasing this positive first impression, they assemble the building blocks to form relationships. Although these visual images are welcoming because minorities and women are represented, universities may need to go further to create a sense of an all-inclusive community by depicting true representations of inclusive relationships. As stated earlier, minorities have been marginalized and do not share in the dominant culture. Universities should consider ways to transcend this situation. After all, universities want all students to identify with their university. Positive image and identification are key to strategic relationships. In the world of new technology, universities need all the positive buzz and word-of-mouth promotion they can get from Generation M's blogs, podcasts, Web sites, and e-mails to continue sto enjoy positive images and good relationships with priority publics.

For consideration

1. What role, if any, should public relations officers play in promoting diversity on campus?
2. What are the implications of depicting diverse images on university Web sites in greater proportion than actually exist on campus? In other words, how ethical is it for a university to promote an image of diversity when the university itself is not diverse?
3. Other than visual images, what other new media techniques or tactics could be used to reflect diversity on college campuses?

4 A popular ad campaign once promoted the idea that "Image is everything." Is it?
5 To what extent does virtual segregation exist beyond college and university Web sites?

For reading

Cook, J., & Finlayson, M. (2005, Summer). The impact of cultural diversity on website design. *SAM Advanced Management Journal, 70*(3), 15–24.

Esrock, S. L., & Leichty, G. B. (1998, Fall). Social responsibility and corporate Web pages: Self-presentation or agenda-setting? *Public Relations Review* (Special Issue on Technology and the Corporate Citizen), *24*(3), 305–319.

Guterman, L. (2000, October 13). Doctoring diversity II. *The Chronicle of Higher Education, 47*(7), p. A12.

Roach, R. (1997, November 27). Website established to showcase higher education diversity initiatives. *Black Issues in Higher Education, 14*(20), 53.

The University of Maryland. (n.d.). *Equity and diversity at the University of Maryland*. Retrieved August 19, 2006, from http://www.umd.edu/diversity/[1]

Note

1. This Web site is an excellent example of how a university promotes diversity on its campus. It provides links to various diversity-related programs and initiatives around the campus. A dedicated Web site of this sort is an invaluable tool for understanding an institution's commitment to diversity.

References

Abate, T. (2006, January 1). Generation M: Are we so immersed in media brine that it's become an environmental health hazard? *San Francisco Chronicle, Chronicle Magazine*, p. 6.

Adelma, C. (1997). Diversity: Walk the walk and talk the talk. *Change, 29*, 34–45.

Admission podcasts tackle what students really want to know. (2005). *Recruitment & Retention in Higher Education, 19*, 1 & 4.

Allen, B. J. (1995). "Diversity" and organizational communication. *Journal of Applied Communication Research, 23*, 143–155.

Allen, R. (2006, February 7). Understanding the Millennial generation. *Daily O'Collegian*. Retrieved July 7, 2006, from http://www.ocolly.com/read_story.php?a_id=29174

Arapan, L. M., Raney, A. A., & Zivnuska, S. (2003). A cognitive approach to understanding university image. *Corporate Communications: An International Journal, 8*, 91–113.

Broom, G. M., Casey, S., & Ritchey, J. (1997). Toward a concept and theory of organization-public relationships. *Journal of Public Relations Research, 9*(2), 83–98.

Bruning, S. D. (2000). Examining the role that personal, professional, and community relationships play in respondent relationship recognition and intended behavior. *Communication Quarterly, 48*, 437–448.

Bruning, S. D. (2002). Relationship building as a retention strategy: Linking relationship attitudes and satisfaction evaluations to behavioral outcomes. *Public Relations Review, 28*(1), 39–48.

Carnevale, A. P. (1999). Diversity in higher education: Why corporate America cares. *Diversity Digest, 3*, 1 & 6.

D'Aprix, R. (2005). The message imperative. *Communication World, 22*, S8.

DeSanto, B. J., & Garner, R. B. (2001). Strength in diversity: The place of public relations in higher education institutions. In R. L. Heath (Ed.), *Handbook of public relations* (pp. 543–549). Thousand Oaks, CA: Sage.

Dichter, E. (1985). What's in an image. *Journal of Consumer Marketing, 2*, 75–81.

Dill, D. (1982). The management of academic culture: Noteson the management of meaning and social integration. *Higher Education, 11*, 303–320.

Druckenmiller, B. (1993). Crises provide insight on image:Preparations necessary to protect goodwill when times turn bad. *Business Marketing, 78*, 40.

Fernando, A. (2004). Creating buzz: New media tactics have changed the PR and advertising game. *Communication World, 21*, 10–11.

Fombrun, C., & Shanley, M. (1990). What's in a name? Reputation building and corporate strategy. *Academy of Management Journal, 33*, 233–258.

Foster, A. L. (2003, May 2). Colleges find more applicants through personalized Web recruiting. *The Chronicle of Higher Education, 49*, pp. A37–A38.

Glass, R. (2004). Marketing your institution effectively: A parent's perspective. *Journal of College Admission, 183*, 2–4.

Goodheart, A. (2004, May/June). The new America. *AARP Magazine*, pp. 8–11.

Green, K. C. (1996). Planning your presence on the Web. *Change*, 28, 67–70.

Grunig, J. E. (1994). Image and substance: From symbolic to behavioral relationships. *Public Relations Review, 19*, 121–139.

Grunig, J. E., & Hung, C. F. (2002, March). *The effect of relationships on reputation and reputation on relationships: A cognitive, behavioral study*. Paper presented to the 5th Educator's Academy Conference, Public Relations Society of America, Miami, FL.

Gurin, P. (1999). New research on benefits of diversity in college and beyond: An empirical analysis. *Diversity Digest, 3*, 15.

Hatch, M. J. (1993). The dynamics of organizational culture. *Academy of Management Review, 18*, 657–693.

Henderson, Z. P. (1992). Educating multicultural groups. *Human Ecology Forum, 20*, 16–19.

Higgins, J., & Mcallaster, C. (2004). If you want strategic change, don't forget to change your cultural artifacts. *Journal of Change Management, 4*, 63–73.

Hon, L. C., and Brunner, B. (2000). Diversity issues and public relations. *Journal of Public Relations Research, 12*, 309–340.

Hon, L. C., & Grunig, J. E. (1999). *Measuring relationships in public relationships*. Manuscript prepared for Relationship Task Force/Measurement Commission for the Institute for Public Relations and Ketchum Public Relations. Gainesville, Florida.

Howard, K. C. (2006, March 6). Millennials spur teaching change. *Las Vegas Review-Journal*, p. 1B.

Immerwahr, J., & Harvey, J. (1995, May 12). What the public thinks of colleges. *The Chronicle of Higher Education, 41*, pp. B1–B2.

Johnston, A. D. (2003, June 2). The brand called "U." *Maclean's, 116*, 50.

Jones, S. (2002, September 15). *The Internet goes to college: How students are living in the future with today's technology*. Washington, DC: Pew Internet & American Life Project. Retrieved July 7, 2006, from http://www.pewinternet.org/pdfs/PIP_College_Report.pdf

Jump, J. (2004, Summer). Admission, heal thyself: A prescription for reclaiming college admission as a profession. *Journal of College Admission, 184*, 12–17.

Kazoleas, D., Kim, Y., & Moffitt, M. A. (2001). Institutional image: A case study. *Corporate Communications: An International Journal*, 6, 205–216.

Klein, A. (2004, March). Does this campaign work? *University Business*, *7*, 66–67.

Klein, A., & Niles, B. (2005). Look to the Web to increase recruitment. *University Business*, 8, 37–38.

Krippendorf, K. (1980). *Content analysis: An introduction to its methodology*. Newbury Park, CA: Sage.

Ledingham, J. A. (2003). Explicating relationship management as a general theory of public relations. *Journal of Public Relations Research*, *15*, 181–198.

Levine, A. (1991). The meaning of diversity. *Change*, *23*, 4–5.

McKay, G. (1997). Internet recruiting. *Executive Report*, *15*(5), 50–51.

Moore, R. M. (2004). The rising tide. *Change*, *36*, 56–61.

National Collegiate Athletic Association—NCAA. (2005). Retrieved August 19, 2006, from http://www.ncaa.org/wps/portal

Padgett, R. (2006). Better public relations on Websites. *Education Digest*, *71*, 54–55.

Pelletier, S. G., & McNamara, W. (1985). To market? *Educational Horizons*, *63*, 54–60.

Pulley, J. L. (2003, October 24). Romancing the brand. *The Chronicle of Higher Education*, 50, pp. A30–A33.

Ralph, S., and Boxall, K. (2005). Visible images of disabled students: An analysis of UK university publicity materials. *Teaching in Higher Education*, *10*, 371–385.

Rowan, C. T. (1997, April 30). The need for diversity in education. *Baltimore Sun*, p. 11A.

Sauerhaft, S., & Atkins, C. (1989). *Image wars: Protecting your company when there's no place to hide*. New York: John Wiley & Sons.

Sax, L. J., Austin, A. W., Korn, W. S., & Mahoney, K. M. (1998). *The American freshman: National norms for fall 1998*. Retrieved March 21, 2002, from http://www.gseis.ucla.edu/heri/norms_pr_98.html

Sevier, R. (2003). The problem with prestige. *University Business*, 6, 16–17.

Stacks, D. W. (2002). *Primer of public relations research*. New York: The Guilford Press.

Stoff, R. (2005). Battle in the Internet age. *St. Louis Journalism Review*, *35*, 9 & 28.

Stoner, M. (2004, April 30). How the Web can speak to prospective students. *The Chronicle of Higher Education*, 50, pp. B10–B11.

Taylor, M., Kent, M. L., & White, W. J. (2001). How activist organizations are using the Internet to build relationships. *Public Relations Review*, *27*(3), 263–284.

Theus, K. T. (1993). Academic reputations: The process of formation and decay. *Public Relations Review*, *19*(3), 277–291.

Toma, J. D., Dubrow, G., & Hartley, M. (2005). The uses of institutional culture. *ASHE Higher Education Report*, *31*, 1–103.

Tucker, P. (2006). Teaching the millennial generation. *The Futurist*, *40*, 7.

University of Idaho, Auburn U. also report incident of doctored photos. (Photographs used in marketing information changes to depict minority students) (Brief Article). (2000, October 26). *Black Issues in Higher Education*, *17*, 12.

Washburn, J. H., & Petroshius, S. M. (2004). A collaborative effort at marketing the university: Detailing a student-centered approach. *Journal of Education for Business*, *80*, 35–40.

Weiner, M. (2005). Marketing PR revolution: How will professionals navigate the new landscape? *Communication World*, *22*, 20–25.

Wilson, L. J. (2001). Relationships within communities. In R. L. Heath (Ed.), *Handbook of public relations* (pp. 521–526). Thousand Oaks, CA: Sage.

Zeller, T. (2006, January 22). A generation serves notice: It's a moving target. *The New York Times*, pp. 1–4.

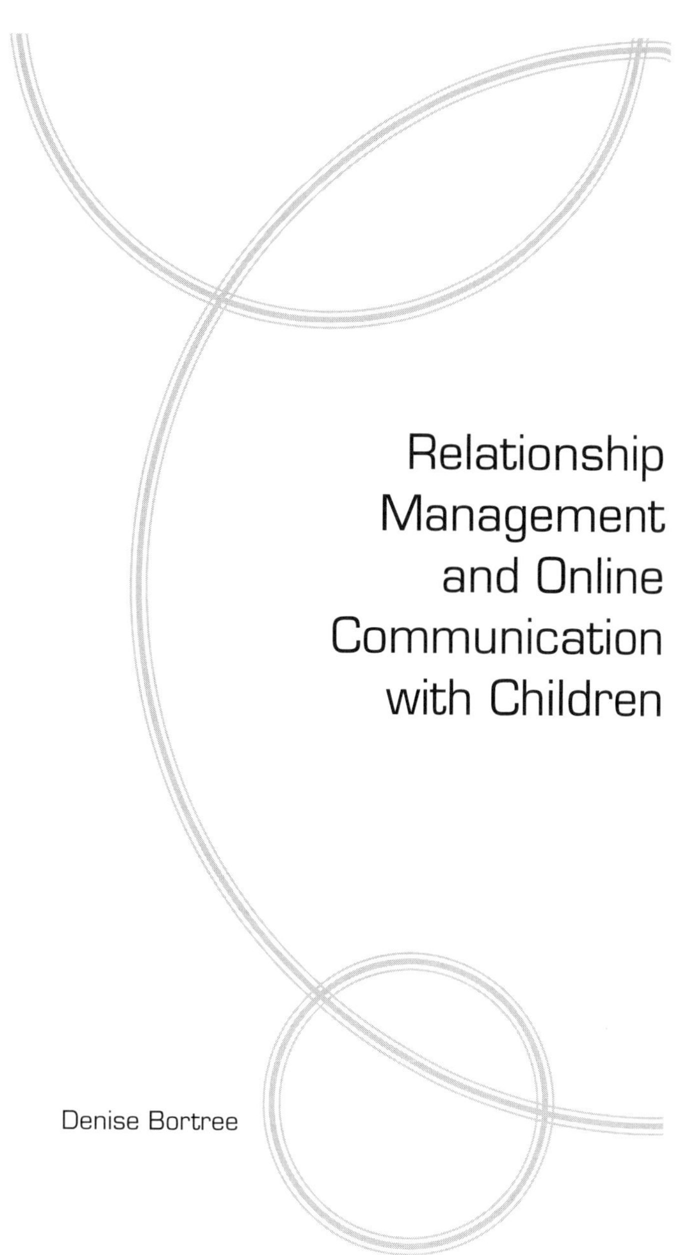

Relationship Management and Online Communication with Children

Denise Bortree

This chapter explores the ways relationship management may be applied to child publics. Because children differ from adults in their social and cognitive abilities, relationship management theory, as it is currently defined, may not apply seamlessly to relationship building with children. Using the framework of Hon and Grunig's (1999) recommendations for maintaining organization-public relationships, the chapter looks at the ways public relations tactics were used by nutrition Web sites to build relationships with child publics. The Web sites demonstrate how interactive content, including games and downloadable material, are used to educate children and encourage them to return to the sites. This chapter includes a discussion of the appropriateness of applying the relationship management recommendations to child publics and offers a modified recommendation for building relationships with these publics.

Many organizations would consider children a valuable public as participants, members, consumers, or activists, and these organizations participate in various types of dialogue with children. In fact, some public relations agencies tout themselves as experts in building campaigns specifically designed to reach children.[1] Organizations with an interest in maintaining relationships with children fall into all sectors, including corporations, government organizations, and nonprofit organizations. Some seek to dialogue with children to build mutual understanding (e.g., libraries or museums), some attempt to influence children's opinions about an issue (e.g., animal rights or human rights organizations), and others attempt to influence children's behaviors (e.g., antismoking or antidrug organizations).

One significant concern that scholars, politicians, and medical practitioners have recently raised about communication between organizations and children is the relationship between advertising, specifically advertising junk food and fast food, and the epidemic of childhood obesity (Strasburger, 2001; N. Wilson, Quigley, & Mansoor, 1999). Many studies have looked at the effects of advertising and general media consumption on children's eating behaviors (Clocksin, Watson, & Ransdell, 2002; Kaiser Family Foundation, 2004; Zuppa, Morton, & Mehta, 2003) and found that media does play a role in children's health. As a result, certain organizations (some in the food industry and some not) have begun to provide nutritional information and guidelines to children. A number of these organizations have chosen the Internet as a point of distribution for the information.

While some organizations provide nutritional information as a way of mitigating the impact of food advertising on children (e.g., Kellogg), others appear to be providing nutritional information as part of an educational effort toward children (e.g., American Diabetes Association). Both types of organizations would be interested in building and maintaining a relationship with children as a way of ensuring the success of their efforts, and ultimately, benefiting their child publics.

Relationship Management

Building and maintaining relationships between organizations and their publics have become the hallmark of public relations. This focus on the relationship as a unit of study was first suggested by Ferguson (1984). Relationships are formed for the benefit of the organization and the public. Ledingham writes that "public relations balances the interests of organizations and publics through the management of organization-public relationships" (2003, p. 181). In other words, through relationship building, organizations hope to benefit both themselves and their targeted publics.

According to Hon and Grunig, public relationships are formed for the following reasons:

> Relationships form because one party has consequences on another party. Organizations have a public relations problem or opportunity and a reason to develop a public relations program when management decisions have consequences on publics inside or outside of the organization or when the behavior of these publics has consequences on the success with which an organizational decision can be implemented. (1999, p. 12)

In many areas, including health and education, the behavior of children can be considered a critical factor in the success of an organization's program. As a result, a long-term relationship is desirable: one that can be used to influence children's behavior in a positive way.

Hon and Grunig (1999) contend that the relationship between an organization and its public can best be measured through four specific outcomes of the relationship:

- **Control Mutuality**—The degree to which parties agree on who has the rightful power to influence one another. Although some imbalance is natural, stable relationships require that organizations and publics each have some control over the other.
- **Trust**—One party's level of confidence in and willingness to open oneself to the other party. There are three dimensions to trust: *integrity:* the belief that an organization is fair and just . . . *dependability:* the belief that an organization will do what it says it will do . . . and, *competence:* the belief that an organization has the ability to do what it says it will do.
- **Satisfaction**—The extent to which each party feels favorably toward the other because of positive expectations about the relationship are reinforced. A satisfying relationship is one in which the benefits outweigh the costs.
- **Commitment**—The extent to which each party believes and feels that the relationship is worth spending energy to maintain and promote. Two dimensions of commitment are continuance commitment, which refers to a certain line of action, and affective commitment, which is an emotional orientation. (p. 3)

These elements, identified for adult relationships, have not been tested with children, so it is not known if they would be important outcomes in an organization-public relationship with children. The following discussion of social and cognitive development in children examines the potential issues of applying relationship management to child publics.

Children's Social and Cognitive Development

Certainly, organizations attempt to engage in relationships with children, but two separate questions remain. At what age do children have the capacity to engage in a relationship with an organization, and how would these relationships look similar and/or different for children?

The answer to the first question may be related to the child's ability to reason abstractly, as the concept of an *organization* is an abstract one. At a fairly young age, a child can understand what a book is. Later, he can understand that the library is a place where books may be checked out for a period of time and then need to be returned. However, it is unlikely that a young child will be able to conceptualize an organization like the National Literacy Association or, for that matter, the American Diabetes Association, which provides information on its Web site for children. If a child cannot conceptualize a specific organization, it is unlikely that he will be able to perceive and maintain a relationship with that entity.

According to Piaget's theory of development (1932/1965), children have all the mental structures in place to think abstractly when they enter the formal operational stage at approximately twelve years old. But, in the concrete operational stage (ages seven to eleven years), children are still developing that capability. Though children in the concrete operational stage would be able to think of themselves as members of the library and users of its services, it is possible that there would be limits to the same children's comprehension or participation in a relationship with an organization that requires abstract conceptualization, such as the National Literacy Association or the American Diabetes Association in the previous example.

However, a "physical" location on the Internet would allow a child to participate in tangible interactions with the organization, and the Web site would create an opportunity for the child to relate to the organization in a concrete way. It is possible that this would increase the likelihood of successful relationship management with children in the concrete operational stage (seven to eleven years of age) (even though the children may perceive the organization as a Web site rather than as an institution). More research is needed in this area.

Assuming that children in the formal operational stage (ages twelve years and older) and even the concrete operational stage (between the ages of seven and eleven) or younger are able to perceive and maintain a relationship with an organization to varying degrees, the next question one might ask is: How would these relationships look similar and/or different for children? Research conducted in the area of children's relationships with peers (Schneider, Smith, Poisson, & Kwan, 2000; von Salisch, 2000) may offer some insight into the way their relationships with organizations are similar to adult relationships with organizations. This research seems to indicate that all four of the outcomes of the organization-public relationship identified by Hon & Grunig (1999) are significant in children's peer relationship. One key aspect of children's relationship building with peers is building trust (Rotenberg, 1991), which they accomplish through keeping and sharing secrets. In addition, elementary school children view their peer relationships as "guided by principles such as *mutual respect* and *reciprocity*" (Ladd, 2005, p. 76), two concepts which have been identified as dimensions of adult organization-public relationships. Like adults, children expect a type of mutuality in their peer relationship; however, the degree to which they would expect this in an organization-public relationship is debatable. In addition, children demonstrate a need for satisfaction and commitment in their peer relationships (Ladd, 2005), also identified as relationship outcome dimensions of organization-public relationships. Considering the complexity of children's interpersonal relationships, it is possible that the dimensions of children's organization-public relationships are similar to those of adults.

On the other hand, children's limitations in relationship skills provide some insight into how the relationships are different. First, relationship management assumes a public with critical skills that children may not possess. Children as young as two or three years of age are aware of another's ability to deceive (Chandler & Hala, 1991) and therefore are in need of developing *trust*, one of the elements of an organization-public relationship. However, children under the age of seven years (those who have not yet reached the concrete operational stage) are only able to perceive messages as true or false (i.e., a lie or a truth) (Piaget, 1932/1965). They lack the critical skills to detect subtle persuasion—a skill needed to enter a relationship that requires both members to engage in negotiation and mutual influence, as would be the case in an organization-public relationship. Therefore, relationship management with children under the age of seven years may not be possible.

Second, children under the age of seven years are not as skilled in interpreting certain types of nonverbal communication (Bugental, Kopeikin, & Lazowski, 1991). This has impact for any relationship management conducted online, particularly in an environment where nonverbal communication is common (Papacharissi, 2002). Children in the concrete operational stage (between seven and eleven years of age) and early in the formal operational stage (twelve years of age and older) are still developing their skills involving trust and perception of nonverbal communication; therefore, organizations should make every effort to be truthful and fair in their dealings with these young publics.

Third, children gradually build a sense of self in a relationship (Barrett, 2000), but a strong sense of self is not in place until late adolescence. The weak sense of self during childhood impacts their ability to participate as an equal partner in the negotiated process of relationship management as it is described by Grunig and Huang (2000); therefore, their participation may be limited, giving the balance of power to the organization. This raises an ethical concern. Public relations relationship building is a dialogue between two entities (organizations or individuals), with each equally participating in the relationship and each mutually influencing the other based on one's interests and concerns. Because of the power imbalance, organizations that seek to engage in a relationship that will not result in a clear benefit for the child should weigh the ethical ramifications of their choice. Grunig and Huang (2000) suggest that, in the case of a severe inequity in power, a third party should be introduced to represent the weaker party and balance the power between the two entities. Engaging a parent or other third party may be a way to address ethical concerns in the case of young children.

Application of Relationship Management

As a starting point, this study used Hon and Grunig's (1999) recommendations for maintaining relationships to examine how the theory can be translated from adults to children. Research on interpersonal relationships, specifically Stafford and Canary's (1991) study of romantic relationships, offers these concepts for relationship maintenance: access, positivity, openness, assurances, networking, and sharing of tasks. These maintenance strategies, when engaged by organizations, are theorized to influence the relationship outcomes previously discussed (control mutuality, trust, satisfaction, commitment, exchange relationship, and communal relationships) (Hon & Grunig, 1999; Grunig & Huang, 2000). This study attempted to link the maintenance strategies to specific tactics used on nutrition Web sites for children. The presence of the tactics was measured, and results are reported in this chapter.

Four of the six maintenance strategies proposed by Hon and Grunig (1999) were used to examine what is currently being done on the Web to engage children in an organization-public relationship.

Positivity is "anything the organization or public does to make the relationship more enjoyable for the parties involved" (Hon & Grunig, 1999, p. 14), and it may play a role in the relationship between child publics and organizations. Online public relations tactics like providing games, interactive content, and downloadable files may fall into this category.

Assurances are "attempts by parties in the relationship to assure the other parties that they and their concerns are legitimate" (Hon & Grunig, 1999, p. 15). Showing concern for a public or demonstrating that the public is valued may be a good way of maintaining a relationship with children. When an organization allows children to contribute to a Web site through message boards and chat rooms or even by assisting in the development of the design of the site, this sends a message to children that they are valued by the organization. These tactics may convey an assurance to children.

Networking is defined as "organizations' building networks or coalitions with the same groups that their publics do, such as environmentalists, unions, or community groups" (Hon & Grunig, 1999, p. 15). The concept of networking with child publics may include types of third-party endorsements: for example, the use of a celebrity or popular cartoon character in materials targeting kids or the provision of lesson plans or other curricula to teachers on a topic of interest to the organization.

Sharing of tasks is defined as "'organizations' and publics' sharing in solving joint or separate problems" (Hon & Grunig, 1999, p. 15). Although this may not translate as easily to relationships with children, possible examples may be encouraging children to get involved in some type of community activity or soliciting donations of time or money from children for a good cause.

The following two concepts were not used in the content analysis because of the lower likelihood that they would impact relationship management. However, future studies should consider their contributions to the organization-child public relationship.

Access is defined as each party giving the other access to significant processes and information within their organization or community. It is possible that this may be done by taking a child behind the scenes of the organization to see how products are produced or how the organization functions.

Openness encourages relationship building through disclosure of information between parties. By disclosing information, an organization can build trust with a skeptical public. Because children tend to build relationships with peers through the sharing of secrets and keeping secrets (Rotenberg, 1991), this may have some impact on relationship building. Future studies should examine the use of this strategy with children.

Impact of Nutritional Information on Children

Because this study looks at nutritional Web sites developed for children, a brief review of health literature related to nutrition was developed as well. According to literature on children's health and nutrition, a number of factors may influence children's health behaviors, including advertising and media consumption levels (Clocksin, Watson, & Ransdell, 2002; N. Wilson, Quigley, & Mansoor, 1999), nutrition of fast food (Mello, Rimm, & Studdert, 2003), poor dietary habits (Kranz, Siega-Riz, & Herring, 2004), and even family environment (Campbell & Crawford, 2001). Medical experts are beginning to stress media education for children and parents as a way of mediating the effects of food advertising and other negative influences (Committee on Public Education, 2001; Rich & Baron, 2001; Hindin, Contento, & Gussow, 2004). Reducing the promotion of specific foods and eliminating the presentation of food as gratification may influence children's perceptions of food and nutrition. In addition, providing good advice on food and nutrition may influence behavior in a positive direction.

Online Privacy of Children

Use of the Internet to reach publics is common practice in public relations today. However, this practice presents unique challenges for organizations that communicate with child publics. Unlike television advertising for children, which is regulated by the Federal Communications Commission, the content of Web sites is not held to a specific standard. Forty-one percent of children are online every day (Livingstone & Bober, 2004), and this frequent exposure to online content raises concerns among both researchers and parents who feel the content should be held to some standard or restricted in some way.

While organizations that communicate with children online are not restricted in the content of their Web sites, they are required to respect the privacy of children. The Children's Online Privacy Protection Act (Federal Trade Commission, 1999) puts a number of restrictions on the type of information that organizations can collect from children online. In general, without parental permission, organizations can collect e-mail addresses for "contests, online newsletters, homework help and electronic postcards" (Exceptions section, para. 1). The organization cannot collect the following information without parental permission: "full name, home address, email address, telephone number or any other information that would allow someone to identify or contact the child" (Personal Information section, para. 1). Furthermore, "The Act and Rule also cover other types of information—for example, hobbies, interests and information collected through cookies or other types of tracking mechanisms—when they are tied to individually identifiable information" (Personal Information section, para. 1). If an organization chooses to collect information from children (under the age of twelve years) online, the organization's Web site must make disclosure of their intent prominent. Organizations are held to more rigorous standards when passing children's personal information to a third party.

The study presented here will examine the types of information that nutrition Web sites attempt to collect from children.

Methodology

This study is based on a content analysis of the universe of kids' nutritional Web sites, as provided by the children's search engine Yahooligans (2006). A total of forty-two nutritional Web sites were listed as sites for children, but an initial evaluation determined that, of those forty-two, only twenty were specifically designed for children. In order for a site to be considered a kids' site for this study, it needed to specifically state "kids" (or "tween" or "children") on the site or in the Web site address, or it needed to have pictures of kids or cartoons on the site. The graphics and content of the site needed to reflect content appropriate for children.

The universe of twenty Web sites included 793 Web pages that were coded by two coders. Intercoder reliability was 90%. The following categories were included in the coding process:

General information was tracked for identification purposes. This category included the name of the Web site; the Web site address; the name of the kids' Web site (if not home page); the kids' Web site address; the extension of the main site (.com, .org, .net, .gov, state, international, or other); the Web site owner; the number of hits or visits; the country of origin for the Web site; the last update of Web site; whether the site was statewide, national, or international; the general topic of Web site; and the total number of pages related to nutrition.

Organization information provided a closer look at the Web site content provider. This category included the type of organization (corporation, health, or other organization; organization representing industry; government; individual; or other), whether the organization was part of a larger organization or project, and the sponsor of the site.

Kid Web site designation was used to determine if the site would be classified as a kid-specific site. The following information was tracked: whether the word "kids" was found on the page, pictures of kids on the site, and the use of cartoons on the site.

Relationship management tactics, based on the literature reviewed herein, included the following: celebrity endorsement, the use of a popular cartoon character, kids' contribution to design or content of site, teacher curricula or lesson plans, a request for donations or other form of participation in the organization, games, downloadable files on site, and interactive features (including chat rooms, message boards, and blogs).

Nutritional information, based on the literature, was tracked: promotion of specific foods, advice on food or nutrition, and presentation of food as gratification.

In light of the general concern about the collection of children's *personal information* by Web sites, the following information was tracked: request for parental permission and collection of personal information (name, age, sex, physical address, e-mail address, phone number, or other).

Results

Of the twenty Web sites, seven were created by corporations, four were created by health or other organizations, five were created by organizations representing industries, three were created by government organizations, and one was created by an individual. Sixteen of the sites were maintained by United States–based organizations, two were maintained by United Kingdom–based organizations, and the country of origin for two of the sites could not be determined. Results for public relations frequencies were as follows.

Positivity. Of the twenty sites reviewed, fifteen (or 75%) included games of some sort. That seemed to be a common option on Web pages directed at kids. Eighteen sites (or 90%) provided some type of interactive content, including interactive coloring pages, mazes, and nutrition calculators. Only five sites (or 25%) provided downloadable files, which were often coloring pages; however, one site allowed users to download a food shopping list.

Assurances. Only one of the twenty Web sites reviewed included some type of content provided by children. That site included both chat rooms and message boards. None of the sites included pages designed by kids.

Networking. Four of the Web sites (or 20%) included some kind of lesson plan or curriculum for teachers. Only one used a popular cartoon character, and none used a celebrity endorsement.

Sharing of tasks. None of the sites asked users for a donation or for participation with the organization.

Regarding nutrition-related frequencies, seventeen sites (or 85%) provided advice on food or nutrition, and eight sites (or 40%) promoted a specific food (whether or not it was produced by the organization). However, no sites promoted food as a form of gratification, reward, or consolation.

An analysis of the data revealed a correlation between public relations strategies (i.e., positivity, assurances, etc.) and nutritional information provided. That is, organizations that used more public relations tactics were more likely to provide nutritional information that encouraged poor behaviors (promoting consumption of specific foods—sometimes unhealthy—rather than providing guidelines for food choices). The study also found that health organizations (and other nonprofits) were less likely than other organizations (corporations, government, and organizations representing industries) to provide information that may have encouraged poor nutritional behaviors. With such a small sample size, these findings may be due to anomalies. However, the questions bear further investigation. If true, this could be an example of negative impact of relationship management on children. As stated previously, organizations that engage with children should be sensitive to the fact that children's interpersonal skills are not fully developed. Therefore, any use of a relationship to influence a child should be entirely for the benefit of the child.

Personal information. Five of the sites (or 25%) requested personal information from the users. This included four requests for the user's name, four requests for the user's age, one request for the user's physical address, five requests for the user's e-mail address and one request for the user's phone number. This raises ethical (and legal) concerns about the personal information collection practices of organizations in this study.

Discussion

As stated earlier, the goal of this study was to examine the types of public relations tactics used in relationship building in children's nutritional Web sites. The study found frequent use of *positivity*, in the form of games, downloadable content, and other interactive technologies. Although the intended purpose of providing these technologies cannot be determined from this type of study, one could assume that the games and interactive content were provided to increase children's enjoyment of the site, encourage their return to the site, and teach them about nutrition through pleasurable interaction with the site. All of these would increase the positive relationship between organizations and child publics.

Infrequent uses of *assurances* and *networking* were found, and no instances of *sharing tasks* were found. Though it is not known whether these three characteristics can improve the relationship between organizations and child publics, prior research on children's relationships with significant others would suggest that they may.

The study presented here is a very small sampling of the large number of Web sites created for children. It is highly likely that further investigation of children's Web sites will find examples of these characteristics. More research should be done to explore the way organizations are currently relating to children. Even more, research needs to be done to examine the way children engage in relationships with organizations.

Recommendations

A number of recommendations can be made based on the findings reported in this chapter and in prior research.

First, the development of content for an organization's Web site is an important part of the relationship-building process. Content should be educational and persuasive in a way that benefits the child. As demonstrated by the study discussed here, messages to children can encourage harmful behaviors if the message creator is not aware of the effect communication can have on children. Organizations that dialogue with children should stay current in their understanding of applicable research, Internet use, and media effects on children.

Second, most of the relationship management strategies proposed by Hon and Grunig (1999) likely will work for child publics; however, organizations need to be creative in their application to this public. Consideration of children's limited cognitive and social abilities should be included in any application of these tactics.

Third, organizations should reconsider their personal information collection practices. As stated earlier, there are specific legal guidelines that organizations in the United States must follow to respect the privacy of children.

Fourth, while most of the content on the Web sites appeared to be for children over the age of seven years old, items like downloadable coloring pages may have been intended for younger children. Organizations that are engaging with very young children need to consider the implications of their behavior on these little ones.

Conclusion

No prior studies have looked at public relations and relationship management between organizations and child members of publics. This study took the first step in that direction by analyzing the content of nutrition Web sites directed toward children. By adapting Hon and Grunig's (1999) recommendations for relationship maintenance, this study looked at four elements of relationship building on the sites.

Analysis found high use of positivity on the Web sites, including interactivity, games, and downloadable files. Future studies should examine whether this concept has significant impact on children's perceptions of an organization.

Finally, organizations that engage in relationship building with children, especially those younger than seven years old, bear the heavy burden of communicating with a disadvantaged public. These organizations should be diligent in their attempts to be truthful and fair in their relationships, as well as keep the best interest of their vulnerable public as a first priority.

For consideration

1. What are some ethical, legal, and practical concerns about conducting relationship management with children?
2. When might it be inappropriate to build or maintain a relationship with children?
3. How might children's experiences with an organization offline influence their relationships online? How might these experiences affect parental relationship building with the organization?
4. What other types of health information might children seek online? What considerations should organizations use when creating content about these topics?
5. In what other ways might the six relationship maintenance concepts (positivity, assurances, networking, sharing of tasks, access, and openness) be used to build relationships online with children?

For reading

Austin, M. J., & Reed, M. L. (1999). Targeting children online: Internet advertising ethics issues. *Journal of Consumer Marketing, 16,* 590–602.

Chandler, M., & Hala, S. (1991). Trust and children's developing theory of mind. In K. J. Rotenberg (Ed.), *Children's interpersonal trust: Sensitivity to lying, deception and promise violation* (pp. 135–159). New York: Springer-Verlag.

Clocksin, B. D., Watson, D. L., & Ransdell, L. (2002). Understanding youth obesity and media use: Implications for future intervention programs. *Quest, 54,* 259–275.

Federal Trade Commission. (1999, November). *How to comply with the children's online privacy protection rule.* Retrieved July 3, 2006, from http://www.ftc.gov/bcp/conline/pubs/buspubs/coppa.htm

Jo, S., & Kim, Y. (2003). The effects of Web characteristics on relationship building. *Journal of Public Relations Research,* 15(3), 199–223.

Note

1. See KidStuff Public Relations at http://www.kidstuffpr.com/ (retrieved August 19, 2006).

References

Barrett, K. C. (2000). The development of self-in-relationship. In R. Mills & S. Duck (Eds.), *The developmental psychology of personal relationships* (pp. 91–108). Chichester, England: John Wiley & Sons, Ltd.

Bugental, D. B., Kopeikin, H., & Lazowski, L. (1991). Children's responses to authentic versus polite smiles. In K. J. Rotenberg (Ed.), *Children's interpersonal trust: Sensitivity to lying, deception and promise violation* (pp. 58–79). New York: Springer-Verlag.

Campbell, K., & D. Crawford. (2001). Family food environments as determinants of preschool-aged children's eating behaviours: Implications for obesity prevention policy. A review. *Australian Journal of Nutrition & Dietetics,* 58(1), 19–26.

Chandler, M. & Hala, S. (1991). Trust and children's developing theory of mind. In K. J. Rotenberg (Ed.), *Children's interpersonal trust: Sensitivity to lying, deception and promise violation* (pp. 135–159). New York: Springer-Verlag.

Clocksin, B. D., Watson, D. L., & Ransdell, L. (2002). Understanding youth obesity and media use: Implications for future intervention programs. *Quest, 54,* 259–275.

Committee on Public Education. (2001). Children, adolescents, and television. *Pediatrics, 107*(2), 423–426.

Federal Trade Commission. (1999, November). *How to comply with the children's online privacy protection rule.* Retrieved July 3, 2006, from http://www.ftc.gov/bcp/conline/pubs/buspubs/coppa.htm

Ferguson, M. A. (1984, August). *Building theory in public relations: Interorganizational relationships.* Paper presented at the annual convention of the Association of Educators in Journalism and Mass Communication, Gainesville, FL.

Grunig, J. E., & Huang, Y. (2000). From organizational effectiveness to relationship indicators: Antecedents of relationships, public relations strategies, and relationship outcomes. In J. A. Ledingham & S. D. Bruning (Eds.), *Public relations as relationship management* (pp. 23–54). Mahwah, NJ: Lawrence Erlbaum Associates.

Hindin, T. J., Contento, I. R., & Gussow, J. D. (2004, February). A media literacy nutrition education curriculum for Head Start parents about the effects of television advertising on their children's food requests: Current research. *Journal of the American Dietetic Association, 104*, 192–198.

Hon, L. C., & Grunig, J. E. (1999). *Measuring relationships in public relations*. Gainesville, FL: Institute of Public Relations.

Kaiser Family Foundation. (2004). *The role of media in childhood obesity*. Retrieved March 10, 2005, from http://www.kff.org/entmedia/entmedia022404pkg.cfm

Kranz, S., Siega-Riz, A. M., & Herring, A. H. (2004). Changes in diet quality in American preschoolers between 1977 and 1998. *American Journal of Public Health, 94*(9), 1525–1530.

Ladd, G. W. (2005). *Children's peer relations and social competence: A century of progress*. New Haven, CT: Yale University Press.

Ledingham, J. A. (2003). Explicating relationship management as a general theory of public relations. *Journal of Public Relations Research, 15*(2), 181–198.

Livingstone, S., & Bober, M. (2004). *UK children go online: Surveying the experiences of young people and their parents*. London: LSE Report. Retrieved August 22, 2006, from http://www.children-go-online.net

Mello, M. M., Rimm, E. B., & Studdert, D. M. (2003). The McLawsuit: The fast-food industry and legal accountability for obesity. *Health Affairs, 22*(6), 207–216.

Papacharissi, Z. (2002). The presentation of self in virtual life: Characteristics of personal home pages. *Journalism and Mass Communication Quarterly, 79*(3), 643–661.

Piaget, J. (1965). *The moral judgment of the child*. New York: Free Press. (Original work published 1932)

Rich, M., & Bar-on, M. (2001). Child health in the information age: Media education of pediatricians. *Pediatrics, 107*(1), 156–162.

Rotenberg, K. J. (1991). The trust-value basis of children's friendship. In K. J. Rotenberg (Ed.), *Children's interpersonal trust: Sensitivity to lying, deception and promise violation* (pp. 160–172). New York: Springer-Verlag.

Schneider, B. H., Smith, A., Poisson, S. E., & Kwan, A. (2000). Connecting children's peer relations with the surrounding cultural context. In R. Mills & S. Duck (Eds.), *The developmental psychology of personal relationships* (pp. 175–198). Chichester, England: John Wiley & Sons, Ltd.

Stafford, L., & Canary, D. J. (1991). Maintenance strategies and romantic relationship type, gender, and relational characteristics. *Journal of Social and Personal Relationships, 8*, 217–242.

Strasburger, V. C. (2001). Children and TV advertising: Nowhere to run, nowhere to hide. *Developmental and Behavioral Pediatrics, 22*, 185–187.

von Salisch, M. (2000). The emotional side of sharing, social support, and conflict negotiation between siblings and between friends. In R. Mills & S. Duck (Eds.), *The developmental psychology of personal relationships* (pp. 49–70). Chichester, England: John Wiley & Sons, Ltd.

Wilson, N., Quigley, R., & Mansoor, O. (1999). Food ads on TV: A health hazard for children? *Australian and New Zealand Journal of Public Health, 23*(6), 647–650.

Yahooligans! (2006). Retrieved November 16, 2004, from http://yahooligans.yahoo.com/Science_and_Nature/Health_and_Safety/Nutrition/ *Note: The site has since moved to http://kids.yahoo.com/directory/Science-and-Nature/Health-and-Safety/Nutrition*

Zuppa, J. A., Morton, H., & Mehta, K. P. (2003). Television food advertising: Counterproductive to children's health? A content analysis using the Australian guide to healthy eating. *Nutrition & Dietetics, 60*, 78–84.

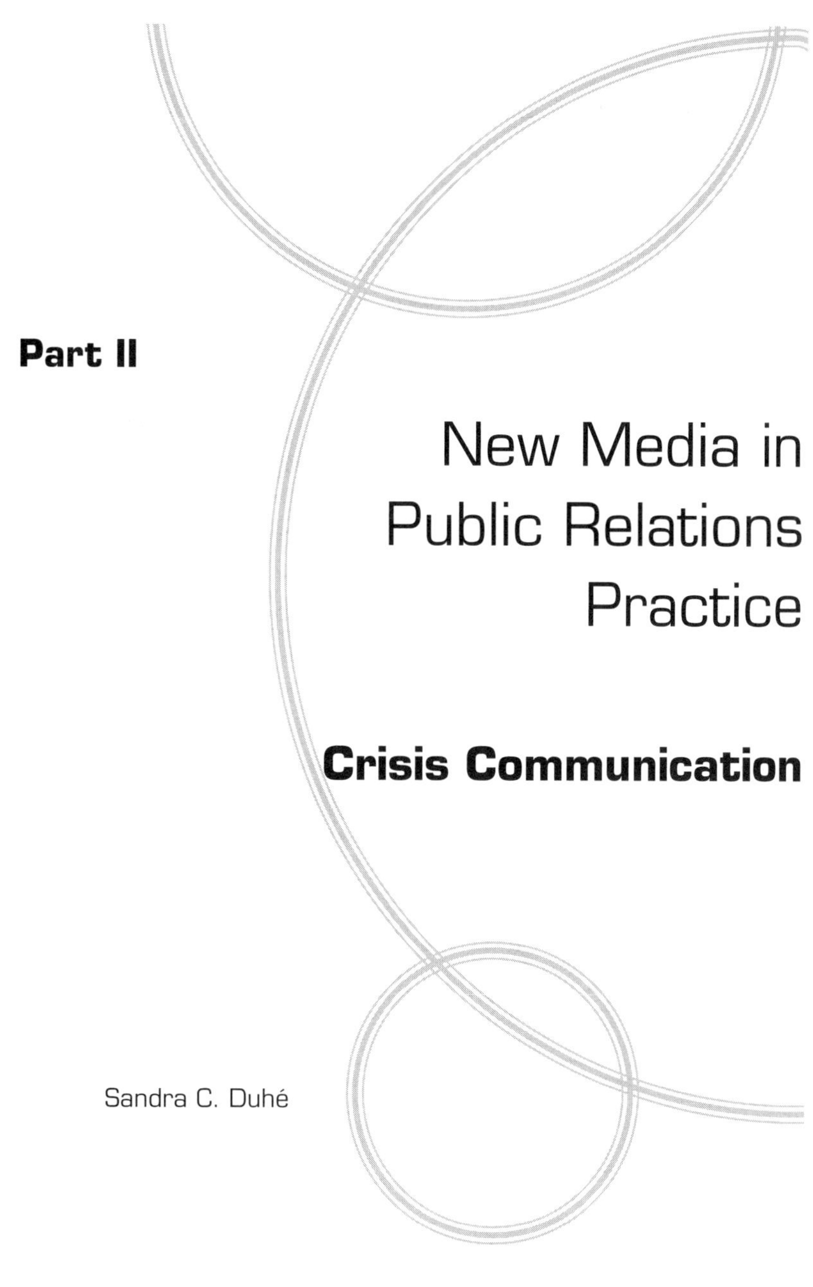

Part II

New Media in Public Relations Practice

Crisis Communication

Sandra C. Duhé

OVERVIEW

Crisis communication is by no means a foreign topic to public relations practitioners and scholars. Crisis responders are pressured to provide thorough, accurate information within minutes of a crisis becoming public. Alongside official statements from the originating site are competing, alternative sources of information being disseminated simultaneously through new media. Although Internet-based channels are attractive for reaching large and dispersed audiences, a natural or man-made disaster can easily render these tools useless. Even if available, audience accessibility to the Internet cannot be assumed.

These final six chapters offer new perspectives and lessons learned in mediated crisis response. We begin with Kenneth Lachlan and Patric Spence's look at both linear and interactive media use during a crisis, along with the roles hazard and public outrage play in the response process. Juliann Scholl, Amy Heuman, Bolanle Olaniran, and David Williams call for the establishment of community-based Crisis Communication Centers to assess and respond to terrorist threats in a way that addresses demographic disparities in access to mediated crisis information. Barbara Gainey highlights the need for organizations to engage in meaningful dialogue with a diverse group of stakeholders to best address the challenges new technologies can bring during a crisis. A big crisis at a small Catholic college serves as the backdrop for Melissa Gibson Hancox (who experienced the crisis firsthand) and Anthony Peyronel's analysis of how a strategic *balance* of channels provides an optimal approach to crisis communication. Liese Hutchison provides an eyewitness account of how the American Red Cross used new and traditional media in the wake of Hurricane Katrina, the largest natural disaster in U.S. history. Lastly, Joe Downing reveals how lessons learned from crisis communication responses to the September 11, 2001, attack and Hurricane Katrina should caution organizations to avoid heavy reliance on new technologies during a crisis.

While the Internet age has ushered in new channels to reach publics during a crisis, it, too, has delivered new technical, timing, and news media challenges that intensify crisis response. Never before have public expectations of response personnel been so demanding. Never before has crisis communication contingency planning been so important.

Hazards, Outrage, and Affected Publics

Crisis Communication in the Internet Age

Kenneth A. Lachlan and Patric R. Spence

Conventional wisdom concerning the relationship between responsible parties and affected publics suggests that crisis messages must be swiftly disseminated through traditional media channels. This allows relationships to be created and managed in the early stages of a crisis, minimizing harm and setting stakeholders at ease in terms of both outrage and perceived hazards associated with the crisis. However, public relations in the Internet age must address interactive media. Affected publics may be prone to seeking additional information and may have access to sources of disinformation, while crisis communication practitioners may have less information control. To this end, we offer a synthesis of recent research in crisis communication, exploring patterns of use and response across linear and interactive media during times of crisis. Using Sandman's model (Sandman, Weinstein, & Hallman, 1998; Sandman, 2003) of risk as a theoretical lens, the chapter explores the interplay of perceived hazard and outrage with both the tendency toward seeking out crisis-related information on the Internet and responses to this content. Lessons learned in this regard from recent crises and implications for crisis communication practitioners are discussed.

Building and maintaining relationships is a central component of public relations. During crisis events, those relationships become important for distributing information, combating misinformation, gaining public trust, and working to reduce the threat, harm, and scope of the crisis. Understanding the needs of the public, and how best to inform the public, will also reduce the harm, threat, and scope of a crisis. Crisis events are becoming more common in frequency and greater in harm, raising concerns about the best ways citizens can obtain information about crises and the connections between the role of public relations and the needs of the public.

Hazard and Outrage in Crisis Communication

At one time in the not-too-distant past, crisis communication was considered only as a post-crisis response. More recently, the role of communication in crisis management has moved away from this perspective and toward a direction in which it is viewed as a continual process, with communication elements that exist before, during, and after crises (Coombs, 1999). As crisis communication has expanded in this direction, the line delineating crisis and risk communication has become somewhat obscured. Take, for example, a flood. In flood-prone locations, risk messages, regarding what steps to take should a flood occur, are continually communicated and reinforced. These risk communication messages originate from multiple locations and, in the precrisis stage, provide tangible behavioral advice. Typical instructions may include preparing a home flood plan and supply kit, eliminating hazards, examining insurance policies, knowing what to do when the water rises, and articulating what to do in the aftermath. If the flood becomes a reality, a more observable transition to crisis communication will take place. Crisis communication messages will then provide information covering the extent of the damage, information about potential threats, information about others that have been affected, and information concerning responses to the crisis with the goal of minimizing the scope, duration, and severity of the crisis.

The role of hazard and outrage. Given the expanding role of crisis communication to include both pre- and postcrisis messages, this information needs to address both the crisis itself and the risks with which it may be associated. Peter Sandman and colleagues (Sandman, Weinstein, & Hallman, 1998; Sandman, 2003) suggest that crisis communication is fundamentally made up of two elements: "scaring people" and "calming people down." Put another way, it is the use of mediated messages for both alerting people to the seriousness of the situation and reassuring them that positive outcomes are imminent. Sandman goes on to suggest that crisis communication should then attempt to steer affected publics toward a level of outrage that is appropriate given the level of hazard. This is loosely expressed as Risk = Hazard + Outrage. The objective of the crisis message is to complete the formula. That is, if an affected public is outraged because there is no clear understanding of the hazard, then messages should educate the public in this regard. If the public understands the hazard, then outrage must be addressed.

However, in most crisis situations, especially exploding crises (Kepplinger, Brosius, Staab, & Linke, 1989), there is little time to compose and refine these messages. This presents the communication practitioner with the challenging task of diagnosing the appropriate degrees of hazard and outrage that must be addressed, crafting the message, and disseminating the message as quickly as possible. As a quick heuristic under these circumstances, hazard can be thought of as the technical seriousness of a risk, while outrage can be thought of as cultural seriousness (Sandman, 2003). The correlation between hazard (how many people are threatened by the risk) and outrage (how many people are upset by the risk) may actually be quite low. Risks may exist that could potentially harm people but fail to upset many and fail to motivate them to take action. For example, a flu outbreak may be harmful, but, as an act of nature, it cannot be attributed to an individual or organization. Thus, under Sandman's model, it would be classified as a high-hazard, low-outrage risk. Other risks may greatly upset people but not pose much of a threat to anyone, leading the public to be fearful and perhaps even take unnecessary precautions (such as the fear that surrounded the Year 2000 (Y2K) computer malfunctions). This constitutes low-hazard, high-outrage risk. In this way, the level of hazard to be addressed may be a product of the degree of outrage. The messages must induce enough fear to motivate the public to take appropriate action yet not induce so much fear as to create panic.

Furthermore, to maximize effectiveness, the messages must meet the public's need for control by outlining the steps that can be taken (outrage) by individuals to reduce their susceptibility to risk (hazard). A message that overemphasizes the public's vulnerability to risk will overproduce fear, reducing the ability of individuals to rationally assess the situation and the appropriate steps to be taken (Aspinwall, 1999). Given these considerations, communication should induce appropriate levels of hazard and outrage and then inform the public of measures they can take to reduce threat. Crisis communication under these circumstances should perform an uncertainty reduction role for the affected public. As such, risk messages *must* then address the concept of outrage in an appropriate manner in order to get and retain the audience's attention.

Applying Sandman's model (Sandman, Weinstein, & Hallman, 1998; Sandman, 2003) of crisis and risk communication may be helpful to crisis communication practitioners by highlighting considerations for appeals to relieve both perceptions of hazards and less specific outrage, anxiety, and grief. A crisis is, however, by its very definition, a nonroutine and low-probability/high-consequence event. Because a crisis is nonroutine, the specific suggestions and lessons that can be learned from previous research may not be appropriate for all crises. The novelty of any specific crisis event may require communication practitioners to think and react in nonprescriptive, incident-specific ways. Therefore, the suggestions offered in this chapter should be viewed as a map from which one might be forced to deviate. Although no two crisis events are the same, lessons learned from previous crises may inform future communication events in ways that meet both the needs of the public and those of first responders, emergency workers, and government agencies.

The role of new media, such as the Internet and e-mail, in times of crisis is largely unknown. As little is known of the intersection between new media and crisis communication, even less has been unearthed concerning the role of new media and capacity of the Internet to meet goals related to hazard and outrage under similar circumstances. The following sections will present an overview of what is known about the role of the Internet in general public relations practice, followed by a brief synopsis of research related to specific crisis events. The chapter concludes with an exploration of the significance of these findings, what might be expected from future affected publics, and suggestions for the best possible use of new media messages in addressing hazard and outrage.

The Internet and Audience Response during Crises

The nature of crisis events inevitably produces a high level of uncertainty in individuals. Weick (1995) posits that these high levels of uncertainty can lead to what he labels a *cosmology episode:* when individuals deeply and suddenly sense that their surroundings no longer constitute a rational, orderly system. Weick believes statements, such as "I've never been here before, I have no idea where I am, and I have no idea who can help me," illustrate a fundamental human reaction to threatening events that create a sense of fear and uncertainty (1993, pp. 634–635). In order to restore reason and stability to their surroundings, people are compelled to seek certainty and predictability. This motivation toward information seeking and uncertainty reduction is an inevitable consequence of crisis (Berlyne, 1960). This drive is especially potent when outcomes associated with the crisis are potentially harmful (Heath & Gay, 1997) or completely uncontrollable (Miller, 1987).

The Internet is obviously a perfect outlet for those seeking information to alleviate their uncertainty. Past research has, however, demonstrated problems with information access and the Internet during times of crisis. For example, on September 11, 2001, many news organizations were not equipped to handle increased Web traffic to their sites; those seeking information on the Internet frequently reported difficulty obtaining access (Jones & Rainie, 2002). CNN.com typically gets about eleven million page views per day; on 9/11, the site was getting nine million hits per hour (Blair, 2002), causing servers to crash and rendering crisis messages unobtainable.

Immediately following a triggering event that signals the beginning of a crisis, the speed at which information is diffused is especially critical. Under these circumstances, the use of new media as an outlet for messages may be especially lucrative, providing the crisis is one that does not inhibit the use of electronic media (such as an accident or natural disaster that could lead to a loss of electricity). For instance, Bracken, Jeffres, Neuendorf, Kopfman, and Moulla (2005) found that audience members responding early to a crisis were those most likely to rely on the Internet and e-mail for information. Typically, these early information seekers then relayed this information to others after obtaining initial crisis information from the Internet. Furthermore, they note that when a crisis erupts during the work day, e-mail and the Internet may emerge as critical information sources due to their availability in the workplace.

Bracken, Jeffres, Neuendorf, Kopfman, and Moulla (2005) do, however, caution that, while new technologies may provide an attractive outlet for obtaining information, affected publics may be less likely to use Internet resources to pass this information along. They argue that, although e-mail is a convenient way of relaying crisis information through interpersonal channels, it may lack the immediacy of a phone call or face-to-face interaction. They argue that, under time-sensitive circumstances, the delayed nature of e-mail exchanges may make it an unattractive option for exchanging crisis information.

Given that the Internet may be most effective as a primary source of crisis information and a less potent resource for interpersonal exchanges, the motives for using the Internet under crisis conditions become an important consideration. A substantive body of research has explored affective and cognitive responses that may be indicative of the motivations for using new media in these circumstances. An understanding of these processes may be informative in assessing what information will be sought by affected publics and, in turn, what impact this may have on audience response.

A long history of research points to the notion that, under crisis conditions, affected publics will turn to the media as a coping mechanism or even as a simple information resource to facilitate interpersonal interactions (Greenberg, 1964). These interpersonal interactions may play a critical role in reducing uncertainty and making people feel better. Crisis events may trigger a wide range of negative emotions, such as sadness, depression, and confusion (Greenberg, Hofschire, & Lachlan, 2002; Seeger, Vennette, Ulmer, & Sellnow, 2002; Spence et al., 2005). Whereas the information component of crisis messages is critical to ensuring appropriate responses by the affected public, it is these basic emotional reactions that may be the primary motivation for information seeking. Understanding these motives and their origins then becomes a key consideration in determining the information that will be sought and any potentially negative responses that may be associated with misinformation or inappropriate treatment of hazard and outrage.

Mood management. Zillmann (1988; 2000) and others argue that the fundamental motivation for media selection, regardless of informational needs, is the management of mood state. In other words, audiences will be motivated to seek out media messages that reduce negative mood states or enhance positive ones (Erber, Wegner, & Therriault, 1996; Knobloch, 2003; Knobloch-Westerwick & Alter, 2006; Zillmann, 1988). Under the extreme duress invoked by some type of crisis, affected publics may be likely to seek out information to alleviate negative mood states, as depression will frequently motivate people toward media consumption (Potts & Sanchez, 1994). Obtaining information that serves to reassure, calm, or alleviate stress may be an especially important function of interactive media under the circumstances.

However, under crisis conditions, it is not guaranteed that mood-based motivations for information seeking will lead affected publics to counteremotional information. The use of media information to manage mood may be more complicated than previously conceptualized (Knobloch, 2003; Knobloch-Westerwick & Alter, 2006). Research suggests that males in a negative mood state may be more likely to seek out bad news than those in a positive mood state (Biswas, Riffe, & Zillmann, 1994) as well as the tendency of female audiences to "enjoy" tragedy (Oliver, 1993; Oliver, Weaver, & Sargent, 2000). Authors go on to claim that perhaps a more complete picture of this process is that under certain socially conscribed circumstances, it is appropriate to experience certain emotional reactions; we then use media as a tool to facilitate these socially appropriate responses, which may, in certain cases, include the facilitation of negative moods (Knobloch, 2003).

Furthermore, there is some evidence that negative mood states may actually improve the ability of a media consumer to process factual information, while this systematic processing may be inhibited by positive mood states (Mitchell, 2000). If this is the case, then any motive toward information seeking—whether to enhance or reduce negative moods—should lead to an increase in the ability of the audience member to understand the information and act in a reasoned manner.

It is here that the induction of outrage is critical in relation to Sandman, Weinstein, and Hallman's (1998) notion of crisis response. Audiences who are outraged to a certain extent or, at the very least, are experiencing a negative emotional reaction to a crisis may be more likely to seek out additional information and understand the directions and implications of the information they receive concerning the crisis. Dutta-Bergman (2005) notes that, in the days following September 11, individuals who reported being depressed, having difficulty sleeping, and having difficulty concentrating were more likely to use the Internet to obtain further information about the attacks and may have been less likely to receive disinformation. Data collected from Hurricane Katrina refugees indicates that those who experienced negative mood states upon hearing of the evacuations were more likely to take remedial steps to protect themselves from the impending crisis (Lachlan & Spence, 2006).

The Internet Crisis Communication Landscape

Given that audiences who are impacted by a crisis, whether at the organizational, community, or national level, will likely experience negative emotions that may motivate their use of new media, it is important to consider what is happening online in terms of crisis messages and availability. A body of research has attempted to address the current state of crisis information on the Internet and the potential of the medium to facilitate these messages.

Of particular concern to researchers in this area is the existence of activist stakeholders who have the potential to exert pressure on organizations or government entities through mediated channels (see Grunig, 1992; Heath, 1997). Some have characterized the traditional relationship between organizations and activist stakeholders as one in which organizations hold the upper hand from a power and resource standpoint (Coombs, 1998; Heath, 1997, 1998).

Of course, the Internet has completely changed this dynamic. Activists no longer have their hands tied in terms of their capacity to exchange information that may be counter to the intended public responses hoped for by organizations and government agencies in times of crisis. Heath notes that

> The Internet and Web have come to be a powerful arena for such discussions which do not allow media reporters, editors, and news directors—or government officials—to be the final power in determining whether issues discussants can have their voices heard. (1998, p. 275)

To this extent, scholars have argued that the Internet, in providing a forum for activist stakeholders, as well as organizations and government, is leading to a leveling of the playing field in terms of the power dynamic between the two sides. Coombs (1998) argues that this leveling of the power dynamic forces crisis practitioners to consider simultaneously the role of activists, the issues management process, and corporate social performance (see Clarkson, 1995), or the broader impact of an organization or business on society.

In this sense, as a resource, the Internet has the capacity to both quell negative public responses and to throw more fuel on the fire, depending on the messages that have been produced by both organizations and activist groups. Whereas organizations and government agencies have traditionally been able to exert information control to a certain extent, this control is greatly limited in the new media environment. Affected publics, who will likely be motivated by both a desire to take appropriate action and a basic need for anxiety reduction, may be prone to seeking out information on the Internet during times of crisis. This presents them with the opportunity to obtain information from a variety of sources, including those that may frame crisis information in a way that is contradictory to the organizations involved.

Furthermore, there is some evidence from research in cognitive consistency and dissonance (see Festinger, 1957; Heider, 1958) that selective exposure processes may drive affected publics to seek information that is cognitively consistent with opinions they already hold. In other words, if a member of an affected public has formed an opinion about the nature of a crisis, those responsible, and what is being done, they may seek out information agreeing with that opinion. Agitated individuals may seek out information supporting the level of outrage and hazard perceptions that they are experiencing, while those that do not hold these opinions may seek information from different channels.

From an information management standpoint, this presents two key issues for crisis communication practitioners when considering the induction of hazard and outrage responses through interactive media. First, given the availability of multiple information sources, the attitudes of affected publics may polarize as individuals seek out cognitively consistent information. Second, from a hazard and outrage standpoint, crisis communication practitioners need to be aware that, for certain members of the affected public, hazard and outrage may fluctuate given the availability of messages that may serve to either reinforce or contradict their cognitive and emotional responses to the crisis.

Hazards, Outrage, and New Media: Implications for Crisis Practitioners

Through careful examination of past research and consideration of audience responses and needs communication, practitioners can identify some predictable patterns that can inform the ways in which crisis messages are crafted and placed using new media. Applying Sandman's model (Sandman, Weinstein, & Hallman, 1998; Sandman, 2003), considerations for appeals to relieve perceptions of hazards and outrage can be helpful to crisis communication practitioners. A brief discussion of the implications of these audience responses for government agencies, officials, municipalities, businesses, and other crisis stakeholders follows.

The Internet is a potential source for many who desire crisis information and one that can provide various degrees of depth of information. As noted, individuals have various motivations for seeking information from new media. Often, individuals first look for information about the scope of the crisis and their susceptibility to harm. For these hazard implications, the Internet can be used to provide information in various depths and through multiple examples to promote the magnitude of the threat. At the same time, Internet messages should provide recommendations about actions that can be taken. Outrage is reduced by providing several actions that people can take; this information can then be distributed through interpersonal networks and e-mail.

The best use of new media for practitioners is to distribute precrisis messages to educate the public about various threats. This, again, is a function of hazard and outrage. The messages can be continually updated to inform individuals about how to make preparations for different types of crises, the likelihood of different occurrences in their geographic area, and the potential magnitude of events. Through providing such information, in the event of a crisis, portions of the public will be better prepared, and the actions taken will have positive social outcomes.

Implications for Relationship Management

Examining how new media are used and how it will evolve causes the crisis practitioner to think about communicating with publics in new ways. First, the distinction between risk and crisis communication needs to be reexamined. That is, they need to be viewed as types of communication with a public that work with each other, rather than as separate forms. This requires that relationships with stakeholders and affected publics begin in the precrisis stage and continue to be maintained regardless of the current conditions. Furthermore, as access to new media expands to larger segments of the population, the use of new media will evolve. As more people have new media available, there will be more opportunities for audience-specific content, requiring further research on the needs and habits of subpopulations. Finally, new media allow for the integration of feedback and communication from the targeted public, which will require continual relationship maintenance between practitioner and public.

For consideration

1. How could crisis communication practitioners go about protecting their organizations from reputational damage that may be the result of online misinformation?

2 Sandman's model posits four types of crisis situations: high hazard–high outrage, high hazard–low outrage, low hazard–high outrage, and low hazard–low outrage. How might computer-mediated messages from crisis practitioners vary across these types of crises? How about messages from activist stakeholders?
3 How can public relations practitioners use new media channels to counter apathy about an issue? That is, how can practitioners effectively create outrage?
4 To what extent is motivation toward cognitively consistent information an important consideration for crisis communication professionals when considering computer-mediated messages?
5 Immediacy and technological unreliability are offered as reasons why the Internet may not be an attractive medium for crisis response messages, despite its ability to provide information tailored to personal goals and needs. What implications does this disadvantage have for crisis response planning purposes?

For reading

Bracken, C. C., Jeffres, L. W., Neuendorf, K. A., Kopfman, J., & Moulla, F. (2005). How cosmopolites react to messages: America under attack. *Communication Research Reports, 22*(1), 47–58.

Coombs, W. T. (1998). The Internet as potential equalizer: New leverage for confronting social irresponsibility. *Public Relations Review, 24*(3), 289–303.

Grunig, L. A. (1992). Activism: How it limits the effectiveness of organizations and how excellent public relations departments respond. In J. E. Grunig (Ed.), *Excellence in public relations and communication management* (pp. 503–530). Hillsdale, NJ: Lawrence Erlbaum Associates.

Heath, R. L. (1998). New communication technologies: An issues management point of view. *Public Relations Review, 24*(3), 273–288.

Jones, S., & Rainie, L. (2002). Internet use and the terror attacks. In B. S. Greenberg (Ed.), *Communication and terrorism* (pp. 27–38). Cresskill, NJ: Hampton Press, Inc.

Seeger, M., Sellnow, T., & Ulmer, R. (2001). Public relations and crisis communication: Organizing and chaos. In R. L. Heath (Ed.), *Handbook of public relations* (pp. 155–166). Sage: Thousand Oaks, CA.

References

Aspinwall, L. G. (1999). Persuasion for the purpose of cancer risk-reduction: Understanding responses to risk communications. *Journal of the National Cancer Institute Monographs, 25*, 88–93.

Berlyne, D. E. (1960). *Conflict, arousal and curiosity*. New York: McGraw-Hill.

Biswas, R., Riffe, D., & Zillmann, D. (1994). Mood influence on the appeal of bad news. *Journalism Quarterly, 71*, 689–696.

Blair, T. (2002, April 2). Internet performs global role, supplementing TV. *USC Annenberg Online Journalism Review*. Retrieved September 21, 2002, from http://www.ojr.org/ojr/technology/1017789404.php

Bracken, C. C., Jeffres, L. W., Neuendorf, K. A., Kopfman, J., & Moulla, F. (2005). How cosmopolites react to messages: America under attack. *Communication Research Reports, 22*(1), 47–58.

Clarkson, M. B. E. (1995). A stakeholder framework for analyzing and evaluating corporate social performance. *Academy of Management Review, 20*, 92–117.

Coombs, W. T. (1998). The Internet as potential equalizer: New leverage for confronting social irresponsibility. *Public Relations Review, 24*(3), 289–303.

Coombs, W. T. (1999). *Ongoing crisis communication: Planning, managing, and responding*. London: Sage.

Dutta-Bergman, M. (2005). Depression and news gathering after September 11: The interplay of affect and cognition. *Communication Research Reports, 22*(1), 7–14.

Erber, R., Wegner, D. M., & Therriault, N. (1996). On being cool and collected: Mood regulation in anticipation of social interaction. *Journal of Personality and Social Psychology, 70*, 757–766.

Festinger, L. (1957). *A theory of cognitive dissonance*. Stanford, CA: Stanford University Press.

Greenberg, B. S. (1964). Diffusion of news of the Kennedy assassination. *Public Opinion Quarterly, 28*, 225–231.

Greenberg, B. S., Hofschire, L., & Lachlan, K. (2002). Diffusion, media use and interpersonal communication behaviors. In B. S. Greenberg (Ed.), *Communication and terrorism* (pp. 3–16). Cresskill, NJ: Hampton Press, Inc.

Grunig, L. A. (1992). Activism: How it limits the effectiveness of organizations and how excellent public relations departments respond. In J. E. Grunig (Ed.), *Excellence in public relations and communication management* (pp. 503–530). Hillsdale, NJ: Lawrence Erlbaum Associates.

Heath, R. L. (1997). *Strategic issues management: Organizations and public policy challenges*. Thousand Oaks, CA: Sage.

Heath, R. L. (1998). New communication technologies: An issues management point of view. *Public Relations Review, 24*(3), 273–288.

Heath, R. L., & Gay, C. D. (1997). Risk communication: Involvement, uncertainty, and control's effect on information scanning and monitoring by expert stakeholders. *Management Communication Quarterly, 10*(3), 342–372.

Heider, F. (1958). *The psychology of interpersonal relations*. New York: John Wiley & Sons.

Jones, S., & Rainie, L. (2002). Internet use and the terror attacks. In B. S. Greenberg (Ed.), *Communication and terrorism* (pp. 27–38). Cresskill, NJ: Hampton Press, Inc.

Kepplinger, H. M., Brosius, H. B., Staab, J., & Linke, G. (1989). Instrumentelle aktualisierung. Grundlagen einer theorie publizistischer konflikte [Instrumental actualization. Fundamentals of a theory of publicistic conflicts]. In M. Kasse & W. Shulz (Eds.), *Massenkommunikation. Theorien, methoden, befunde*. Wiesbaden, Germany: Westdeutscher Verlag.

Knobloch, S. (2003). Mood adjustment via mass communication. *Journal of Communication, 53*, 233–250.

Knobloch-Westerwick, S., & Alter, S. (2006). Mood adjustment to social situations through mass media use: How men ruminate and women dissipate angry moods. *Human Communication Research, 32*(1), 58–73.

Lachlan, K. A., & Spence, P. R. (2006). *The impact of medium choice on perceived crisis message adequacy and hazard and outrage responses during Hurricane Katrina*. Unpublished manuscript, Boston College, MA.

Miller, S. M. (1987). Monitoring and blunting: Validation of a questionnaire to assess styles of information seeking under threat. *Journal of Personality and Social Psychology, 52*, 345–353.

Mitchell, M. (2000). Able but not motivated? The relative effects of happy and sad mood on persuasive message processing. *Communication Monographs, 67*, 215–227.

Oliver, M. B. (1993). Exploring the paradox of the enjoyment of sad films. *Human Communication Research, 19*, 315–342.

Oliver, M. B., Weaver, J. B., & Sargent, S. (2000). An examination of factors related to sex differences in the enjoyment of sad films. *Journal of Broadcasting and Electronic Media, 44*, 282–300.

Potts, R., & Sanchez, D. (1994). Television viewing and depression: No news is good news. *Journal of Broadcasting & Electronic Media, 38*(1), 79–90.

Sandman, P. (2003, April). Four kinds of risk communication. *The Synergist*, pp. 26–27.

Sandman, P., Weinstein, N. D., & Hallman, W. K. (1998). Communications to reduce risk underestimation and overestimation. *Risk Decision and Policy, 3*(2), 93–108.

Seeger, M. W., Vennette, S., Ulmer, R. R., & Sellnow, T. L. (2002). Media use, information seeking and reported needs in post crisis contexts. In B. S. Greenberg (Ed.), *Communication and terrorism* (pp. 53–63). Cresskill, NJ: Hampton Press, Inc.

Spence, P. R., Westerman, D. K., Skalski, P. D., Seeger, M., Ulmer, R. R., Venette, S., & Sellnow, T. L. (2005). Proxemic effects on information seeking after the September 11 attacks. *Communication Research Reports, 22*(1), 39–46.

Weick, K. (1993). The collapse of sensemaking in organizations: The Mann Gulch disaster. *Administrative Science Quarterly, 38*, 628–652.

Weick, K. (1995). *Sensemaking in organizations*. Thousand Oaks, CA: Sage.

Zillmann, D. (1988). Mood management through communication choices. *American Behavioral Scientist, 31*, 327–340.

Zillmann, D. (2000). Mood management in the context of selective exposure theory. *Communication Yearbook, 23*, 103–123.

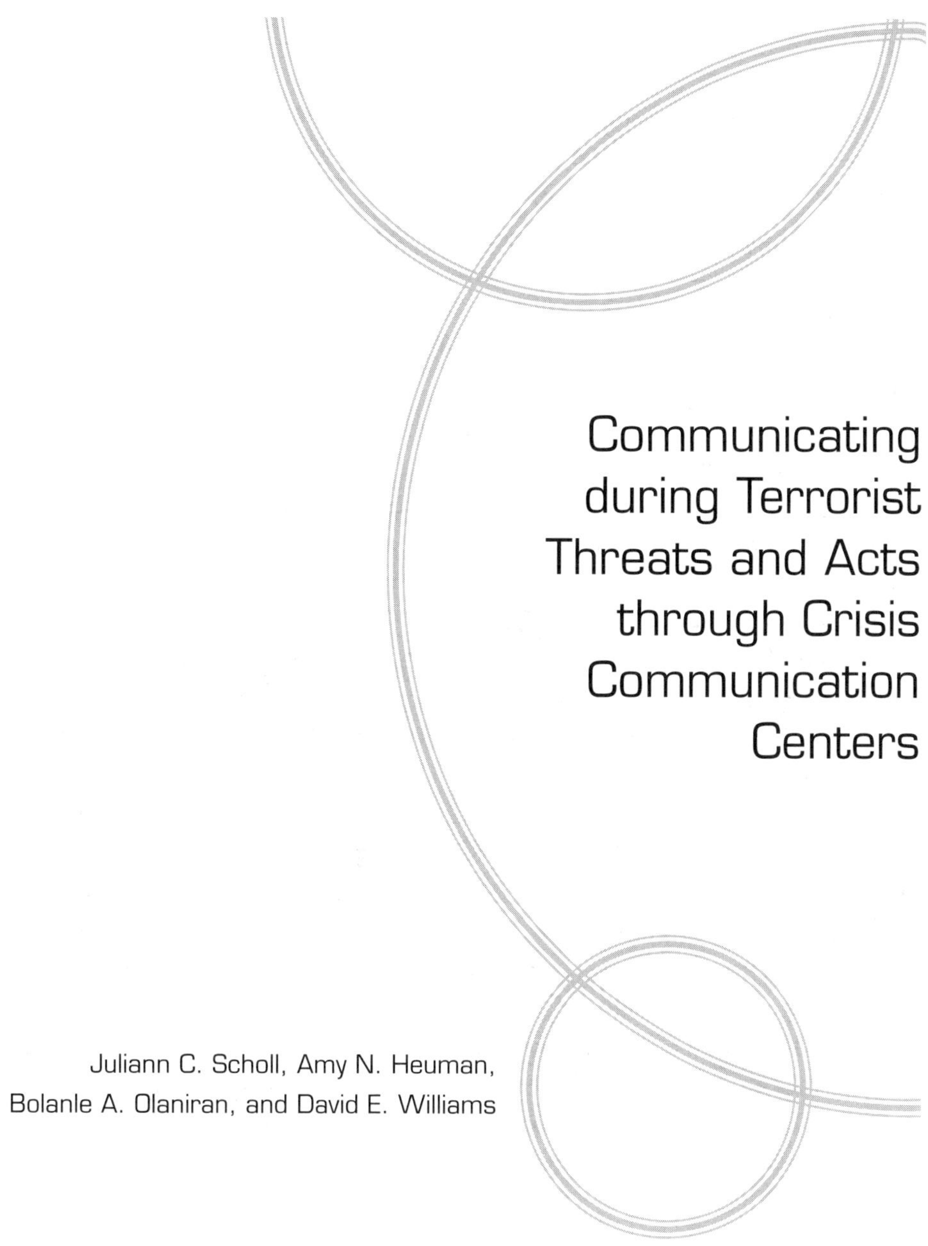

Communicating during Terrorist Threats and Acts through Crisis Communication Centers

Juliann C. Scholl, Amy N. Heuman,
Bolanle A. Olaniran, and David E. Williams

Crisis communication changed forever on September 11, 2001. The need to interface with the public regarding issues of health and safety was never more prominent than during the moments following the terrorist acts that took place that day. This chapter will introduce a mechanism for enhancing mediated public relations as it pertains to crisis communication regarding terrorist threats and acts.

Given the need for community-wide access to information regarding terrorist threats and acts, government agencies at the local and state levels should consider development of Crisis Communication Centers (CCCs). Although the United States Department of Homeland Security and other agencies hold a macroscopic view of national security, CCCs would be developed to serve the unique terrorist risk and public communication needs of the local public.

CCCs are locally planned and developed agencies that would be staffed by needs according to the size of the community they represent. These centers would be locally governed but connected to the Department of Homeland Security. The purpose of such centers would be to monitor regularly the community for vulnerability to terrorist threats and acts and facilitate communication with the public regarding those concerns. While CCCs would be local government agencies, their success would depend greatly on local community and industry support and the availability of some media, communication, and public relations expertise.

CCCs are based on the crisis response principle of anticipation (Scholl, Williams, & Olaniran, 2005). The anticipatory model of crisis management (Olaniran & Williams, 2000) holds that crisis managers optimize their success in responding to crisis when they proactively anticipate the possibility of crisis. This process involves a detailed analysis of all factors related to a particular organization (e.g., environmental factors, technological risks, and financial vulnerabilities). The anticipatory model encourages public relations and crisis management professionals to engage in a systematic review of all crisis concerns, develop response scenarios, and maintain relationships with key stakeholders who will assist the organization in responding to a crisis. Crisis managers using the anticipatory model must work to maintain a thorough understanding of all factors related to the organization that might signal the onset of a crisis. This understanding suggests a thorough familiarity with elements of the organization and how the organization interacts with its stakeholders and environment. The greater the level of understanding, the greater control crisis managers can employ in responding to a crisis.

For CCCs, anticipation is key in signal detection of terrorist threats and acts. CCCs would be locally developed and maintained units that function to assess terrorist vulnerabilities and the optimal means of communicating terrorist threats and acts to their specific communities. CCCs would be staffed by community members and municipal employees who have specific knowledge and expertise within the community. CCC staff would be able to share their expertise about the community in determining what terrorists would most likely target. For example, in a farming community, the CCC would need input from agricultural experts in the area who could assess how terrorists might attack their industry. Risks, such as destroying crops in order to attack the economic base of the community, tampering with livestock to spread illness, or stealing crop-dusting planes in order to use them to spread dangerous chemicals, would all be possible vulnerabilities that the local agricultural experts could address.

Members of health professions would be key representatives on the CCC staff and stakeholders with whom the CCC staff would maintain contact. As most potential terrorist threats ultimately involve a concern for health and safety, these stakeholders must be aware of their role in crisis response and how to manage their communication with the public. Their roles would move beyond that of treating the aftermath of a terrorist attack. They would also play a critical role in information dissemination and community awareness. As they will be consulted for input, the CCC would need to work closely with members of health professions to create messages that both alert the public to health concerns and help alleviate concerns so that public panic does not escalate unnecessarily. Public safety concern for communication from health professionals reaches beyond the hospital spokesperson to individuals working with those on the site of an attack in the hospital.

Upon appointing a CCC staff adequate to access the terrorist risk in their community, the team would begin the process of community assessment. Guidelines from the Department of Homeland Security would provide a starting point for this process as the team identifies targets, risks, and vulnerabilities. However, team members' local expertise and connections in the community will provide the most valuable insight. The CCC would develop and maintain a list of areas of concern for the community and would maintain relationships with key individuals who can alert the CCC to changes or activities that might signal a terrorist threat. For example, early insight from the community hospital of otherwise unspectacular, but unexplained, injuries due to minor chemical burns or illness from an unknown vapor could signal the onset of a terrorist act that would otherwise go undetected.

Whereas public relations crisis management personnel are advised to have regular crisis response training, CCC staff should also have regular reviews of the vulnerabilities and periodic communication with their contacts within the community. The development of a community assessment instrument has been suggested elsewhere as a means for facilitating this process (Scholl, Williams, & Olaniran, 2005). This instrument would include a generic inventory of known terrorist targets and then would include a list of local sites of interest.

Communication with the public is one of the most important tasks for the CCC staff. The CCC must select a member who could fulfill the role of spokesperson. This is a critical decision, as the person must connect with the broad spectrum of citizens represented in the community. One consideration is to select a recognized and respected member of the health profession. This individual could have a broad range of appeal to the community and would be able to speak with credibility on the health and safety concerns of the community during a crisis. Prominent members of the local government, education system, or the military might also serve in this capacity with similar benefits of community appeal.

Selection of a medium for communicating with the public is another vital role of the CCC. Qualified staff members should survey the community to answer some basic questions about their information-gathering preferences. Foremost among concerns over information dissemination is who should present the information, what types of media community members prefer, and whether there is an individual (other than the CCC) spokesperson to whom the community would be more attentive. Is there a person whom the community would prefer to deliver news about terrorist threats or acts in their community? The CCC must also determine what level of potential threat warrants public communication. Answers to these questions can vary greatly, thus highlighting the need for locally developed and operated CCCs.

In some communities, a local news anchor might be the preferred source for information, while, in other areas, a prominent member of the police department or a member of the clergy would be preferred. A survey of the community will reveal some of these answers. Such a survey might also give insights into the need for non-English spoken or written information.

Technological elements of public communication will be another component of the CCCs' work. Use of the Internet to host a Web site, send electronic mail, and host bulletin boards will be central to the work of information dissemination. The primary purpose of such Web sites would be to facilitate communication with the public, by both receiving and sending information. For example, local CCCs could sponsor Web sites that would allow citizens to respond to inquiries regarding from whom they would like to receive crisis information and how they would like to receive those messages. This could prove useful in the early stages of the development of each CCC as it works to determine how to present information to the public and who should present it.

During times of crisis or uncertainty, electronic communication could be used to alert the public to what is happening, if anything; what the public should do; and what the perceived risks might be. Access to this information would be an issue of concern for each local team. Electronic communication could also be used for internal purposes by members of the team.

Public relations and crisis management professionals can enter a new realm of service to the public with the creation of local CCCs. However, coordinating the communicative elements to serve the diversity of interests and abilities represented in each local community would take extensive work on the part of CCC teams. Particular attention must be given to the role of mediated public relations as it interfaces with health professionals; members of society who have less influence, input, and access to public information; and those who understand the complexities of technological issues in computer-mediated communication.

The Role of Health Information in Crisis

The interchange between behavioral, biological, and societal influences on health and behavior ultimately shapes individuals' health outcomes. In particular, the emergence of a crisis (e.g., a terrorist threat or attack or a natural disaster) within a community can have large-scale and long-term influences on community members' physical and psychological health. Community-wide crises have informational impacts, as they disrupt "communication practices at the individual, relational, and societal levels, which are closely related to . . . health" (Becker & Thompson, 2005, p. 56). Public health campaigns must utilize the most effective strategies for enhancing health behaviors, beliefs, and outcomes among targeted communities, especially in the wake of far-reaching crises. This understanding can provide insight to health professionals regarding the construction of successful intervention campaigns that improve health through behavioral modification.

Community-wide crises bring about many unnoticed effects on psychosocial and physical health. For example, when individuals experience trauma, they feel disconnected from their world and have difficulty making sense of it (Niederhoffer & Pennebaker, 2002). Many people affected by crises have otherwise seemingly normal lives, marriages, and jobs. Unfortunately, it is only after the trauma that many people elect to seek help, even after their jobs, health, and relationships fall apart (Sheely, 2003). After terrorist attacks, individuals may come to the hospital presenting unexplained physical symptoms, which are often exacerbated by psychological stress (Taintor, 2003). Researchers have yet to explore "the less visible—yet important—impacts of these attacks on victims' interpersonal relationships and psychosocial health" (Becker & Thompson, 2005, p. 47).

The availability of various media outlets allows community members to obtain health information quickly as well as engage in public discourse about certain events influencing all aspects of their health. Of all media available, the Internet has probably most significantly altered the way decision makers and public relations experts can create communities through the dissemination of important health information (Leaffer & Mickelberg, 2006). However, public health and public relations leaders need to ensure that technological advances in science and health do not contribute to the unbalanced rise of poorer health outcomes for selected American communities.

Despite their unintended and sometimes negative effects, media can play an important role in how individuals obtain reliable and appropriate health information, especially that which can aid in coping with crises. By understanding how individuals typically respond to crises, community leaders and public relations experts may take advantage of these tendencies to make their health campaigns more accessible, believable, and effective. For instance, public relations experts may use various media forms to open more dialogue among community members. Mehl and Pennebaker (2003) found that people seemed more motivated to communicate with others, especially through the telephone, after the September 11 terrorist attacks. As time went on, people showed more tendencies to engage in face-to-face communication. When community members are in open public dialogue with each other, they have the opportunity to help propagate important health information that may originate from mediated sources.

Public relations personnel need to be mindful of the types of images they incorporate into their mediated messages. For example, frequent viewing of disturbing images has been associated with posttraumatic stress disorder and depression (Ahern et al., 2002; Saylor, Cowart, Lipovsky, Jackson, & Finch, 2003). Although mediated messages can reach wide audiences and can be an efficient way to disseminate vast amounts of information, excessive viewing of disturbing images can thwart positive health outcomes.

Becker and Thompson (2005) suggest that social support coupled with viewing of such images will help individuals cope with a tragedy. Social support is central to the coping and healing process (Kulic, 2003), and talking about traumas and devastating events may reduce and improve health (Becker & Thompson). Self-disclosure about traumatic events has been associated with decreases in doctor visits, increased immune functioning, and positive behavioral changes (Pennebaker, 1997). However, some interpersonal communication about the trauma may only enable victims to relive it, thereby doing more harm than good. Also, through emotional contagion, victims or relief workers who talk about their distress or anger may inadvertently spread those negative feelings to others around them (Kulic, 2003). Those who utilize media are encouraged to minimize these negative impacts and to disseminate messages that provide access to social support networks and groups.

One cannot talk about mediated messages without acknowledging cultural- and community-based disparities in access to mediated health information. "While the trends toward increased consumer use of online health information are truly encouraging, they also reveal several key demographic differences in consumer usage of the Internet" (Leaffer & Mickelberg, 2006, p. 54). In turn, these discrepancies can affect who obtains the greatest benefit from relevant health information. Disseminators of mediated health messages should learn who is using the Internet and who has greater access to it. The University of California, Los Angeles, Digital Future Report (as cited in Leaffer & Mickelberg) reported that those who have greater access to mediated health information (e.g., via the Internet) report good or excellent health. Also, younger, employed, white, suburban and urban Americans are more likely to go online for medical information, compared to other demographic groups.

Experts in health campaigns and public relations should find ways to reach those demographic groups who have limited access to medical information and simultaneously report lower levels of health. Such interventions may include training targeted populations in the use of publicly accessible information as well as being willing and able to honor the time commitments required to make such projects successful (Ruffin, Cogdill, Kutty, & Hudson-Ochillo, 2005). Leaffer and Mickelberg (2006) further argue that the Internet can serve as an arena in which people from different political ideologies, cultural backgrounds, and educational levels can bond as they share personal experiences and anecdotal advice.

Speaking to Unspoken-to Populations during a Crisis

Whereas the Internet may provide a forum for diverse discussions and rich information, it becomes difficult for certain segments of the population to engage this medium when there is a lack of access, a desire for an alternative medium, or illiteracy in technology usage. In developing a crisis plan, CCCs need to be aware of the potential for certain communities to be ignored, overlooked, or not spoken to during times of crisis. Given the events of Hurricane Katrina in 2005 and the lack of a timely, well-executed response, CCCs need to take measures to better plan for ways to get the message out to impoverished, aging, and rural populations who may or may not have access to forms of technology, such as the Internet, or even telephones and television.

Scholars argue that the inability to access and use technology is leading to a new form of social exclusion (Facer & Furlong, 2001). This "information poverty" (Chatman, 1996; Norris, 2001) or what some have called the divide between the "haves" and the "have-nots" is most popularly referred to as the *digital divide*. Information have-nots most often comprise underrepresented populations and are often linked to those from nondominant racial and ethnic groups, lower economic classes, and the aging population (Keefe, 2005; Martin & Nakayama, 2006). Information poverty excludes certain sectors of society from information loops. What results is a lack of access to information technology designed to connect those in need to jobs, housing, banking, social networks, and information in times of crises (Katz & Aspden, 1997).

In terms of computer usage and Internet connectivity, 51% of all U.S. households (53.7 million households) own a personal computer; 41.5% have Internet access (U.S. Department of Commerce, 2001). Yet, of these, only 23.5% are African American households, and 23.6% are Latino households. Along class lines, only 21% of households earning less than $30,000 per year have high-speed Internet access (Keefe, 2005). Each of these statistics points to a need to pay attention to underrepresented populations in times of crisis.

Equally important, but often unmentioned, are the voices of disenfranchised groups, such as Native American tribal communities, the homeless, and non-English speakers. Within these cultures, affordability might be one concern, but access to basic technology media (e.g., television and telephones), language, and Internet connectivity are others. According to the 1990 U.S. census report, "just 53% of Native households on reservation and trust lands have telephones, compared to 94% for all American homes" (as cited in Dorr & Akeroyd, 2001, p. 38). In terms of access to higher forms of technology, only 15% of all Native American households have a computer, and only 10% have Internet access.

Homeless populations also need to be considered when developing a plan for a crisis response. In the United States, approximately two million adults are homeless (Hershberger, 2003). Hershberger goes on to state that the two main causes of homelessness are poverty and the lack of affordable housing. Other factors are a lack of affordable healthcare, job loss, domestic violence, mental illness, and addictions. No singular reason fully explains why persons become homeless. Rather, a combination of reasons, or what is referred to as "multi-crisis" homelessness, is what leads to their condition (*National Coalition for the Homeless*, n.d.). Assisting homeless persons in times of crisis is a major concern. Ensuring that CCCs include the homeless populations in their plans of action is vital. Planning in advance for how CCC team members will provide access to technology and the dissemination of information will be a complex process requiring much time and forethought.

Those who speak English as a second language and those who are non-English speakers in the United States also need to be given attention. Graddol (1997) notes that, as recently as 1996, over 80% of Web pages still privileged the English language. Although Warschauer (2000) and Graddol (1997) suggest the potential for a substantial shift in this percentage, we also must be aware of the importance of utilizing multiple languages in times of crisis. For a case in point, we can refer to the growing Latino population and note that, for some, a Spanish language–based identity remains important, while, for others, an immersion in English language is desired as a vehicle to American middle-class life (Warschauer, 2000). Language as a link to identity cannot be understated as a cultural value that deserves consideration.

A positive trend for technology usage can be found in research that suggests that the homeless, Native American, African American, and Latino communities are using library facilities and services more than other groups as a means of getting in touch with technology (Dorr & Akeroyd, 2001; Hershberger, 2003; Oder, 2002). This provides opportunities for disseminating information in times of crisis via public library outlets. Mossberger, Tolbert, and Stansbury (2003) find that, in addition to libraries, community members are seeking out public schools, recreational facilities, churches, and even senior centers as a means of gaining access. CCCs might consider these resources when suggesting a plan of action in the event of a major crisis situation.

Programs, such as the Gates Foundation's (2006) Native American Access to Technology Program (NAATP) and communication connection programs for the homeless being sponsored by nonprofit organizations across the country, along with Internet technology providers working to ensure a surge in bilingual and multilingual Web sites, can provide needed access to public information and crisis communication messages. Using computer literacy education and training for various forms of technology and connecting the homeless to voicemail services can be considered as part of a preventative plan for managing crisis (Dorr & Akeroyd, 2001; Hershberger 2003; Taglang, 2001). Finding ways to implement such programs into a CCC mission might also be beneficial in long-term planning.

Importantly, relating to the community and recognizing the cultural values of community members is essential in reaching those who often go unspoken to in times of crisis. The selection of team members and representatives who are well-respected and valued within a community is necessary for a local CCC to be effective. Spending time communicating with community members will afford insights into the communication channels or mediums they prefer as a means of disseminating critical information. If barriers to access (i.e., Internet, telephones, and language) are of concern, or if communities prefer verbal communication to mediated communication, then the CCCs' plans should reflect these preferences. Of main concern is getting community members involved in the dialogue and planning for times of crisis, so that their voices are included in the efforts to prepare for crisis.

Concerns with Technological Communication during a Crisis

CCCs can benefit greatly by incorporating technology-oriented media in their plans. However, a broader understanding of the benefits and pitfalls of technology-based mediated communication is essential. Communication technologies offer tremendous opportunity to address concerns of victims, employees, and other stakeholders during crisis management.

With more organizations realizing that they are no longer in control, they are forced to explore communication technologies in public relations campaigns and conflict management. This lack of control brought about by technology-mediated channels emphasizes the need for organizations to take a proactive rather than a reactionary stance when it comes to issues of control in crisis management (Olaniran & Williams, 2001; Williams & Olaniran, 1994). The proactive stance involves anticipating implications of consequences of certain operating policies and behaviors and putting in place systems to address or minimize negative consequences. Technology-mediated communication channels offer a way to achieve greater control of information dissemination in crisis management.

Implications for Relationship Management

The advent of CCCs suggests that public relations professionals must invoke social scientific methods of research in determining the best means to address their publics. Long before crisis messages can be created, CCC staff members must conduct original research in their community to determine how to speak to the various segments of the public, who should speak to them, and how best to reach all who need to know about terrorist threats and acts.

New media use by public relations professionals is being driven by fear and anger in the American public as people demand a return to a feeling of safety and security. This sentiment is a strong force in the political process at the national level. The advent of CCCs would further drive the sentiment into local communities and give citizens of those communities that develop CCCs a sense of being proactive in the quest to return to a secure homeland.

The development of CCCs can help foster a sense of togetherness in communities where that togetherness may not exist. CCCs unite a community in an attempt to share important information in preparing for the possibility of terrorist threats and in responding in the case of an actual terrorist act. As CCCs utilize new media to disseminate information, members of the same community who previously had nothing in common will now share a common source for critical information.

For consideration

1. What criteria could be used to determine who would be a credible spokesperson(s) for a community in times of crisis?
2. What specific actions could a public relations practitioner take to help a CCC identify the different cultures in a community?
3. What challenges might exist in how a public relations practitioner would need to communicate information about a terrorist threat to various cultural groups? What role could new media and interpersonal contact play in effectively communicating these challenges?
4. To what extent does the potential for a terrorist threat change the crisis planning and preparedness process?
5. What can be done, from a public relations perspective, to counter community apathy about establishing a CCC?

For reading

Holtz, S. (2002). *Public relations on the Net* (2nd ed.). New York: AMACOM.

Horton, J. L. (2001). *Online public relations: A handbook for practitioners*. Westport, CT: Quorum.

Middleberg, D. (2000). *Winning PR in the wired world: Powerful communication strategies for the noisy digital space*. New York: McGraw-Hill.

Scholl, J. C., Williams, D. E., & Olaniran, B. A. (2005). Preparing for terrorism: A rationale for the crisis communication center. In H. D. O'Hair, R. L. Heath, & G. R. Ledlow (Eds.), *Community preparedness and response to terrorism: Vol. 3* (pp. 243–268). Westport, CT: Praeger.

Tuman, J. S. (2003). Communicating terror: The rhetorical dimensions of terrorism. Thousand Oaks, CA: Sage.

References

Ahern, J., Galea, S., Resnick, H., Kilpatrick, D., Bucuvalas, M., Gold, J., & Vlahov, D. (2002, Winter). Television images and psychological symptoms after the September 11 terrorist attacks. *Psychiatry:* Interpersonal & Biological Processes, 65(4), 289–300.

Becker, J. A. H., & Thompson, S. (2005). The intersection of terrorism, interpersonal communication, and health. In J. A. Johnson, G. R. Ledlow, & M. A. Cwiek (Eds.), *Community preparedness and response to terrorism [Volume III]: Communication and the media* (pp. 47–64). Westport, CT: Greenwood Publishing Group.

Chatman, E. (1996). The impoverished life world of outsiders. *Journal of the American Society for Information Science*, *47*(3), 193–206.

Dorr, J. & Akeroyd, R. (2001). Bridging the digital divide. *Computers in Libraries*, *10*, 37–42.

Facer, K., & Furlong, R. (2001). Beyond the myth of the "Cyberkid": Young people at the margins of the Information revolution. *Journal of Youth Studies*, 4(4), 451–469.

Gates Foundation. (2006). *Native American access to technology (1999–2003)*. Retrieved August 19, 2006, from http://www.gatesfoundation.org/

Graddol, D. (1997). *The future of English*. London: The British Council.

Hersberger, J. (2003). Are the economically poor information poor? Does the digital divide affect the homeless and access to information? *The Canadian Journal of Information and Library Science*, *27*(3), 46–63.

Katz, J., & Aspden, P. (1997). Motivations for and barriers to Internet usage: Results of a national public opinion survey. *Internet Research*, *7*, 170–188.

Keefe, B. (2005, October 6). Survey finds 1 in 5 Americans say they've never used the Web. *Cox News Service*. Retrieved August 19, 2006, from http://www.daytondailynews.com/search/content/shared/news/nation/stories/10/06_COXINTERNET_STUDY.html

Kulic, K. R. (2003). An account of group work with family members of 9/11. *J Spec Group Work*, *28*, 195–198.

Leaffer, T., & Mickelberg, L. (2006, May/June). The digital health care revolution: Empower health consumers. *The Futurist*, 53–57.

Martin, J. N., & Nakayama, T. K. (2006). *Intercultural communication in contexts* (4th ed.). Boston, MA: McGraw-Hill.

Mehl, M. R., & Pennebaker, J. W. (2003). The social dynamics of a cultural upheaval: Social interactions surrounding September 11, 2001. *Psychological Science*, *14*, 579–585.

Mossberger, K., Tolbert, C. J., & Stansbury, M. (2003). *Virtual inequity: Beyond the digital divide*. Washington, DC: Georgetown University Press.

National Coalition for the Homeless. (n.d.). Retrieved June 11, 2006, from http://www.nationalhomeless.org

Niederhoffer, K. G., & Pennebaker, J. W. (2002). Sharing one's story: On the benefits of writing or talking about emotional experience. In C. R. Snyder & S. J. Lopez (Eds.), *Handbook of positive psychology* (pp. 573–583). London: Oxford University.

Norris, P. (2001). *Digital divide: Civic engagement, information poverty, and the Internet worldwide*. Cambridge, UK: Cambridge University Press.

Oder, N. (2002). Benton: Feds retreat from fighting the digital divide. *Library Journal, 127*(5), 16.

Olaniran, B. A., & Williams, D. E. (2000). Anticipatory model of crisis management: A vigilant response to technological crises. In R. L. Heath (Ed.), *Handbook of public relations* (pp. 581–594). Newbury Park, CA: Sage.

Pennebaker, J. W. (1997). Writing about emotional experiences as a therapeutic process. *Psychological Science*, 8, 162–166.

Ruffin, A. B., Cogdill, K., Kutty, L., & Hudson-Ochillo, M. (2005). Access to electronic health information for the public: Analysis of fifty-three funded projects. *Library Trends*, 53(3), 434–453.

Saylor, C. F., Cowart, B. L., Lipovsky, J. A., Jackson, C., & Finch, A. J., Jr. (2003). Media exposure to September 11. *American Behavioral Science*, 46, 1622–1642.

Scholl, J. C., Williams, D. E., & Olaniran, B. A. (2005). Preparing for terrorism: A rationale for the crisis communication center. In H. D. O'Hair, R. L. Heath, & G. R. Ledlow (Eds.), *Community preparedness and response to terrorism: Vol. 3* (pp. 243–268). Westport, CT: Praeger.

Sheely, G. (2003). *Middletown America: One town's passage from trauma to hope*. New York: Random House.

Taglang, K. (2001, December 10). A low-tech, low cost tool for the homeless. *Digital Divide Network*. Retrieved August 19, 2006, from http://www.digitaldivide.net/articles/view.php?ArticleID=152

Taintor, Z. (2003). Assessing mental health needs. In B. Levy & V. Sidel (Eds.), *Terrorism and public health: A balanced approach to strengthening systems and protecting people* (pp. 49–68). Oxford: Oxford University Press.

U.S. Department of Commerce. (2001). *A nation online*. Retrieved June 10, 2006, from http://www.ntia.doc.gov/ntiahome.dn/

Warschauer, M. (2000). Language, identity, and the Internet. In B. E. Kolko, L. Nakamura, & G. B. Rodman (Eds.), *Race in cyberspace* (pp. 151–170). New York: Routledge.

Williams, D.E. & Olaniran, B.A. (1994). Exxon's crisis decision making flaws: The hypervigilant response to the Valdez grounding. *Public Relations Review*, 20, 5–18.

Crisis Communication Evolves in Response to New Public Engagement Constructs in Educational Settings and Beyond

Barbara S. Gainey

This chapter addresses organizational relationship management and new media in the context of crisis management and crisis communication. The chapter examines the constructs of strategic public engagement (through such channels as advisory councils, focus groups, and town hall–type meetings) and stakeholder networks, changes in demographics of organizational publics, crisis communication challenges posed by new constructs of public engagement, and the impact of 24/7 media coverage and new media on crisis management. The chapter explores relevant theoretical perspectives, evolving communication models, and relevant case studies. Organizations that can dig deeper for opportunities to have meaningful dialogue about topics of concern to the organization and key publics position themselves to strengthen relationships in a precrisis environment. Engaging in meaningful communication with a broader, more inclusive, stakeholder audience has implications for stronger organizations, stronger communities, and more effective crisis management.

Effective Crisis Management Depends on Relationship Building

To paraphrase a well-known expression, "Crisis makes for strange bedfellows." Indeed, the traumatic events of recent years—from the Columbine High School shootings in Colorado; to the September 11, 2001, terrorist attacks; to more recent terrorist alerts, warnings, and incidents around the globe—have spurred unexpected connections of individuals and organizations to confront common threats. *The Atlanta Journal-Constitution* took note of a new "Security R Us" approach to managing crises through the involvement of unusual security partners, including clam diggers working in the shadow of Boston's Logan International Airport and private-sector truck drivers reporting suspicious activities, all in the name of homeland security (Malone, 2005, A1).

As Kruckeberg and Starck observed, "The sense of community that existed a century ago is no longer common" (1988, p. 37). There is a new sense of urgency to connect with and meaningfully engage diverse and perhaps disconnected publics to meet the demands of a seemingly more fragile existence in communities small and large. With the new bonds of electronic communication, or new media, organizations are realizing that, to survive and flourish, building human communication channels around the globe is increasingly important.

Organizations also must come to terms with the realities and pressures of a 24/7 news cycle, citizens who are instantaneously on top of the latest breaking story through electronic wonders on their desks or in their hands, and the sheer size and complexities of a global society. As public school district leaders discovered after the traumatic shootings at Columbine High School in April 2001, "Everything has the potential for going national" (Nora Carr, as cited in Cook, 2001, p. 16). Around-the-clock news organizations, such as CNN, the rapid development of the Internet as a source of news, and competition for breaking news work together to create a media frenzy at the site of any crisis.

Such was the experience of Rick Kaufman, Jefferson County Public School District's executive director of public engagement and communications in Golden, Colorado, when two shooters entered Columbine High School in Littleton, Colorado. Kaufman said he was "floored by the speed with which the media was able to create this mini, thriving operation" (Cook, 2001, p. 17). In the first twenty-four hours of the crisis, the school district logged 1,500 telephone calls from the media, in addition to a media encampment at the school site. Inquiries continued for at least fourteen months.

The prospect of encountering a crisis that could threaten the success of an organization *and* capture the attention of a mass audience has never been more real. This chapter will examine the problems and opportunities of organizations as they attempt to engage diverse publics and respond to the news media in a crisis-rich environment.

Organizations that can dig deeper for opportunities to have meaningful dialogue about topics of concern to the organization *and* key stakeholders position themselves to strengthen relationships before a crisis happens. Engaging in meaningful communication with a broader, more inclusive, stakeholder audience has implications for stronger organizations, stronger communities, and more effective crisis management.

Developing Relationships through Strategic Public Engagement

In the first Extra-Strength Tylenol product-tampering case more than twenty years ago, Johnson & Johnson set the standard for effective crisis management and safeguarding good relationships with key stakeholders—employees, customers, the media, and the greater community. Developing relationships between organizations and their publics has long been recognized as central to the public relations function in an organization. Edward Bernays, a pioneer in the practice, wrote,

> Improving public relations for an individual or an institution is not a matter of using this or that tool or technique to bring about the desired effect. *The total person or institution needs to be brought into a better relationship or adjustment with the environment upon which he or it depends* [italics added]. (1952, p. viii)

According to L. A. Grunig, J. E. Grunig, and Ehling, "The nature of relationships between organizations and stakeholders . . . emerges, then, as a *central concept in a theory of public relations* [italics added] and organizational effectiveness" (1992, p. 81).

The nature of relationships between organizations and their stakeholders has been a subject of increasing study and discussion among public relations scholars. Although the definition of relationships between organizations and publics is far from complete, one aspect of this relationship is when "one or both parties perceive mutual threats from an uncertain environment" (Broom, Casey, & Ritchey, 2000, p. 17).

Both public relations and crisis management literature point to building meaningful relationships with—or *reengaging*—the organization's stakeholders as essential to achieving the organization's mission. Developing and maintaining "mutually beneficial relationships" between organizations and key stakeholders clearly has implications for crisis management (Cutlip, Center, & Broom, 2000, p. 6).

Creating New Partnerships

Effective crisis management requires that organizations identify key stakeholders: those publics that may be important in the event of a crisis. These key communicators often function as opinion leaders within their spheres of influence. Most organizations have at least rudimentary communication networks in place with some of their key stakeholders. Effective crisis management, however, requires that meaningful two-way communication channels be established with a more diverse group of key stakeholders *before* a crisis event happens. Relationships with the media and others important to an organization's survival "are something you need to create and maintain, not just in the midst of the water rising," says Barry Gaskins, public information officer, Pitt County Schools, North Carolina (Cook, 2001, p. 19). Credibility with an organization's publics must be earned over time; the positive relationships that have been built can be valuable when confronting a crisis.

How Are Stakeholder Networks Changing?

Crises that have the potential to unfold in new media will require new partnerships and collaborations with stakeholders. Traditional stakeholder groups are changing, and organizational communication maps should reflect these societal shifts. Demographics are changing inside and outside of organizations; older Americans are quickly outnumbering the young, and minority populations are growing. The first wave of the baby boomer generation is retiring and has different interests and demands than the smaller young-adult population, sometimes called the *millennial generation*. In some U.S. communities, majority populations have been replaced by a combination of multiple minority populations.

Minority populations of African Americans, Hispanics, and Asians are gaining a stronger voice. Larger cities are often a patchwork of different nationalities, races, and cultures, creating new communication challenges in crisis situations. For example, Montgomery County Public Schools in Rockville, Maryland, publishes an emergency preparedness information brochure in multiple languages—English, Chinese, French, Korean, Spanish, and Vietnamese—to reach a more diverse population (Montgomery County Public Schools, 2006).

At the same time that the demographics of stakeholders are evolving, organizations also are experiencing dramatic technological changes. Organizations are in the midst of an information explosion, and new technologies are revolutionizing the way we communicate. The millennial generation, those born between 1980 and 2000, has grown up in this new world of digital technology and media convergence. Communication through new media ranges from using Internet blogs and chat rooms, to text messaging on cell phones, to downloading the latest news and videos to cell phones, iPods, or personal digital assistants (PDAs) (Zeller, 2006). Organizations also find themselves facing increased competition for leaders who can meet the new demands of the marketplace, employees, stockholders, and other key stakeholders (often including the media). In tandem, societal unrest continues to spill into the workplace (Marx, 2000).

To be crisis ready, organizations must (a) evaluate the communications climate (Is the organization an open system, willing to share and receive information, internally and externally? Or is the organization a closed system, reluctant to share information and not willing to seek and respond to information from external audiences?); (b) identify the stakeholders that are crucial to the organization's survival and success; (c) create a written communication network of stakeholders that are potentially important to the organization in a crisis; (d) develop systematic (regular and ongoing), planned, two-way communication between the organization and key stakeholders; (e) ensure that this planned communication program incorporates a mix of traditional and new media; (f) continue to develop new ways of meaningfully engaging stakeholders in dialogue about topics that the organization and its publics have in common, that is, areas of shared interest and concern; (g) evaluate new media as channels for engaging stakeholders in new and different ways; and (h) anticipate the demands that users of new media will place on the organization in the event of a crisis.

How Do We Reengage Our Publics? Theory and Practice

Wadsworth argues that

> Engaging the public means just that. The process requires a constituency that is broader and more inclusive than the "usual suspects" with whom leaders and experts are accustomed to working. And sharing responsibility with this broader constituency is necessary in order to move from a critical or confrontational debate to meaningful participation. (1997, p. 752)

To position organizations to more effectively respond to crises requires that organizations move beyond the mere identification of important stakeholders. It is no longer enough to mail or e-mail the perfunctory monthly newsletter to a mass mailing list or schedule the traditional, quarterly, all-employee meeting. Organizations must seek to *engage* key constituencies in meaningful conversation. *Engagement* and *meaningful conversation* demonstrate that the organization is seriously participating in a dialogue, sharing information, listening, and responding to feedback. The organization is willing to consider modifying its behavior based on this dialogue. Stakeholders have an opportunity to influence the organization, bring about change, and even share in the responsibility for making decisions that may affect them.

The systems approach stipulated by Cutlip, Center, and Broom (2000) provides for an open system that responds and adjusts to change pressures from the environment (such as a crisis) in order to achieve and maintain goal states (or a central mission). Open systems attempt to explain the organization to external audiences. Open systems exchange information. The open system is most likely to be responsive to internal and external pressures in a crisis, enabling it to continue to focus on its central mission.

J. E. Grunig (1976) discusses two types of organizational communication systems. Fatalistic systems are closed systems, with information flowing out but not in: an ineffective system for responding to crises. Problem-solving or open systems face few constraints as they monitor the environment, attempting to understand and communicate external attitudes and opinions to management.

Reengaging the public has been cited by one researcher as one of the critical issues facing schools of the twenty-first century. Marx defines public engagement as

- Building public understanding and support
- Developing a common culture
- Building a sense of community
- Creating legitimate partnerships and collaborations
- Capitalizing on the community as a source of support
- Developing parent participation
- Building a sense of "we" versus "us and them." (2000, p. 88)

Public relations and public engagement both emphasize the importance of building mutually beneficial relationships. Communicating with the community should be viewed as an asset rather than as an expense. Former director of communications for the Rockwood

School District in Glenco, Missouri, and consultant on public engagement projects with the Annenberg Institute for School Reform at Brown University, J.S. Arnett said, "Like an old-fashioned barn raising, confidence is an organization's strength, and the ability to withstand the elements is often proportionate to the number of people who participate in its construction" (1999, p. 27).

Rich Bagin, executive director of the National School Public Relations Association, wrote that public engagement participants are "engaged *with* people and issues, not *by* them," indicating a deeper level of involvement (1998, p. 8). In addition, public engagement enables

> schools and communities to build trust and mutual accountability for school programs. . . . At a time when many parents . . . and the majority of taxpayers are no longer connected (*let alone engaged*) in any tangible fashion to our public schools, it may be time to create that direct link of public education to our democratic society. (p. 7)

Educational systems are not alone in recognizing the need to reengage the public. An international trend toward reengagement of stakeholders in decision making and policy setting is taking place (Rowe & Frewer, 2005).

Organizations can engage stakeholders through a number of traditional and nontraditional channels. Rowe and Frewer (2005) have created an extensive list of traditional strategies for engaging stakeholders. These include citizen advisory committees, citizen panels, community forums, focus groups, town meetings, neighborhood planning, and task forces. Nontraditional methods might include online discussion groups through organizational Web sites and blogs. Public relations practitioners need to note, however, that use of new media may vary according to age and other demographics.

For example, according to the Pew Internet and American Life Project (as cited in Zeller, 2006), among those with access to the Internet, e-mail services are used nearly equally by retirees (90%) and teenagers (89%), but use of other new media varies. Although only 9% of those in their thirties have created blogs, about 40% of teenage and twenty-something Internet users have done so. Thirty percent of adults age twenty-nine to forty report visiting blogs, while nearly 80% of teenagers and young adults (twenty-eight years of age and younger) regularly visit blogs. Text messaging by cell phone users also varies: 44% for the older audience compared to 60% for the younger group.

New Constructs of Public Engagement

Expanded use of the Internet is offering organizations new ways of engaging a diverse and sometimes global public. Some of the opportunities offered by the World Wide Web include (a) a channel for reaching traditionally isolated publics; (b) communication free of filters and traditional media gatekeepers; (c) feedback opportunities or opportunities to solicit and respond to concerns, questions, opinions, complaints, and praise; (d) information "to allow [stakeholders] to engage an organization in dialogue as an informed partner" (Kent & Taylor, 1998, p. 328); (e) constant updates, including revised frequently asked questions (FAQs) sections, text, graphics, and audio/video that can be downloaded, along with new interactive and searchable features that can motivate stakeholders to return to the sites and promote engagement; (f) speed of delivery or access to information; (g) opportunity for one stakeholder to network with other stakeholders on shared concerns or issues; and (h) a democratization of issues discussion, whereby nonprofit, smaller, or activist organizations can have the same access to stakeholders and discussion of timely issues as large corporations with more financial resources (Heath, 1998; see also, Coombs, 1998; Kent & Taylor, 1998; Kent, Taylor, & White, 2003; Taylor, Kent, & White, 2001; Ryan, 2003).

Unfortunately, many organizations are not yet using the full capabilities of the Internet to engage stakeholders in meaningful dialogue. Some studies indicate that access to the Web may mean organizations are creating a "presence" on the Web to enhance visibility or image, rather than emphasizing content and interactivity to facilitate two-way communication to engage and build relationships with targeted stakeholders (Kent, Taylor, & White, 2003).

In fact, in one study of activist organizations' Web sites, fewer than 30% of the activist organizations responded to publics when contacted directly through e-mail (Kent, Taylor, & White, 2003). In another study, however, public relations practitioners seem to recognize the potential of the Web for engaging stakeholders in dialogic communication (Ryan, 2003). Most practitioner respondents indicated that links were important to enable stakeholders to submit comments, suggestions, or complaints or to contact the public relations department. Practitioners also recognized that Web sites are useful for crisis information.

New media also offer technologies to communicate with internal audiences, such as employees, in a crisis. For example, American Airlines used mediated channels, including the employee intranet, hotlines, Internet kiosks, a proprietary software system, and fax-on-demand technologies, to supplement supervisory communications after 9/11 (Downing, 2004).

A study by Taylor and Perry compared the online use of traditional tactics with the online use of "innovative media tactics" (2005, p. 212) by organizations in crisis. Findings showed that most of the organizations that responded were most comfortable using familiar, traditional tactics—news releases, fact sheets, FAQs, and news conference transcripts—online. Ninety-eight percent of the organizations used at least one of the traditional media tactics online, and 34% used only traditional tactics in their online crisis response.

It is encouraging to note that 66% of the organizations used at least one of the innovative media tactics online, and one organization relied only on the innovative media tactics online. The innovative online media tactics were (a) dialogic communication that encouraged visitors to respond to issues via the Internet; (b) connecting links or hot buttons to provide additional resources at other Internet sites; (c) real-time monitoring to provide hour-by-hour updates to monitor the crisis; (d) multimedia effects, such as taped or live video, photographs, or audio effects; and (e) online chats to involve stakeholders in the situation (Taylor & Perry, 2005, p. 212).

The study found that the most popular innovative media tactics were connecting links (46%) and opportunities for two-way communication (44%). Sixty-four percent of the organizations used a mixed-media approach (i.e., a combination of traditional and innovative online techniques) in responding to crisis (Taylor & Perry, 2005).

Communication Challenges Posed by Public Engagement and New Media

According to Dr. Rita R. Colwell (2001), former director of the National Science Foundation, the terrorist attacks of September 11, 2001, and subsequent incidents of anthrax poisonings through the mail have set the nation on a new course. "As we incorporate the phrase 'homeland security' into our national lexicon, every sector of society . . . will be in the business of preparedness" (para. 60).

As Columbine High School motivated educators to plan for the unthinkable, the events of September 11, 2001, drove home the need for crisis planning in corporate America. According to a chief executive officer (CEO) reputation survey by *PR Week* / Burson-Marsteller, 21% of 194 CEO respondents said they had no crisis plan in place on 9/11 (Schoenberg, 2005). In the aftermath, however, 63% said they started to address crisis planning.

Other studies have found that more progress is needed. For example, according to a study by the International Profit Associates Small Business Research Board, 79% of American small businesses indicate that they do not have a disaster-recovery plan in place ("Study," 2005). Other organizations are evaluating if their crisis plans cover enough possible scenarios, such as responding to major flu outbreaks or the avian flu (Brickey, 2005).

One of the demands affecting crisis planning is the need to communicate with and engage a more diverse audience. Two crises illustrate this point. When severe acute respiratory syndrome (SARS) struck Canada in spring 2003, officials found that using diverse voices—multiple spokespersons with differing types of expertise—helped reach a diverse public with a more credible message (Duhé, 2005).

A similar experience was found by Griffin Hospital, a small community hospital in Derby, Connecticut, when it faced a single case of anthrax poisoning in 2001. The hospital made two important decisions. The first was to have different areas of expertise reflected through multiple hospital spokespersons, rather than relying on a single spokesperson, as is often encouraged in crisis management literature. The second was to share information openly and frequently, as was the nature of the open communication culture of the hospital. According to the hospital president, "This is an open, transparent organization. It has been what has made us successful" (Wise, 2003, p. 467).

In the United States, the terrorist attacks on September 11, 2001, provided an early lesson in the use of new media in a crisis. Although television is generally credited as the dominant source for news on 9/11, a significant number of viewers also turned to news Web sites, such as CNN.com, and government sites. For example, the Port Authority of New York and New Jersey had a 7,000% increase in traffic on its Web site (Baron, 2003; Carey, 2003). People had difficulty in the early hours of the crisis in accessing Web sites because of demand; broadband Web access was generally lower in 2001, and the quality could not compete with television. As broadband access and quality improve, Internet usage in a major crisis will grow. In the meantime, organizations are turning to Web pages as an important communication link in times of crisis.

The Internet became an important tool for Montgomery County Schools in Rockville, Maryland, to communicate with publics through frequent Web page updates in the wake of the sniper shootings in the Washington, D.C. / Maryland area.[1] Today, school districts and other organizations may have links on their home Web sites to direct visitors to specific Web pages with emergency-related information, including resources in the languages of diverse stakeholders.

In the midst of a crisis, use of new media can supplement and, at the same time, stymie official channels of communication. Officials responding to the July 2005 terrorist bombings in London experienced firsthand the challenges of citizens reporting through new media. Images captured by cell phone users and blogs posted by eyewitnesses competed with traditional journalists for attention from the news-hungry public. In an effort to battle misinformation, officials tried to focus attention on official statements, rather than devote precious time to responding to the many reports that were circulating (Murphy, 2005).

There are other concerns prompted by the increase in numbers of participants on the new media public stage. American Association for the Advancement of Science (AAAS) President Shirley Ann Jackson worries about a "communication environment that features more sources of information and more opposing perspectives from a host of 'new experts,' making it more difficult for the public to decipher what is 'fact'" ("Jackson," 2005, p. 1893). As society faces security threats and controversial issues related to scientific advances, the AAAS is calling for scientists and engineers to become a more proactive voice in this new public arena. Jackson says that "respectful engagement" (p. 1893) with the public can help promote scientific discovery.

As public relations practitioners work to keep up with new media and their ramifications for crisis response, the technology revolution continues unabated. According to Leichtman Research Group, "69% of all U.S. households now subscribe to an online service at home, and high-speed Internet services now account for about 60% of all online subscribers" (Center for Media Research, 2006, para. 1). In May 2006, Nielsen / Net Ratings found that 72% of active U.S. home Web users connected via broadband, compared with 57% just one year earlier (Gaither, 2006).

The end result of the diffusion of high-speed Internet connections will be an increase in the use of video and audio downloads, both a blessing and a curse, depending on your perspective. The June 2006 posting of a photo of a Dell notebook computer bursting into flames in Japan was widely distributed via Web sites and blogs (Darlin, 2006). Remember the old adage, "A picture is worth a thousand words?" The flaming notebook image just added fuel to the fire of blog and Web site complaints about the personal computer (PC) manufacturer's customer service response in the previous year. This online assault on Dell's image and reputation, which did lead to mainstream press coverage, prompted the company to respond with a $100 million program to improve service (Darlin, 2006). (Dell concluded that the laptop fire was caused by a faulty lithium-ion battery cell. The faulty batteries have since resulted in a battery recall by Dell and Apple Computer, users of the Sony Corporation–manufactured batteries.) The impact of such images creates multiple challenges for organizations and the public relations professionals who advise them. Organizations must keep up with how they are portrayed in new media and decide when and how to respond.

Summary

Whether we get our news from the business pages of the newspaper or the blogosphere, the lesson is the same. Technology and new media will continue to advance, and the way organizations and stakeholders use these new communication tools also will evolve. New media and its consumers must be recognized as integral to building and maintaining organization-stakeholder relationships and effective crisis management.

Organizations must recognize the realities of the new century: more diverse stakeholders who need to be reconnected and reengaged, a more competitive news media environment, new media that offer opportunities and challenges, and an often crisis-laden environment. Organizations that step up to address these realities *before* the next crisis occurs will strengthen their ability to respond to crises effectively and to solidify important relationships with stakeholders and communities.

For consideration

1. What new media strategies could be used to facilitate public engagement in a *pre-crisis* environment?
2. What specific factors will encourage organizations to adopt more new media tactics as part of their planned crisis response?
3. What internal and external challenges might a public relations practitioner confront when attempting to adopt new media strategies to respond to crises? To what extent can these challenges be overcome?
4. How can new media tools be used to research the changing demographics of publics?
5. How can more traditional means of communication be used during a crisis if electricity is unavailable? What does this imply for crisis preparedness?

For reading

Baron, G. R. (2003). *Now is too late: Survival in an era of instant news*. Upper Saddle River, NJ: Prentice Hall.

Coombs, W. T. (2007). *Ongoing crisis communication: Planning, managing, and responding* (2nd ed.). Thousand Oaks, CA: Sage Publications.

Fearn-Banks, K. (2007). *Crisis communications: A casebook approach* (3rd ed.). Mahwah, NJ: Lawrence Erlbaum Associates.

Holtz, S. (1999). *Public relations on the Net*. New York: AMACOM.

Noll, A. M. (Ed.). (2003). *Crisis communications: Lessons from September 11*. Lanham, MD: Rowman & Littlefield.

Note

1. The Montgomery County Public Schools: Emergency Information (2006) Web site provided frequent updates regarding the sniper shootings in October 2002; information on the site has since been changed.

References

Arnett, J. S. (1999, September). From public enragement to engagement. *The School Administrator*, 24–27.

Bagin, R. (1998, May). A practical look at public engagement. *NSPRA Network*, 7–8.

Baron, G. R. (2003). *Now is too late: Survival in an era of instant news*. Upper Saddle River, NJ: Prentice Hall.

Bernays, E. L. (1952). *Public relations*. Norman, OK: University of Oklahoma Press.

Brickey, H. (2005, November 23). Few firms in area have plans for a crisis. *The Blade (OH)*. Retrieved February 15, 2006, from EBSCO database.

Broom, G. M., Casey, S., & Ritchey, J. (2000). Concept and theory of organization-public relationships. In J. A. Ledingham & S. D. Bruning (Eds.), *Public relations as relationship management: A relational approach to the study and practice of public relations* (pp. 3–22). Mahwah, NJ: Lawrence Erlbaum Associates.

Carey, J. (2003). The functions and uses of media during the September 11 crisis and its aftermath. In A. M. Noll (Ed.), *Crisis communications: Lessons from September 11* (pp. 1–16). Lanham, MD: Rowman & Littlefield.

Center for Media Research. (2006, July 11). *Research brief: High speed Internet in 60% of online households*. Retrieved July 11, 2006, from http://www.centerformediaresearch.com/cfmr_brief.cfm?fnl=060710

Colwell, R. R. (2001, November 7). *Science: Before and after September 11*. Director's Forum Lecture, Woodrow Wilson International Center for Scholars, Washington, DC. Retrieved July 27, 2006, from http://www.nsf.gov/news/speeches/colwell/rc011107wodrowilson.htm

Cook, G. (2001, June). The media and the message. *American School Board Journal*, 16–22.

Coombs, W. T. (1998). The Internet as potential equalizer: New leverage for confronting social irresponsibility. *Public Relations Review*, *24*(3), 289–303.

Cutlip, S. M., Center, A. H., & Broom, G. M. (2000). *Effective public relations* (8th ed.). Upper Saddle River, NJ: Prentice-Hall.

Darlin, D. (2006, July 10). Dell's exploding computer and other image problems. *The New York Times*. Retrieved July 11, 2006, from http://www.nytimes.com/2006/07/10/technology/10dell.html?ex=1310184000&en=6e111c

Downing, J. R. (2004). American Airlines' use of mediated employee channels after the 9/11 attacks. *Public Relations Review*, *30*(1), 37–48.

Duhé, S. F. (2005, Spring). The sources behind the first days of the anthrax attacks: What can practitioners learn? *Public Relations Quarterly*, *50*(1), 7–13.

Gaither, C. (2006, July 10). Use of Internet video is growing at a faster clip. *Los Angeles Times*, *C1*. Retrieved July 11, 2006, from http://www.latimes.com/

Grunig, J. E. (1976, November). Organizations and public relations: Testing a communication theory. *Journalism Monographs*, *46*, 1–59.

Grunig, L. A., Grunig, J. E., & Ehling, W. P. (1992). What is an effective organization? In J. E. Grunig (Ed.), *Excellence in public relations and communication management* (pp. 65–90). Hillsdale, NJ: Lawrence Erlbaum Associates.

Heath, R. L. (1998). New communication technologies: An issues management point of view. *Public Relations Review*, *24*(3), 273–288.

Jackson: Scientists should work to solve problems and build trust (AAAS News and Notes). (2005, March 25). *Science*, *307*(5717), 1893.

Kent, M. L., & Taylor, M. (1998). Building dialogic relationships through the World Wide Web. *Public Relations Review*, *24*(3), 321–334.

Kent, M. L., Taylor, M., & White W. J. (2003). The relationship between Web site design and organizational responsiveness to stakeholders. *Public Relations Review, 29*(1), 63–77.

Kruckeberg, D., & Starck, K. (1988). *Public relations and community: A reconstructed theory*. New York: Praeger.

Malone, J. (2005, March 6). More eyes to combat terrorism. *The Atlanta Journal-Constitution*, pp. A1, A10.

Marx, G. (2000). *Ten trends: Educating children for a profoundly different future*. Vienna, VA: Center for Public Outreach, Educational Research Service.

Montgomery County Public Schools. (2006, January 11). *Emergency information*. Retrieved June 10, 2006, from http://www.mcps.k12.md.us/info/emergency/

Murphy, C. (2005, September 30). 7 July: Putting crisis theory into practice. *PR Week* (London edition), pp. 25–28.

Rowe, G., & Frewer, L. J. (2005, Spring). A typology of public engagement mechanisms. *Science, Technology, & Human Values, 30*(2), 251–290.

Ryan, M. (2003). Public relations and the Web: Organizational problems, gender, and institution type. *Public Relations Review, 29*(3), 335–349.

Schoenberg, A. (2005, Spring). Do crisis plans matter? A new perspective on leading during a crisis. *Public Relations Quarterly*, 50(1), 2–7.

Study: Most firms unprepared for disaster. (2005, November 18). *The Central New York Business Journal*, p. 11.

Taylor, M., Kent, M. L., & White, W. J. (2001). How activist organizations are using the Internet to build relationships. *Public Relations Review, 27*(3), 263–284.

Taylor, M., & Perry, D. C. (2005). Diffusion of traditional and new media tactics in crisis communication. *Public Relations Review, 31*(2), 209–217.

Wadsworth, D. (1997). Building a strategy for successful public engagement. *Phi Delta Kappan, 78*(10), 749–752.

Wise, K. (2003). The Oxford incident: Organizational culture's role in an anthrax crisis. *Public Relations Review*, 29(4), 461–472.

Zeller, T., Jr. (2006, January 22). A generation serves notice: It's a moving target. *The New York Times*. Retrieved January 23, 2006, from http://www.nytimes.com/2006/01/22/business/yourmoney/22youth.html?ei=5088&en=d23

Balancing Traditional Channels of Public Relations with New Media Technologies in Crisis Communication

The Case of a Catholic College

Melissa K. Gibson Hancox and Anthony C. Peyronel

This chapter explores the changing role of traditional channels of public relations with new media technologies in crisis communication. Using a case study approach, the chapter focuses on a 2004 crisis in which the president of a small Catholic college was accused of sexual misconduct with minors, demonstrating that institutions of higher education are prime ground for crisis events. The case provides a framework for understanding the need for a balanced public relations strategy during times of crisis. The chapter concludes by offering advice to aspiring public relations professionals on mediated public relations in crisis communication management.

"The Monster on the Hill." At first glance, the title sounds like a low-quality horror flick complete with bad lighting, costumes, and plot lines. But to the public relations practitioners at a small Northeastern Catholic college, this was no movie—it was the description used by bloggers to describe their president during a crisis that shook the organization to its roots.

Founded in 1926, Mercyhurst College is a private liberal arts college with approximately four thousand students and is situated on a sloping hillside in Erie, Pennsylvania. On October 10, 2004, Mercyhurst College students, faculty, and administrators woke to find a six-page story on the front page of the *Erie Times News* accusing then President Dr. William P. Garvey, sixty-eight years of age, of physically and sexually abusing boys in the 1960s, 1970s, and 1980s (Palattella, 2004a).

The allegations were made by four adult men who claimed they were molested by Garvey when they were basketball players on a grade-school basketball team he coached. One accuser, Dr. Chuck Rosenthal (2004), now a professor at Loyola Marymount University in California, detailed his accusations in a fictionalized book titled *Never Let Me Go*. Rosenthal alleged that Garvey lured him into having sex with him at the age of thirteen when Garvey coached the grade-school team at St. John the Baptist Catholic Church in Erie. The three other alleged victims included Rosenthal's two younger brothers.

The men recounted instances when Garvey would punish them for poor performance on the court or in the classroom by telling them to strip naked in the locker room and spanking them barehanded. The men alleged Garvey would hold their testicles in his hand while he spanked them. The former players also described an exercise technique that Garvey forced the players to perform. They alleged that Garvey told the players to jump up and down naked in the shower while he would squeeze their testicles, an exercise he told the players would make them jump higher by tightening their pubic-area muscles (Palattella, 2004a).

Rosenthal's description of the relationship in the newspaper story was even more lurid. Rosenthal claimed that his sexual relationship with Garvey continued until he was nineteen years old. Rosenthal said the molestation at the hands of a trusted adult caused him to develop emotional, relational, and substance abuse problems. As a young adult, he later tried to kill himself (Palattella, 2004a).

The story also included claims by two male prostitutes who said Garvey solicited them for sex in the mid-1980s when they were sixteen years old (Palattella, 2004a). According to the men, Garvey paid them for performing oral sex acts. After the encounters, the two men said they devised a scheme to extort money, blackmailing Garvey and claiming to receive approximately $1,000 from him.

Crisis Warnings Ignored

Like many organizational crises, the Mercyhurst College crisis was not entirely unexpected but rather had been smoldering under the surface for several years. Mercyhurst College officials, including the board of trustees, knew of the impending crisis as early as 2002 when Rosenthal "recounted his claims of being molested by his grade-school basketball coach Garvey to a *Los Angeles Times* columnist" (P. Howard, 2005, p. 2A). Reporters who were investigating the claims contacted the College. Marlene Mosco, president of the board of trustees, refused to acknowledge any merit to the sexual molestation accusations, calling Rosenthal a "liar" (P. Howard).

The thirty-five members of the board of trustees were so confident in Garvey's leadership that they extended his presidential contract in May 2003 through 2006; then, in July 2004, they extended his contract through 2007. Although the statute of limitations had expired, and Garvey could not be criminally charged with any crimes related to the accusations, the crisis threatened the image of the Catholic college at a time when the Catholic Church was already under fire nationwide for molestation issues with clergy (Dixon, 2004). Furthermore, Garvey had become closely identified with the organization, having served as a professor at Mercyhurst College since 1962, and then as president since 1980. The crisis was one that would redefine the history of the small Catholic college on the hill.

Crisis Defined

According to Fearn-Banks (2001), a crisis is "a major occurrence with a potentially negative outcome affecting an organization as well as its publics, services, products, and/or good name" (p. 480). Typical organizational crises include man-made or natural disasters, organizational downsizing, product tampering, product recalls, job-related death of an employee, hostile takeovers, insider trading, and employee misconduct, to name just a few. Crisis management, from a public relations approach, is the strategic planning that the public relations practitioner and the organization itself undertake to handle the crisis. All crises run the risk of escalating, catching the eye of the media, disrupting normal operations in the organization, and jeopardizing the image of the organization to its publics (Fink, 1986).

Communication, however, "can be used to influence how stakeholders interpret a crisis and the organization in the crisis" (Coombs & Holladay, 1996, p. 280). This is where public relations practitioners can help to manage the crisis for their organization. The way in which the crisis is managed is crucial in determining how, and if, the organization will survive the crisis. "Making the right strategic moves will determine just how long a conflict will persist and how much damage or benefit it will cause" to the organization (Wilcox & Cameron, 2006, p. 243). Ineffective management of a crisis can result in lost profits, reduced employee morale, and poor organizational image. "In some cases, the company gains admiration and a better reputation" as the result of a crisis handled well (Fearn-Banks, 2001, p. 485).

All organizations are susceptible to crises. In fact, "most organizations will receive unfavorable coverage at some point" in their organization's history (Lyon & Cameron, 2004, p. 213). Crises are an inevitable part of organizational life for public relations practitioners. As Duke and Masland explain, "No matter how busy you are, and no matter where you work in private business, large corporation, non-profit organization, college or university one day you too may face a crisis" (2002, p. 35).

Like the case of Mercyhurst College, colleges and universities are prime sites for potential crises, including campus crimes, natural disasters, administrative scandals, and student injuries or deaths. "College communities with their large population of single young adults, experiencing their first taste of freedom away from parental supervision, can become an environment susceptible to any number of crisis situations" (Duke & Masland, 2002, p. 30). The irony is that the very college campus where students study public relations is just as susceptible to organizational crises as any other organization. "Such crisis communication concepts are taught in college public relations classes. And the same concepts are practiced by public relations professionals at these colleges which, despite their bucolic settings, are not immune to crises" (Duke & Masland, 2002, p. 30).

Effective crisis management is especially critical at colleges and universities. No longer regarded as sheltered ivory towers, higher education institutions have been subjected to increasing levels of scrutiny for the last twenty-five years. More than ever, as the Mercyhurst College case demonstrates, colleges and universities must deal with crises strategically, especially in the face of scandal and controversy.

Ironically, credibility is largely regarded as the key component of effective crisis management, especially in handling a campus controversy (Jones, 2000; Ross & Hallstead, 2001). Indeed, how a college or university handles bad news is often more important than how it promotes good news, although this fact is often lost on college and university presidents:

> Leaders of institutions who actively seek positive coverage from the media need to understand that when they do so, they are opening up the institution in bad times as well as good times. The same education editor who did such a fine story on your institution's student volunteer programs will expect you to respond just as quickly when she calls about citizens' complaints to city council about noisy parties in student apartments. You can't welcome the media with open arms for good stories and brush them off for the bad ones. If you do, you will suddenly find there is no more interest in your stories. (Jones, 2000, p. 193)

Effective crisis management is predicated upon a policy of open and honest communication in the face of difficult circumstances.

The Impact of Technology

Public relations practitioners must constantly engage in what is called *relationship management*. Relationship management refers to public relations activities that "initiate, build, and maintain mutually beneficial relationships with internal and external publics" (Bruning, Castle, & Schrepfer, 2004, p. 435). Over the past twenty years, increases in technology have changed the way public relations practitioners engage in relationship management, particularly during an organizational crisis. Technology has changed the way public relations practitioners function and strategize (Lordan, 2001). This dawning of new technologies has forced the field of public relations to respond differently to crisis events. Because of technological advances, practitioners must (a) increase two-way communication with publics, (b) reduce reliance on traditional print media, (c) manage increased information clutter, and (d) respond to more active information-seeking publics during times of crisis (Sparks Fitzgerald & Spagnolia, 1999, p. 12).

Technology has altered and broadened the way organizations (and public relations departments) respond during a crisis. Take, for example, public media, including newspapers, magazines, radio, television, and others. These mass communication outlets are inherently non-personal, are controlled by the media organizations themselves, and reach large audiences (Hallahan, 2001). In a crisis situation, public media are a traditional outlet for information distribution.

But because of new technologies, the demand for information from public media has only been heightened. "As a result of the Internet, every organization is now a 24-hour business with customers and journalists expecting instant accessibility and immediate response" (C. M. Howard, 2000, p. 9). In a crisis event, practitioners must be accessible to public media and aware of how to use these channels to not only distribute important information about the crisis, but also communicate key messages to key publics. "In general, because of technology, practitioners can provide and receive information more quickly and gain almost instant response time, even internationally" (Sparks Fitzgerald & Spagnolia, 1999, p. 13).

Another change for public relations practitioners has been the growth of interactive media. Interactive media include the Internet, intranets, databases, e-mail, blogs, newsgroups, chat rooms, and other sources (Hallahan, 2001). During a crisis event, publics turn to interactive media channels as a way of seeking information or processing the event due to a lack of communication by the organization (Perry, Taylor, & Doerfel, 2003). In this way, technological advances have changed the way that publics get information. Instead of passively watching a news report or reading a story in print, publics can now solicit information directly from the organization through means like e-mail or ask others for information on blogs or chat rooms.

New technologies, especially Internet-based peer-to-peer sites, are likely to exacerbate this trend. For example, blogging is becoming an increasingly popular way for college students to communicate with each other and, in doing so, often allows them to bypass traditional campus media sources (Stoner, 2006). Conversely, blogs are "a democratic medium bespeaking quick communication, informality, and transparency" and providing the perfect opportunity for "direct, person-to-person communication" between students (Stoner, 2006, p. 27). Moreover, opinions expressed through student blogs often carry more credibility than information communicated through official campus channels (Stoner).

Another traditional public relations channel that has been impacted by technology is controlled media or media that are "physically produced and delivered to the recipient by the sponsor" (Hallahan, 2001, p. 467). Controlled media include brochures, newsletters, sponsored magazines, annual reports, and books produced by the organization. Because there are no gatekeepers, organizations have complete freedom to design the message and distribute the communication piece strategically.

In a crisis event, controlled media offer advantages in terms of time and space. With controlled media, such as an organizational newsletter, the public relations practitioner can spend a great deal of time designing and placing the message, even detailed and complex messages (Hallahan, 2001). Due to technological advancements, many controlled media pieces available to organizations offer great advantages during a crisis event. For example, newsletters and magazines, once only in the domain of printed material, are often found in online formats. Public relations practitioners can easily create and disseminate controlled media with greater ease and reach than any time in the past.

The preceding discussion has focused on what are considered to be more mass communication channels. It is essential to also consider more interpersonal communication channels when strategically responding to a crisis. For example, speeches, exhibits, meetings, conferences, demonstrations, rallies, and other special events offer a means to communicate in a semipersonal manner with key publics (Hallahan, 2001).

Similarly, the most interpersonal of communication channels one-on-one communication offers important opportunities in crisis management situations. Examples of one-on-one communication include personal visits, lobbying, personalized letters, and telephone calls (Hallahan, 2001). From a crisis management perspective, these one-on-one contacts may appear simplistic in nature but often form the basis of effective relationship management (Bruning, Castle, & Schrepfer, 2004). One-on-one communication in public relations is often used in unique situations that are "highly interactive and sometimes volatile" in nature (Hallahan, 2001, p. 469). Crisis events could be characterized as this kind of special situation. In times of crisis, public relations practitioners can use one-on-one communication to align allies and negotiate perceptions with hostile members of a public.

However, even these interpersonal communication channels have not gone untouched by the hand of technology. During crisis situations, public relations practitioners can use technology to enhance even the most traditional of public relations practices. For example, an organization could hold a video teleconference and discuss issues of importance during a crisis as easily with employees in India as with employees in Idaho (Kotcher, 1992).

Crisis Management in the Mercyhurst College Case

As discussed above, public relations practitioners faced with managing a contemporary crisis face a myriad of decisions regarding how to balance traditional communication channels with new media technologies in the most strategic fashion. So, how did Mercyhurst College manage its own crisis?[1]

When the story about President William Garvey finally did break in October 2004, students on campus woke to find the front pages of all the local *Erie Times News* newspapers on campus missing. President Garvey's Personal Assistant and Secretary to the Board of Trustees Mary Daly removed the front pages of all copies of the newspaper from student residence halls in an attempt to keep the story from students (Palattella, 2004b). This action only further angered students who felt their rights had been violated and cast a veil of suspicion over the college administration.

Marlene Mosco, speaking for the board of trustees, maintained Garvey's innocence and refused to speak with reporters immediately after the story ran (P. Howard, 2005). The day after the story broke, Garvey and Mosco each released a one-page statement regarding the crisis, which was sent via e-mail to students, faculty, and staff, as well as local public media. The college also posted these statements to the Mercyhurst College intranet.

Not once after the crisis erupted did college officials meet with college staff members, thus losing an opportunity to engage in valuable interpersonal communication with a key public. Only after the faculty demanded a face-to-face meeting with officials did Mosco meet with faculty three days after the story. Mosco's demeanor only angered many faculty when she cut the meeting short, telling faculty, "If you can't say something in support of William Garvey, then keep your mouth shut."[2]

Meanwhile, the campus was inundated with local and national media attention. After Chuck Rosenthal's appearance on a national talk show that week, print and broadcast stories soared. Within days, the story had been picked up by national outlets and, later, online publications (Lipka, 2004). Officials held one public forum with students several weeks after the incident, but the forum turned contentious as students felt Mosco and other officials were handling the crisis poorly; the forum was closed to the public and to the media.

In the case of Mercyhurst College, officials appeared to be unprepared for the demand of information by public media sources. Officials repeatedly refused to meet individually with journalists, instead issuing brief responses that amounted to name calling of the alleged victims. Journalist Pat Howard critiqued the college's handling of the crisis in a newspaper editorial:

> When the newspaper contacted Mosco [President of the Board of Trustees] for the college's response, she labeled Rosenthal [an alleged victim] a liar which remains her and the college's position of record to this day. More striking, however, was her reaction to being informed that the newspaper had uncovered a great deal of information [about Garvey] beyond Rosenthal's memoir. Mosco declined to meet with the newspaper to review that information. (2005, p. 2A)

As this comment indicates, local journalists were given very little information after the crisis erupted, a strategy that only served to anger the media and the community at large.

Mercyhurst College used little or no controlled media during its management of the crisis event. Even though the college produced various internal newsletters and sponsored magazines, no messages regarding the crisis were presented in these channels.

For the most part, College officials appeared stunned by the impact that new technologies had on the diffusion of the crisis. While officials attempted to control information being distributed, the amount of discussion by students, alumni, and community members on blogs, chat rooms, and other electronic sites flourished.

Blog sites served as a place for students and community members to foster rumors and banter about conspiracy theories about the crisis ("*Would You Want Your Child*," 2004). That student blogs played a key role in the Mercyhurst College case is not surprising, especially since college officials, in an autocratic approach, made every effort to block the flow of information through traditional channels.

Mercyhurst College discovered the rapid and aggressive use of interactive media during its crisis. Officials naively assumed the crisis would attract relatively little attention and would have a short life span. But this was far from the reality of what occurred. Instant messaging by students carried rumors and speculations at a rapid pace. Blog sites detailed conspiracies about the event and proliferated information. Alumni also used e-mail and phone calls to inquire about the crisis.

In the case of Mercyhurst College, officials chose to overlook the impending crisis and were caught unprepared for the national attention it received. In fact, two months prior to the story breaking, the director of public relations issued a plea to a professional listserv asking fellow public relations practitioners for advice on how to handle the crisis—a plea which was later discovered and posted by bloggers to point out the lack of crisis preparation on the college's part, even though the crisis was not entirely unexpected (*Gennifer Biggs*, 2005).

Unprepared in handling the crisis, the college eventually hired Simpson Communications, a Virginia-based public relations firm specializing in crisis communication, to manage the crisis (Palattella, 2004b). Unfortunately, as in many crisis events, valuable time immediately after the story broke had been wasted for positioning the college's messages and maintaining a positive image to key publics. Even the college itself later admitted that "mistakes were made" in the handling of the Garvey crisis (Palattella, 2005b).

Arguably, the public relations response by Mercyhurst College, its public relations practitioners, and its administrators failed on many counts, including the lack of understanding about the role that mediated public relations would play in the crisis event. In addition, too much time was focused on denying and attempting to control the amount of information that was disseminated, rather than focusing on relationship management with its key publics students, alumni, faculty, staff, and community members.

An Integrated Approach

New media technologies have changed the playing field of modern day crisis communication. Organizations that rely on narrow-minded approaches to responding to a crisis face criticism and disapproval from key publics. However, effective relationship management practices during a crisis cannot discount the role of traditional public relations channels, including one-on-one communication with key publics. New media technologies have altered these traditional models and call for more expanded and broad-based approaches to crisis communication.

Even though the landscape of public relations (and crisis communication) has inherently changed due to new technologies, ignoring other traditional communication channels is unwise. Just as it would be unwise for a practitioner to overlook new technologies like blogging or e-mail, it is equally unwise to ignore traditional channels like television, newspapers and radio, or public meetings. Balance is the key, even in an increasingly mediated world. According to Devere Logan of D. E. Logan Public Relations, "Relationships are not developed by machines. High tech will never replace high touch" (as cited in Knowles, 2006, p. 28). Overreliance on new technologies is a mistake (Knowles). Rather, practitioners should use a blended approach when creating their own crisis communication plans.

Practical Advice for Practitioners

The case of Mercyhurst College demonstrates lessons in managing the demands of traditional relationship management with more modern complexities brought on by new technologies. Some lessons offer sound advice for crisis situations in any organization.

Lesson #1: Be proactive and develop a crisis plan before a crisis ever occurs. Sometimes crises can be predicted; sometimes they cannot. The best advice is to be prepared. Have a plan for how your organization will respond when faced with various kinds of crises (Olaniran & Williams, 2001). Effective public relations practitioners practice or have mock crisis drills with staff to prepare for crisis events.

Lesson #2: Use an integrated approach when deciding how to respond to the crisis event. Remember that approaches to crisis communication are best managed through an integrated approach, one that balances traditional and new media approaches. So, while you might be monitoring blogs and posting to your organization's Web site, do not overlook the need to engage in traditional public relations tactics, like maintaining relationships with key publics.

Lesson #3: Don't panic when stories about the crisis disseminate across communication channels. Instead, marshal your communication channels to your advantage. Realize that much of the coverage of the story is out of your hands and focus on what you can control responding to the crisis. "During and after a crisis, companies are judged not by the original negative actions but by how they handle them" (Pines, 2000, p. 16).

Lesson #4: Select a spokesperson who is comfortable communicating across a variety of channels and who will best represent the organization. Although during the worst crises there is an expectation that the chief executive officer or president of an organization will step forward as the spokesperson (Farmer & Tvedt, 2005), at other times, the best representative for the organization might be someone else (Pines, 2000).

Lesson #5: Be honest, be forthright, and be accountable not only to the publics but also to yourself. As the professional hallmarks of the Public Relations Society of America (PRSA) suggest, as a public relations practitioner, you must be accountable to yourself and to your profession above all else, including your organization. You should never compromise your ethics to accommodate the organization for which you are employed at the moment. Effective crisis management is predicated on a policy of open and honest communication in the face of difficult circumstances. The notion that you can postpone the truth is always seductive but ultimately wrong.

Lesson #6: Understand the 24/7 news cycle. Because of advances in technology, news media now operate on a tweny-four-hour, seven-days-a-week news cycle. Unlike the past, when news was constrained by publication or broadcast times, today's journalists seek and post news around-the-clock (Brown, 2003). If you don't supply information, journalists will just find a way to get it; by offering information, you have a better chance of positioning your organization's message.

Lesson #7: Embrace technology before it strangles you. Public relations practitioners in the twenty-first century must not only tolerate technology, but also embrace it. Technology has changed public relations practices forever, but practitioners can use technology to their benefit. For example, instead of resisting new technologies, organizations should "designate a PR pro or department to constantly monitor online communications like blogs and message boards to prevent crisis situations" (Knowles, 2006, p. 20).

Conclusion

Although the realm of public relations is inherently muddied and made more complex by the influx of new technologies, "experienced PR practitioners agree that newer communication technologies can and will bring higher efficiency to the profession if approached conscientiously" (Knowles, 2006, p. 28). Public relations practitioners will continue to be challenged by balancing new communication technologies with traditional channels during crisis events.

In a crisis, relationships with key publics must be managed, be it through blog sites or personal telephone calls, e-mail, or face-to-face meetings. Public relations practitioners can plan and respond more effectively by using a strategic balance of communication channels, ranging from mass communication to interpersonal communication. Organizations that fail to do so face uncertain outcomes when a crisis hits.

So, how can a poorly handled crisis affect an organization? What happens when an organization fails during times of crisis to balance traditional relational management with the increasing demands of new technologies? Just follow the Mercyhurst College example and see.

A month after the story first broke, the board of trustees hired a retired county judge and an ex Pennsylvania state trooper to investigate the claims against Dr. William P. Garvey. On December 15, 2004, the investigators sent a memo to the board of trustees indicating that there was "merit in the allegations that Garvey had serially abused teenagers" (P. Howard, 2005, p. 2A) and that other individuals were ready to come forward with similar allegations.

On December 16, 2004, Garvey resigned as president of Mercyhurst College after twenty-five years in office (Palattella, 2005a). At that time, the board of trustees terminated any further investigation into the claims made by the alleged victims. Because the board of trustees had recently renewed his contract, the college was obligated to continue to pay him two-thirds of his salary, approximately $120,000 per year through 2007.

In September 2005, the Mercyhurst College Board of Trustees issued a 412-word statement saying that "mistakes were made" in how the college handled the crisis (Palattella, 2005b). After this public statement, other officials at the college left in the wake of the crisis, including the director of public relations. Soon after, a member of the board of trustees publicly resigned in protest at the handling of the crisis (Palattella, 2005d).

Faculty members, angry over the way the crisis was managed, demanded that the name of a park on campus in Garvey's honor be changed (Palattella, 2005c). In addition, the Mercyhurst College Faculty Senate threatened to take a vote of no confidence in the leadership of Marlene Mosco as president of the board of trustees. Up to eighteen months after the crisis broke, hostile letters to the editor about Mercyhurst College continued to appear in local newspapers. And, even today, bloggers continue to banter about the "Monster on the Hill" and discuss when the scandal will be made into its own Hollywood movie.

For consideration

1. What public relations challenges could be the most difficult to manage in a higher education setting? What constraints could a practitioner face that would not otherwise be an issue in the private sector?
2. To what extent would new media channels be effective in reaching students, faculty, staff, and alumni during a college or university crisis? What advance planning would be required?
3. How should blogs be used during a crisis event, if at all?
4. What direction does PRSA's code of ethics and professional conduct provide to public relations professionals trying to decide whether it is appropriate to post favorable messages about their organizations on blogs during a crisis?
5. What risks are inherent to an organization's openness to two-way communication with journalists? What types of benefits make risk taking worthwhile?

For reading

Barton, L. (2001). *Crisis in organizations II*. Cincinnati, OH: South-Western.

Coombs, W. T. (1999). *Ongoing crisis communication: Planning, managing, and responding*. Thousand Oaks, CA: Sage Publications.

Fearn-Banks, K. (2001). *Crisis communications: A casebook approach*. Mahwah, NJ: Lawrence Erlbaum.

Griese, N. L. (2002). *How to manage organizational communication during crisis: A primer of best practices for public relations professionals*. Tucker, GA: Anvil Publishers.

Seeger, M. W., Sellnow, T. L., & Ulmer, R. R. (2003). *Communication and organizational crisis*. Westport, CT: Greenwood Publishing.

Notes

1. The lead author's firsthand experience at Mercyhurst College provided the background to this case study. In the three years during which she was a faculty member and, later, department chair in

the Communication Department, she witnessed the organization's management of the crisis prior to, during, and after the crisis erupted. Her experiences as an administrator and a faculty member during that time provided insight into the organization's actions and crisis management approach.

2. The lead author was at this meeting and witnessed this statement being made.

References

Brown, T. S. (2003). Powerful crisis communications lessons: PR lessons learned from Hurricane Isabel. *Public Relations Quarterly, 48*(4), 31–34.

Bruning, S. D., Castle, J. D., & Schrepfer, E. (2004). Building relationships between organizations and publics: Examining the linkage between organization-public relationships, evaluations of satisfaction, and behavioral intent. *Communication Studies, 55*(3), 435–446.

Coombs, W. T., & Holladay, S. J. (1996). Communication and attributions in a crisis: An experimental study in crisis communication. *Journal of Public Relations Research, 8*(4), 279–295.

Dixon, M. A. (2004). Silencing the lambs: The Catholic Church's response to the 2002 sexual abuse scandal. *Journal of Communication and Religion, 27*, 63–86.

Duke, S., & Masland, L. (2002). Crisis communication by the book. *Public Relations Quarterly, 47*(3), 30–35.

Farmer, B., & Tvedt, L. (2005). Top management communication during crises: Guidelines and a "perfect example" of a crisis leader. *Public Relations Quarterly, 50*(2), 27–31.

Fearn-Banks, K. (2001). Crisis communication: A review of some best practices. In R. L. Heath (Ed.), *Handbook of public relations* (pp. 479–485). Thousand Oaks, CA: Sage Publications.

Fink, S. (1986). *Crisis management: Planning for the inevitable*. New York: AMACOM.

Gennifer Biggs Preparing for the Garvey stuff to hit the fan. (2005, August 6). Retrieved July 12, 2006, from http://community.livejournal.com/_gryphon/2162.html?thread=59250#t59250

Hallahan, K. (2001). Strategic media planning: Toward an integrated public relations media model. In R. L. Heath (Ed.), *Handbook of public relations* (pp. 461–470). Thousand Oaks, CA: Sage Publications.

Howard, C. M. (2000). Technology and tabloids: How the new media world is changing our jobs. *Public Relations Quarterly, 45*(1), 8–12.

Howard, P. (2005, August 28). Mosco turns bad to worse by covering up Garvey probe. *Erie Times News*, p. 2A.

Jones, D. (2000). Media relations: Successfully positioning your institution. In P. M. Buchanan (Ed.), *Handbook of institutional advancement* (pp. 187–194). Washington, DC: Council for the Advancement and Support of Education.

Knowles, S. L. (2006). Married to technology, for better or for worse? PR pros discuss pros and cons of new communication methods. *Public Relations Tactics, 13*(3), 20, 28.

Kotcher, R. L. (1992). The technological revolution has transformed crisis communications. *Public Relations Quarterly, 37*(3), 19–21.

Lipka, S. (2004, October 22). College president accused of past sexual abuse. *The Chronicle of Higher Education*. Retrieved November 21, 2005, from http://chronicle.com/weekly/v51/i09/09a04002.htm

Lordan, E. J. (2001). Cyberspin: The use of new technologies in public relations. In R. L. Heath (Ed.), *Handbook of public relations* (pp. 583–589). Thousand Oaks, CA: Sage Publications.

Lyon, L., & Cameron, G. T. (2004). A relational approach examining the interplay of prior reputation and immediate response to a crisis. *Journal of Public Relations Research, 16*(3), 213–241.

Olaniran, B. A., & Williams, D. E. (2001). Anticipatory model of crisis management: A vigilant response to technological crises. In R. L. Heath (Ed.), *Handbook of public relations* (pp. 487–500). Thousand Oaks, CA: Sage Publications.

Palattella, E. (2004a, October 10). Garvey past questioned. New memoir claims Garvey molested grade-school boy. *Erie Times News*, pp. 1A-6A.

Palattella, E. (2004b, October 15). College wants law firm to probe claims. *Erie Times News*, p. 1A.

Palattella, E. (2005a, February 24). Garvey era ends. *Erie Times News*, pp. 1A, 4A.

Palattella, E. (2005b, September 9). Mistakes were made: Mercyhurst offers explanation of how it handled Garvey probe. *Erie Times News*, pp. 1A, 4A.

Palattella, E. (2005c, October 1). College faculty: Rename Garvey park. *Erie Times News*, pp. 1B-2B.

Palattella, E. (2005d, December 8). Mercyhurst trustee resigns over Garvey probe. *Erie Times News*, pp. 1A, 6A.

Perry, D. C., Taylor, M., & Doerfel, M. L. (2003). Internet-based communication in crisis management. *Management Communication Quarterly, 17*(2), 206–232.

Pines, W. L. (2000). Myths of crisis management. Public *Relations Quarterly, 45*(3), 15–17.

Rosenthal, C. (2004). *Never let me go*. Granada Hills, CA: Red Hen Press.

Ross, J. E., & Hallstead, C. P. (2001). Feeding the media monster. In J.E. Ross & C.P. Hallstead (Eds.), *Public relations and the presidency* (pp. 119–126). Washington, DC: Council for the Advancement and Support of Education.

Sparks Fitzgerald, S., & Spagnolia, N. (1999). Four predictions for PR practitioners in the new millennium. *Public Relations Quarterly, 44*(3), 12–14.

Stoner, M. (2006, January). The real world. *CASE Currents*, pp. 26–31.

Wilcox, D. L., & Cameron, G. T. (2006). *Public relations strategies and tactics* (8th ed.). Boston: Pearson Education.

Would you want your child to go to Mercyhurst College? (2004, October 21). Retrieved July 12, 2006, from http://www.livejournal.com/community/_gryphon/2162.html

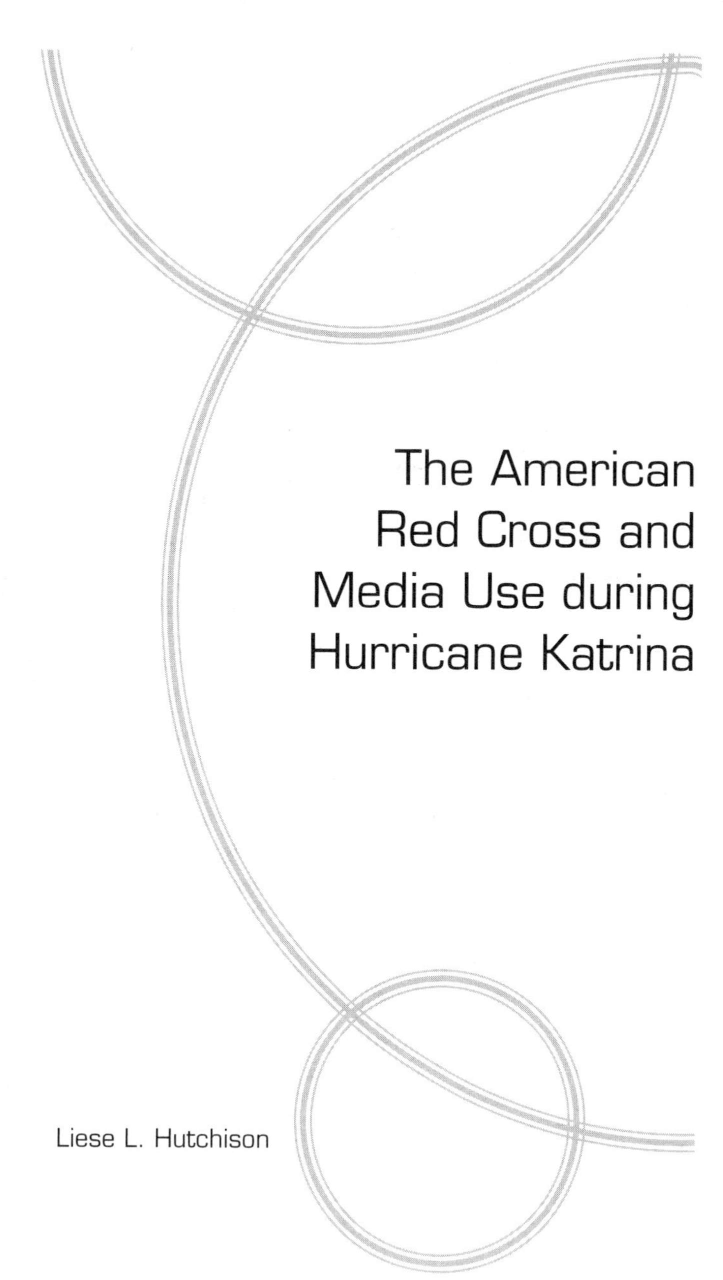

The American Red Cross and Media Use during Hurricane Katrina

Liese L. Hutchison

Media use during a crisis is obligatory but fraught with peril. Why do organizations rely on mass and new media to communicate during disasters? How should they utilize the media? What happens to their relationships with publics before, during, and after a crisis? This chapter focuses on mediated public relations during times of disaster and how one organization, the American Red Cross, utilized media during the largest natural disaster in U.S. history—Hurricane Katrina.

Mediated Public Relations during Crises

Even though public relations professionals may prefer to distribute information to their organizational publics by controlled methods, such as newsletters, memos, or Internet postings, during a crisis, most publics receive information from the media. Fearn-Banks (2002) agrees that "information about a crisis reaches publics through the media more than any other means" (p. 65). Heath and Millar (2003) define *crisis* as "an untimely but predictable event that has actual or potential consequences for stakeholders' interests as well as the reputation of the organization suffering the crisis" (p. 2). They continue by arguing that, if the organization responds well to the crisis, it will survive, but, if it responds poorly, the organization's reputation could be ruined. Borda and Mackey-Kallis (2003) propose that, for an organization to survive a crisis, communication is the key, and that the organization must gather information, package it, and deliver it successfully. Gathering information involves researching the background of the crisis, designating a spokesperson, and identifying short- and long-term goals; packaging information includes disclosing all that is known, telling the truth, and showing concern; and delivering the information includes getting the information out quickly and being assertive with the media by getting the information out first.

The relationship an organization has with its publics before a crisis may play a role in how the public perceives the organization's response to the crisis (Coombs, 2000). It may be safe to conclude that a favorable perception bodes well for the organization before a crisis, but there are two potential downfalls: The organization may be held to a higher standard, or its reputation could be shattered if the core of the relationship is at stake. For the first point, this may mean that an organization that always performed well in the public's mind would have to be exceptional in its response to a crisis. Regarding the second potential pitfall, if the organization is hypocritical during a crisis (e.g., claims to be environmentally sound but then is shown to pollute), the core relationship could be broken apart.

Sellnow, Seeger, and Ulmer (2002) argue that crisis communication focuses on organizational or political crises. Natural disasters, such as floods, tornados, earthquakes, and hurricanes, have received less attention, even though disasters "have profound effects when combined with established human structures in communities" (p. 274). These disasters, Brandon (2002) notes, also involve more than one organization, making communication potentially confusing. He points out that, during disasters, all levels of government may be involved, as are police and other emergency agencies and numerous nongovernmental organizations, such as the United Nations or the Red Cross. This causes communication challenges, because disaster scenes "would see each of the responding organizations flocking independently to the scene, each with its own native communications chosen independently and based on analysis of its own role and operations" (p. 137).

Disasters are also of high interest to the public as well as the media. Sood, Stockdale, and Rogers (1987) state that "news media personnel see disasters as a special opportunity to provide a valuable service to local victims, to increase broadcast ratings and newspaper circulation" (p. 29). In a study conducted on how the media operated during five disasters from 1979 to 1984, the authors found that the "reporters who were most successful in filing their news stories about disasters were the most resourceful in circumventing access and mobility problems" (p. 31). Thus, mediated public relations during a disaster offers more complications to public relations professionals. With competing organizations responding to the disaster and vying for the public's attention, news media trying to increase viewership and readership, and reporters going out on their own to find stories, public relations professionals must utilize their skills to help their organizations survive before, during, and after disasters.

One communication vehicle used increasingly by organizations during times of crisis and disaster is the Internet. Perry, Taylor, and Doerfel (2003) conducted an eighteen-month study of organizational use of the Internet during crises and found that most organizations used the Internet to communicate with all publics, including the media. Of those organizations incorporating the Internet into crisis communications, most utilized online technology by posting traditional public relations tools, such as news releases and fact sheets. New media tactics, such as video or audio clips, two-way interaction, and real-time monitoring of the crisis, were used less often. They concluded that "an organization's attempt to maintain relationships with its various publics via the Internet while under intense scrutiny may minimize the potential damage of a crisis with its stakeholders and maximize recovery" (p. 231).

Organizations can manage relationships using mediated means during crises and disasters as long as these organizations communicate quickly and honestly, work closely with other responding organizations, realize that the media is probably trying to make its own mark through its reporting of the situation, and fully understand their relationships with publics before the crisis occurs. To illustrate one organization's use of mediated public relations during a disaster, this chapter focuses on the American Red Cross and its response to Hurricane Katrina.

Hurricane Katrina Facts

Hurricane Katrina struck the Louisiana coast and the city of New Orleans on Monday morning, August 29, 2005. According to the National Oceanic and Atmospheric Administration's National Hurricane Center (2005), it formed six days earlier as a tropical storm off the coast of the Bahamas, slowly built to hurricane strength before hitting southern Florida, and then quickly mushroomed to a category 5 hurricane. Katrina was the eleventh named storm of the 2005 hurricane season and the second category 5 storm to form. As it bore its way toward the Gulf Coast, evacuation orders were given to residents from the Florida Panhandle to the Texas coastline. The storm weakened to a category 3, but catastrophic damage still ensued. Storm surges from Mobile to Biloxi drowned hundreds, but initial media reports indicated that New Orleans escaped major damage and that the city's levees held. Unfortunately, this good news did not last. Later in the morning of August 29, citizens of New Orleans reported flooding. Suddenly, tens of thousands of residents were in harm's way, and hundreds subsequently died.

Hurricane Katrina became the largest natural disaster in U.S. history. It was not only the costliest hurricane in terms of dollars and cents, but it also was the deadliest hurricane in almost eighty years, killing approximately 1,600 people along the Gulf Coast. Hundreds of thousands of residents from Alabama, Louisiana, and Mississippi sought shelter, and the American Red Cross launched its largest relief effort in its 124-year history (*American Red Cross*, 2006a).

American Red Cross Responds

Founded on May 21, 1881, by Clara Barton, the American Red Cross responds to more than seventy thousand disasters a year (*American Red Cross*, 2006a). From single house fires to flooding, from tornados to terrorist attacks, from wildfires to earthquakes, its mission is to aid people during their most desperate time of need with shelter, food, water, mental health counseling, and financial assistance. The 2005 hurricane season tested the organization's disaster response effort like never before. Hurricane Katrina required the American Red Cross to perform a series of firsts: Shelter more than four hundred thousand people in twenty-seven states, serve a million meals in a single day, and provide financial assistance to more than one million families. Hurricane Rita hit less than a month later in Texas, with Wilma soon following in October in southern Florida. The American Red Cross response to Katrina expanded to include these two storms as well. The unprecedented effort produced eye-popping statistics (see Table 1). During this time, the Red Cross also raised and spent the most in its history: $2.1 billion.

American Red Cross Media Use[1]

The American Red Cross's media objectives before, during, and after a disaster are to communicate with its publics about what the American Red Cross is doing, what help it needs, and how people can support the organization. As a nonprofit organization chartered by the U.S. Congress to respond to disasters, the American Red Cross communicates with numerous publics, including federally elected officials and up to the U.S. president; governmental agencies, such as the Department of Homeland Security; corporate partners; media; staff; volunteers; donors; and the people it helps. The corporate communication staff includes a senior vice president of communication, who reports directly to the president, and two vice presidents, one for internal communication and the other for external communication.

American Red Cross disaster communicators are overseen by the senior director of disaster response, preparedness, and international services and fall into three categories: (a) the public affairs staff at the Disaster Operations Center (DOC) in Washington, D.C.; (b) chapter communicators at the approximately eight hundred local chapters across the United States and its territories; and (c) Rapid Response Team (RRT) members. The American Red Cross media plan for hurricanes covers two areas—before and after landfall. Before landfall, the DOC staff alerts the RRT that they may be deployed. RRT members, both paid staff and volunteers, are public relations professionals, former journalists, and others who are trained to go to the potential disaster areas before the storm hits. In 2002, the Public Relations Society of America (n.d.) and the American Red Cross created The Power of Two partnership to recruit and train public relations professionals to aid in communication during disasters.

RRT members work with national and international media who are typically on the scene, while local chapter public affairs officers continue to work with their local media.

Table 1: American Red Cross (ARC) Hurricane Relief Statistics

ARC Hurricane Relief Effort (Hurricanes Katrina, Rita, & Wilma)	Statistics
Overnight stays in shelters	3.4 million
Number of states sheltering evacuees	27 and the District of Columbia
Number of evacuees	Nearly 450,000 in 1,300 shelters
Number of meals served	30 million (plus 28 million snacks)
Mental health professionals deployed	4,600
Disaster relief workers deployed	More than 245,000 (95% of whom were volunteers)
New volunteers signed up	Approximately 60,000
Hits on Family Links Registry	More than 350,000
Hits in September 2005 on http://www.redcross.org	9.6 million, five times more than in the previous month

Note. Data from *American Red Cross*, 2006, retrieved June 23, 2006, from http://www.redcross.org

The rationale for this approach is that chapter communicators know their local media and are better equipped to work with them. RRT only arrives when national media arrive. At the same time, the DOC public affairs staff prepares messaging, press releases, question and answer documents, fact sheets, and stories for the American Red Cross Web site (http://www.redcross.org), while handling interviews with national and international media not physically located at the disaster area.

Before the storm, media use involves disseminating preparedness messages to media at all levels. These messages reach American Red Cross publics through media interviews and public service announcements. They include encouraging residents to create a family calling plan, telling them what to gather before evacuating (e.g., birth certificates and prescriptions) and what should go in a disaster preparedness kit (e.g., food and water for three days), and urging them to seek the closest shelter. For publics outside the disaster area, messaging may include how and where to give blood, volunteer, and donate money.

The American Red Cross has an extensive array of prepared public service announcements for radio, print, television, and online media. To maximize the use of media interviews to relay information, RRT members typically work in teams of two and may be deployed, for example, to the largest shelter in a given area, at a feeding/distribution center, or as a roaming team seeking satellite trucks in the field to aid producers and reporters looking for stories.

The size of the storm determines media interest. Before Katrina made landfall, RRT members were deployed to south Florida, Mobile, Biloxi, Baton Rouge, and numerous towns in between. Armed with cell phones and BlackBerries, they conducted hundreds of print, radio, online, and television interviews over the phone, in person, during press conferences, and via satellite before the August 29 landfall. Chapter communicators and the DOC public affairs staff did the same. By communicating before landfall, the American Red Cross employed a proactive media strategy that helped it to establish itself as the major nonprofit agency responsible for disaster relief operations.

After a typical storm, RRT members work with national and international media in the disaster area who are interested in doing stories on feeding operations, human interest pieces on shelter life, and distribution stories on financial assistance or clean-up supplies, for example. The DOC public affairs staff releases daily fact sheets with operational statistics, such as number of shelters open, meals served, and funds donated, along with press releases to the media, chapter communicators, and RRT members. Table 2 includes a sampling of American Red Cross press release headlines for Hurricanes Katrina and Rita.

Table 2: Selected American Red Cross (ARC) Hurricane Press Release Headlines

Release Date	Headline
Aug. 26, 2005	ARC Urges Preparedness as Katrina Approaches Gulf Coast States
Aug. 27, 2005	ARC Encourages Every Household to Develop a "Family Communication Plan"
Aug. 29, 2005	ARC Launches Largest Mobilization Effort in History for Hurricane Katrina
Sept. 2, 2005	ARC Announces Family Linking Available via http://www.redcross.org or by Calling 1-877-LOVED-1S
Sept. 5, 2005	One Week after Katrina's Gulf Coast Assault, the ARC Is Setting a Record Relief Pace
Sept. 10, 2005	ARC Launches Toll-Free Call Center Expanding Ability to Meet Essential Needs of Hurricane Evacuees
Sept. 12, 2005	ARC Extending Assistance for Evacuees in Hotel Accommodations
Sept. 13, 2005	ARC Asks for Patience in Financial Assistance Effort
Sept. 19, 2005	ARC Projects Hurricane Katrina Most Costly Relief Effort in its History
Sept. 21, 2005	ARC and Microsoft Announce New Family Linking Web site KatrinaSafe.Org
Sept. 29, 2005	ARC Committed to Fighting Fraud
Sept. 30, 2005	ARC Tops $1 Billion in Hurricane Response
Sept. 30, 2005	ARC Providing Financial and Housing Assistance to Victims of Hurricane Rita

Note. Headlines from *Press Room: Archives 2005*, by American Red Cross, 2006, retrieved August 19, 2006, from http://www.redcross.org/press/archive/0,1081,0_489_,00.html

The DOC staff also distributes daily talking points to chapter communicators and RRT members. This enables the organization to stay on message regardless of how many communicators are working with the media. The DOC then posts the fact sheets and press releases on the American Red Cross Web site (http://www.redcross.org) for the general public and handles calls from national and international media not present in the disaster zone.

A typical national media time line for hurricanes is less than a week. That is, one or two days before the storm are spent reporting on preparedness, three or four days after the storm are spent writing relief stories, and then national media are ready to move on to the next news event. Hurricane Katrina changed that time line, however. National media focused on Katrina and, subsequently, Rita and Wilma for months, putting a strain on American Red Cross communicators at all levels. For example, during the first five days after Katrina's landfall, the DOC public affairs staff received between two hundred and three hundred media calls per hour, twenty-four hours a day; RRT members stayed deployed for up to three months, versus the usual deployment of one week; and chapter communicators in the affected areas worked around the clock through the end of the year.

In addition to working with print, radio, and television stations from around the world, the American Red Cross continuously posted stories on relief efforts on its Web site, http://www.redcross.org. The site received 9.6 million hits in September 2005, five times as many as in August. The organization also launched a Spanish language Web site, Cruz Roja Americana (2006), four days before Hurricane Katrina hit the Gulf Coast.

The American Red Cross's intranet site was constantly updated with staff and volunteer information to ensure all knew what the American Red Cross was doing 24/7. Talking points for chapter communicators and RRT members were posted on this internal site as well.

One Web-related tool that the national American Red Cross did not offer was a blog. Several of the organization's more than eight hundred chapters had launched local blogs at the time of this publication, but a national blog was not expected to be launched until the end of 2006.

The American Red Cross utilized media conferences extensively after Katrina. With major operations in Alabama, Mississippi, Louisiana, and Texas, American Red Cross communicators participated in dozens of media conferences in conjunction with local, state, and federal officials in those states. For example, after the Houston Astrodome was set up as an American Red Cross shelter for New Orleans evacuees, the American Red Cross set up a roped-off area for media in the parking lot, a media conference room in the adjacent Reliant Center, and a media work room where reporters could file stories and enjoy the air conditioning.

More than one hundred reporters, camera people, sound crew, and producers from as far away as Japan and Great Britain were credentialed, allowing them to attend regularly scheduled morning media conferences that typically included elected officials, medical personnel, police and fire officials, and representatives from the American Red Cross. The media received color-coded armbands that allowed them to be escorted into the shelter by an American Red Cross communicator to conduct interviews or shoot images. American Red Cross communicators also participated in media conferences in Washington, D.C., and in the other twenty-plus states where sheltering operations took place.

In addition to media relations and Internet postings, the American Red Cross utilizes other communication tools during disaster response. It employs still photographers to capture images for online and media use. Camera crews shoot pool footage to share with television stations when the disaster location is in close quarters and inaccessible by a large number of people or when the American Red Cross wants to ensure the privacy of its shelter clientele.

The organization is increasingly using podcasts and webcasts as a way to reach its publics on the local chapter level. Another high-tech tool employed by the organization is a fleet of emergency communication response vehicles. These mobile trucks are equipped with satellite phones, cell phones, two-way radios, and wireless Internet access, and they are deployed into disaster areas to aid in communication efforts between relief organizations, emergency officials, and clients.

The American Red Cross media strategy strives to be proactive. Hurricane devastation exposed around the world brought heightened media interest, unprecedented support, calls to action, and, of course, controversies that put the American Red Cross in a reactive media position after the storm.

Controversies

Generally, the American Red Cross receives positive media attention before and after disasters. The media understand that the organization's only goal is to help. For typical storms, occasional negative stories may appear; for example, there may be coverage about emergency response vehicles not going into one particular neighborhood for mobile feeding or the American Red Cross policy of not taking pets in shelters. Even with these challenges, by mid-September, unaided recall of the American Red Cross as a responding organization to Hurricane Katrina was 77%. By late September, 89% of the public agreed that the American Red Cross was doing its best to help hurricane-affected victims under the most difficult conditions it had ever faced.[2]

Because of Katrina's destructive prowess, American Red Cross controversies grew in proportion to the size, scope, and length of disaster operations. The first involved the Louisiana Superdome in New Orleans. The American Red Cross determines where it opens shelters based on the structural soundness of a building and whether it is outside of the U.S. Army Corps of Engineers' mapped flood zones. Unfortunately, the entire city of New Orleans is in a flood zone, so all American Red Cross shelters were north toward Baton Rouge and beyond. Residents of New Orleans expected the American Red Cross to be at the Superdome, and it was not. The media also questioned the lack of American Red Cross presence. The American Red Cross responded by explaining its rationale, which did little to console the thousands who initially felt abandoned in New Orleans ("Red Cross: State Rebuffed," 2005).

As most who work in media relations know, the media want to report numbers. With evacuees spread out over so many states, it was difficult for the American Red Cross to release exact numbers of people in shelters or those being fed on a timely basis. In addition, many shelters were operating without electricity and in areas without cell phone service. Shelter managers were unable to relay numbers to the DOC, and, in turn, the DOC was not able to compile accurate numbers for the media. As the situation improved in the Gulf states, so did the reporting of numbers.

Families losing touch with each other also caused concern. Tens of thousands of families were evacuated from New Orleans to the Houston Astrodome. In Houston, the American Red Cross opened the largest shelter in its history in less than twenty-four hours after Texas agreed to take thousands from the New Orleans Convention Center and the Superdome. As the busses drove across the state line, several were diverted to Dallas and San Antonio, causing family members to be split up. Families were also separated immediately after the flooding, with some members being at home while others were at work and school.

Busses arrived at the Houston Astrodome at about midnight on Wednesday, August 31. Each evacuee was given a medical review, during which basic information was written down, before being brought inside for a hot shower, food and water, and a cot. Because evacuees came in so quickly, an electronic system was not in place to register each person. They were exhausted and simply wanted to eat, clean up, and rest. And, as it turned out, they wanted to find out where their family members were as well.

The American Red Cross immediately recognized that, with thousands of evacuees seeking shelter throughout the country, it needed to quickly find a way to link family members together. For the media (and those looking for loved ones), however, the American Red Cross did not move fast enough. Although the organization is geared toward sheltering and feeding, it was not, at that time, prepared to quickly create a method for family members to find each other. On September 2, it launched the toll-free number 1–877-LOVED-1S and set up a Family Links Registry through its Web site. The phone number received more than 350,000 calls, and more than 340,000 people registered online in the weeks following the storm. Slowly but surely, families were reconnected, turning a negative story positive as images of reunited families flooded the airways.

On Friday, September 2, 2005, NBC Universal Television Group aired a live benefit, *A Concert for Hurricane Relief*, on its NBC, MSNBC, and CNBC networks. The American Red Cross benefit raised $50 million, along with a few eyebrows, when recording artist Kanye West stated that President George W. Bush "doesn't care about black people." The American Red Cross (2005) quickly released the following statement late that night:

> The American Red Cross is incredibly grateful for the support we're receiving in the wake of the catastrophic events caused by Hurricane Katrina. We want to acknowledge the ongoing support of NBC-Universal, which aired a telethon tonight on behalf of the victims of this tragedy. During the telecast, a controversial comment was made by one of the celebrities. We would like the American public to know that our support is unwavering, regardless of political circumstances. We are a neutral and impartial organization, and support disaster victims across the country regardless of race, class, color or creed. We cannot, and we do not endorse any comments of a political nature. (para. 1 and 2)

As part of its response efforts, the American Red Cross provided Client Assistance Cards for families to purchase basic items. The cards operated like a debit card, and their monetary value was determined by the amount of damage a person's or family's home received and how many members were in that family. At the Houston Astrodome, these cards were distributed in less than one week after evacuees arrived. The American Red Cross had seventy thousand of these cards stockpiled, which proved to be far fewer than what was needed. Because the Federal Emergency Management Agency (FEMA) also distributed funds this way, no more cards were to be found in the United States, resulting in delays. Fraud occurred because many evacuees had no identification when they left home quickly without driver's licenses and Social Security cards. Erring on the side of helping, the American Red Cross distributed some funds to nonevacuees and to some evacuees more than once. The media quickly questioned the American Red Cross's commitment to safeguarding donor money ("Despite Katrina Efforts," 2005).

Other controversies included community and faith-based organizations questioning why the American Red Cross was not supporting their efforts, donated items being stolen by volunteers, the American Red Cross president resigning, and Senator Grassley of Iowa questioning the American Red Cross's efficient use of donor money.

In the year following Katrina, the American Red Cross rededicated itself to solving the issues raised by the unprecedented storm and issued a report in June 2006 indicating the following corrective actions:[3]

- One million Client Assistance Cards were stockpiled to ensure that no shortage will ever occur.
- Background checks are conducted on all volunteers to minimize fraud.
- Partnerships were created with hundreds of community and faith-based organizations to assist in their local relief efforts.[4]
- A best practices approach to fully utilize and account for donations was employed.
- A National Shelter System was created with FEMA that helps the American Red Cross and other humanitarian organizations track details about shelter locations, daily populations, and availability.
- Infrastructure upgrades have been installed to ensure that the American Red Cross can handle all calls to its 1–800-RED-CROSS information and donation line as well as visits to its Web site (http://www.redcross.org).

In terms of communicator improvements, the American Red Cross held extensive retraining for RRT members in February 2006 with the awareness that the media will continue to ask tough questions and probe American Red Cross practices more closely in future disasters. The crisis media training included mock press conferences with hostile media members, satellite interviews, and ambush interviews. In addition, the American Red Cross held numerous chapter-level media training sessions throughout spring 2006 to improve the responses and expertise of chapter communicators. At the national level, the DOC public affairs staff reorganized to handle massive media inquiries if a Katrina-type disaster struck again.

The American Red Cross's unprecedented response to the largest national disaster in U.S. history came with a few missteps, but, without mediated public relations, the organization would never have been able to explain to a global audience what it was doing and how concerned citizens could help. Without the media and the Internet, the American Red Cross could not have raised awareness of relief needs and, therefore, relief dollars to help the more than one million families in need after Hurricanes Katrina, Rita, and Wilma.

For consideration

1. Is the American Red Cross strategy of sending in RRT members before a storm to do preparedness messaging an advisable one, or is this work best handled by local, state, and federal emergency officials? What are the advantages and potential drawbacks of the American Red Cross's approach?
2. Some of the controversies discussed involved highly emotional issues, such as lost family members or no American Red Cross presence at the New Orleans Superdome. To what extent can an organization effectively respond to emotion by using facts?
3. How could American Red Cross better leverage mediated communication tools in its response efforts?
4. To what extent does disaster response require a mixture of traditional and new media channels? That is, is either means of communicating sufficient on its own?
5. Are media reports an effective means of evaluating disaster response efforts? How else could a disaster response organization evaluate its effectiveness?

For reading

Fearn-Banks, K. (2002). *Crisis communications: A casebook approach*. Mahwah, NJ: Lawrence Erlbaum Associates.

Ledingham J., & Bruning, S., (Eds.). (2000). *Public relations as relationship management: A relational approach to the study and practice of public relations*. Mahwah, NJ: Lawrence Erlbaum Associates.

Millar, D., & Heath, R. (2003). *Responding to a crisis: A rhetorical approach to crisis communication*. Mahwah, NJ: Lawrence Erlbaum Associates.

Mitroff, I. (2005). *Why some companies emerge stronger and better from crisis: 7 essential lessons for surviving disaster.* New York: AMACOM.

Sood, R., Stockdale, G., & Rogers, E. (1987). How the news media operates in natural disasters. *Journal of Communication*, *37*(3), 27–40.

Notes

1. The author is a member of the American Red Cross Rapid Response Team. Information for this section came from personal experience during her three-year volunteer effort.
2. Data gathered from omnibus telephone surveys of one thousand U.S. adults age eighteen and older (margin of error, +/- 3.1%), commissioned by the American Red Cross and conducted by ORC International and Bruskin Goldring on September 14 and 22, 2006.
3. See American Red Cross (2006d). The full report was retrieved June 23, 2006, from http://www.redcross.org/static/file_cont5448_lang0_2006.pdf
4. See American Red Cross (2006c) for the May 2, 2006, press release: American Red Cross, NAACP and Faith-Based Groups Partner to Provide Disaster Training Nationwide.

References

American Red Cross. (2005, September 2). *Statement: September 2 NBC-Universal Telethon*. Retrieved August 19, 2006, from http://www.redcross.org/pressrelease/0,1077,0_489_4527,00.html

American Red Cross. (2006a). Retrieved June 1, 2006, from http://www.redcross.org/

American Red Cross. (2006b). *Press Room: Archives 2005*. Retrieved August 19, 2006, from http://www.redcross.org/press/archive/0,1081,0_489_,00.html

American Red Cross. (2006c, May 2). *American Red Cross, NAACP and Faith-Based Groups Partner to Provide Disaster Training Nationwide*. Retrieved August 19, 2006, from http://www.redcross.org/pressrelease/0,1077,0_489_5355,00.html

American Red Cross. (2006d, June). *From challenge to action: American Red Cross actions to improve and enhance its disaster response and related capabilities for the 2006 hurricane season and beyond*. Retrieved June 23, 2006, from http://www.redcross.org/static/file_cont5448_lang0_2006.pdf

Borda, J., & Mackey-Kallis, S. (2003). A model for crisis management. In D. Millar & R. Heath (Eds.), *Responding to a crisis: A rhetorical approach to crisis communication*. Mahwah, NJ: Lawrence Erlbaum Associates.

Brandon, W. (2002). An introduction to disaster communications and information systems. *Space Communications*, *18*(3/4), 133–139.

Coombs, W. T. (2000). Crisis management: Advantages of a relational perspective. In J. Ledingham & S. Bruning (Eds.), *Public relations as relationship management: A relational approach to the study and practice of public relations* (pp. 73–94). Mahwah, NJ: Lawrence Erlbaum Associates.

Cruz Roja Americana. (2006). Retrieved June 1, 2006, from http://www.cruzrojaamericana.org/

Despite Katrina efforts, Red Cross draws criticism. (2005, September 28). *USAToday.com*. Retrieved July 17, 2006, from http://www.usatoday.com/news/nation/2005–09–28-katrina-red-cross_x.htm

Fearn-Banks, K. (2002). *Crisis communications: A casebook approach*. Mahwah, NJ: Lawrence Erlbaum Associates.

Heath, R., & Millar, D. (2003). A rhetorical approach to crisis communication: Management, communication processes, and strategic responses. In D. Millar & R. Heath (Eds.), *Responding to a crisis: A rhetorical approach to crisis communication* (pp. 1–17). Mahwah, NJ: Lawrence Erlbaum Associates.

National Oceanic and Atmospheric Administration's National Hurricane Center. (2005, October 31). *Post storm data acquisition*. Retrieved August 19, 2006, from http://www.nws.noaa.gov/om/data/pdfs/KatrinaPSDA.pdf

Perry, D., Taylor, M., & Doerfel, M. (2003). Internet-based communication in crisis management. *Management Communication Quarterly*, *17*(2), 206–232.

Public Relations Society of America. (n.d.). Retrieved June 1, 2006, from http://www.prsa.org

Red Cross: State rebuffed relief efforts. (2005, September 9). *CNN.com*. Retrieved July 17, 2006, from http://www.cnn.com/2005/US/09/08/katrina.redcross/

Sellnow, T., Seeger, M., & Ulmer, R. (2002). Chaos theory, informational needs, and natural disasters. *Journal of Applied Communication Research*, *30*(4), 269–292.

Sood, R., Stockdale, G., & Rogers, E. (1987). How the news media operates in natural disasters. *Journal of Communication*, *37*(3), 27–40.

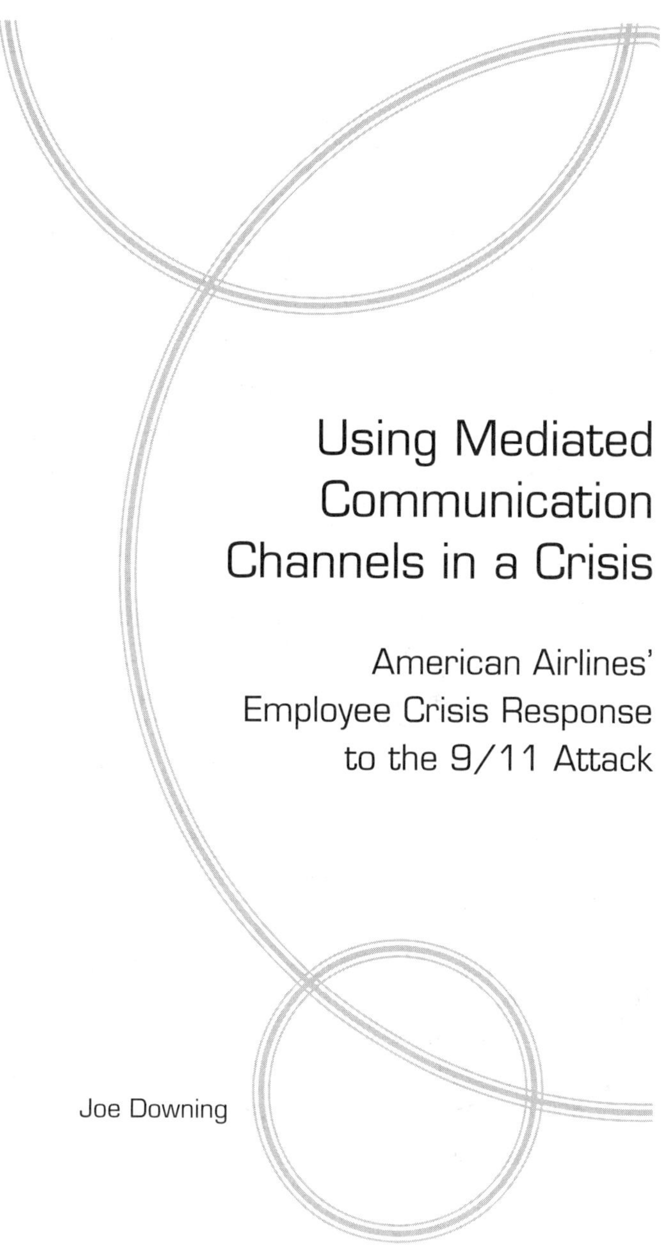

Using Mediated Communication Channels in a Crisis

American Airlines' Employee Crisis Response to the 9/11 Attack

Joe Downing

Increasingly, public relations professionals use mediated channels to communicate with their employees after a crisis. This chapter illustrates how communicators from American Airlines used both one-way and two-way/interactive media in their employee crisis response to the September 11, 2001, attack on the United States. One lesson from 9/11, which was reinforced during Hurricane Katrina, is that organizations must not rely too heavily on these technologies during a crisis. Three new technologies—the Wi-Fi Mesh, Voice over Internet Protocol (VoIP), and mass notification services—are offered as effective ways to strengthen an organization's communication network in the event of a crisis.

In this chapter, I argue how professional communicators can use new technologies to communicate with employees more effectively after a crisis. The chapter is structured as follows. First, I explain how the birth of CNN and its rival twenty-four-hour news networks have changed stakeholders' perceptions of how quickly organizations should release information following a crisis. Second, I observe the significant effect that the Internet has had on crisis communication practices in general and on employee crisis response tactics in particular. Third, by way of a case study, I show how the interaction of these two developments influenced American Airlines' use of mediated channels in its employee crisis response to the September 11, 2001, U.S. attack. Fourth, I use lessons from 9/11, then Hurricane Katrina, to demonstrate the vulnerability of relying too heavily on electronic channels like e-mail, the telephone, and the Internet after a crisis. Last, I look toward the future of mediated employee crisis communication. In this last section of the chapter, I introduce three new technologies—the Wi-Fi Mesh, VoIP, and mass notification technologies—that professional communicators can adopt as part of their overall crisis response plan.

Crisis Management Efforts

The birth of the twenty-four-hour news network. In the early 1980s, CNN changed how professional communicators dealt with the media. Before CNN, communicators worked on fixed deadlines to get their organization's message to reporters. With the birth of the twenty-four-hour news network, though, the public no longer had to wait to watch the evening news or to read about events in their morning newspaper (Holtz, 2002). In short, the cable news channels changed stakeholders' opinions of how quickly organizations were expected to communicate information. This is especially true when a crisis struck the organization.

When covering a crisis, news anchors were also under pressure to keep new information flowing into their newsroom. If the company involved in the crisis was not prepared to provide that information, the media often used outsiders—pundits and other speculators—to fill in the missing gaps (Calloway, 1991; Sweetman, 2000).

While CNN and its rival networks had shortened the news reporting cycle, the publics' widespread adoption of the Internet in the mid- to late 1990s challenged the media's traditional role as *gatekeeper* of the news. The Internet, combined with the accelerated news reporting cycle, meant that public relations professionals played a more important role in shaping a company's public image after a crisis. Suddenly, a new class of citizen journalist emerged on the scene. Targeted and niche publications began to flourish, and the public enjoyed more sources to get their news.

In response, organizations had to develop innovative methods to reach their stakeholders during a crisis. At first, these communicators relied on traditional, one-way channels to disseminate this information. For example, communicators posted press releases, transcripts of news conferences, question and answer documents, and fact sheets about the company and the crisis (Taylor & Perry, 2005).

The interactive nature of the Internet, however, now allows communicators to experiment with more interactive, two-way communication channels (Taylor & Perry, 2005). Companies, for instance, now communicate with their stakeholders through e-mail (Taylor

& Perry), instant messaging (IM) services (Boorstin, Lashinsky, Mehta, Sellers, & Stires, 2005), blogs (Marken, 2005), and organizations' intranet and Internet sites (Adams & Clark, 2001; Holtz, 2002; Horton, 2001; Middleberg, 2000). Innovations in wireless and cellular phone networks also allow organizations to send text messages (also known as short message service [SMS]) to stakeholders' cell phones or personal digital assistants (PDAs).

Incorporating Internal (Employee) Audiences into the Organization's Crisis Response

Professional communicators' use of these new technologies, of course, is a recent phenomenon. Before 2001, most organizations focused their crisis response efforts on external stakeholders like victims' families, the financial community, and the media. In this tradition, the primary goal of an effective crisis response was to maintain a positive image with the company's external stakeholders (Coombs, 2000; Pearson & Mitroff, 1993; Seeger, Sellnow, & Ulmer, 2001).

As at least an indirect result of the September 11, 2001, attack, crisis managers began to recognize the importance of using these new technologies to communicate with their employees as well as with their external stakeholders (Argenti, 2002; Wright, 2002).

If they are to be successful in this endeavor, organizations must recognize the information and emotional needs employees have during and after a crisis. Employees' immediate concern is knowing that their colleagues are safe and that the crisis itself is under control (Pincus & Acharya, 1988). Once the organization communicates this information, it can begin to address employees' secondary information needs. These include issues like how employees can access their benefits (Boorstin, Lashinsky, Mehta, Sellers, & Stires, 2005; Cagle, 2006) and counseling services available to employees.

Once the initial shock of the crisis begins to wear off, employees' attention may then turn to whether their organization can financially survive the crisis (Bierck, 2002). The financial impact of the crisis may also force the organization to lay off employees. Leeper (2004) has argued that those employees who remain with the company may experience lower organizational commitment levels, decreased productivity, and morale problems because of layoffs.

Communicators, then, face a daunting task. Initially, they must allow employees the space to grieve their lost colleagues. At some point, though, the organization has to move forward (Seeger & Ulmer, 2002), and employees must return their focus to the job at hand.

In the next section, I draw upon American Airlines' employee crisis response to 9/11 to show how the airline used mediated channels to address these two types of messages. However, as I will later show, overreliance on a particular communication technology can cause its own set of issues.

American Airlines Case Study

In February 2003, American Airlines management granted me unrestricted access to proprietary documents that the airline had sent its employees after September 11. This chapter focuses mainly on how professional communicators at American Airlines responded to employees after the attack. Readers interested in how the airline responded to its external stakeholders should consult Greer and Moreland (2003).

Background information. Before September 11, AMR Corporation, which runs American Airlines, American Eagle, and Trans World Airlines (TWA), employed over 122,000 people. Of this group, most employees (about 70%) belonged to a union (Downing, 2004).

Historically, the relationship between American Airlines management and its employee unions has been difficult. Tensions between the two groups were especially high during Bob Crandall's tenure from 1980 to 1998, first as president, and then as chief executive officer (CEO) (Ray, 1999). Donald Carty, who replaced Crandall as CEO, worked hard to strengthen the ties between these two groups. Carty also led the airline through the September 11 crisis, but he later resigned amid allegations that he had planned to give his senior executives retention bonuses—while simultaneously asking for wage concessions from the airline's unions (Zellner, 2003). Carty's successor, Gerard Arpey, continues to serve as the airline's CEO and has thus far kept the airline out of bankruptcy court (Trottman, 2006).

Overview of employee communication practices within American Airlines. Like most major organizations, American Airlines uses various methods (or channels) to communicate with its employees. Much of the daily interaction within the airline takes place face-to-face between individual employees and their supervisors (Larkin & Larkin, 1996). The supervisor is the link between the employees and the frontline communicator, who writes employee newsletter articles, runs a local intranet site, and supports the communication efforts between frontline supervisors and their employees.

On the corporate level, three departments support the efforts of these frontline communicators: Human Resources, Corporate Communications, and Information Technology. Each of these departments uses both print and mediated channels to communicate with airline employees.

In a company like American Airlines, communicators face challenges to disseminate information efficiently, especially in crisis situations, because employees have varying access to media. An internal study conducted within the airline before the attack found that 40% of employees shared online access at work. Employees often accessed the Internet through kiosks. Eleven percent of employees had a corporate e-mail account, and only 10% had individual access to the Internet at work. Gate and ticket agents only had access to text-only messages on their computer terminal, while mechanics and ramp workers received most of their information directly from their frontline supervisor. Pilots and flight attendants, when not in flight, had access to fax-on-demand technologies and Internet kiosks. In an emergency, the airline's System Operations Control group could also disseminate information to airborne pilots.

Description of different communication channels. American Airlines, like most companies, has both external and internal audiences to which they must quickly deliver information. At the time of the 9/11 crisis, communicators used the AMR Corporation Web site (http://www.amrcorp.com)[1] as the primary online resource to post information to external stakeholders like the media and the financial community. Amrcorp.com, which was launched in 1995, included press releases, statistics about the airlines' fleet, executive biographies, and information targeted at those who invested in the company (AMR Corporation, 2002). Airline passengers used a second Web site, http://www.aa.com, to buy tickets, get flight schedules, and check arriving and departing gate information (Downing, 2004).

On the internal side, the airline used six primary communication channels to reach its employees: *Flagship News*, the Don Carty Hotline, *HeAAds Up!*, *Jetwire*, *Special Jetwire*, and Jetnet. The first channel, *Flagship News*, was the airline's primary employee print newsletter. This publication was important since relatively few employees had continuous Internet access. The second communication channel, the Don Carty Hotline, provided employees or journalists with a prerecorded telephone message from Carty.

The third and fourth communication channels, *HeAAds Up!* and *Jetwire*, were distributed electronically. *HeAAds Up!* was the electronic publication sent to nearly five thousand American Airlines managers. Its purpose was to notify managers of information before it was officially released to employees. Oftentimes, the publication included talking points that managers could use to handle employee inquiries.

Jetwire was the company-wide electronic newsletter. *Jetwire* was published daily and sent to employees who had a corporate e-mail account and was available to virtually all employees on Sabre, the airline's internal reservations and operations computer system. *Jetwire* included standard features that reported the previous day's on-time arrival and departure information, how many lost baggage claims the company incurred the previous day, and the current value of AMR stock. Other relevant employee information was also included in the airline's daily *Jetwire* message. During the 9/11 crisis, the transcribed text of Mr. Carty's hotline messages was included in *Special Jetwire* e-mails, as were any updates or breaking news events that were not included in that day's *Jetwire*.

The fifth communication channel, Jetnet, was the airline's Web portal for internal communications. Employees managed their payroll services, employee benefits, and travel privileges through Jetnet. From 1996 to 2002, the airline maintained IntrAAnet, a rather simple intranet site that contained mostly press releases, links to outside news about the company, and some investor relations material. In late April 2002, Jetnet replaced IntrAAnet.

In the next section, I discuss how American Airlines used these different channels to communicate with its employees after the September 11 attack.

September 11, 2001. On the morning of September 11, American Airlines and United Airlines each lost two planes. Seventeen American Airlines crew members died that morning; sixteen crew members from United Airlines also were killed. Excluding the hijackers, 213 passengers died as a result of the four plane crashes ("9/11 Commission Report," 2004; McCartney & Carey, 2001).

Immediately after the attack, American Airlines management dispatched 350 Customer Assistance Relief Effort (CARE) volunteers to help victims' families at the two crash sites. Another 300 CARE volunteers in Dallas were available by phone to aid these families, as well as to help American Airlines employees cope with the aftermath of the day's events (Fearn-Banks, 2002). By September 14, American Airlines' Employee Assistance Program (EAP) had set up a Comfort Room in the airline's Fort Worth headquarters. Three more rooms were available to employees at the Dallas–Fort Worth International Airport. The purpose of these rooms was to provide a space where airline employees could talk to an EAP representative or just take a break and relax.

Airline communicators used mediated channels to support the interpersonal efforts of CARE team and EAP personnel. Tim Wagner worked in Corporate Communications and, at the time, served as one of the airline's two Webmasters. Wagner began to update Amrcorp.com five minutes after he learned about the second plane crash.

Wagner faced two difficult challenges: (a) providing timely updates to various stakeholders of the airline and (b) avoiding a server overload because of the dramatic increase of visits to the site. Immediately, Wagner decided to replace the normal Web site content with two simple messages. First, Wagner acknowledged what he knew, which was that two planes had crashed. He wrote that the airline would post more information as it became available. Second, Wagner posted phone numbers that victims' families could call. Wagner also minimized the use of graphics on the Web site and contacted his colleagues in Information Technology to divert job-related tasks to a different server (Price, 2002, para. 2). While Wagner was updating Amrcorp.com, other communicators were working to push information quickly to employees.

One of the first messages the airline sent its employees that morning was from Corporate Security. The message asked employees to limit outgoing phone calls. Shortly after that, Corporate Communications sent an e-mail that urged employees to limit their Internet usage.

Mick Doherty, managing editor of employee publications, sent a *HeAAds Up!* message to managers. This was the first detailed message airline management had sent its employees after the attack. Corporate Communications then sent that message to *Special Jetwire*. The text of this message is included in Figure 1.

Figure 1: American Airlines Message to Employees on September 11, 2001

—AMERICAN CONFIRMS TWO AIRCRAFT LOST—

American Airlines confirmed today that it lost two aircraft in tragic incidents this morning. American said the flights were Flight 11, a Boeing 767 en route from Boston to Los Angeles with 81 passengers, nine flight attendants and two pilots; and Flight 77, a Boeing 757 operating from Washington Dulles to Los Angeles with 58 passengers, four flight attendants and two pilots.

> Because of the heightened security due to the nature of today's events, American said it is working closely with U.S. government authorities and will not release more information at this time.
>
> The government has shut down the entire air traffic system in the United States. American, TWA and American Eagle will not operate.
>
> "We are horrified by these tragic events," said Chairman Don Carty. "Our thoughts and prayers go out to the families of all involved."
>
> Media calls should be directed to the FBI. Customers who wish to receive information about relatives should call American's response number at 1-800-245-0999.

Recall that only 40% of American Airlines employees had continuous Internet and e-mail access. Those employees could call the Don Carty Hotline to receive updated information. Airline communicators forwarded Mr. Carty's messages to employees' voicemail boxes. The transcripts of Don Carty Hotline messages were faxed to field offices and crew lounges and were also posted to gate agents' computer terminals.

The company's servers became less burdened twenty-four hours after the attack. However, Wagner, Doherty, and their colleagues faced another problem. Immediately after the two plane crashes, the Federal Bureau of Investigation (FBI) placed a gag order on American Airlines and United Airlines. Government officials banned officials from both airlines from speaking to the press. The gag order lasted almost two weeks. Airline officials, however, were allowed to post certain information, including updates on how the attack affected passengers' travel plans.

Wagner posted what he could on Amrcorp.com, Aa.com, and IntrAAnet. Posts included information about colleagues who were killed in the attack, counseling services that were available to employees, and transcripts of Carty's hotline messages. Most of the information targeted at external stakeholders (victims' families and the media) was posted on Amrcorp.com. One of the principal goals here was to include enough detail so that stakeholders would look to this Web site for information and not call the airline directly.

In the weeks following the attack, American Airlines received hundred of condolence letters, prayers, and other messages from customers, competitors, and vendors. On September 24, 2001, Doherty began to intersperse these inspirational messages among the grim operational and financial news that the airline was sending its employees. Later, Doherty asked employees to contribute their own reactions to 9/11. Employees began to send their stories to Doherty. These stories, plus letters and prayers from those outside the airline, became a formal communications campaign called Good Words. This online campaign represented a shift from the use of one-way communication to a more interactive, two-way exchange between the airline and its employees.

Lessons learned. American Airlines managers learned three important lessons from their employee crisis response to the September 11 attack. The first lesson was the importance of internal employee campaigns like Good Words, which included employees' reactions to the crisis. Good Words represented the community-building and interactivity features that mediated communication channels can deliver following a crisis.

A second lesson airline managers learned was that lines of authority between the different communications functions (in this case, Human Resources, Corporate Communications, and Information Technology) must be settled prior to a major crisis. Otherwise, coordinating information and other resources can be difficult. Airline communicators learned a third, and final, lesson, which was how to update the company's Web site during future crises. When a later crisis occurred, Wagner grayed out the entire Amrcorp.com site but kept the links on the site active. This use of a "dark site" (Davis & Gilman, 2002) can effectively reflect the somber mood within an organization.

The Vulnerability of Communication Networks during a Crisis

American Airlines' response to 9/11 shows how corporate communicators have several online channels available to them after a crisis. The demands of the crisis situation, including the demographics of employees who work for the organization, should guide communicators' decisions about their use.

An equally important lesson to be learned from 9/11 was how vulnerable traditional communication channels, like landline phones, cellular phones, and proprietary corporate networks, are during a crisis. Regrettably, this same issue surfaced again almost four years later when Hurricane Katrina struck the Gulf Coast in late August 2005.

During a crisis, heavy use and equipment damage can shut down landline and/or cellular phone services. Traditional landline phones use centralized switches to handle all phone calls (Thompson, 2005), and electricity is needed to power the switches that keep communication networks working (Stuver, Keene, & Carlisle, 2004).

Data can be transferred through either wired (landline) or wireless (cellular) networks. Regardless of the network used, each requires vast amounts of electricity to work effectively. Phone networks are designed so that no more than 10% of the population speaks at the same time (Thompson, 2005). However, speaking on the phone is precisely what people tend to do in a crisis. As a result, both landline and cellular phone circuits may become jammed.

Substantial increases in incoming and outgoing messages can also cripple internal corporate networks. For instance, in the twenty-four hours after the attack, American Airlines Reservations received up to forty thousand phone calls.

The physical buildings that house these communication switches can also be damaged in a crisis. For instance, when the Twin Towers of the World Trade Center fell on September 11, a building near the site that housed switches was damaged. This combination of overloaded circuits and damaged switch boxes caused blackouts in parts of New York City (Thurm, 2001).

Similar physical communication network problems surfaced four years later when Hurricane Katrina struck the Gulf Coast (Block, Schatz, & Fields, 2005). High water damaged above-the-ground and underground phone lines. Strong winds also blew down cellular towers in the region (Thompson, 2005). Before Hurricane Katrina, Marriott International, the hotel chain, had the foresight to move the company's e-mail servers out of New Orleans and up to Washington, D.C. (Boorstin, Lashinsky, Mehta, Sellers, & Stires, 2005). As a result, Marriott experienced few network problems during the hurricane.

Alternatives to Traditional Communication Networks

Satellite phones. During both 9/11 and Katrina, some organizations used satellite phones to bypass these traditional communication networks altogether. Satellite phones have their own set of drawbacks, however. Satellite phones are expensive and, like cell phones, need electricity to operate. In general, cell phone batteries cost less and last longer than batteries designed for satellite phones (Thompson, 2005). Satellite phones are also expensive and require line-of-sight access to the sky to operate (Drucker, 2005). This limits their use inside a building.

Wi-Fi mesh. Some cities—and larger corporations—have begun to develop a new communication network called Wi-Fi mesh. A Wi-Fi mesh works much like the Internet, which relies on a vast set of disparate communication networks to send and receive data. For example, when a person sends an e-mail, electronic data typically pass through many communication networks. If one or more of these networks is damaged, the Internet is set up to bypass these particular networks. As a result, the recipient usually receives the e-mail.

A Wi-Fi mesh operates in much the same manner as the Internet. Instead of relying on traditional computer servers, a Wi-Fi mesh uses a network of nodes that are placed atop individuals' roofs. These nodes are small (about the size of a small shoebox) and, therefore, less susceptible to damage from the elements than, for example, a large cellular tower (Thompson, 2005). Each node is also inexpensive for a company to purchase: They cost roughly $500 apiece and use little electricity. In fact, a car battery can keep each node working should the city's power supply go out (Thompson).

To work, a Wi-Fi mesh requires a clear line of sight between these nodes (Thompson, 2005). Even if fire, wind, water, or some other force of man or nature destroys one or more nodes, the decentralized nature of a Wi-Fi mesh will allow data to continue to flow through the network.

Voice over Internet Protocol (VoIP). Individuals who use a Wi-Fi mesh enjoy the same functionality as their counterparts who rely on a wired network to access the Internet. One

Internet technology that shows promise is called Voice over Internet Protocol (VoIP), which allows individuals to make and receive phone calls through their computer. Companies like Vonage and Skype market VoIP services to consumers directly. VoIP shows special promise during a crisis situation because, if a crisis disrupts traditional landline and/or cellular phone service, users can use their computers as a phone (Thompson, 2005). (This assumes, of course, that the individual has the needed electricity to run his or her computer.)

What the Future Holds for Mediated Employee Crisis Response

Organizations affected by 9/11 and Hurricane Katrina have learned not to rely too heavily on a single communication channel during a crisis. For example, during Katrina, some Gulf Coast employers had a difficult time reaching their employees by telephone. Without this contact, employers had no way of knowing if their employees were accounted for and safe (Boorstin, Lashinsky, Mehta, Sellers, & Stires, 2005).

However, each communication network, including the Wi-Fi mesh, can be rendered useless during a crisis. Yet, rarely do communication networks shut down all at once. For instance, landline and cellular phone services were disrupted in New York City on 9/11, but Internet and text messaging services were largely unaffected (Thurm, 2001). Sadly, a number of employees trapped in the Twin Towers used text messaging services to say their final goodbyes to their families (Zuckerman & Cowan, 2001).

Mass notification technologies. In the future, organizations must simultaneously use multiple communication channels to reach employees. This is critical, especially in the event that one communication channel fails (Stuver, Keene, & Carlisle, 2004). A recent technology called *mass notification system* has been created to address this problem. Currently, two mass notification systems dominate the U.S. market: Send Word Now[2] and 3n.[3] Although each service offers slightly different features, the underlying method used by both companies is the same.

Communicators register one or more of an organization's stakeholder groups (employees, the media, and so forth) with the service. Either the organization or its individual stakeholders use a Web-based interface to input their contact information (e-mail address, cellular phone, landline phone, fax machine, IM address, and so on) into the mass notification service's database. When a crisis strikes the organization, the communicator designs a message and then uses the service's Web-based interface to deliver that message. Alternatively, the communicator can use a landline or cellular phone to contact the service. Send Word Now or 3n automatically sends the message to registered stakeholders who can receive the message as a text message or through a number of alternative communication channels (Jones, 2005). Organizations like the U.S. Postal Service (Harwood, 2003) and PriceWaterhouseCoopers (Gurliacci, 2005) have already signed up with a mass notification service provider.

Conclusion

In this chapter, I have stressed how professional communicators can use electronic channels to respond to employees after a crisis. American Airlines' use of its Good Words campaign illustrated how the airline incorporated employee reactions into the broader crisis response strategy of the organization.

Relying too heavily on these mediated communication channels during a crisis, however, is a mistake. Organizations affected by 9/11 and Hurricane Katrina learned how vulnerable traditional communication networks can be when disaster strikes. Savvy communications will incorporate new technologies like the Wi-Fi Mesh, VoIP, and mass notification systems into their crisis response plans.

For consideration

1. How can organizations communicate care and concern for affected publics in the aftermath of a crisis and avoid appearing insincere in doing so?
2. To what extent can organizations effectively communicate care and concern through new media channels?
3. For what occasions should only interpersonal channels be used to communicate with employees in the aftermath of a crisis?
4. Which is more important in the rush to communicate in a crisis: accuracy or speed? To what extent does the answer depend on the audience?
5. What impact could a government-mandated gag order have on the reputation of a company responding to a major crisis?

For reading

9/11 Commission report: Final report of the National Commission on Terrorist Attacks upon the United States. (2004). New York: Norton.

Argenti, P. (2002, December). Crisis communication: Lessons from 9/11. *Harvard Business Review, 80(12)*, 80(12), 103–109. Retrieved November 12, 2005, from the EBSCO database.

Coombs, W. T. (1999). Ongoing crisis communication: Planning, managing, and responding. Thousand Oaks, CA: Sage.

Pearson, C. M., & Mitroff, I. I. (1993). From crisis prone to crisis prepared: A framework for crisis management. *Academy of Management Executive*, *7*(1), 48–59.

Seeger, M. W., Sellnow, T. L., & Ulmer, R. R. (2001). Public relations and crisis communication: Organizing and chaos. In R. L. Heath (Ed.), *Handbook of public relations* (pp. 155–166). Thousand Oaks, CA: Sage.

Notes

1. This Web site is no longer active. Viewers are now redirected to http://www.aa.com
2. Web site retrieved August 19, 2006, from http://www.sendwordnow.com/default.aspx
3. Web site retrieved August 19, 2006, from http://www.3nonline.com/

References

9/11 Commission report: Final report of the National Commission on Terrorist Attacks upon the United States. (2004). New York: Norton.

Adams, T. L., & Clark, N. E. (2001). *The Internet: Effective online communication*. Dallas: Harcourt College Publishers.

AMR Corporation. (2002, January). *Internet sites* [Web site no longer available]. Retrieved December 27, 2002, from http://www.amrcorp.com/facts/sheets/Internet.com

Argenti, P. (2002, December). Crisis communication: Lessons from 9/11. *Harvard Business Review, 80*(12), 103–109. Retrieved November 12, 2005, from the EBSCO database.

Bierck, R. (2002, May). What will you say when disaster strikes? *Harvard Management Communication Letter, 5*(5), 1–4. Retrieved June 12, 2005, from the EBSCO database.

Block, R., Schatz, A., & Fields, G. (2005, September 6). Power failure: Behind Katrina response, a long chain of weak links. *The Wall Street Journal*, p. A1.

Boorstin, J., Lashinsky, A., Mehta, S. N., Sellers, P., & Stires, D. (2005, October 3). New lessons to learn. *Fortune*, pp. 87–88. Retrieved April 2, 2006, from the EBSCO database.

Cagle, J. (2006, March/April). Internal communication during a crisis pays dividends. *Communication World*, pp. 22–23. Retrieved May 2, 2006, from the EBSCO database.

Calloway, L. J. (1991). Survival of the fastest: Information technology and corporate crises. *Public Relations Review, 17*(1), 85–92. Retrieved November 3, 2005, from the Elsevier Science Direct database.

Coombs, W. T. (2000). Designing post-crisis messages: Lessons for crisis response strategies. *Review of Business, 21*(3), 37–41. Retrieved December 17, 2005, from the EBSCO database.

Davis, S. C., & Gilman, A. D. (2002, August). Communications coordination. *Risk Management*, pp. 38–42. Retrieved December 17, 2006, from the ABI Proquest database.

Downing, J. (2004). American Airlines' use of mediated employee channels after the 9/11 attacks. *Public Relations Review, 30*(1), 37–48.

Drucker, E. (2005, October 15). Refining disaster strategies. *Wireless Week, 11*(21), p. 15.

Fearn-Banks, K. (2002, September). A snapshot of how organizations responded to tragedy. *PR Tactics*, pp. 30–31. Retrieved November 12, 2005, from the EBSCO database.

Greer, C. F., & Moreland, K. D. (2003). United Airlines' and American Airlines' online crisis communication following the September 11 terrorist attacks. *Public Relations Review, 29*(4), 427–441. Retrieved November 3, 2005, from the Elsevier Science Direct database.

Gurliacci, D. (2005, December 5). Business alert system set up to make emergencies easier. *Fairfield County Business Journal*, p. 3.

Harwood, M. (2003, December 29). Business report. *Federal Times*, p. 18.

Holtz, S. (2002). *Public relations on the Net* (2nd ed.). New York: AMACOM.

Horton, J. L. (2001). Online public relations: A handbook for practitioners. Westport, CT: Quorum.

Jones, K. (2005, May 24). Casting a wide net. *PC Magazine*, p. 24.

Larkin, T. J., & Larkin, S. (1996, May/June). Reaching and changing frontline employees. *Harvard Business Review, 74*, 95–104.

Leeper, K. (2004). Downsizing or reduction-in-force: A crisis residual. In D. P. Millar & R. L. Heath (Eds.), *Responding to crisis: A rhetorical approach to crisis communication* (pp. 299–310). Mahwah, NJ: Erlbaum.

Marken, G. (2005). To blog or not to blog, that is the question? *Public Relations Quarterly*, 50(3), 31–33. Retrieved May 27, 2006, from the EBSCO database.

McCartney, S., & Carey, S. (2001, October 15). Flying blind. *The Wall Street Journal*, p. A1. Retrieved December 17, 2005, from the ABI Inform Newspaper database.

Middleberg, D. (2000). Winning PR in the wired world: Powerful communications strategies for the noisy digital space. New York: McGraw-Hill.

Pearson, C. M., & Mitroff, I. I. (1993). From crisis prone to crisis prepared: A framework for crisis management. *Academy of Management Executive*, 7(1), 48–59.

Pincus, J. D., & Acharya, L. (1988). Employee communication strategies for organizational crises. *Employee Responsibilities and Rights Journal*, *1*(3), 181–199.

Price, T. (2002) *Portals in a storm: Crisis communication online* (3rd ed.). Washington, DC: Foundation for Public Affairs.

Ray, S. J. (1999). Strategic communication in crisis management: Lessons from the airline industry. Westport, CT: Quorum.

Seeger, M. W., Sellnow, T. L., & Ulmer, R. R. (2001). Public relations and crisis communication: Organizing and chaos. In R. L. Heath (Ed.), *Handbook of public relations* (pp. 155–166). Thousand Oaks, CA: Sage.

Seeger, M. W., & Ulmer, R. R. (2002). A post-crisis discourse of renewal: The cases of Malden Mills and Cole Hardwoods. *Journal of Applied Communication Research*, 30(2), 126–142.

Stuver, P., Keene, J., & Carlisle, J. (2004). *Using mass notification in a disaster.* White paper. Retrieved May 25, 2006, from http://www.3nonline.com/

Sweetman, B. (2000). Managing a crisis. *Air Transport World*, 5, 71–74. Retrieved December 17, 2002, from the EBSCO database.

Taylor, M., & Perry, D. C. (2005). Diffusion of traditional and new media tactics in crisis communication. *Public Relations Review*, *31*(2), 209–217. Retrieved November 3, 2005, from the Elsevier Science Direct database.

Thompson, C. (2005, September 18). The way we live now: Idea Lab. *New York Times Magazine*, p. 24.

Thurm, S. (2001, September). A day of terror: Net proves more reliable than phones in effort to contact friends, loved ones. *The New York Times*, p. A3.

Trottman, M. (2006, April 17). Airline CEO's novel strategy: No bankruptcy. *The Wall Street Journal*, pp. B1, B4.

Wright, D. K. (2002, March). Examining how the 11th September, 2001 terrorist attacks precipitated a paradigm shift advancing communications and public relations into a more significant role in corporate America. *Journal of Communication Management*, 6(3), 280–292. Retrieved March 29, 2005, from the ABI Proquest database.

Zellner, W. (2003, May 5). What was Don Carty thinking? *Business Week*, p. 32.

Zuckerman, G., & Cowan, L. (2001, September 13). At the top: Cantor Fitzgerald took pride in its position at the World Trade Center. *The Wall Street Journal*, p. C1.

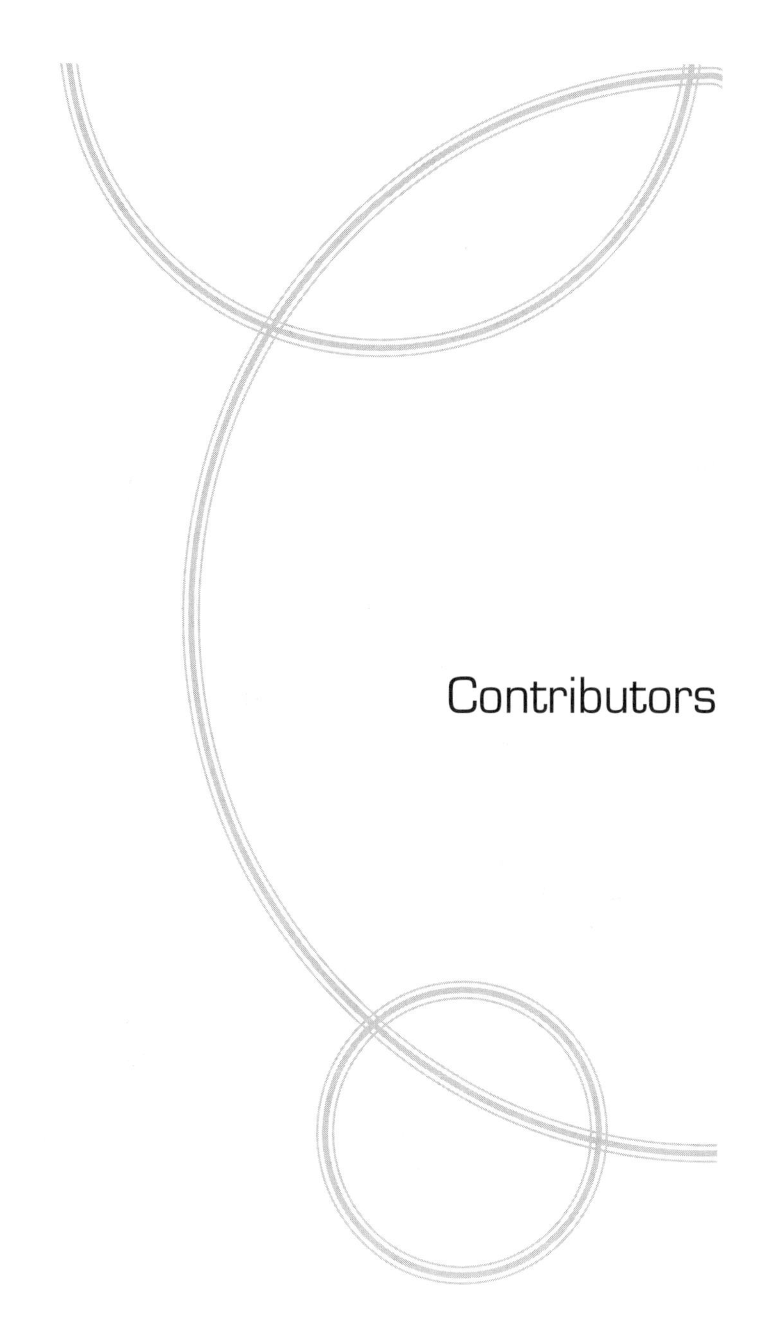

Contributors

JENNIFER L. BARTLETT teaches and researches in public relations and corporate communication in the School of Advertising, Marketing, and Public Relations at the Queensland University of Technology, Australia. Her research focuses on institutional theory, legitimacy, corporate social responsibility and reputation, and the implications for public relations practice. In industry, she has worked in public affairs within the banking and energy sectors and retains links through ongoing industry consulting.

DENISE BORTREE, M.A.M.C., M.Ed. (University of Florida), is currently a doctoral candidate in the College of Journalism and Communication at the University of Florida. She is studying children and Internet usage, Internet communication, and public relations. Her research has been published in *Education, Communication and Information (ECi)* and *Journal of Applied Communications*.

MARY HELEN BROWN, Ph.D. (University of Texas at Austin), is an associate professor and acting chair of the Department of Communication and Journalism at Auburn University. Her special areas of interest are organizational communication, storytelling, and sense making.

BRIGITTA R. BRUNNER, Ph.D. (University of Florida), is an assistant professor in the Department of Communication and Journalism at Auburn University. Her research focuses on public relations and diversity issues.

ERIN C. BRYAN, M.A. (University of Delaware), is visiting instructor in the Communication Department at West Virginia Wesleyan College and teaches public relations and communication studies courses. Bryan's research experience spans organizational culture and outcomes of organizational practices.

IDIL CAKIM is a vice president of interactive media at GolinHarris, focusing on social and user-generated media. Formerly, she was a director of knowledge development at Burson-Marsteller, USA where she specialized in interactive marketing and technology-related custom research. She managed Burson-Marsteller's Internet-related proprietary studies, including the *e*-fluentials® and the online crisis audit. Cakim is a frequent industry speaker on emerging technology issues and has been quoted widely regarding research and online communications in publications including the *Financial Times*, *USA Today*, *The New York Times*, *San Jose Mercury News*, *Business Week*, and CBS *Market Watch*. Cakim has a B.A. in sociology from Bryn Mawr College and an M.A. from the Annenberg School of Communication at the University of Pennsylvania.

JEFFREY R. CARLSON, M.A. (Purdue University), is instructor in the Communication Department at West Virginia Wesleyan College and teaches communication studies, persuasion, and research methods courses. Carlson has also served as a marketing representative and as human performance consultant in industry. Carlson's research focuses on leadership in nonprofit organizations, marketing communication, and small group dynamics.

W. TIMOTHY COOMBS, Ph.D. (Purdue University), is an associate professor in communication studies at Eastern Illinois University. His crisis communication research focuses on the development and testing of the Situational Crisis Communication Theory (SCCT). His crisis communication work was awarded the 2002 Jackson, Jackson, and Wagner Behavioral Research Prize from the Public Relations Society of America, and he is cowinner of the 2002 PRIDE Award for Best Article from the Public Relations Division of the National Communication Associations.

ZORAIDA R. COZIER, Ph.D. (Purdue University), worked as an assistant professor of public relations at Georgia Southern University for two years (2004 to 2006) prior to joining Concorde Career College in San Diego, California. Her research interests include organizational public relations, discourse publics, structurationist perspective of public relations, and mental illness communities. A related published work by Cozier is the "The Development of a Structuration Analysis of New Publics in an Electronic Environment" in the 2001 *Handbook of Public Relations* by Heath and Vasquez. Cozier has presented numerous competitive papers at international and national conferences on case studies that draw on structuration theory and public relations.

MARCIA WATSON DISTASO is an assistant professor in the College of Communications at Penn State University. Her dissertation at the University of Miami explores the use of impression management in earnings releases. Her professional background in finance led to her research focus on investor relations.

JOE DOWNING, Ph.D. (Rensselaer Polytechnic Institute), investigates the role that emerging communication technologies play in employee communication practices in for-profit organizations. His work has appeared in *Public Relations Review*, *Journal of Business Communication*, *American Communication Journal*, *Communication Education*, and the *Journal of Technical Writing and Communication*. Joe Downing is an assistant professor of communication at Penn State-York campus. Prior to earning his doctoral degree, Downing worked in human resource capacities in two organizations in Denver, Colorado. From 1998 to 2004, he consulted for General Electric's (GE) Global Research Center, where he analyzed the internal and cross-cultural communication processes between research and development scientists in Bangalore, India; Shanghai, China; and Niskayuna, New York.

SANDRA C. DUHÉ, Ph.D., is assistant professor and coordinator of the public relations program at the University of Louisiana at Lafayette and associate director of communication at the University of Louisiana Center for Business and Information Technologies. Prior to joining academia in 2004, she was a public affairs manager for three multinational corporations with extensive experience in media relations, corporate brand management, crisis response, risk communication, and community coalition building. She holds master's degrees in public relations and applied economics and received her Ph.D. in political economy from the University of Texas at Dallas. Her research focuses on corporate public relations with particular interests in political economy, complexity science, and new media perspectives in public relations practice and theory.

MOHAN J. DUTTA, Ph.D. (Minnesota), is associate professor of health communication and public relations in the Department of Communication at Purdue University, where he is a member of the Center for Education and Research in Information Assurance and Security, Asian Studies, and the E-Enterprise Centers. He also serves as an affiliate faculty member for the Discovery Learning Center, the Regenstreif Center for HealthCare Engineering, and the Burton D. Morgan Enterpreneurship Center. Dutta's work explores the intersections of critical theory and cultural studies in the context of public relations.

MAHMOUD EID is an assistant professor at the Department of Communication, University of Ottawa, Ontario, Canada. He previously taught in the University of Regina's School of Journalism in Regina, Saskatchewan, and in Carleton University's School of Journalism and Communication in Ottawa, Ontario. His professional expertise lies in quantitative and qualitative research regarding the effects of mass media and social development. His teaching experience, research interests, and publications concentrate on international communication, media studies, communication research methods, terrorism, crisis management and conflict resolution, modernity, and the political economy of communication. He received a B.A. (1991) in public relations and advertising and an M.A. (1997; thesis titled *The Role of Arabic Mass Media as Tools in Conflict Resolution: The 1991 Gulf War as a Case Study*) in international communication from Cairo University's Faculty of Mass Communication in Egypt and a Ph.D. (2004; thesis titled *Interweavement—Building a Crisis Decision-Making Model for Rational Responsibility in the Media: International Communication, Political Crisis Management, and the Use of Mathematics*) in communication from Carleton University's School of Journalism and Communication in Canada. Eid is the editor of *Cybercultures* and a co-editor of *Introduction to Media Studies*. He serves as an editorial board member for several academic journals and as an organizing committee member for various international conferences. He has presented numerous papers in global conferences and has published several book chapters as well as journal articles in *Communications: The European Journal of Communication Research, International Journal of the Humanities, First Monday, INFORMATION, Journalism Ethics for the Global Citizen*, and elsewhere.

GLEN FEIGHERY, Ph.D. (University of North Carolina, Chapel Hill), is an assistant professor in the Department of Communication at the University of Utah.

SARAH BONEWITS FELDNER, Ph.D. (Purdue University), is an assistant professor of communication studies at Marquette University. Her research focuses on critical and rhetorical approaches to organizational communication, specifically organizational identity. Her most recent work looks at the relationship between organizational mission and individual identity.

ROMY FRÖHLICH, Ph.D. (University for Music and Theatre, Hannover, Germany), is full professor at Ludwig-Maximilians-University of Munich (Germany) and former president of the German Communication Association (DGPuK). Her major research and teaching interests include public relations (theory and practice), women in mass communication, journalism education, and news content / content analysis.

BARBARA S. GAINEY, Ph.D., is an assistant professor in the Department of Communication's public relations track at Kennesaw State University. Gainey has more than twenty years of professional communication experience, starting in newspaper work. Most of her experience has been in public relations in public and corporate sectors, including substantial administrative experience. At Kennesaw State University, Gainey is the faculty advisor for the university's new Public Relations Student Society of America chapter. She holds a Ph.D. in mass communications (public relations emphasis) from the University of South Carolina. Her research interests are in the areas of public relations, crisis management / crisis communication (particularly in educational settings), leadership, and public engagement.

PETER W. GALARNEAU, JR., M.A. (West Virginia University), is assistant professor in the Communication Department at West Virginia Wesleyan College and teaches public relations and media studies courses. Prior to entering full-time teaching, Galarneau served in various public relations positions, such as sports information director, Webmaster, assistant director of marketing, and director of public relations. Galarneau's research includes media convergence and effects in computer-mediated communication.

MELISSA K. GIBSON HANCOX, Ph.D. (Ohio University), is an assistant professor and undergraduate program coordinator in the Department of Communication and Media Studies at Edinboro University of Pennsylvania. In addition to prior teaching appointments at Gannon University and Western Michigan University, she served as department chairperson at Mercyhurst College. She has also worked professionally in journalism and public relations. In addition to a textbook and previous book chapters, she has publications in *Journal of Applied Communication Research*, *Management Communication Quarterly*, *Journal of Management Consulting*, *Southern Communication Journal*, and *Mid-American Journal of Business*.

CALIN GURAU, Ph.D., is professor in marketing at Groupe Sup De Co Montpellier, France. He is a junior fellow of the World Academy of Art and Science, Minneapolis, USA. He worked as marketing manager in two Romanian companies, and he has received degrees and distinctions for studies and research from the University of Triest, Italy; University of Vienna, Austria; Duke University, United States of America; University of Angers, France; and Oxford University, United Kingdom. His present research interests are focused on the marketing strategies of high-technology firms and marketing strategies on the Internet.

JILL HARRISON-REXRODE is a doctoral student in sociology at Virginia Tech. Her research interests include social capital and culture consumption across the life course. Jill's past research projects include an exploration of the relationship between perceptions of communicative abilities and ratings of social capital by college students. Her current work uses Richard Peterson's omnivore hypothesis to explore cultural consumption patterns of Americans.

VINCENT HAZLETON, Ph.D., APR, Fellow of the Public Relations Society of America, is professor of communication at Radford University. His teaching and research interests include public relations theory, public relations competence, public relations strategies, and social capital.

AMY N. HEUMAN, Ph.D. (Bowling Green State University), is assistant professor in the Department of Communication Studies at Texas Tech University, where her research areas include inter/cultural communication, multicultural identity negotiation, critical Latino/a studies, ethnography, and communication pedagogy and instruction.

SHERRY J. HOLLADAY, Ph.D. (Purdue University), is a professor at Eastern Illinois University. Her research interests include crisis communication, the effects of the Internet on corporate communication, and aging employees. She was the cowinner of the 2002 PRIDE Award for Best Article from the Public Relations Division of the National Communication Associations.

LIESE L. HUTCHISON is an associate professor in the Department of Communication at Saint Louis University, with a secondary appointment at the Center for International Studies. She teaches public relations principles and practices, public relations writing, public relations ethics, cases in public relations, international public relations, and integrated communication campaigns. She has taught at Saint Louis University's Madrid campus, at the Maastricht Center for Transatlantic Studies in the Netherlands, and as a Fulbright scholar at Concordia International University in Tallinn, Estonia. Her research interests include studying the consequences of globalization, international public relations, crisis communication, social responsibility and ethics, and public relations education. She has more than fifteen years of professional public relations experience. An active public relations consultant and volunteer, her recent volunteer work includes serving as a national media spokesperson for the American Red Cross during disasters such as Hurricanes Charlie, Ivan, and Katrina.

CASSANDRA IMFELD, Ph.D. (University of North Carolina, Chapel Hill), is an instructor at Oglethorpe University.

RIC JENSEN, Ph.D. (Texas A & M), is an adjunct instructor in the Journalism Department at Northwestern State University in Natchitoches, Louisiana, where he teaches Principles of Public Relations and Impact of the Mass Media on Society with distance education. He also serves as an adjunct instructor for the Mass Communication Department at Sam Houston State University, where he teaches Online Journalism, and works as an adjunct lecturer for the Texas A & M University Sports Management Program, where he teaches Sports Public Relations. Jensen works as a public relations professional at the Texas Water Resources Institute at Texas A & M University, where he develops white papers about science communication of environmental issues. His research interests include public relations and the environment, public understanding of science, online public relations and journalism, and sports public relations and sports.

WILLIAM R. KENNAN, Ph.D., is professor of communication at Radford University. He conducts research on social capital formation as it impacts organizational performance. His teaching areas include research methods, change management, and organizational communication at both the undergraduate and graduate levels.

NETE NØRGAARD KRISTENSEN, Ph.D. (University of Copenhagen), is associate professor in the Department of Media, Cognition, and Communication, Film and Media Studies Section at the University of Copenhagen, Denmark. Her research interests include strategic communication and public relations, media science, cultural journalism, and news and war.

KENNETH A. LACHLAN, Ph.D. (Michigan State University), is assistant professor of communication at Boston College. His research interests include the psychological processing of both linear and interactive media, crisis communication, and media effects. His recent research has been published in *American Journal of Public Health*, *Journal of Broadcasting and Electronic Media*, and *Human Communication Research*.

KATHLEEN M. LONG, Ph.D. (University of Connecticut), is professor and chair of the Communication Department at West Virginia Wesleyan College. Long has taught public relations and communication courses for twenty-one years and served in various public relations positions for ten years before entering full-time teaching. Long's research spans intercultural, interpersonal, and instructional communication and, more recently, best practices in public relations.

REBECCA J. MEISENBACH, Ph.D. (Purdue University), is an assistant professor of communication at the University of Missouri-Columbia. Her research primarily focuses on qualitative and rhetorical approaches to nonprofit and gendered organizing. Her most recent work considers the identity negotiations of higher education fundraisers.

MARCUS MESSNER is an assistant professor in the School of Mass Communications at Virginia Commonwealth University. His dissertation at the University of Miami explores the intermedia agenda-setting effects between traditional news media and Web logs. He holds a master's degree in print journalism and worked as a political and business editor for a German daily newspaper.

BOLANLE A. OLANIRAN, Ph.D. (University of Oklahoma), is a professor in the Department of Communication Studies at Texas Tech University. His research and work in crisis management focuses on the preparation for and the anticipatory process of crisis management.

MARIA DE FÁTIMA OLIVEIRA, M.S. (Temple University), is a doctoral student in the School of Communications and Theater at Temple University. Her research interests focus on organizational discourse, reputation and issues management, social networks, and the impact of new media on public relations.

KATIE DELAHAYE PAINE is the founder of KDPaine & Partners, LLC, and publisher of the first blog and the first newsletters for marketing and communications professionals dedicated entirely to measurement and accountability. She writes KDPaine's Measurement Blog (http://kdpaine.blogs.com) and publishes *The Measurement Standard* newsletter. Her book, *Measures of Success: KDPaine's Guide to Measuring Your Public Relationships*, will be released in the spring of 2007. Prior to launching KDPaine & Partners in 2002, Paine was the founder and president of the Delahaye Group, which she sold to Medialink in 1999. For the past seventeen years, Paine has been providing marketers and communications professionals with tools, data, and information to help them make better business decisions. She and her firms have read and analyzed millions of news articles, Internet postings, and internal communications and have conducted hundreds of thousands of interviews in the relentless pursuit of quantitative and qualitative measures of her client's marketing success. She works with some of the world's most admired companies, including Raytheon, Hewlett-Packard, and Southwest Airlines. Most recently, her endeavors have been focused on providing cost-effective measurement programs for nonprofits, small businesses, and government agencies. Paine was an initial founder of the Institute for Public Relations Measurement Commission. She served as the U.S. liaison to the European Standards Task Force to set international standards for media evaluation. She writes a regular column for *PRNews* on corporate image and crisis communications and contributes to *PRNews*, *Communications World*, *PR Week*, *Business Marketing*, and *New Hampshire Magazine*. Prior to founding the Delahaye Group, Paine was the director of corporate communications for Lotus Development Corporation and previously was manager of merchandising for Hewlett-Packard Personal Computer Group. An accomplished speaker, Paine frequently lectures to conferences and universities, including the Conference Board, the American Strategic

Management Institute, the Public Relations Society of America, the International Association of Business Communicators, the Institute for International Research, the International Public Relations Research Conference, Ragan Communications Conferences, the PR Executive Forum, the International Public Relations Association, the University of New Hampshire, and Southern New Hampshire University. Paine was named Entrepreneurial Venture Creator Person of the Year by the University of New Hampshire's Whittemore School of Business and is the 2006 recipient of New Hampshire Business Review's Business Excellence Award for Media & Marketing. A cum laude graduate of Connecticut College's class of 1974, Paine majored in history and Asian studies. She received an honorary doctorate of laws from New Hampshire College in May 1996. She is an Athena award winner and a Board member of the New Hampshire Political Library. Her life is featured in Mark Albion's books *Making a Life, Making a Living*, and *True to Yourself*.

MAHUYA PAL is a doctoral candidate in the Department of Communication at Purdue University and specializes in public relations and organizational communication.

ANTHONY C. PEYRONEL, Ed.D. (University of Pittsburgh), is an associate professor and department chairperson in the Department of Communication and Media Studies at Edinboro University of Pennsylvania. Prior to his current academic employment, he was the director of communication at Westminster College and a print and broadcast reporter at various news organizations. His prior publications have appeared in *Public Relations Society of America Counselors in Higher Education Monograph Series*, *The CASE International Journal of Educational Advancement*, *Quill Magazine*, and *Journal of Marketing for Higher Education*.

JULIANN C. SCHOLL, Ph.D. (University of Oklahoma), is assistant professor of communication studies at Texas Tech University. Her research specialty is in health and safety problematics. Recent projects have focused on crisis communication centers, deception between patients and providers, the effects of humor in patient-provider interactions, and end-of-life care among minority populations.

GLENN W. SCOTT, Ph.D. (University of North Carolina, Chapel Hill), is an assistant professor in the School of Communications at Elon University.

BEY-LING SHA, Ph.D., APR, is an assistant professor in the School of Journalism and Media Studies at San Diego State University. Previously, she served on the faculties of the University of Maryland at College Park and the American University of Paris. For five years, Sha was a public affairs officer for the U.S. Census Bureau. In recent years, she also served as research consultant to Microsoft Corporation, Perrier-Vittel, the Urban Institute, the Chesapeake Bay Foundation, and other organizations. Sha's primary research program combines theories of mathematical physics with public relations scholarship. Her other research areas include international public relations, activism, cultural identity, gender, and health communication. Her research has been published in *Journal of Public Relations Research*, *Public Relations Review*, and *Journal of Promotion Management*. She holds a Ph.D. in mass communication from the University of Maryland at College Park and is accredited in public relations by the Universal Accreditation Board.

PATRIC R. SPENCE, Ph.D. (Wayne State University), is an assistant professor in the Department of Communication at Western Kentucky University. His research interests include crisis communication, organizational communication, and quantitative research methods. His recent work has appeared in *Communication Research Reports* and *Journal of Modern Applied Statistical Methods*.

DON W. STACKS, Ph.D. (University of Florida), is professor and director of the public relations program in the School of Communication at the University of Miami. He has written more than 150 scholarly articles and papers and has authored or coauthored seven books on communication topics. His awards include the Ralph Nichols Award for Research in Listening; the Institute for Public Relations Research and Education Pathfinder Award; the Public Relations Society of America Outstanding Educator Award; and the Jackson, Jackson, and Wagner Behavioral Science Prize.

MIHAELA VORVOREANU, Ph.D. (Purdue University), is an assistant professor in the Department of Communication Studies at Clemson University. Her research interests include online public relations, new communication technologies, and the internationalization of U.S. academe. She is originally from Bucharest, Romania.

RICHARD D. WATERS is a former fundraiser and consultant for healthcare and social service nonprofit organizations. Currently, he is an assistant professor in the department of communication in North Carolina State University's College of Humanities and Social Sciences.

DAMION WAYMER received his Ph.D. from Purdue University and is assistant professor of communication at the University of Houston School of Communication. His dissertation research used frame analysis to explore Cincinnati's gentrification and race relation initiatives; his thesis research used both quantitative and qualitative methodologies to explore the ways that corporate social performance (CSP) and organizational social responsibility can be used as effective employee recruitment and retention tools. His teaching and research interests include, generally, public relations and organizational rhetoric. More specifically, his research uses public relations and communication theories to explore the ways that marginalized or underrepresented publics can and do gain access to voice, as well as what strategies are available to them to challenge various issues they might encounter.

DAVID E. WILLIAMS, Ph.D. (Ohio University), is a professor in the Department of Communication Studies at Texas Tech University. His research interests include crisis management and crisis communication. He was worked with coauthors on the development of the Anticipatory Model of Communication and the development of Crisis Communication Centers. He has also authored analyses of public relations response strategies to various crises.

JEFFREY WIMMER, Ph.D. (Ludwig-Maximilians-University of Munich, Germany), is assistant lecturer and scientific employee at the Free University of Berlin (Germany). His primary research interests include public relations (theory and practice), international communication, and theories of public and counterpublic spheres.

DIANE F. WITMER, Ph.D. (University of Southern California), APR, Fellow of the Public Relations Society of America, is a professor of communications at California State

University, Fullerton. Her research interests include computer-mediated communication, organizational communication, and public relations. Her work includes articles in *Communication Monographs*, *Communication Education*, and *Journal of Computer-Mediated Communication*. Additional published work includes a peer-reviewed text, entitled *Spinning the Web: A Handbook for Public Relations on the Internet*, and contributions in a variety of texts and theoretical volumes. Her practical experience includes both general public relations and Web development for corporate and nonprofit organizations.

DEBRA A. WORLEY, Ph.D., APR, is associate professor of communication and coordinator of the public relations program at Indiana State University. Her primary research/teaching interests are in public relations, organizational communication, and communication ethics.

JORDI XIFRA, Ph.D. (University Autonomous of Barcelona, Spain), law degree (University of Barcelona, Spain), is professor of public relations in the Faculty of Tourism and Communication at the University of Girona, Spain. Since 1998, he has published several Spanish-language books on the theories, strategies, and tactics of public relations.